Industrial Network Security

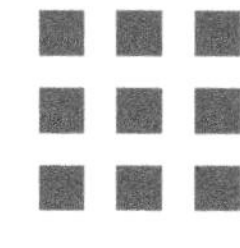

Industrial Network Security

Securing Critical Infrastructure Networks for Smart Grid, SCADA, and Other Industrial Control Systems

Third Edition

Eric D. Knapp

Leader and Visionary in Industrial Control Systems Cyber Security,
Boston, MA, United States

ELSEVIER SYNGRESS®

Syngress is an imprint of Elsevier
50 Hampshire Street, 5th Floor, Cambridge, MA 02139, United States

ISBN: 978-0-443-13737-2

For information on all Syngress publications visit our website at
https://www.elsevier.com/books-and-journals

Publisher: Mara Conner
Acquisitions Editor: Chris Katsaropoulos
Editorial Project Manager: Toni Louise Jackson
Production Project Manager: Selvaraj Raviraj
Cover Designer: Vicky Pearson Esser

Typeset by TNQ Technologies

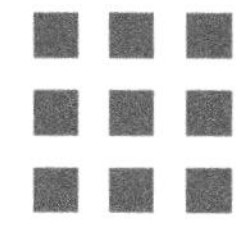

Contents

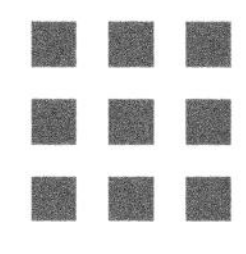

Biography

Eric D. Knapp is a globally recognized expert in industrial control systems (ICS) cyber security. He is the original author of *Industrial Network Security: Securing Critical Infrastructure Networks for Smart Grid, SCADA, and Other Industrial Control Systems* (First Edition) and the coauthor of *Applied Cyber Security and the Smart Grid.* Eric has held senior technology positions at NitroSecurity, McAfee, Wurldtech, and Honeywell, where he has consistently focused on the advancement of end-to-end ICS cyber security in order to promote safer and more reliable automation infrastructures. Eric has decades of experience in OT cybersecurity and holds multiple patents in the areas of risk management, asset protection, and secure data transfer. Eric's research and development efforts are the result of his never-ending quest to improve the field of industrial cyber security.

In addition to his work in information security, Eric is also an award-winning fiction author, best known for the cult hit *Cluck: Murder Most Fowl.* He studied English and Writing at the University of New Hampshire and the University of London and holds a degree in communications.

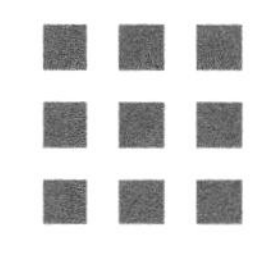

Acknowledgments

While there are so many who deserve acknowledgement for helping in so many ways through the years—my amazing wife, my friends, and colleagues—I dedicate this third edition solely to the one friend who helped start it all. Many decades ago, Mohan Ramanathan and I set off on a journey together to understand the strange new world of industrial cyber security, back in a time when "OT" wasn't even a term yet, long before any actual cyber attack against an industrial control system had ever been seen. Out of necessity, we started off by learning. "Industrial Cyber Security" was at best an emergent field. It was something very few people fully understood. It was a time when cyber attacks against industrial control systems were something to be imagined and feared, but had never been realized.

A lot has changed in the world since the second edition was published. Those attacks that were previously only imagined were proven real, first in labs and then through an increasing cadence of incidents around the world.

To my readers, I owe you thanks. Without the success of the first and second editions, the third would not be possible. But without my friend Mo, the first edition would never even have been written. So all of my thanks go to you, my friend, may you rest in peace.

The world is dimmer in the absence of your brightness, but I'll keep fighting to keep the lights on.

1 Introduction

Information in this chapter

- Book overview and key learning points
- Book audience
- Diagrams and figures
- The smart grid
- OT, IoT, IIoT, and xIoT
- How this book is organized
- Changes made to the third addition

Book overview and key learning points

This book is now in its third edition, published over a decade since the first edition's release in 2011. In some ways, a lot has changed during that time, and yet in others ways, very little has changed. In 2011, the entire concept of industrial cyber security was relatively new. Today, it is possible to specialize in this discipline and dedicate one's career to it. Formal education is offered by many universities as well as organizations such as the Cybersecurity and Infrastructure Security Agency (CISA), the International Information System Security Certification Consortium (ISC2), the SANS Technology Institute, and others. At the same time, for many readers coming from backgrounds in both industrial control ("OT") and information technology ("IT"), the idea of industrial cyber security will be entirely new.

One thing that has definitely changed: it is no longer optional to ignore the subject of securing industrial automation and process control environments from the ever rising threat of a cyberattack.

Since the first edition, this book has attempted to define an approach to industrial network security that considers the unique network, protocol, and application characteristics of an **industrial control system** (**ICS**) while also taking into consideration a variety of common compliance controls. For the purposes of this book, a common definition of ICS will be used in lieu of the more specific **supervisory control and data acquisition** (**SCADA**) or **distributed control system** (**DCS**) terms. Note that these and many other specialized terms are used extensively throughout the book. While we have made an effort to define them all, an extensive glossary has also been included to provide a quick reference if needed. If a term is included in the glossary, it will be printed in bold type the first time that it is used.

Industrial Network Security. https://doi.org/10.1016/B978-0-443-13737-2.00016-6

One term that is new is the aforementioned "**OT,**" or "**operational technology.**" The acronym "OT" is widely used today to discuss any and all aspects of industrial cyber security. While it provides a simple and convenient way to reference an otherwise complex and nuanced subject, it is often misleading, and so it will be discussed in more detail in Chapters 2 and 3.

Although many of the topics described herein—and much of the general guidance provided by regulatory standards organizations—are built upon common enterprise security methods and reference readily available information security tools, there remains little information available about how to implement these tools in an industrial network. This book attempts to rectify this by providing deployment and configuration guidance where possible, and by identifying why security controls should be implemented, where they should implemented, how they should be implemented, and how they should be used.

Book audience

To adequately discuss industrial network security, the basics of two very different underlying communication systems need to be understood: the Ethernet and Internet Protocol (IP) networking communications used ubiquitously in the enterprise, and the control and field bus protocols are used to manage and/or operate automation systems.

As a result, this book possesses a bifurcated audience. For the plant operator with an advanced engineering degree and a decade of programming for process controllers, the basics of industrial network protocols in Chapter 4 have been presented within the context of security in an attempt to not only provide value to such a reader but also to get that reader thinking about the subtle implications of cyber security. For the information security professional, familiar tenant of information security and basic information security practices have been provided within the new context of an OT environment.

There is an interesting dichotomy between the two that provides a further challenge. IT security typically strives to protect digital information by securing the users and **hosts** on a network, while at the same time enabling the broad range of open communication services required within modern business. OT, on the other hand, strives for the efficiency and reliability of a fine-tuned OT system, while always addressing the safety of the personnel, plant, and environment in which they operate. While there has been a long-standing friction between these two groups, only by giving the necessary consideration to both sides can the true objective be achieved: a secure industrial network architecture that supports safe and reliable operation, minimizes risk, and provides business value to the larger enterprise. This latter concept is referred to as "operational integrity."

While earlier editions focused on introducing each audience to the basics of the other's field, this edition aims to provide more nuanced information and guidance for a third audience: the OT Cybers Security Professional, who already understands the basics of both fields, but who is struggling to actually establish an OT security program, or to improve on the program that is currently in place.

Diagrams and figures

The network diagrams used throughout this book have been intentionally simplified and have been designed to be as generic as possible while adequately representing ICS architectures and their industrial networks across a very wide range of systems and suppliers. As a result, the diagrams will undoubtedly differ from real ICS designs and may exclude details specific to one particular industry while including details that are specific to another. However, they will provide a high-level understanding of the specific industrial network security controls being discussed.

The smart grid

Although the smart grid is of major concern and interest, for the most part, it is treated as any other industrial network within this book, with specific considerations being made only when necessary (such as when considering available **attack vectors**). As a result, there are many security considerations specific to the smart grids that are unfortunately not included. This is partly to maintain focus on the more ubiquitous ICS security requirements; partly due to the relative immaturity of smart grid security and partly due to the specialized and complex nature of these systems. Although this means that specific measures for securing synchrophasers, meters, etc. are not provided, the guidance and overall approach to security that is provided herein is certainly applicable to smart grid networks. For more in-depth reading on smart grid network security, consider *Applied Cyber Security and the Smart Grid* by Eric D Knapp and Raj Samani (ISBN: 978-1-59749-998-9, Syngress).

OT, IoT, IIoT, and xIoT

As mentioned above, "OT" or "operational technology" is a convenient way to reference "industrial automation and control systems, industrial networks, and the assets that comprise them. While "OT" and " Internet of Things (IoT)" are often conflated, they are actually very different.

According to Gartner, "The **Internet of Things** (**IoT**) is the network of physical objects that contain embedded technology to communicate and sense or interact with their internal states or the external environment."[1] The term is widely attributed to a Kevin Ashton in 1999. Mr Ashton was MIT's Executive Director of Auto-ID Labs, and a pioneer in the use of RFID for automated inventory tracking.[2] IoT became extremely popular in the early 2000s as more and more smart devices were introduced commercially. By the early 2020s, IoT devices are commonplace and include smart doorbells, surveillance cameras, sensors, and actuators used in industrial applications, medical devices, and more. As IoT evolved, so did the nomenclature. For example, IoT used for industrial applications is typically referred to as **Industrial Internet of Things** (**IIoT**). When referencing connected devices that could span many uses and specializations, the term **xIoT** (where the "x" represents a variable).

For purposes of industrial network security, IoT is something that needs to be acknowledged and understood. However, the focus on this book is on industrial networks, rather than IoT: that is, on structured networks used for industrial automation and control, rather than the Internet. IoT will not be discussed in depth, although the ubiquity of interconnected devices that are available across many industries needs to be acknowledged. An industrial automation and control system *could* include distributed devices that are interconnected via the Internet, either deliberately or accidently, and the introduction of such interconnected devices could have severe impact on the proper implementation of zones and conduits (see Chapter 9: Establishing zones and conduits). This is especially true in industrial systems that are highly distributed by nature, such as smart cities and smart grid applications.

How this book is organized

This book is divided into a total of 11 chapters, followed by three appendices guiding the reader where to find additional information and resources about industrial protocols, standards and regulations, and relevant security guidelines and best practices (such as **NIST**, **ChemITC,** and **ISA**).

The chapters begin with an introduction to industrial networking, and what a cyberattack against an ICSs might represent in terms of potential risks and consequences, followed by details of how industrial networks can be assessed, secured, and monitored in order to obtain the strongest possible security, and conclude with a detailed discussion of various compliance controls, and how those specific controls map back to network security practices.

It is not necessary to read this book cover to cover, in order. The book is intended to offer insight and recommendations that relate to both specific security goals as well as the cyclical nature of the security process. That is, if faced with performing a **security assessment** on an industrial network, begin with Chapter 6; every effort has been made to refer the reader to other relevant chapters where additional knowledge may be necessary.

Chapter 2: About Industrial Networks

In this chapter, there is a brief primer of ICSs, industrial networks, **critical infrastructure**, common cyber security guidelines, and other terminology specific to the lexicon of industrial cyber security. The goal of this chapter is to provide a baseline of information from which topics can be explored in more detail in the following chapters (there is also an extensive Glossary included to cover the abundance of new acronyms and terms used in OT networks). Chapter 2 also covers some of the basic misperceptions about industrial cyber security, in an attempt to rectify any misunderstandings prior to the more detailed discussions that will follow.

Chapter 3: Industrial Cyber Security, History, and Trends

Chapter 3 is a primer for industrial cyber security. It introduces industrial network cyber security in terms of its history and evolution, by examining the interrelations between "general" networking, industrial networking, and potentially critical infrastructures. Chapter 3 covers the importance of securing industrial networks, discusses the impact of a successful industrial attack, and provides examples of real historical incidents—including a discussion of the **advanced persistent threat** and the implications of cyber war.

Chapter 4: Introduction to ICS Systems and Operations

It is impossible to understand how to adequately secure an OT environment without first understanding the fundamentals of ICS systems and operations. These systems use specialized devices, applications, and protocols because they perform functions that are different than enterprise networks, with different requirements, operational priorities, and security considerations. Chapter 4 discusses control system **assets**, operations, protocol basics, how control processes are managed, and common systems and applications with special emphasis on smart grid operations.

Chapter 5: ICS Network Design and Architecture

Industrial networks are built from a combination of Ethernet and TCP/IP networks (to interconnect general computing systems and servers) and at least one real time network or fieldbus (to connect devices and process systems). These networks are typically nested deep within the enterprise architecture, offering some implied layers of protection against external threats. In recent years, the deployment of remote access and wireless networks within industrial systems offers new entry points into these internal networks. Chapter 5 provides an overview of some of the more common industrial network designs and architectures, the potential risk they present, and some of the methods that can be used to select appropriate technologies and strengthen these critical industrial systems.

Chapter 6: Industrial Network Protocols

This chapter focuses on industrial network protocols, including **Modbus**, **DNP3**, **OPC**, **ICCP**, **CIP**, **Foundation Fieldbus**, **Wireless HART**, **Profinet** and **Profibus**, **Zigbee**, and others. This chapter will also introduce vendor-proprietary industrial protocols, and the implications they have in securing industrial networks. The basics of protocol operation, frame format, and security considerations are provided for each, with security recommendations being made where applicable. Where properly disclosed vulnerabilities or exploits are available, examples are provided to illustrate the importance of securing industrial communications.

Chapter 7: Hacking Industrial Systems

Understanding effective cyber security requires a basic understanding of the threats that exist. Chapter 6 provides a high-level overview of common attack methodologies, and how industrial networks present a unique **attack surface** with common attack vectors to many critical areas.

Chapter 8: Risk and Vulnerability Assessments

Industrial control systems are often more susceptible to a cyberattack, yet they are also more difficult to patch due to the extreme uptime and reliability requirements of operational systems. Chapter 8 focuses on risk and vulnerability assessment strategies that specifically address the unique challenges of assessing risk in industrial networks, in order to better understand—and therefore reduce—the vulnerabilities and threats facing these real-time systems.

Chapter 9: Establishing Zones and Conduits

A strong cyber security strategy requires the isolation of devices into securable groups. Chapter 9 looks at how to separate functional groups and where functional boundaries should be implemented, using the Zone and Conduit model originated by the Purdue Research Foundation in 1989 and later adapted by ISA 99 (now known as ISA/**IEC** 62,443). Specifics are then provided on how to secure not only the interior zones but also the conduits used to interconnect these zones, including common security products, methods, and policies that may be implemented.

Chapter 10: OT Attack and Defense Lifecycles

To successfully protect an industrial environment against cyberattacks, it is necessary to understand the basics of attack and defense lifecycles. This chapter includes discussion of attack lifecycles including the MITER ATT&CK framework, in order to understand how to best utilize defensive capabilities, cybersecurity controls, countermeasures, policies, and procedures.

Chapter 11: Implementing Security and Access Controls

With a new understanding of how specific controls can influence attack and defensive lifecycles from Chapter 10, this chapter goes into more specifics about the various cybersecurity controls that are commercially available, and how they can be implemented. This chapter discusses various controls that are required to obtain cybersecurity data that are necessary for broader cybersecurity monitoring efforts.

Chapter 12: Exception, Anomaly, and Threat Detection

Industrial control network monitoring and analysis tools have become increasingly popular: enough so to justify their won chapter. These tools can provide valuable insight to the OT security teams responsible for cybersecurity monitoring and analysis. However, they can also bring new challenges, including unique deployment and operationalization considerations.

Chapter 13: Security Monitoring of Industrial Control Systems

Completing the cycle of situational awareness requires further understanding and analysis of the threat indicators that you have learned how to detect in previous chapters. Chapter 13 discusses how obtaining and analyzing broader sets of information can help you better understand what is happening and make better decisions. This includes recommendations of what to monitor, why, and how. Information management strategies—including **log** and **event** collection, direct monitoring, and correlation using **security information and event management** (**SIEM**) and other tools—are discussed here, including guidance on data collection, retention, and management.

Chapter 14: Standards and Regulations

There are many regulatory compliance standards applicable to industrial network security, and most consist of a wide range of procedural controls that are not easily resolved using IT. On top of this, there is an emergence of a large number of industrial standards that attempt to tailor many of the general-purpose IT standards to the uniqueness of ICS architectures. There are common cyber security controls (with often subtle but importance variations), however, which reinforce the recommendations put forth in this book. Chapter 12 attempts to map those cyber security–related controls from some common standards—including **NERC CIP**, **CFATS**, NIST 800–53, **ISO**/IEC 27,002:2005, ISA 62,443, **NRC** RG 5.71, and NIST 800–82—to the security recommendations made within this book, making it easier for security analysts to understand the motivations of compliance officers, while compliance officers are able to see the security concerns behind individual controls.

Chapter 15: Common Pitfalls and Mistakes

Back by popular demand, this chapter highlights some common pitfalls and mistakes—including errors of complacency, common misconfigurations, and deployment errors. By highlighting the pitfalls and mistakes, it is easier to avoid repeating those mistakes.

Changes made to the third edition

For readers of previous editions of industrial network security, securing critical infrastructure networks for smart grid, SCADA and other ICSs, you will find new and updated content throughout the book. However, the largest changes that have been made include.

- Revised diagrams, designed to provide a more accurate representation of industrial systems so that the lessons within the book can be more easily applied in real life.
- Better organization of topics, including major revisions to both introductory chapters (Chapter 2, that are intended to provide a more effective introduction of topics.
- The separation of "hacking methodologies" and "risk and vulnerability assessment" into two chapters, expanding each to provide significantly more detail to each very important subject.
- The expansion of "risk and vulnerability assessment" to expand further beyond network-scan-based assessments and include more system-level assessment guidance, including safety considerations, cyber-physical threat modeling, and cybersecurity HAZOPs discussions.
- The inclusion of wireless networking technologies and how they are applied to industrial networks, including important differences between general-purpose IT and specific ICS technology requirements.
- Much greater depth on the subjects of industrial firewall implementation and industrial protocol filtering—important technologies that were in their infancy during the first edition but are now commercially available.
- The inclusion of real-life vulnerabilities, exploits, and defensive techniques throughout the book to provide a more realistic context around each topic, while also proving the reality of the threat against critical infrastructure.
- An entirely new chapter on "OT Defense Lifecycle & Defensive Methods," which discusses the OT cyber defensive lifecycle, from detection to response to recovery. This chapter acts as a precursor to the previous chapter on "Implementing Security and Access Controls," which has been expanded to include newer security controls and to provide more specific guidance where available.
- For readers of earlier editions, or who have established careers or responsibilities in OT cyber security, new material has been added. Discussions of cyber security posture, cyber security maturity, and the lifecycle of a cybersecurity incident have been added, as well as expanded discussions of when, where, and how to implement OT cyber security controls.
- The closing chapter on "Pitfalls and Mistakes" is back! Since the first edition was published, the industry has made its share of new blunders. Do not be embarrassed if you have made these mistakes; learn from them; and maybe laugh a little along the way.

Conclusion

Writing the first edition of this book was an education, an experience, and a challenge. In the months of research and writing, several historic moments occurred concerning ICS security, including the first ICS-targeted cyber weapon: Stuxnet. At the time, Stuxnet was the most sophisticated cyberattack to date. Since then, its complexity and sophistication have been surpassed more than once, and the frequency of new threats continues to rise. There is a growing number of attacks, more relevant cyber security research (from both **blackhats** and **whitehats**), and new evidence of advanced persistent threats, cyber espionage, nation-based cyber privacy concerns, and other socio-political concerns on what seems like a daily basis.

Hopefully, this book will be both informative and enjoyable, and it will facilitate the increasingly urgent need to strengthen the security of our industrial networks and automation systems. Even though the attacks themselves will continue to evolve, the methods provided herein should help to prepare against the inevitable advancement of industrial network threat.

Endnotes

1. Gartner, Inc. "Gartner Glossary, Information Technology". Document from the web, cited January 2023. https://www.gartner.com/en/information-technology/glossary/internet-of-things
2. Keith D. Foote, "A Brief History of the Internet of Things" January 14, 2022. Document from the web. https://www.dataversity.net/brief-history-internet-things/

2 About Industrial Networks

Information in this chapter

- The Use of Terminology Within This Book
- Understanding "OT" versus "IT"
- Common Industrial Security Recommendations
- Advanced Industrial Security Recommendations
- Common Misperceptions About Industrial Network Security

It is important to understand some of the terms used when discussing industrial networking and industrial control systems (ICSs), as well as the basics of how industrial networks are architected and how they operate before attempting to secure an industrial network and its interconnected systems. It is also important to understand some of the common security recommendations deployed in business networks and why they may or may not be truly suitable for effective industrial network cyber security.

What is an industrial network? Because of a rapidly evolving socio-political landscape, the definition of a cyber threat has become blurred. Terms such as "critical infrastructure," "APT," "IoT," "SCADA," and "Smart Grid" are used freely and often incorrectly. More recently, "operational technology" (OT) has become a popular way to describe ICSs and other industrial environments (see "Understanding "OT" vs. "IT," below). It can be confusing to discuss a distributed metering system or an oil refinery in the same general terms. Without some generalization, however, the diversity of industrial networks and the markets they serve would require a stack of books each catering to slight nuances of both technology and vernacular. Many regulatory agencies and commissions have also been formed to help secure different types of industrial networks for different industry sectors—each introducing their own specific nomenclatures and terminology.

In hopes that common security needs, practices, and guidance can be provided across all applications within a single text, this chapter will attempt to provide a baseline for talking about industrial network cybersecurity, introducing the reader to some of the common terminology, issues, and security recommendations that will be discussed throughout the remainder of this book.

The use of terminology within this book

The authors have witnessed many discussions on industrial cybersecurity fall apart due to disagreements over terminology. There is a good deal of terminology specific to both cybersecurity and to ICSs that will be used throughout this book. Some readers may be cybersecurity experts who are unfamiliar with ICSs, while others may be industrial system professionals who

Industrial Network Security. https://doi.org/10.1016/B978-0-443-13737-2.00014-2

are unfamiliar with cybersecurity. For this reason, a conscience effort has been made by the authors to convey the basics of both disciplines and to accommodate both types of readers.

Some of the terms that will be used extensively include.

- Assets (including whether they are physical or logical assets, and if they are classified as cyberassets, critical assets, and critical cyberassets)
- Enclaves, zones, and conduits
- Enterprise or business networks
- Industrial control systems—DCS, PCS, SIS, SCADA
- Building control systems
- Industrial networks
- Plants, mills, refineries, lines
- Industrial protocols
- Network perimeter or electronic security perimeter (ESP)
- Critical infrastructure

Some cyber security terms that will be addressed include.

- Attacks
- Breaches
- Vulnerabilities
- Exploits
- Incidents
- Risk
- Security measures, security controls, or countermeasures

These will be given some cursory attention here, as a foundation for the following chapters. There are many more specialized terms that will be used, and so an extensive glossary has been provided at the back of this book. The first time a term is used; it will be printed in bold to indicate that it is available in the glossary.

Note

The book title "*Industrial Network Security: Securing Critical Infrastructure Networks for Smart Grid, SCADA, and Other Industrial Control Systems*" was chosen because this text discusses all of these terms to some extent. "Industrial cybersecurity" is a topic relevant to many industries, each of which differs significantly in terms of design, architecture, and operation. An effective discussion of cybersecurity must acknowledge these differences; however, it is impossible to cover every nuance of distributed control system (DCS), supervisory control and data acquisition (SCADA), smart grids, critical manufacturing, etc. This book will focus on the commonalities among these industries, providing a basic understanding of industrial automation, and the constituent systems, sub-systems and devices that are used. Every effort will also be made to refer to all industrial automation and control systems (DCS, PCS, SCADA, etc.) as simply ICSs or just ICS. It is also important to understand that industrial networks are one link in a much larger chain comprising: fieldbus networks, process control networks, supervisory networks, business networks, remote access networks, and any number of specialized applications, services, and communications infrastructures that may all be interconnected and therefore must be assessed and secured within the context of cybersecurity. A smart grid, a petroleum refinery, and a city skyscraper may all utilize ICS systems, yet each represents unique variations in terms of size, complexity, and risk. All are built using the same technologies and principles making the cyber security concerns of each similar and the fundamentals of industrial cyber security equally applicable.

Note

This book does not go into extensive detail on the architecture of smart grids due to the complexity of these systems. Please consult the book "*Applied Cyber Security and the Smart Grid*"[1] if more detail on smart grid architecture and its associated cyber security is desired.

Attacks, breaches and incidents; malware, exploits, and APTs

The reason that you are reading a book titled "*Industrial Network Security*" is likely because you are interested in, if not concerned, about unauthorized access to and potentially hazardous or mischievous usage of equipment connected to an industrial network. This could be a deliberate action by an individual or organization, a government backed act of cyberwar, the side effect of a computer virus that just happened to spread from a business network to an ICS server, the unintended consequence of a faulty network card or—for all we know—the result of some astrological alignment of the sun, planets and stars (aka "solar flares"). While there are subtle differences in the terms "incident" and "attack"—mostly to do with intent, motivation, and attribution—this book does not intend to dwell on these subtleties. The focus in this book is how an attack (or breach, or exploit, or incident) might occur and subsequently how to best protect the industrial network and the connected ICS components against undesirable consequences that result from this action. Did the action result in some outcome—operational, health, safety, or environment, which must be reported to a federal agency according to some regulatory legislation? Did it originate from another country? Was it a simple virus or a persistent rootkit? Could it be achieved with free tools available on the Internet or did it require the resources of a state-backed cyber espionage group? Do such groups even exist? The author of this book thinks that these are all great questions, but ones best served by some other book. These terms may therefore be used rather interchangeably herein.

Assets, critical assets, cyberassets, and critical cyberassets

An asset is simply a term for a component that is used within an ICS. Assets are "physical" such as a workstation, server, network switch or PLC. Physical assets also include the large quantity of sensors and actuators used to control an industrial process or plant. There are also "logical" assets that represent what is contained within the physical asset such as a process graphic, a database, a logic program, a firewall rule set, or firmware. When you think about it, cybersecurity is actually focused on the protection of "logical" assets and not the "physical" assets that contain them. Physical security is that which tends to focus more on the protection of a physical asset. Security from a general point-of-view can therefore effectively protect a "logical" asset, a "physical" asset, or both. This will become more obvious as we develop the concept of security controls or countermeasures later in this book.

The Critical Infrastructure Protection (CIP) standard by the North American Electric Reliability Corporation (NERC) through version 4 has defined a "critical cyber asset" or "CCA" was defined as any device that uses a routable protocol to communicate outside the electronic security perimeter (ESP), uses a routable protocol within a control center, or is dial-up accessible.[2] This changed in version five of the standard by shifting from an individual asset approach, to one that addresses groupings of CCA's called bulk electric system (BES) cyber "systems."[3] This approach represents fundamental shift from addressing security at the component or asset level, to a more holistic or system-based one.

A broad and more generic definition of "asset" is used in this book: where any component—physical or logical; critical or otherwise—is simply referred to as an "asset." This is because most ICS components today, even those designed for extremely basic functionality, are likely to contain a commercial microprocessor with both embedded and user-programmable code that most likely contains some inherent communication capability. History has proven that even single-purpose, fixed-function devices can be the targets, or even the source of a cyberattack by specifically exploiting weaknesses in a single component within the device (See Chapter 3, "Industrial Cyber Security History and Trends"). Many devices ranging from ICS servers to PLCs to motor drives have been impacted in complex cyberattacks—as was the case during the 2010 outbreak of **Stuxnet** (see "Examples of Advanced Industrial Cyber Threats" in Chapter 7, "Hacking Industrial Control Systems"). Regardless of whether a device is classified as an "asset" for regulatory purposes or not, they will all be considered accordingly in the context of cybersecurity.

Security controls and security countermeasures

The term "security controls" and "security countermeasures" are often used, especially when discussing compliance controls, guidelines, or recommendations. They simply refer to a method of enforcing cybersecurity: either through the use of a specific product or technology, a security plan or policy, or other mechanism for establishing and enforcing cyber security, in order to reduce risk.

Firewalls and intrusion prevention systems

While there are many other security products available—some of which are highly relevant to industrial networks—none have so been so broadly used to describe products with such highly differing sets of capabilities. The most basic "firewall" must be able to filter network traffic in at least one direction, based on at least one criterion such as Internet protocol (IP) address or communication service port. A firewall may or may not also be able to tracking the "state" of a particular communication session, understanding what is a new "request" versus what is a "response" to a prior request.

A "deep packet inspection" (DPI) system is a device that can decode network traffic and look at the contents or payload of that traffic. DPI is typically used by intrusion detection systems (IDS), intrusion prevention systems (IPS), advanced firewalls, and many other specialized cybersecurity products to detect signs of attack. IDS can detect and alert but do not block or reject bad traffic. IDS can block traffic. Industrial networks support high availability making most general IPS appliances less common on critical networks—IPS is more often applied at upper-level networks where high availability (typically >99.99%) is not such a high priority. The result is that good advice can lead to inadequate results, simply through the use of overused terms when making recommendations.

Note

Most modern IPS can be used as IDS by configuring the IPS to alert on threat detection but not to drop traffic. Because of this the term "IPS" is now commonly used to refer to both IDS and IPS systems. One way to think about IDS and IPS is that an IPS device that is deployed in-line (a "bump in the wire") is more capable of "preventing" an intrusion by dropping suspect packets, while an IPS deployed out-of-band (e.g., on a span port) can be thought of as an IDS because it is monitoring mirrored network traffic and can detect threats but is less able to prevent them. It may be the same make and model of network security device, but the way it is configured and deployed indicates whether it is a "passive" IDS or an "active" IPS.

Consider that the most basic definition of a firewall, given above, fails to provide the basic functionality recommended by NIST and other organizations, which advise filtering traffic on both the source and destination IP address and the associated service port, bi-directionally. At the same time, many modern firewalls are able to do much more: looking at whole application sessions rather than isolated network packets; by filtering application contents; and then enforcing filter rules that are sometimes highly complex. These unified threat management (UTM) appliances are becoming more common in protecting both industrial and business networks from today's advanced threats. Deploying a "firewall" may be inadequate for some installations while highly capable at others, depending upon the specific capabilities of the "firewall" and the particular threat that it is designed to protect the underlying system against. The various network-based cyber security controls that are available and relevant to industrial networks are examined in detail in Chapter 10, "Implementing Security and Access Controls" and Chapter 11, "Anomaly and Threat Detection."

Industrial control system

An **industrial control system** is a broad class of automation systems used to provide control and monitoring functionality in manufacturing and industrial facilities. An ICS actually is the aggregate of a variety of system types including process control system (PCS), DCS, SCADA system, safety instrumented system (SIS), and many others. A more detailed definition will be provided in Chapter 4, "Introduction to Industrial Control Systems and Operations."

Figure 2.1 is a simplified representation of an ICS consisting of two controllers and a series of inputs and outputs connecting to burners, valves, gauges, motors, etc. that all work in a tightly integrated manner to perform an automated task. The task is controlled by an application or logic running inside the controller, with local panels or human–machine interfaces (HMI) used to provide a "view" into the controller allowing the operator to see values and make changes to how the controller is operating. The ICS system typically includes toolkits for creating the process logic that defines the task, as well as toolkits for building custom operator interfaces or graphical user interfaces (GUI) implemented on the HMI. As the task executes, the results are recorded in a database

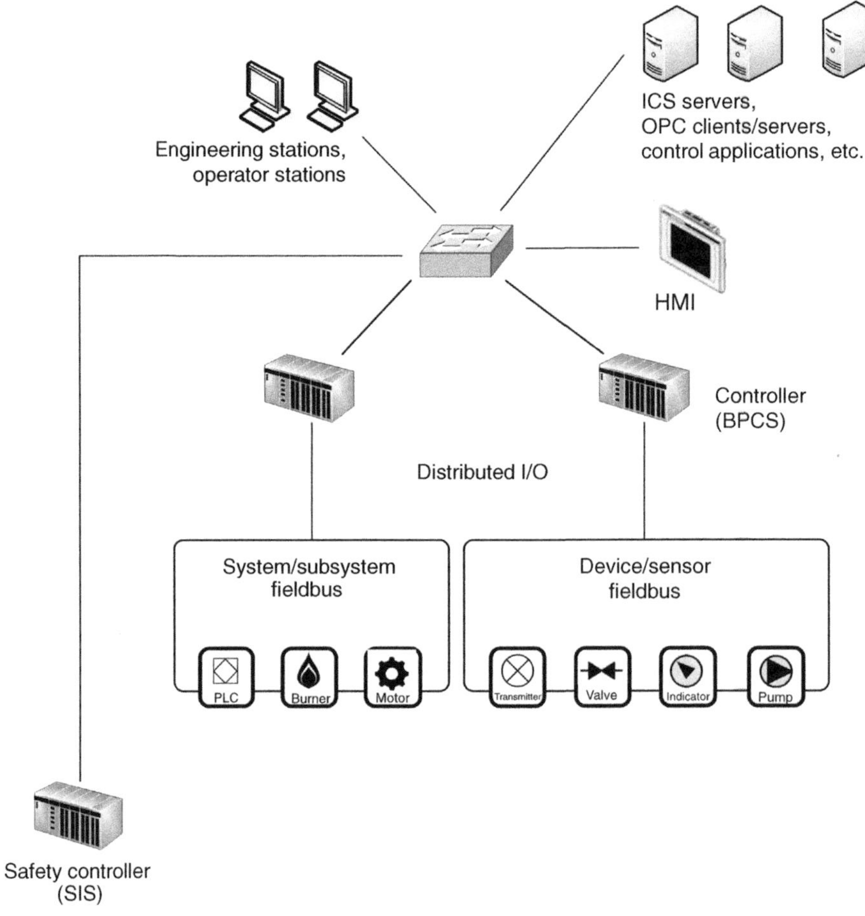

FIGURE 2.1 Sample industrial automation and control system.

called an Historian (see Chapter 4, "Introduction to Industrial control Systems and Operations" for more information and detail on how such a system operates).

Building control systems

Building control systems (**BCS**) are specialized industrial control systems that focus on aspects of building automation and control, including physical access and entry, heating, ventilation, and air conditioning (HVAC) and other climate controls, lighting, safety, and other systems. As with any industrial system, any digitally controllable mechanical aspect of a building can be connected, monitored, and controlled. Much like a "Smart Home" that consists of home automation, a BCS makes many aspects of building management intelligent and automated.

For readers who are responsible for securing a BCS, there is good news: a BCS is essentially just a focused application of the industrial automation and control systems that are discussed extensively in this book. Unless noted otherwise herein, the information and guidance offered for securing OT devices applies equally to a BCS.

DCS or SCADA?

Originally, there were significant differences between the architectures of a DCS versus that of an SCADA system. As technology has evolved, these differences have diminished, and there can often be a blur between whether a particular ICS is in fact classified as DCS or SCADA. Both systems are designed to monitor (reading data and representing it to a human operator and possibly to other applications such as historians and advanced control applications) and to control (defining parameters and executing instructions) manufacturing or industrial equipment. These system architectures vary by vendor, but all typically include the applications and tools necessary to generate, test, deploy, monitor, and control an automated process. These systems are multifaceted tools, meaning that a workstation might be used for purely supervisory (read only) purposes by a quality inspector, while another may be used to optimize process logic and write new programs for a controller, while yet a third may be used as a centralized user interface to control a process that requires more human intervention, effectively giving the workstation the role of the HMI.

It should be noted that ICSs are often referred to in the media simply as "SCADA," which is both inaccurate and misleading. Looking at this another way—an SCADA system is in fact an ICS, but not all ICS are SCADA! The authors hope to help clarify this confusion in Chapter 4, "Introduction to Industrial Control Systems and Operations."

Plants, mills, refineries, and lines

Every industry has specific terminology for industrial control environments. Certain aspects of power generation and chemical manufacturing, for example, refer to their systems as "plants," whereas other manufacturing facilities might refer to their industrial systems as "lines." The paper industry operates "mills" for the processing of materials into paper and may also operate secondary "plants" for producing products from that paper. Certain chemical processes may occur in "refineries," while others might occur in a "plant." Within a plant, mill, refinery, *et. al.*, many different automation processes may occur: discreet versus continuous manufacturing; cracking versus coking; pressurization versus filtration; effluence versus emissions; etc.

Again, there is good news in that all of these industries and processes can be effectively generalized when discussing the basics of industrial network cybersecurity. As this book delves into more advanced topics, the specifics of one industry might have some influence, but the fundamentals behind them should be applicable to all industrial environments.

Industrial networks

The various assets that comprise an ICS are interconnected over an industrial network. While the ICS represented in Figure 2.1 is accurate, in a real deployment, the management and supervision of the ICS will be separated from the controls and the automation system itself. Figure 2.2 shows how an ICS is actually part of a much larger architecture, consisting of plant areas that contain common and shared applications, area-specific control devices, and associated field equipment, all interconnected via a variety of network devices and servers. In large or distributed architectures, there will be a degree of local and remote monitoring and control that is required (i.e., in the plant), as well as centralized monitoring and control (i.e., in the control room). This is covered in detail in Chapter 5, "Industrial Network Design and Architecture." For now, it is sufficient to understand that the specialized systems that comprise an ICS are interconnected, and this connectivity is what we refer to as an industrial network.

Industrial protocols

Most ICS architectures utilize one or more specialized protocols that may include vendor-specific proprietary protocols (such as Honeywell CDA, General Electric SRTP or Siemens S7, and many others) or nonproprietary and/or licensed protocols including OPC, Modbus, DNP3, ICCP, CIP, PROFIBUS, and others. Many of these were originally designed for serial communications but have been adapted to operate over standard Ethernet link layer using the Internet Protocol with both UDP and transmission control protocol (TCP) transports and are now widely deployed over a variety of common network infrastructures. Because most of these protocols operate at the application layer, they can be accurately (and often are) referred to as applications. They are referred to as protocols in this book to separate them from the software applications that utilize them—such as DCS, SCADA, EMS, historians, and other systems.

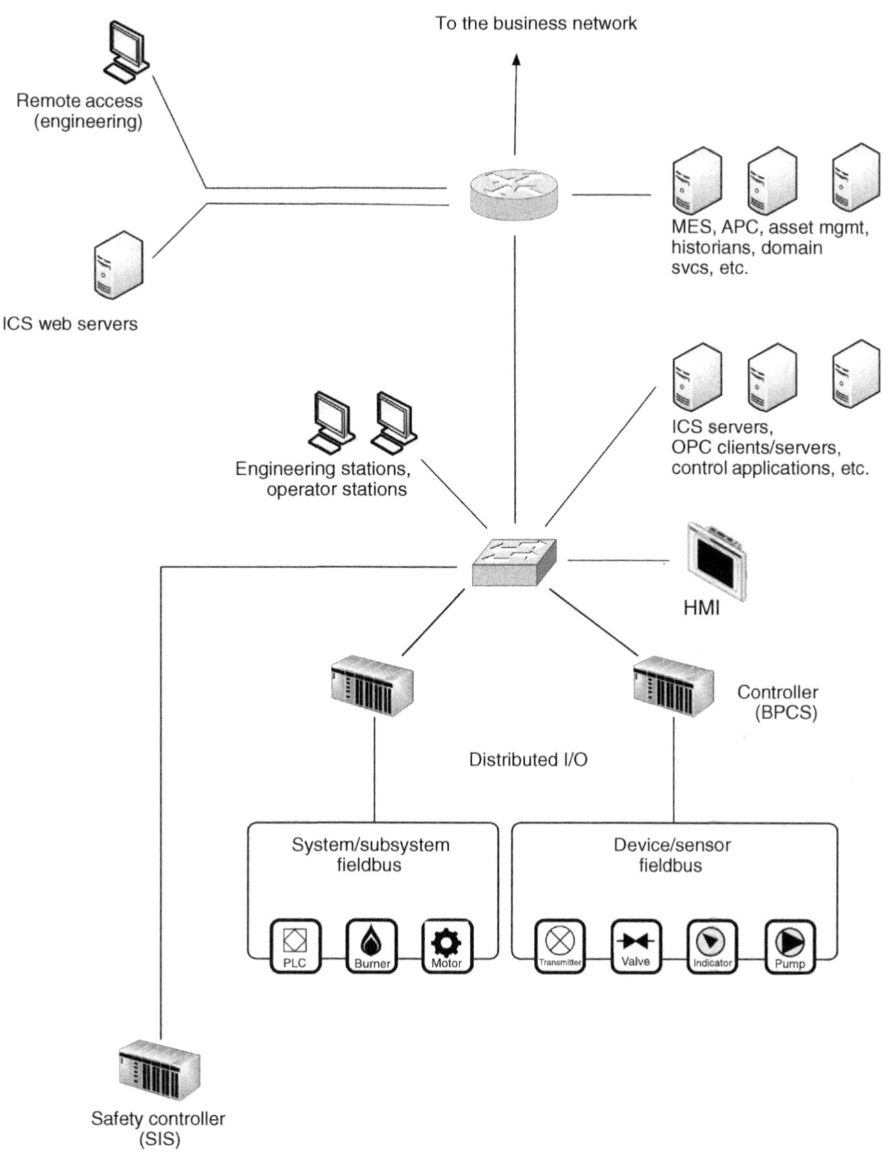

FIGURE 2.2 Sample network connectivity of an industrial control system.

Call out

The open systems interconnection (OSI) model

The OSI model defines and standardizes the function of how a computing system interacts with a network. Each of seven layers is dependent upon and also serves the layers above and below it, so that information from an application (defined at the topmost or application layer) can be consistently packaged and delivered over a variety of physical networks (defined by the bottommost or physical layer). When one computer wants to talk to another on a network, it must step through each layer: Data obtained from applications (layer 7) are represented to the network (layer 6) in defined sessions (layer 5), using an established transport method (layer 4), which in turn uses a networking protocol to address and route the data (layer 3) over an established link (layer 2) using a physical transmission mechanism (layer 1). At the destination, the process is reversed in order to deliver the data to the receiving application. With the ubiquity of the Internet Protocol, a similar model called the TCP/IP model is often used to simplify these layers. In the TCP/IP model, layers five through 7 (which all involve the representation and management of application data), and layers one and 2 (which define the interface with the physical network) are consolidated into a single application layer and network interface layer. In this book, we will reference the OSI model in order to provide a more specific indication of what step of the network communication process we are referring to (Figure 2.3).

Because these protocols were not designed for use in broadly accessible or public networks, cybersecurity was seen as compensating control and not an inherent requirement. Now, many years later, this translates to a lack of robustness that makes

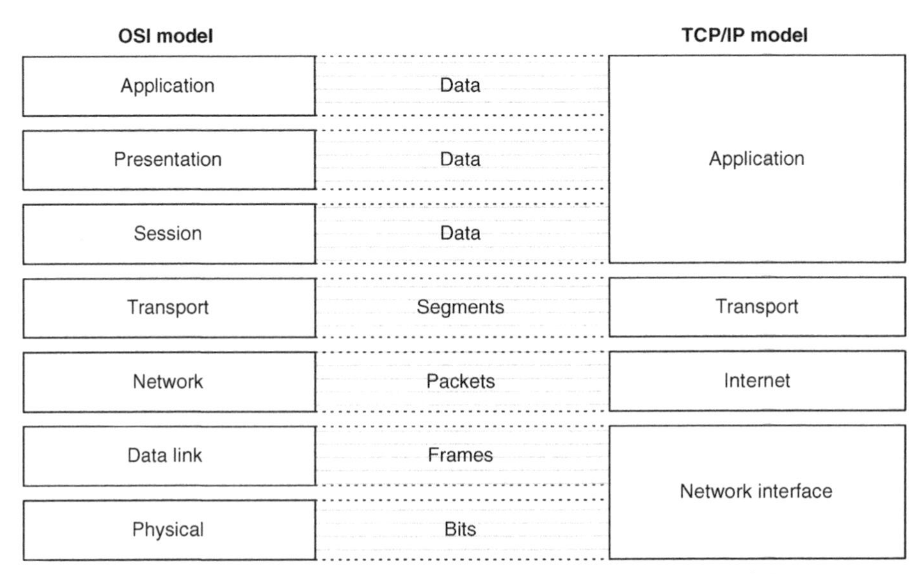

FIGURE 2.3 The Open Systems Interconnection (OSI) and transmission control protocol (TCP)/Internet protocol (IP) models.

the protocols easily accessed—and in turn they can be easily broken, manipulated, or otherwise exploited. Some are proprietary protocols (or open protocols with many proprietary extensions such as Modbus-PEMEX), and as such, they have benefited for some time by the phenomena of "security by obscurity." This is clearly no longer the case with the broader availability of information on the World Wide Web, combined with an increasing trend of industrial-focused cyber security research. Many of the concerns about industrial systems and critical infrastructure stem from the growing number of disclosed vulnerabilities within these protocols. One disturbing observation is that in the few years following the Stuxnet attack, many researchers have found numerous vulnerabilities with open protocol standards and the systems that utilize them. Little attention has been given to the potential problem of vulnerabilities in the proprietary products that are often times too cost prohibitive for traditional researchers to procure and analyze. These proprietary systems and protocols are at the core of most critical industry and represent the greatest risk should they be compromised. See Chapter 6, "Industrial Network Protocols" and Chapter 7, "Hacking Industrial Systems" for more detail on these protocols, how they function, and how they can/have been compromised.

Networks, routable networks and non-routable networks

The differentiation between routable and nonroutable networks is becoming less common as industrial communications become more ubiquitously deployed over IP. A "nonroutable" network refers to those serial, bus, and point-to-point communication links that utilize **Modbus/RTU**, DNP3, fieldbus, and other networks. They are still networks: they interconnect devices and provide a communication path between digital devices and in many cases, are designed for remote command and control. A "routable" network typically means a network utilizing the Internet Protocol (TCP/IP or UDP/IP), although other routable protocols such as AppleTalk, DECnet, Novell IPX, and other legacy networking protocols certainly apply. "Routable" networks also include routable variants of early "nonroutable" ICS protocols that have been modified to operate over TCP/IP, such as **Modbus over TCP/IP**, **Modbus/TCP**, and **DNP3 over TCP/UDP**. ICCP represents a unique case in that it is a relative new protocol developed in the early 1990s, which allows both a point-to-point version and a wide-area routed configuration.

Routable and nonroutable networks would generally interconnect at the demarcation between the control and supervisory control networks, although in some cases (depending upon the specific industrial network protocols used) the two networks overlap. This is illustrated in Figure 2.4 and is discussed in more depth in Chapter 5, "Industrial Control System Network Design and Architecture" and Chapter 6, "Industrial Network Protocols."

These terms were popularized through NERC CIP regulations, which implies that a routable interface can be easily accessed by the network either locally or remotely (via adjacent or public networks) and therefore requires special cyber security consideration,

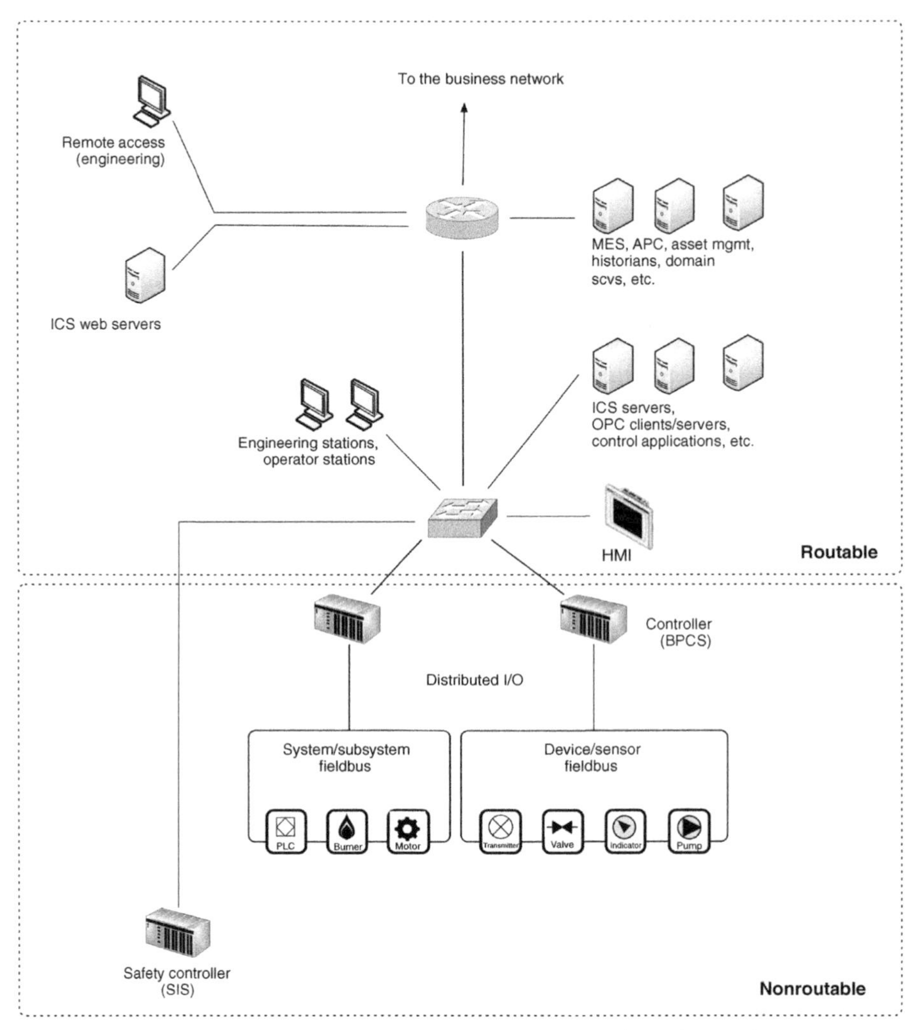

FIGURE 2.4 Routable and nonroutable areas within an industrial control system.

and inversely that nonroutable networks are "safer" from a network-based cyberattack. This is misleading and can prevent the development of a strong cyber security posture. Today, it should be assumed that *all* industrial systems are connected either directly or indirectly to a "routable" network, whether or not they are connected via a routable protocol. Although areas of industrial networks may still be connected using serial or bus networks that operate via specific proprietary protocols, these areas can be accessed via other interconnected systems that reside on a larger IP network. For example, a PLC may connect to discrete I/O over legacy fieldbus connections. If considered in isolation, this would be a nonroutable network. However, if the PLC also contains an Ethernet uplink to connect to a centralized ICS system, the PLC can be accessed via that network and then manipulated to alter communications on the "nonroutable" connections. To further complicate things, many devices have remote access capabilities such as modems, infrared receivers, radio, or other connectivity options that may not be considered "routable" but are just as easily accessed by a properly equipped attacker. Therefore, the distinction between routable and nonroutable—though still widely used—is no longer considered a valid distinction by the authors. For the purposes of strong and cohesive cybersecurity practices, all networks and all devices should be considered potentially accessible and vulnerable. See Chapter 8, "Risk and Vulnerability Assessments" for more detail on determining accessibility and identifying potential attack vectors.

Enterprise or business networks

An ICS is rarely an isolated system (in years of ICS design, we have found only a handful of examples of control systems that had no connectivity to any network). For every factory floor, electric generator, petroleum refinery, or pipeline, there is a corporation or organization that owns and operates the facility, a set of suppliers that provides raw materials, and a set of customers that receive the manufactured products. Like any other corporation or organization, these require daily business functions: sales, marketing, engineering, product management, customer service, shipping and receiving, finance, partner connectivity, supplier access, etc. The network of systems that provide the information infrastructure to the business is called the business network.

There are many legitimate business reasons to communicate between the enterprise systems and industrial systems, including production planning and scheduling applications, inventory management systems, maintenance management systems, and manufacturing execution systems to name a few. The business network and the industrial network interconnect to make up a single end-to-end network.

Figure 2.5 illustrates this end-to-end functional network, as well as the separation of the business networks from the industrial networks, which consist of plant, supervisory, and functions. In this example, there is a high degree of redundancy in all areas, which is intended to make a point: the network infrastructure may be designed using the same "enterprise" switches and routers as those used in the business network. In some areas of an industrial network, "industrial" switches and routers may be used, which support harsher environments, offer higher availability, eliminate moving parts such as fans and are otherwise engineered for "industrial" and sometimes "hazardous" use. In this book, the industrial network is defined by its function, not by the marketing designation provided by a product vendor, and so the supervisory network in Figure 2.4 is considered an industrial network even though it uses enterprise-class networking gear.

It should also be noted that there are several systems and services that exist in both business and industrial networks such as directory services, file servers, and databases. These common systems should not be shared between business and industrial networks but rather replicated in both environments in order to minimize the interconnectivity and reduce the potential attack surface of both the ICS and enterprise infrastructure.

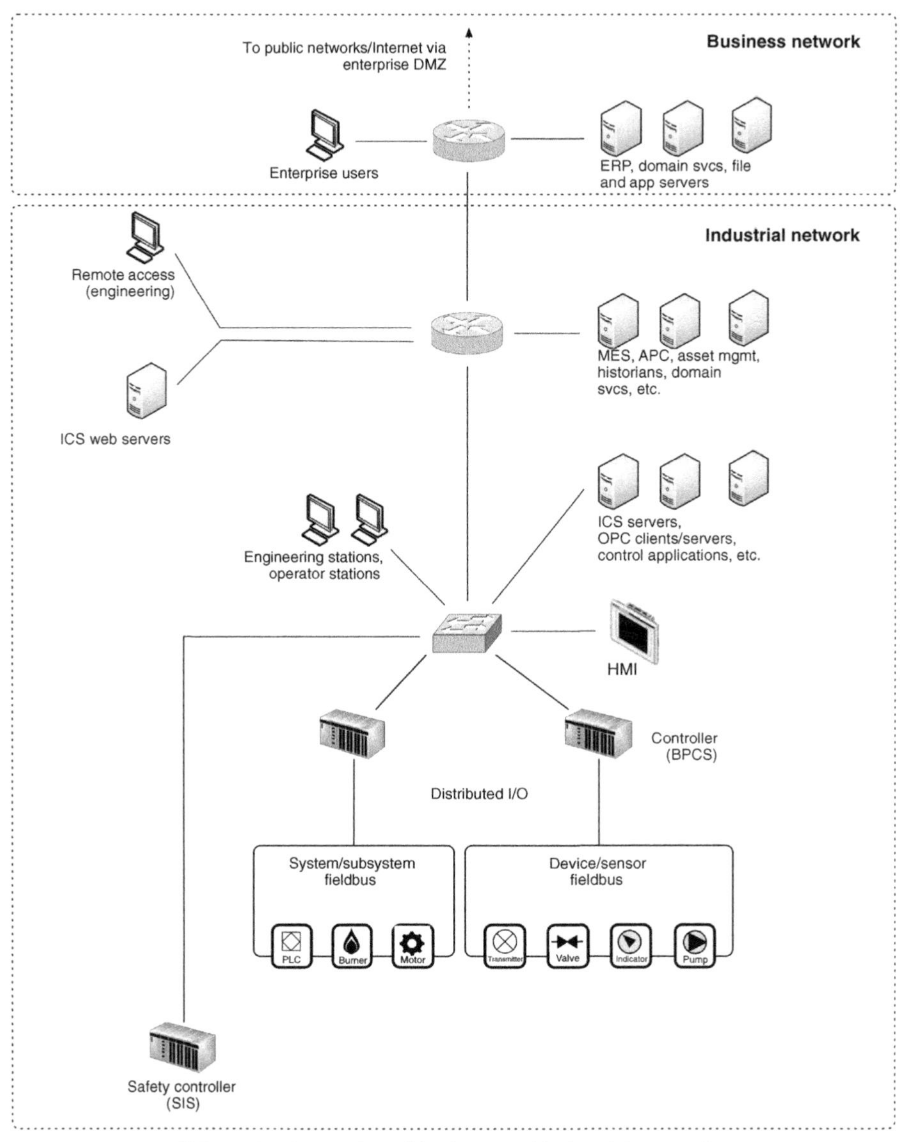

FIGURE 2.5 Separation of business and industrial networks.

This book does not focus on the business network or its systems except where they might be used as an attack vector into the ICS. There are numerous books available on enterprise cybersecurity if more information is required on this subject. This book will also not focus on how internal attacks originating from the industrial network might be used to gain unauthorized access to business networks (this is a legitimate concern, but it is outside of the scope of this book).

Zones and enclaves

The terms "enclave" and "zone" are convenient for defining a closed group of assets or a functional group of devices, services, and applications that make up a larger system. While the term "enclave" is often used in the context of military systems, the term "zone" is now becoming more recognized because it is referenced heavily within the widely adopted industry standards: ISA-62,443 (formerly ISA-99). Originally developed from the Purdue Reference Model for Computer Integrated Manufacturing,[4] the concept of zones and conduits has now become widely adopted.

Within this model, communications are limited to only those devices, applications, and users that should be interacting with each other legitimately in order to perform a particular set of functions. Figure 2.6 shows zones as illustrated within IEC-62,443, where Figure 2.7 then shows the same model applied to the sample network architecture used throughout this book.

The term "zone" is actually not new but in fact has been used for many years in describing a special network that is created to expose a subset of resources (servers, services, applications, etc.) to a larger, untrusted network. This "demilitarized zone" or DMZ is typically used when enterprises want to place external-facing services like web servers, email servers, B2B portals, etc. on the Internet while still securing their more trusted business networks from the untrusted public Internet networks. It is important to note that at this point in the book, Figure 2.7 has been simplified and omits multiple DMZ's that would typically be deployed to protect the plant and enterprise zones.

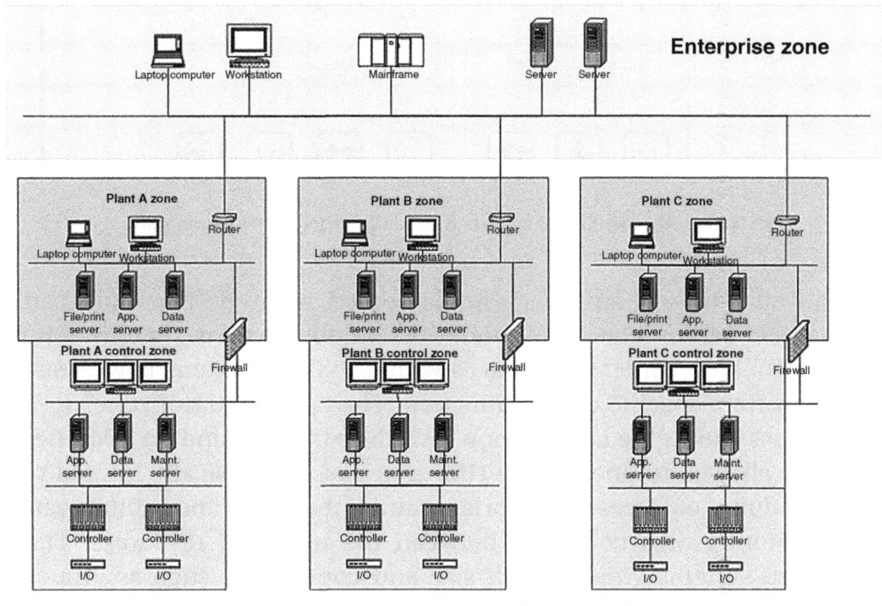

FIGURE 2.6 The ISA-62,443 zone and conduit model (block diagram).

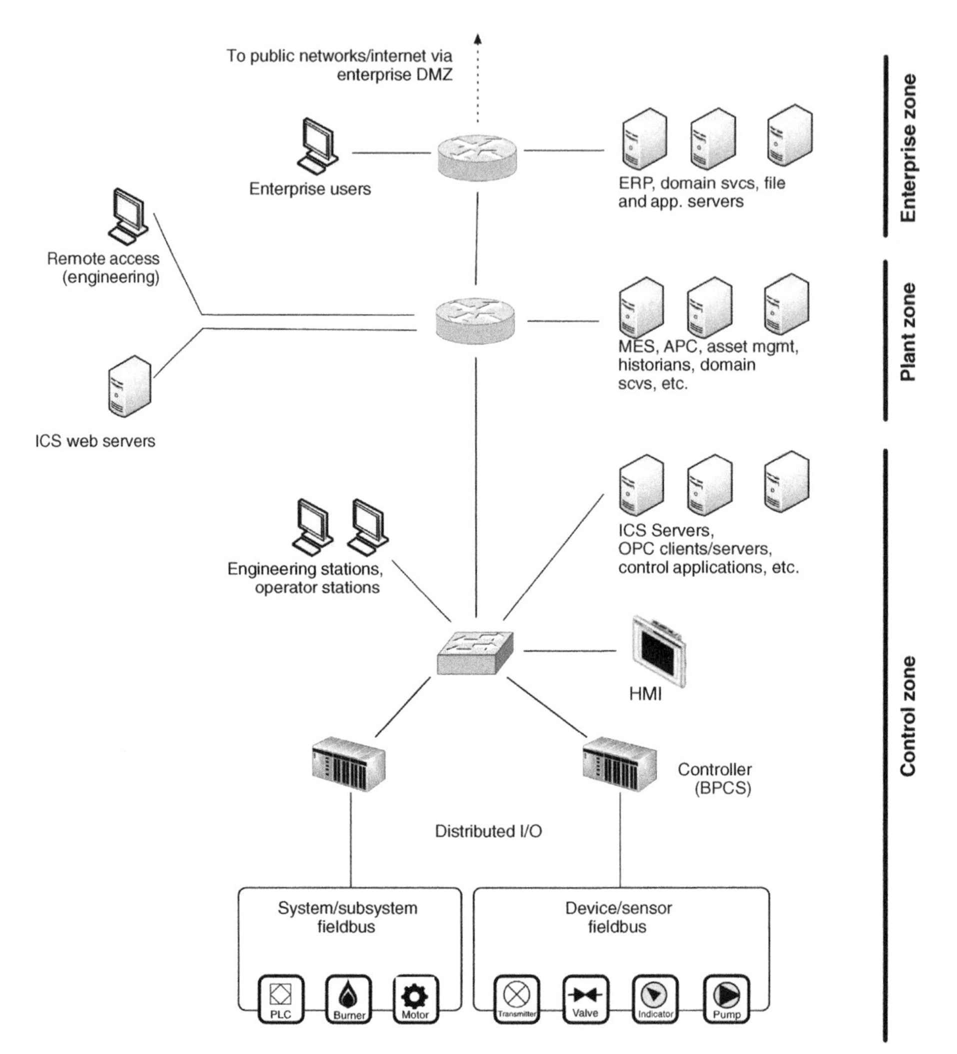

FIGURE 2.7 The ISA-62,443 zone and conduit model (network diagram).

While highly effective when properly implemented, zones and conduits can become difficult to engineer and to manage in more distributed, complex systems. For example: in a simple control loop, an HMI interfaces with a PLC that interacts with sensors and actuators to perform a specific control function. The "plant control zone" in Figure 2.6 includes all devices within the control loop including the PLC and an HMI. Because the authorized users allowed to operate the HMI may not be physically located near these devices, a "conduit" enforces appropriate authentication and authorization (and potentially monitoring or accounting) between the user and resources. This can be exasperated when systems grow in both size and complexity, such as in a smart grid architecture. Smart grids are highly complex and highly interconnected, as evident in

NISTIR 7628 Guidelines for smart grid cyber security v1.0–Aug 2010

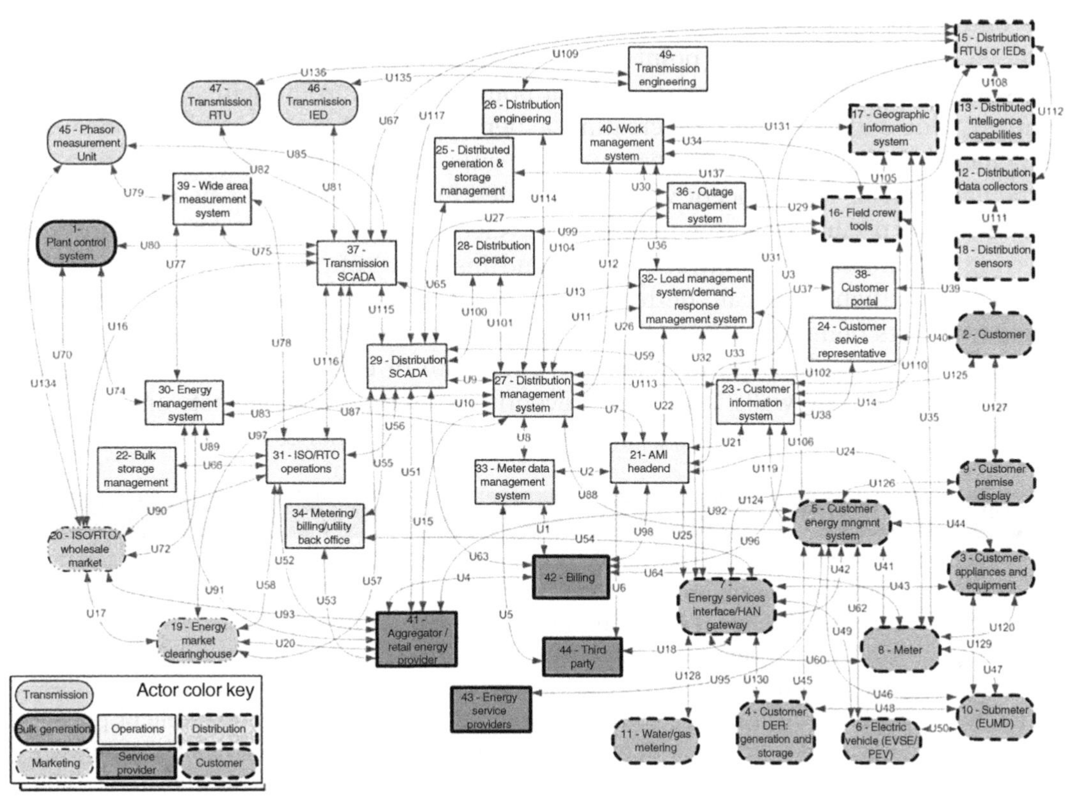

FIGURE 2.8 The challenge of applying zones to the smart grid (from NISTIR 7628).

Figure 2.8, making it difficult to adequately separate systems into security zones. For more on the zone and conduit model and how to apply it to real industrial control environments, see Chapter 9, "Establishing Zones and Conduits."

Note

Zone and conduits are a method of **network segregation** or the separation of networks and assets in order to enforce and maintain access control. A zone does not necessarily require a physical boundary, but it does require a logical delineation of systems (i.e., assets combined with the communication conduits that exist between them). Zones are an important aspect of cybersecurity as they define acceptable versus unacceptable access to the various systems and sub-systems that comprise an ICS that are placed within a particular zone. Though many standards may not specifically mention zones, most describe the concept of segmentation as one of the fundamental network security controls. Zones and conduits are typically the outcome of this network segmentation activity. The mapping and management of zones can become confusing because a single asset could exist in multiple logical zones. The concept of zones is expanded further in Chapter 9, "Establishing Zones and Conduits," but for now, it is enough to understand the term and how it will be used.

Network perimeters or "electronic security perimeters"

The outermost boundary of any closed group of assets (i.e., a "zone") is called the perimeter. The perimeter is a point of demarcation between what is outside of a zone and what is inside. A perimeter is a logical point at which to implement cybersecurity controls. One hidden aspect of creating a perimeter is that it provides a means to implement controls on devices that may not support the direct implementation of a particular control. This concept will be explained further later in this book.

NERC CIP popularized the terminology "electronic security perimeter" or "ESP" referring to the boundary between secure and insecure zones.[5] The perimeter itself is nothing more than a logical "dotted line" around that separates the closed group of assets within its boundaries from the rest of the network. "Perimeter defenses" are the security defenses established to police the entry into or out of the different zones and typically consist of firewalls, intrusion prevention system, or similar network-based filters. This is discussed in depth in Chapter 9, "Establishing Zones and Conduits."

Note: Perimeter security and the cloud

When dealing with well-defined, physically segmented, and demarcated networks, perimeters are easily understood and enforced. However, as more and more remote systems become interconnected, often relying on shared resources stored in a central data center, a perimeter becomes more difficult to define and even more difficult to enforce. A smart grid, for example, may utilize broadly distributed measurement devices throughout the transmission and distribution grid, all of which interact with a centralized service. This is an example of private cloud computing, and it comes with all of the inherent risks and concerns of cloud-based computing. For more information about cloud computing, please refer to the "CSA Guide to Cloud Computing" by Raj Samani, Brian Honan, and Jim Reavis, published by Elsevier.

Critical infrastructure

For the purposes of this book, the terms "industrial network" and "critical infrastructure" are used in somewhat limited contexts. Herein, "industrial network" is referring to any network operating some sort of automated control system that communicates digitally over a network, and "critical infrastructure" is referring to the critical *systems and assets* used within a networked computing infrastructure. Confusing? It is, and this is perhaps one of the leading reasons that many critical infrastructures remain at risk today: many ICS security seminars have digressed into an argument over semantics, at the sake of any real discussion on network security practices.

Luckily, the two terms are closely related in that the defined critical *national* infrastructures, meaning those systems listed in the **Homeland Security Presidential Directive Seven** (**HSPD-7**), typically utilizes some sort of industrial control systems. In its own words, "HSPD-7 establishes a national policy for Federal departments and agencies to identify and prioritize [the] United States critical infrastructure[s] and key resources and to protect them from terrorist attacks." HSPD-7 includes public safety, bulk electric energy, nuclear energy, chemical manufacturing, agricultural and pharmaceutical manufacturing and distribution, and even aspects of banking and finance—basically, anything whose disruption could impact a nation's economy, security, or health.[6] While financial services, emergency services, and healthcare are considered a part of our critical national infrastructure; they do not typically directly operate industrial control networks and so are not addressed within this book (although many of the security recommendations will still apply, at least at a high level).

Utilities

Utilities—water, wastewater, gas, oil, electricity, and communications—are critical national infrastructures that rely heavily on industrial networks and automated control systems. Because the disruption of any of systems associated with these infrastructures could impact our society and our safety, they are listed as critical by HSPD-7. They are also clear examples of industrial networks because they use automated and distributed PCSs. Of the common utilities, electricity is often separated as requiring more extensive security. In the United States and Canada, it is specifically regulated to standards of reliability and cybersecurity. Petroleum refining and distribution are systems that should be treated as both a chemical/hazardous material and as a critical component of our infrastructures, but at the time, this book was published were not directly regulated by federal authorities for cybersecurity compliance in a manner similar to NERC CIP.

Nuclear facilities

Nuclear facilities represent unique safety and security challenges due to their inherent danger in fueling and operation, as well as the national security implications of the raw materials used. These plants typically comprise a base load contribution to the national electric grid. This makes nuclear facilities a prime target for cyberattacks and makes the consequences of a successful attack more severe. The **Nuclear Regulatory Commission** (**NRC**), as well as NERC and the Federal Energy Regulatory Commission (FERC), heavily regulate nuclear energy in the United States when it comes to supplying electricity to the grid. Congress formed the NRC as an independent agency in 1974 in an attempt to guarantee the safe operation of nuclear facilities and to protect people and the environment. This includes regulating the use of nuclear material including by-product, source, and special nuclear materials, as well as nuclear power.[7]

Bulk electric

The ability to generate and distribute electricity in bulk is highly regulated. Electrical energy generation and distribution is defined as critical infrastructures under HSPD-7 and is heavily regulated in North America by **NERC**—specifically via the NERC CIP reliability standards—under the authority of the Department of Energy (DoE). The DoE is also ultimately responsible for the security of the production, manufacture, refining, distribution, and storage of petroleum, natural gas, and nonnuclear electric power.[8]

It is important to note that energy generation and distribution are two distinct industrial network environments, each with its own nuances and special security requirements. Energy generation is primarily concerned with the safe manufacture of a product (electricity), while energy distribution is concerned with the safe and balanced distribution of that product. The two are also highly interconnected, obviously, as generation facilities directly feed the power grid that distributes that energy, since bulk energy must be carefully measured and distributed upon production. For this same reason, the trading and transfer of power between power companies is an important facet of an electric utility's operation and the stability of the grid at large.

The smart grid—an update to traditional electrical transmission and distribution systems to accommodate digital communications for metering and intelligent delivery of electricity—is a unique facet of industrial networks that is specific to the energy industry that raises many new security questions and concerns.

Although energy generation and distribution are not the only industrial systems that need to be defended, they are often used as examples within this book. This is because NERC has created the CIP reliability standard and enforces it heavily throughout the United States and Canada. Likewise, the NRC requires and enforces the cyber security of nuclear power facilities. Ultimately, all other industries rely upon electric energy to operate, and so the security of the energy infrastructure (and the development of the smart grid) impacts everything else. Talking about securing industrial networks without talking about energy is practically impossible.

Is bulk power more important than the systems used in other industry sectors? That is a topic of heavy debate. Within the context of this book, we assume that all control systems are important, whether or not they generate or distribute energy, or whether they are defined that way by HSPD-7 or any other directive. A speaker at the 2010 Black Hat conference suggested that ICS security is overhyped because these systems are more likely to impact the production of cookies than they are to impact our national infrastructure.[9] Even the production of a snack food can impact many lives—through the manipulation of its ingredients or through financial impact to the producer and its workers and the communities in which they live. What is important to realize here is that the same industrial systems are used across designated "critical" and "noncritical" infrastructures—from making cookies to making electrical energy.

Smart grid

The smart grid is a modernization of energy transmission, distribution, and consumption systems. A smart grid improves upon legacy systems through the addition of monitoring, measurement and automation—allowing many benefits to energy producers (through accurate demand and response capabilities for energy generation), energy providers (through improved transmission and distribution management, fault isolation and recovery, metering and billing, etc.), and energy consumers (through in-home energy monitoring and management, support for alternate energy sources such as home generation or electric vehicle charge-back, etc.). The specific qualities and benefits of the smart grid are far too extensive and diverse to list them all herein. The smart grid is used extensively within this book as an example of how an industrial system—or in this case a "system of systems"—can become complex, and as a result become a large and easy target for a cyberattacker.

This is partly because by becoming "smart," the devices and components that make up the transmission, distribution, metering, and other components of the grid infrastructure have become sources of digital information (representing a privacy risk), have been given distributed digital communication capability (representing a cyber security risk), and have been highly automated (representing a risk to reliability and operations should a cyberattack occur). In "Applied Cyber Security and the Smart Grid", the smart grid is described using an analogy of human biology: the increased monitoring and measurement systems represents the eyes, ears, and nose, as well as the sensory receptors of the brain; the communication systems represent the mouth, vocal chords, eyes, and the ears, as well as the communicative center of the brain; and the automation systems represent the arms, hands, and fingers, as well as the motor functions of the brain. The analogy is useful because it highlights the common participation of the brain: if the smart grid's brain is compromised, all aspects of sensory perception, communication, and response can be manipulated.

The smart grid can be thought of within this book as a more complex "system of systems" that is made up of more than one industrial network, interconnected to provide end-to-end monitoring, analytics, and automation. The topics discussed herein apply to the smart grid even though they may be represented in a much simpler form. Some of the differences in smart grid architecture and operations are covered in Chapter 5, "Industrial Network Design and Architecture," and in more detail in the complimentary publication "Applied Cyber Security and the Smart Grid."

Chemical facilities

Chemical manufacture and distribution represent specific challenges to securing an industrial manufacturing network. Unlike the "utility" networks (electric, water, wastewater, natural gas, fuels), chemical facilities need to secure their intellectual property as much as they do their control systems and manufacturing operations. This is because the product itself has a tangible value, both financially and as a weapon. For example, the formula for a new pharmaceutical could be worth a large sum of money on the black market. The disruption of the production of that pharmaceutical could be used as a social attack against a country or nation, by impacting the ability to produce a specific vaccine or antibody. Likewise, the theft of hazardous chemicals can be used directly as weapons or to fuel illegal chemical weapons research or manufacture. Chemical facilities need to also focus on securing the storage and transportation of the end product for this reason.

Understanding "OT" versus "IT"

While the term "industrial network environment" is probably more accurate, it is far easier to use the term "operational technology" or "OT" to quickly refer to industrial control systems, and the assets, networks, and protocols that they are built from. As such, the use of the term OT has become ubiquitous.

However, it is also misleading.

According to ISA, "information technology (IT) is defined as hardware, software, and communications technologies that focus on the storage, recovery, transmission, manipulation, and protection of data. Operations technology (OT) is defined as hardware and software that detects or causes a change through the direct monitoring and control of physical devices, processes, and events".[10] To simply this, IT focuses on "cyber," or digital information, while OT focuses on what is "physical." This is why attacks that utilize digital systems (such as a network) to achieve a physical outcome (such as production failures or the production of lower quality goods) are sometimes referred to as a "**cyber-physical attack**," or alternatively a "**kinetic attack**." Please refer to Chapter 7, "Hacking industrial systems" for more information about hacking industrial networks, and to Chapter 3, "Industrial Cyber Security History and Trends" for a broader discussion of IT, OT, and the convergence of the two.

Common Industrial Security Recommendations

Many of the network security practices that are either required or recommended by the aforementioned organizations are consistent between many if not all of the others. Although all recommendations should be considered, these common "best practices" are extremely important and are the basis for many of the methods and techniques discussed within this book. They consist of the following steps.

(1) identifying what systems need to be protected,
(2) separating the systems logically into functional groups,
(3) implementing a defense-in-depth strategy around each system or group,
(4) controlling access into and between each group,
(5) monitoring activities that occur within and between groups, and
(6) limiting the actions that can be executed within and between groups.

Identification of critical systems

The first step in securing any system is determining what needs to be protected, and this is reflected heavily in NERC CIP, NRC 10 CFR 73.54, and ISA-62,443. Identifying the assets that need to be secured, as well as identifying their individual importance to the reliable operation of the overall integrated system, is necessary for a few primary reasons. First, it tells us what should be monitored and how closely. Next, it tells us how to logically segment the network into high-level security zones. Finally, it indicates where our point security devices (such as firewalls and intrusion protection systems) should be placed. For North American electric companies, it also satisfies a direct requirement of NERC CIP and therefore can help to minimize fines associated with noncompliance.

Identifying critical systems is not always easy. The first step is to build a complete inventory of all connected devices in terms of not only the physical asset itself but also the logical assets that reside within. Remember that in the end, cybersecurity controls will be applied to protect specific logical assets, so it is important to adequately define them at this early stage. For example, an active directory server that performs the file and storage services role and therefore contains the "files" as a logical asset is different from another AD server that is assigned the domain services roles and contains "credentials" as one of its logical asset. Each of these devices should be evaluated independently. If it performs a critical function, it should be classified as critical. If it does not, consider whether it could impact any other critical devices or operations. Could it impact the network itself, preventing another device from interacting with a critical system and therefore causing a failure? Finally, does it protect a critical system in any way?

The NRC provides a logic map illustrating how to determine critical assets, which are adapted to more generic asset identification in Figure 2.9. This process will help to separate devices into two categories.

- Critical assets
- Noncritical assets

In many larger operations, this process may be over simplified. There may be different levels of "criticality" depending upon the individual goals of the operational process, the operating company, and even the nation within which that company is incorporated. A general rule to follow once the basic separation of critical versus noncritical has been completed is as follows. Are there any critical assets that are not functionally related to other critical assets? If there are, next ask if one function is more or less important than the other. Finally, if there is both a functional separation *and* a

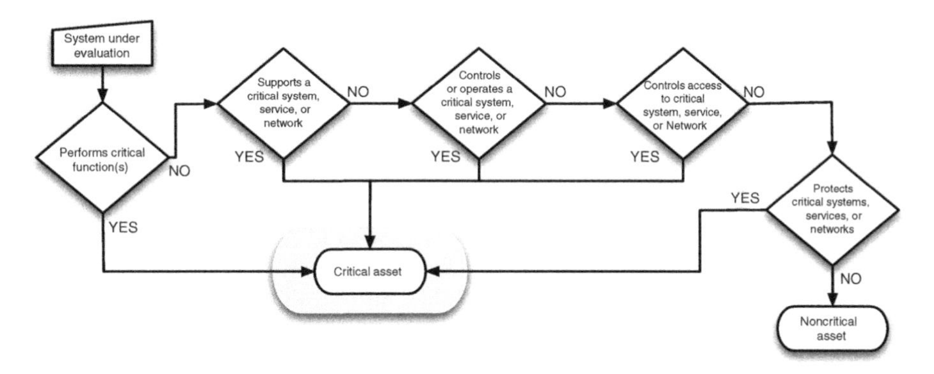

FIGURE 2.9 Nuclear regulatory commission (NRC) process diagram for identifying critical cyberassets.

difference in the criticality of the system, consider adding a new logical "tier" to your network. Also remember that a device could potentially be critical *and* also directly impact one or more other critical assets. Consider ranking the criticality of devices based on their total impact to the overall system as well. Each layer of separation can then be used as a point of demarcation, providing additional layers of defense between each group.

Network segmentation/isolation of systems

The separation of assets into functional groups allows specific services to be tightly locked down and controlled and is one of the easiest methods of reducing the attack surface that is exposed to potential threat actors. It is possible to eliminate most of the vulnerabilities—known or unknown—that could potentially allow an attacker to exploit those services simply by disallowing all unnecessary services and communication ports.

For example, if several critical services are isolated within a single functional group and separated from the rest of the network using a single firewall, it may be necessary to allow several different traffic profiles through that firewall (see Figure 2.10). If an attack is made using an exploit against web services over port 80/tcp, that attack may compromise a variety of services including email services, file transfers, and patch/update services.

However, if each specific service is grouped functionally and separated from all other services, as shown in Figure 2.11—that is, all patch services are grouped together in one group, all database services in another group, etc.—the firewall can be configured to disallow anything other than the desired service, preventing an update server using HTTPS from being exposed to a threat that exploits a weakness in SQL on the database servers. Applying this to the reference design, it is easy to see how additional segmentation can protect attacks from pivoting between centrally located services. This is the fundamental concept behind the design of what are called "functional DMZs."

In an industrial control system environment, this method of service segmentation can be heavily utilized because there are many distinct functional groups within an industrial network that should not be communicating outside of established parameters. For

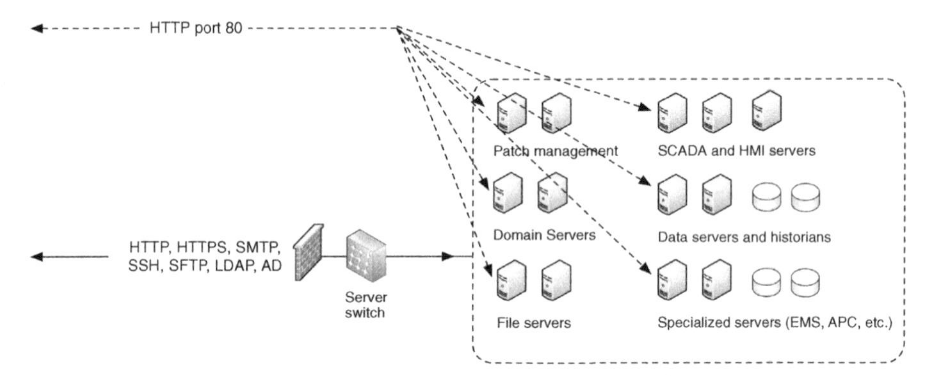

FIGURE 2.10 Placing all services behind a common defense provides a broader attack surface on all systems.

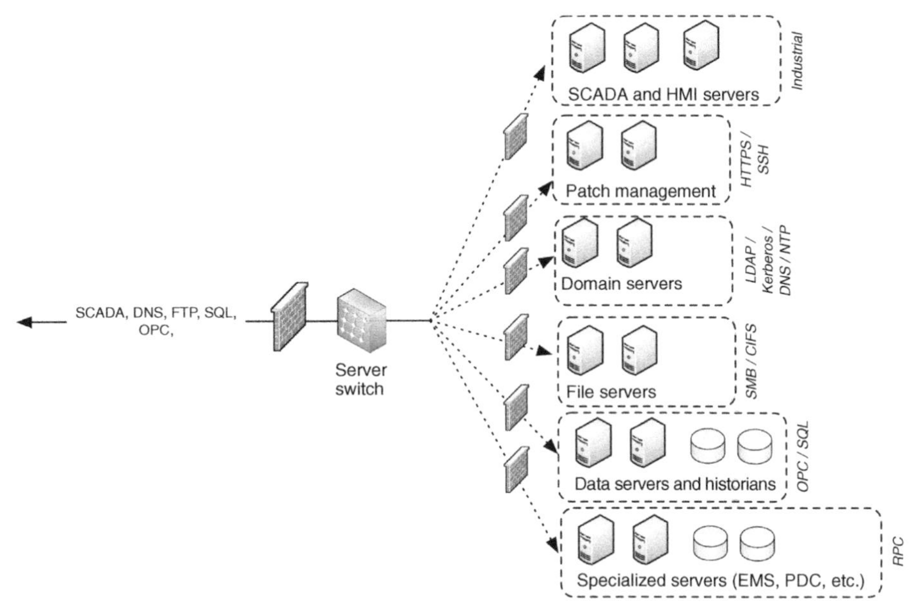

FIGURE 2.11 Separation into functional groups reduces the attack surface to a given system.

example, protocols such as Modbus or DNP3 (discussed in depth in Chapter 6, "Industrial Network Protocols") are specific to ICS systems and should never be used within the business network, while Internet services such as HTTP, IMAP/POP, FTP, and others should never be used within supervisory or control network areas. In Figure 2.12, it can be seen how this layered approach to functional and topological isolation can greatly improve the defensive posture of the network.

These isolated functional zones are often depicted as being separated by a firewall that interconnects them by conduits with other zones within this book. In many cases, a separate firewall may be needed for each zone. The actual method of securing the zone can vary and could include dedicated firewalls, intrusion protection devices, application content filters, access control lists, and/or a variety of other controls. Multiple zones can be supported using a single firewall in some cases through the careful creation and management of policies that implicitly define which hosts can connect over a given protocol or service port. This is covered in detail in Chapter 9, "Establishing Zones and Conduits."

> **Caution**
>
> Do not forget to control communications in both directions through a firewall. Not all threats originate from outside to inside (less trusted to more trusted networks). Open, outbound traffic policies can facilitate an insider attack, enable the internal spread of malware, enable outbound command and control capabilities, or allow for data leakage or information theft.

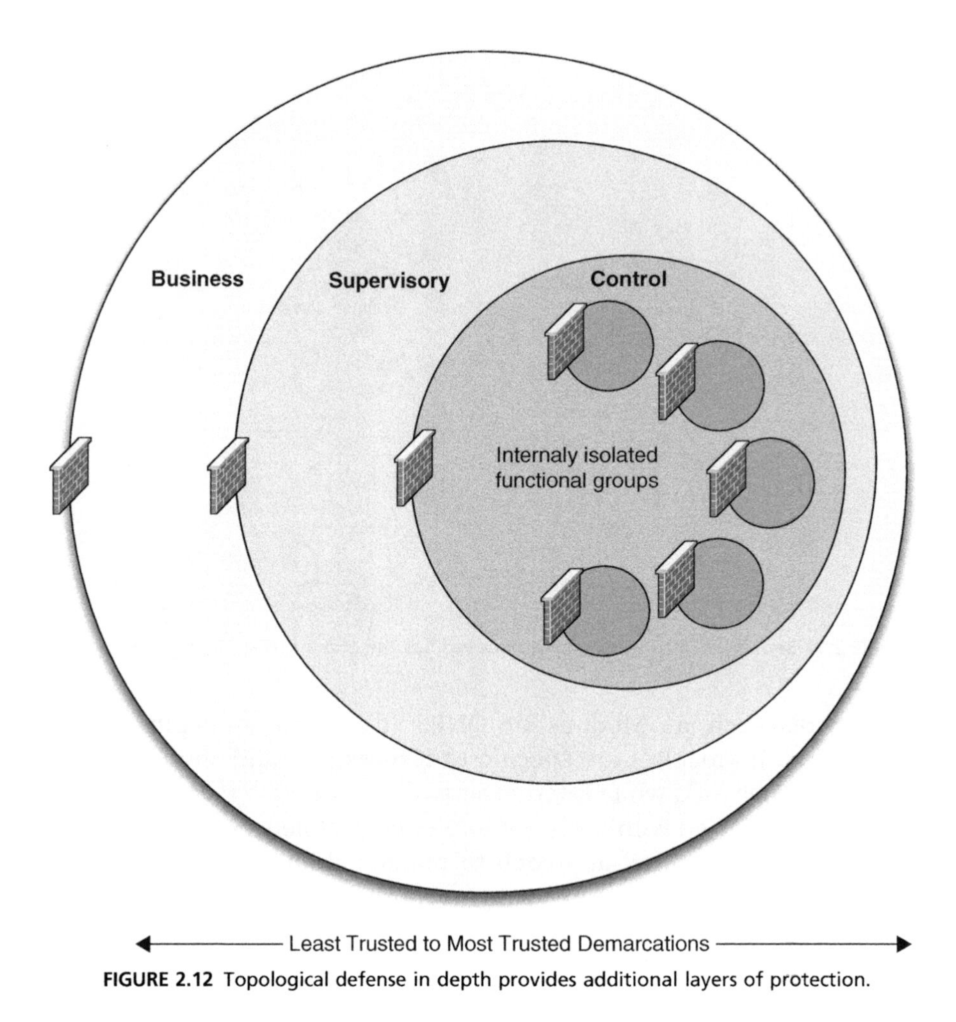

FIGURE 2.12 Topological defense in depth provides additional layers of protection.

Defense in depth

All standards organizations, regulations, and recommendations indicate that a defense-in-depth strategy should be implemented. The philosophy of a layered or tiered defensive strategy is considered a best practice even though the definitions of "defense in depth" can vary somewhat from document to document. Figure 2.13 illustrates a common defense-in-depth model, mapping logical defensive levels to common security tools and techniques.

The term "defense in depth" can and should be applied in more than one context because of the segregated nature of most industrial systems, including

- The layers of the open systems interconnection (OSI) model, from physical (layer 1) to application (layer 7).

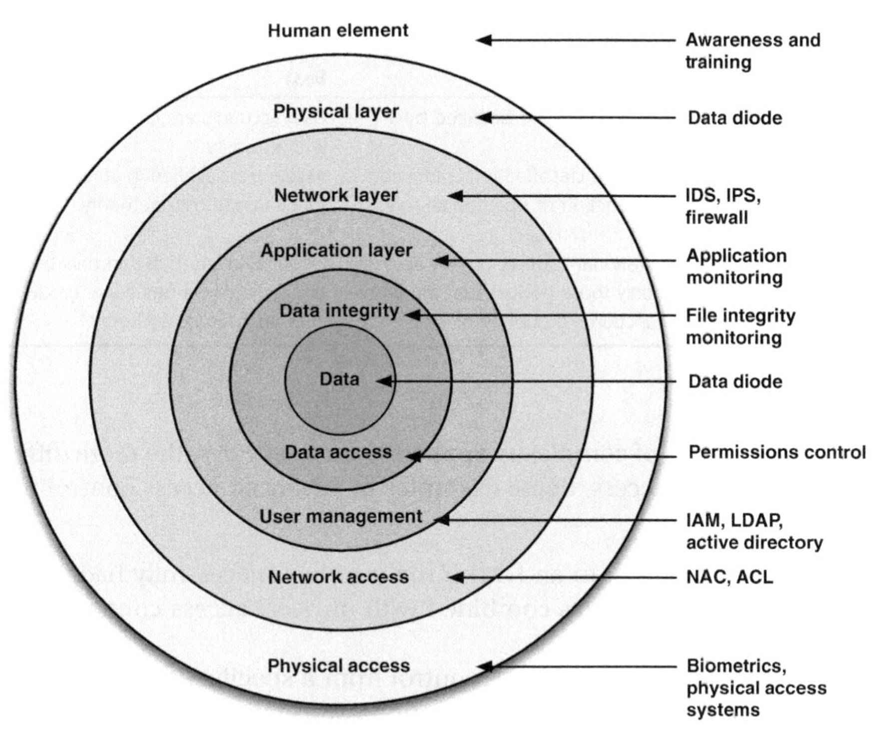

FIGURE 2.13 Defense in depth with corresponding protective measures.

- Physical or topological layers consisting of subnetworks and/or functional zones.
- Policy layers, consisting of users, roles, and privileges.
- Multiple layers of defensive devices at any given demarcation point (such as implementing a firewall and an intrusion prevention system).

Access control

Access control is one of the most difficult yet important aspects of cybersecurity. Access control considers three very important aspects of how a user interacts with resources (e.g., local application, remote server, etc.). These aspects are: identification, authentication, and authorization. It becomes more difficult for an attacker to identify and exploit systems by locking down services to specific users or groups of users accessing specific resources. The further access can be restricted, the more difficult an attack becomes. Although many proven technologies exist to enforce access control, the successful implementation of access control is difficult because of the complexity of managing users and their roles and their mapping to specific devices and services that relate specifically to an employee's operational responsibilities. As shown in Table 2.1, the strength of access control increases as a user's identity is treated with the additional context of that user's roles and responsibilities within a functional group.

Table 2.1 Adding Context to User Authentication to Strengthen Access Control

Good	Better	Best
User accounts are classified by authority level	User accounts are classified by functional role	User accounts are classified by functional role and authority
Assets are classified in conjunction with user authority level	Assets are classified in conjunction with function or operational role	Assets are classified in conjunction with function and user authority
Operational controls can be accessed by any device based on user authority	Operational controls can be accessed by only those devices that are within a functional group	Operational controls can only be accessed by devices within a functional group by a user with appropriate authority

Again, the more layers of complexity applied to the user rules, the more difficult it will be to gain unauthorized access. Some examples of advanced access control include the following.

- Only allow a user to log in to an HMI if the user has successfully badged into the control room (user credentials combined with physical access controls - station-based access control)
- Only allow a user to operate a given control from a specific controller (user credentials limited within a security zone—area of responsibility)
- Only allow a user to authenticate during that user's shift (user credentials combined with personnel management—time-based access control)

Tip

Authentication based on a combination of multiple and unrelated identifiers provides the strongest access control, for example, the use of both a digital and a physical key, such as a password and a biometric scanner. Another example may include the use of dedicated hosts for specific functions. The specific purpose of each ICS component under evaluation must be considered and account for unique operational requirements of each. It may be possible to implement strong, multifactor authentication at an engineering workstation, where this may not be acceptable at an operator HMI that depends on shared operator accounts.

Advanced Industrial Security Recommendations

The cybersecurity industry evolves rapidly, and newer security products and technologies are being introduced every day—certainly faster than they can be referenced or recommended by standards and other industry organizations. Some advanced security recommendations include: real-time activity and event monitoring using a security information and event management system (SIEM); network-based anomaly detection tools; policy whitelisting using an industrial firewall or industrial protocol filter; end-system malware protection using application whitelisting; and many others. There are undoubtedly new security products available since the time of this writing—it is good advice to always research new and emerging security technology when designing, procuring, or implementing new cybersecurity measures.

Security Monitoring

Monitoring an information technology system is a recognized method of providing situational awareness to a cybersecurity team and monitoring tools such as SIEM and log management systems are heavily utilized by enterprise IT departments for this reason. Improved situational awareness can also benefit industrial networks, although special care needs to be taken in determining what to monitor, how to monitor it, and what the information gathered means in the context of cybersecurity. For more detail on how to effectively monitor an industrial network, see Chapter 12, "Security Monitoring of Industrial Control Systems."

Policy whitelisting

"Blacklists" define what is "bad" or not allowed—malware, unauthorized users, etc. A "whitelist" is a list of what is "good" or what is allowed—authorized users, approved resources, approved network traffic, safe files, etc. A policy whitelist defines the behavior that is acceptable. This is important in ICS architectures, where an industrial protocol is able to exhibit specific behaviors such as issuing commands, collecting data, or shutting down a system. A policy whitelist, also referred to as a protocol whitelist, understands what industrial protocol functions are allowed and prevents unauthorized behaviors from occurring. Policy whitelisting is a function that is available to newer and more advanced industrial firewalls. This is discussed in more detail in Chapter 11, "Anomaly and Threat Detection."

Application whitelisting

Application whitelisting defines the applications (and files) that are known to be "good" on a given device and prevents any other applications from executing (or any other file from being accessed). This is an extremely effective deterrent against malware, since only advanced attacks directly against resident memory of an end system have the ability to infect systems with properly implemented application whitelisting. This also helps improve resilience of those systems that are not actively patched either due to operational issues or vendor specifications. This is discussed in more detail in Chapter 11, "Anomaly and Threat Detection."

Common Misperceptions About Industrial Network Security

In any discussion about industrial cybersecurity, there is always going to be objections from some that are based on misperceptions. The most common are.

- **Cybersecurity of industrial networks is not necessary.** The myth remains that an "air gap" separates the ICS from any possible source of digital attack or infection. This is simply no longer true. While network segmentation is a valuable method for establish security zones and improving security, the absolute separation of networks promised by the air gap is virtually impossible to obtain. "Air" is not an adequate defense against systems that support wireless diagnostics ports, removable media that can be hand-carried, and so on. This myth also assumes that all threats originate from outside the industrial network and fails to address the risk from the insider and the resulting impact of a cyber event on the ICS from an authorized user. This is a religious debate to some. To the authors of this book, the air gap is a myth that must be dispelled if cybersecurity is to be taken seriously.
- **Industrial security is an impossibility.** Security requires patching. Devices need to be patched to protect against the exploitation of a discovered vulnerability, and antivirus systems need regular updates. Control environments cannot support adequate patch cycles, making any cyber security measures moot. While it is true that these are challenges faced in ICSs, it does not mean that a strong security posture cannot be obtained through other compensating controls. Industrial security requires a foundation of risk management and an understanding of the security lifecycle.
- **Cybersecurity is someone else's responsibility.** This comment is typically heard from plant operational managers hoping that IT managers will adopt responsibility (and budget) for cybersecurity. It is more often than not in operations' benefit to take responsibility for cybersecurity. Cybersecurity will have ownership at the highest executive levels in a properly structured organization, and appropriate responsibilities will trickle down to both IT and operations as needed, so that they can work in concert—as can be seen in this book (and already within this chapter), cybersecurity is an end-to-end problem that requires an end-to-end solution.
- **It is the same as "regular" cybersecurity.** This is another common misperception that can sometimes divide IT and plant operations' groups within an organization. "You have an Ethernet network, therefore my UltraBrand Turbo-charged Firewall with this state-of-the-art unified threat management system will work just as well in the ICS as it does in the enterprise! After all, the vendor said it supported SCADA protocols, and all SCADA protocols the same!" One thing that will become abundantly clear as you read this book is that industrial and business networks are different and require different security measures to adequately protect them.

For examples of how these and other misperceptions can manifest in the real world, see Chapter Appendix A-1, "Examples of Industrial Control System Cyber Events."

Assumptions made in this book

The security practices recommended within this book aim for a very high standard and in fact go above and beyond what is recommended by many government and regulatory groups. So which practices are really necessary, and which are excessive? It depends upon the nature of the industrial system being protected and the level of risk mitigation desired. What are the consequences of a cyberattack? The production of energy is much more important in modern society than the production of a Frisbee (unless you happen to be a professional Ultimate Frisbee champion!). The proper manufacture and distribution of electricity can directly impact our personal safety by providing heat in winter or by powering our irrigation pumps during a drought. The proper manufacture and distribution of chemicals can mean the difference between the availability of flu vaccines and pharmaceuticals and a direct health risk to the population. Most ICSs are by their nature important regardless of an ICS's classification, and any risk to their reliability holds industrial-scale consequences. These consequences can be localized to a particular manufacturing unit or spread to larger regional and national levels. While not all manufacturing systems hold life-and-death consequences, it does not mean that they are not potential targets for a cyberattack. What are the chances that an extremely sophisticated, targeted attack will actually occur? The likelihood of an incident diminishes as the sophistication of the attack—and its consequences—grow, as shown in Figure 2.14. By implementing security practices to address these uncommon and unlikely attacks, there is a greater possibility of avoiding the devastating consequences that correspond to them.

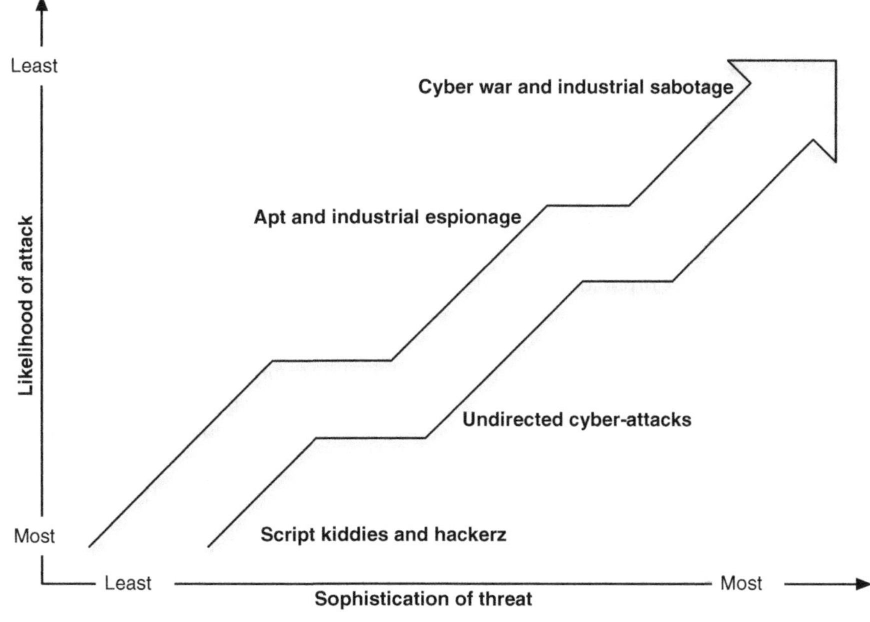

FIGURE 2.14 Likelihood versus consequence of a targeted cyberattack.

The goal of this book is to secure any industrial network. It focuses on critical infrastructure in particular and will reference various standards, recommendations, and directives as appropriate. It is important to understand these directives regardless of the nature of the control system that needs to be secured, especially NERC CIP, Chemical Facility Anti-Terrorism Standards (CFATS), Federal Information Security Management Act (FISMA), ISA, and the control system security recommendations of National Institute of Standards and Technology (NIST). Each has its own strengths and weaknesses, but all provide a good baseline of best practices for industrial network security. References are given when specific standards, best practices, and guidance are discussed. It is however, difficult to devote a great deal of dedicated text to these documents due to the fact that they are in a constant state of change. The industrial networks that control critical infrastructures demand the strongest controls and regulations around security and reliability, and accordingly there are numerous organizations helping to achieve just that. The Critical Infrastructure Protection Act of 2001 and HSPD-7 define what they are, while others—such as NERC CIP, NRC, CFATS, and various publications of NIST—help explain what to do.

Summary

Understanding industrial network security first requires a basic understanding of the terminology used, the basics of industrial network architectures and operations, some relevant cybersecurity practices, the differences between industrial networks and business networks, and why industrial cybersecurity is important. By evaluating an industrial network, identifying and isolating its systems into functional groups or "zones," and applying a structured methodology of defense-in-depth and strong access control, the security of these unique and specialized networks will be greatly improved. The remainder of this book will go into further detail on how industrial control systems operate, how they can be exploited, and how they can be protected.

Endnotes

1. Eric D. Knapp and Raj Samani, "Applied Cyber Security and the Smart Grid", Elsevier, 2013.
2. North American Electric Corporation, Standard CIP–002–4, Cyber Security, Critical Cyber Asset Identification, North American Electric Corporation (NERC), Princeton, NJ, approved January 24, 2011.
3. North American Electric Corporation, Standard CIP–002–5.1, Cyber Security, Critical Cyber Asset Identification, North American Electric Corporation (NERC), Princeton, NJ, approved January 24, 2011.
4. Purdue Research Foundation (Theodore J. Williams, Editor); A Reference Model For Computer Integrated Manufacturing (CIM), A Description from the Viewpoint of Industrial Automation; Instrument Society of America, North Carolina, 1989.
5. North American Electric Corporation, Standard CIP–002–4, Cyber Security, Critical Cyber Asset Identification, North American Electric Corporation (NERC), Princeton, NJ, approved January 24, 2011.
6. Department of Homeland Security, Homeland security presidential directive 7: critical infrastructure identification, prioritization, and protection. <http://www.dhs.gov/xabout/laws/gc_1214597989952.shtm>, September 2008 (cited: November 1, 2010).

7. U.S. Nuclear Regulatory Commission, The NRC: who we are and what we do. <http://www.nrc.gov/about-nrc.html> (cited: November 1, 2010).
8. Department of Homeland Security, Homeland security presidential directive/HSPD-7. Roles and responsibilities of sector-specific federal agencies (18) (d). <http://www.dhs.gov/xabout/laws/gc_1214597989952.shtm>, September 2008 (cited: November 1, 2010).
9. J. Arlen, SCADA and ICS for security experts: how to avoid cyberdouchery. in: Proc. 2010 BlackHat Technical Conference, July 2010.
10. Halsey, Jennifer. ISA. 12/7/2020. Document from the Internet. https://blog.isa.org/what-is-the-difference-between-it-and-ot-security.

3 Industrial Cybersecurity History and Trends

Information in this chapter

- The convergence of OT and IT
- Importance of securing industrial networks
- Evolution of the cyber threat
- Insider threats
- Hacktivism, cybercrime, cyberterrorism, and cyberwar

Securing an industrial network and the assets connected to it, although similar in many ways to standard enterprise information system security, presents several unique challenges. While the systems and networks used in industrial control systems (ICSs) are highly specialized, they are increasingly built upon common computing platforms using commercial operating systems. At the same time, these systems are built for reliability, performance, and longevity. A typical integrated ICS may be expected to operate without pause for months or even years, and the overall life expectancy may be measured in decades. Attackers, on the contrary, have easy access to new exploits and can employ them at any time. In a typical enterprise network, systems are continually managed in an attempt to stay ahead of this rapidly evolving threat, but these methods often conflict with an industrial network's core requirements of reliability and availability.

Doing nothing is not an option. Because of the importance of industrial networks and the potentially devastating consequences of an attack, new security methods need to be adopted. Industrial networks are being targeted as can be seen in real-life examples of industrial cyber sabotage (more detailed examples of actual industrial cyber events will be presented in Chapter 7, "Hacking Industrial Systems"). They are the targets of a new threat profile that utilizes more sophisticated and targeted attacks than ever before. An equally disturbing trend is the rise in accidental events that have led to significant consequences caused when an authorized system user unknowingly introduces threats into the network during their normal and routine interaction. This interaction may be normal local system administration or via remote system operation.

The convergence of OT and IT

The overall trend of industrial digitization is often referred to more simply as the "convergence of information technology (IT) and operational technology (OT)."

Industrial Network Security. https://doi.org/10.1016/B978-0-443-13737-2.00009-9

In the early days of industrial automation, the control of a process was done manually; mechanical controls were used to manage analog systems. As technology advanced, electronic controls were introduced, and a need to communicate between these controls led to the introduction of various proprietary protocols.[1] Industrial networks were comprised of proprietary connectivity between assets built upon proprietary hardware with proprietary software. Figure 3.1 shows a clear separation between these early industrial controls systems and IT systems.

The concept of "OT" as we know it today did not arise until after a more recent shift toward standardized systems using commercial off-the-shelf (COTS) components. When industrial control assets started to be built using standard technologies using commercial operating systems, Ethernet networking, USB interfaces, etc., it became very easy to think of industrial computing hardware and software as "the same" as commercial computing hardware and software used in IT. In Figure 3.2, we now see IT and OT begin to overlap. However, while the technology used is the same in many cases, the way that it is used differs significantly between IT and OT.

Information technology continued to advance, IT and OT systems continued to converge. With enough relevant data, businesses were able to drive advancements in

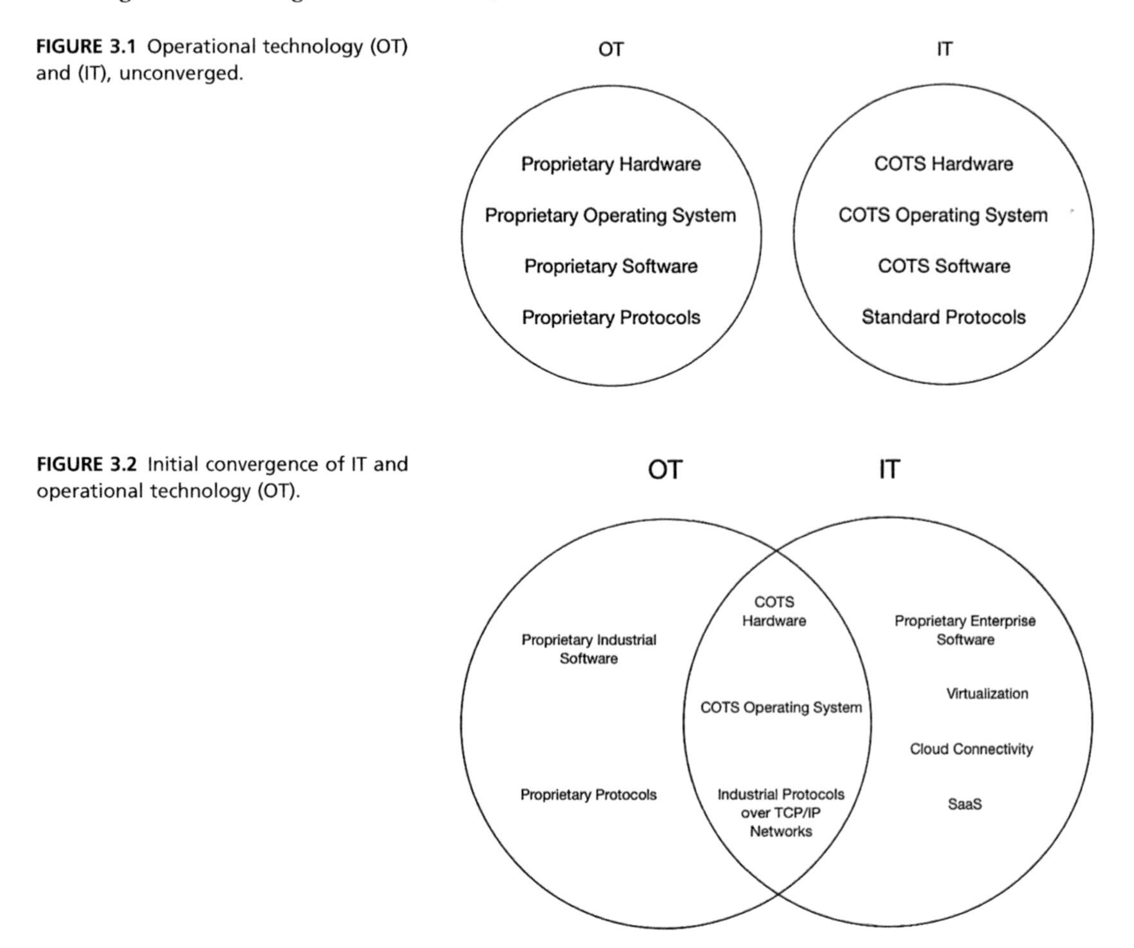

FIGURE 3.1 Operational technology (OT) and (IT), unconverged.

FIGURE 3.2 Initial convergence of IT and operational technology (OT).

everything from operational efficiencies, customer engagement, market intelligence, supply chain management, and more. With increased availability of Internet connected "smart" devices (from wearable technology, smart phones, fitness trackers, smart fire protection systems, smart refrigerators, *et. al.*), the availability of new and valuable data further fueled a race toward big-data analytics. Increased amounts of data and more advanced analytics seemed to correlate directly to the bottom line, which increased the computational and data storage needs, driving additional advancements in cloud infrastructures. Traditional data centers began to feel the upper limitations of scalability and performance, resulting in a shift to **cloud computing** and **software as a service (SaaS)** applications that proved to be both effective and efficient for businesses.

Industrial organizations also saw benefits to an operating model built upon real-time analytics. The once-simple sensors and actuators that make up ICSs also became "smart" digital devices. ICSs shifted further toward digitization, with smart and connected devices creating an Industrial Internet of Things (IIoT) that promised the same operational benefits to industrial operators that general businesses had realized. In 2011, the "Industrial Internet" was introduced at the Hannover Messe trade fair in Germany.[2]

Like in business, the increased demand for large amounts of industrial data and advanced analytics of that data required the scalability of cloud computing. Cloud computing became an increasingly popular technology for ICS.

However, the introduction of cloud computing poses a unique challenge to industrial environments. How do you enable connected technologies while maintaining an isolated "air gapped" system? The use of cloud computing in ICS comes with an increased risk as well. By connecting systems and assets to the Internet, the attack surface increases considerably. The introduction of this network connectivity in and of itself introduces a direct vector in the ICS. Therefore, it is important to carefully consider when and how cloud computing is used. When cloud-based systems and applications are used, the connectivity to the cloud must be carefully controlled and closely monitored (see Chapters 10, 11, 12, and 13 for more about how to control and monitor systems).

At this point, the convergence of IT and OT is even more pronounced. In Figure 3.3, we can see how the introduction of cloud computing models and SaaS further blur the line between IT and OT, including instances where proprietary industrial control software (which is exclusive to OT) is connected to proprietary enterprise software.

When attempting to secure an industrial control environment, it is important to remember that despite this convergence, fundamental differences remain, including how these systems are used and the potential consequences of an incident. While not all inclusive, Figure 3.4 shows how different the impact can be between an IT and an OT system. Impact or consequence is a fundamental consideration of risk, and therefore plays an important role in differentiating IT and OT.

Importance of securing industrial networks

The need to improve the security of industrial networks cannot be overstated. Most critical manufacturing facilities offer reasonable physical security preventing

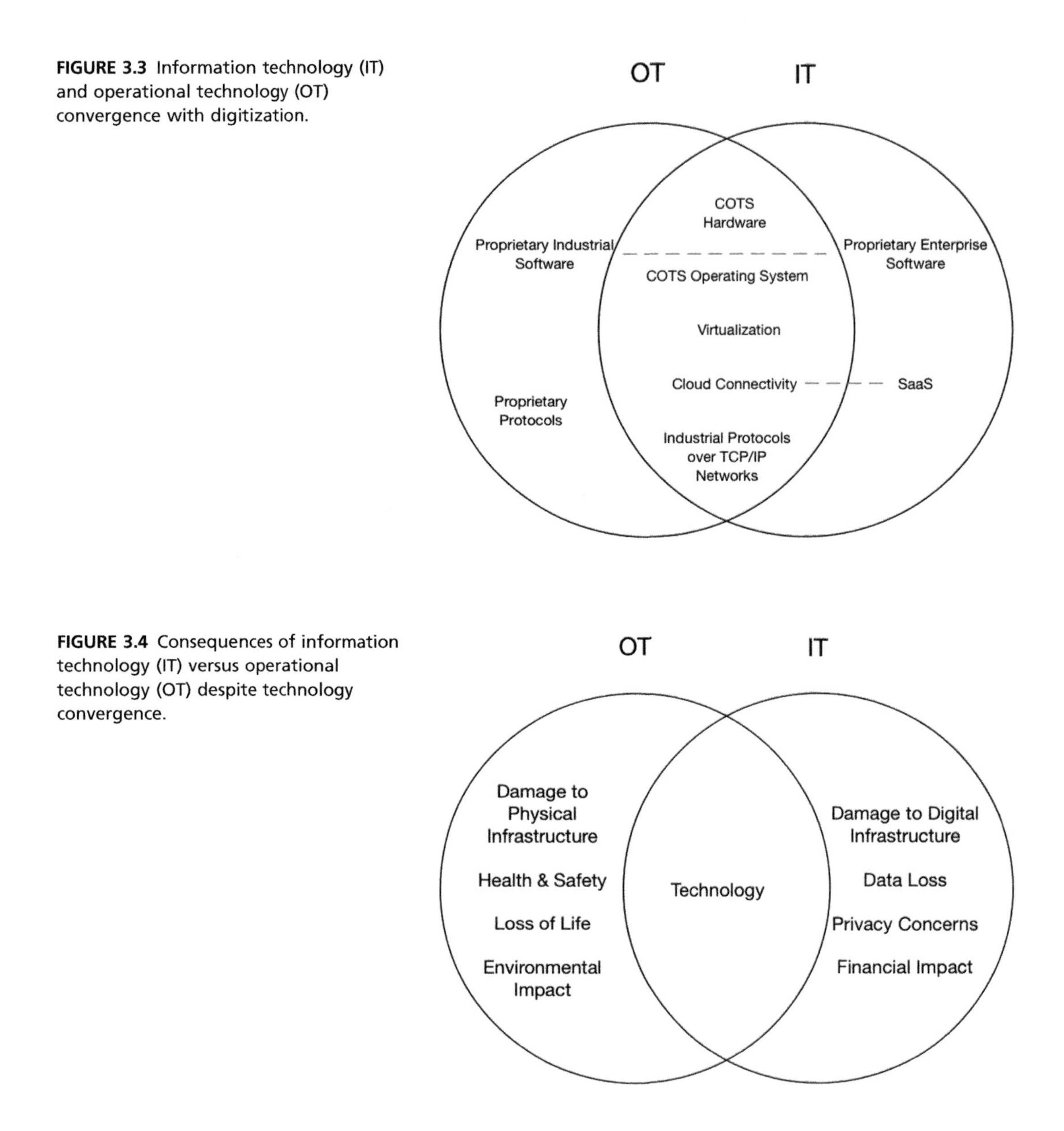

FIGURE 3.3 Information technology (IT) and operational technology (OT) convergence with digitization.

FIGURE 3.4 Consequences of information technology (IT) versus operational technology (OT) despite technology convergence.

unauthorized local access to components that form the core of the manufacturing environment. This may include physically secured equipment rack rooms, locked engineering work centers, or restricted access to operational control centers. The only method by which an ICS can be subjected to external cyber threats is via the industrial networks and the connections that exist with other surrounding business networks and enterprise resources.

Many industrial systems are built using legacy devices, and in some cases run legacy protocols that have evolved to operate in routable networks. Automation systems were built for reliability long before the proliferation of Internet connectivity, web-based applications, and real-time business information systems. Physical security was always a concern, but information security was typically not a priority because the control systems were air-gapped—that is, physically separated with no common system (electronic or otherwise) crossing that gap, as illustrated in Figures 3.1 and 3.5.

Ideally, the air gap would still remain and would still apply to digital communication, but in reality, it does rarely exists. Many organizations began the process of re-engineering their business processes and operational integration needs in the 1990s. Organizations began to perform more integration between not only common ICS applications during this era but also the integration of typical business applications like production planning systems with the supervisory components of the ICS. The need for real-time information sharing evolved as well as these business operations of industrial networks evolved. A means to bypass the gap needed to be found because the information required originated from across the air gap. In the early years of this integration "wave," security was not a priority, and little network isolation was provided. Standard

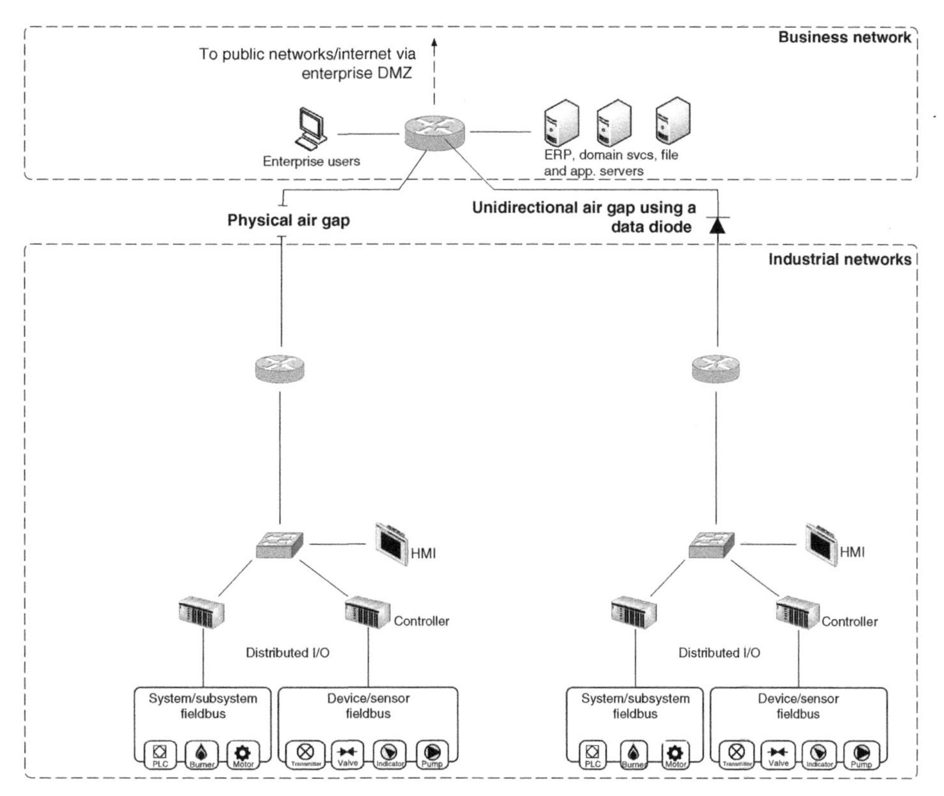

FIGURE 3.5 Air gap separation.

routing technologies were initially used if any separation was considered. Firewalls were then sometimes deployed as organizations began to realize the basic operational differences between business and industrial networks, blocking all traffic except that which was absolutely necessary in order to improve the efficiency of business operations.

The problem is that—regardless of how justified or well intended the action—the air gap no longer exists, as seen in Figure 3.2. There is now a path into critical systems, and any path that exists can be found and exploited (Figure 3.6).

Security consultants at Red Tiger Security presented research in 2010 that indicates the current state of security in industrial networks. Penetration tests were performed on approximately 100 North American electric power generation facilities, resulting in more than 38,000 security warning and vulnerabilities.[3] Red Tiger was then contracted by the U.S. Department of Homeland Security (DHS) to analyze the data in search of trends that could be used to help identify common attack vectors and, ultimately, to help improve the security of these critical systems against cyberattack.

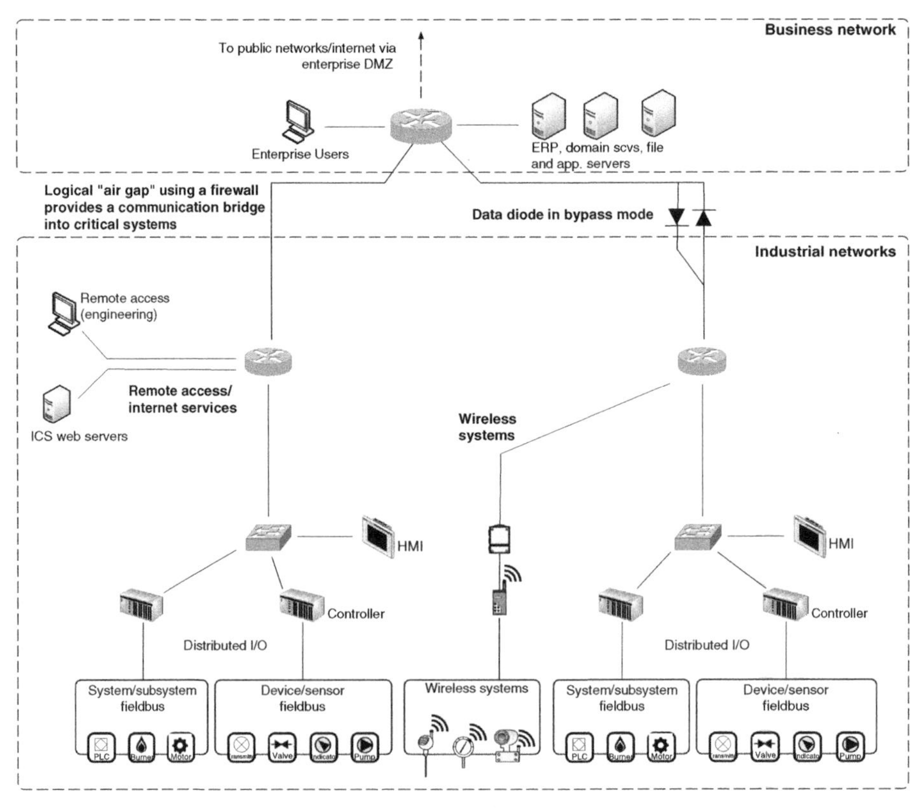

FIGURE 3.6 Crossing the the air gap.

The results were presented at the 2010 Black Hat USA conference and implied a security climate that was lagging behind other industries. The average number of days between the time a **vulnerability** was disclosed publicly and the time the vulnerability was discovered in a control system was 331 days: almost an entire year. Worse still, there were cases of vulnerabilities that were over 1100 days old, nearly 3 years past their respective "0-day."[4]

What does this mean? It says that there are known vulnerabilities that can allow hackers and cybercriminals entry into control networks. Many of these vulnerabilities are converted into reusable modules using open source penetration testing utilities such as **Metasploit** and Backtrack, making exploitation of those vulnerabilities fairly easy and available to a wide audience. This says nothing of the numerous other testing utilities that are not available free-of-charge and that typically contain exploitation capabilities against 0-day vulnerabilities as well. A more detailed look at ICS exploitation tools and utilities will be discussed in Chapter 7, "Hacking Industrial Systems."

It should not be a surprise that there are well-known vulnerabilities within control systems. Control systems are by design very difficult to patch. By intentionally limiting (or even better, eliminating) access to outside networks and the Internet, simply obtaining patches can be difficult. Actually applying patches once they are obtained can also be difficult and restricted to planned maintenance windows because reliability is paramount. The result is that there are almost always going to be unpatched vulnerabilities. Reducing the window from an average of 331 days to a weekly or even monthly maintenance window would be a huge improvement. A balanced view of patching ICS will be covered later in Chapter 10, "Implementing Cyber Security Controls."

The evolution of the cyber threat

It is interesting to look at exactly what is meant by a "cyberthreat." Numerous definitions exist, but they all have a common underlying message: (a) unauthorized access to a system and (b) loss of confidentiality, integrity, and/or availability of the system, its data, or applications. Records dating back to 1902 show how simple attacks could be launched against the Marconi Wireless Telegraph system.[5] The first computer worm was released just over 25 years ago. Cyberthreats have been evolving ever since: from the Morris worm (1988), to Code Red (2001), to Slammer (2003), to Conficker (2008), and to Stuxnet (2010) and beyond. When considering the threat against industrial systems, this evolution is concerning for three primary reasons. First, the initial attack vectors still originate in common computing platforms—typically within level 3 or 4 systems. This means that the initial penetration of industrial systems is getting easier through the evolution and deployment of increasingly complex and sophisticated malware. Second, the industrial systems at levels 2, 1, and 0 are increasingly targeted. Third, the threats continue to evolve, leveraging successful techniques from past malware while introducing new capabilities and complexity. A simple analysis of Stuxnet reveals that one of the propagation methods used included the exploitation of the same vulnerabilities used by the Conficker worm that was identified and supposedly patched in 2008. These systems are extremely vulnerable and can be considered a decade or more behind typical enterprise systems in terms of cybersecurity maturity. This means that, once breached, the result is most likely a *fait accompli*. The industrial systems as they stand today simply do not stand a chance against the modern attack capability. Their primary line of defense remains the business networks that surround them and network-based defenses between each security level of the network. Twenty percent (20%) of incidents are now targeting energy, transportation, and critical manufacturing organizations according to the 2013 Verizon Data Investigations Report.[6]

Note

It is important to understand the terminology used throughout this book in terms of "levels" and "layers." Layers are used in context of the open systems interconnection (OSI) 7-layer model and how protocols and technologies are applied at each layer.[7] For example, a network MAC address operates at layer 2 (data link layer) and depends on network "switches," while an IP address operates at layer 3 (network layer) and depends on network "routers" to manage traffic. The TCP and UDP protocols operate at layer 4 (transport layer) and depend on "firewalls" to handle communication flow.

Levels on the other hand are defined by the ISA-95[8] standard for the integration of enterprise and production control systems, expanding on what was originally described by the Purdue Reference Model for Computer Integrated Manufacturing (CIM)[9] most commonly referred to as the "Purdue Model." Here, the term Level 0 applies to field devices and their networks; level 1 basic control elements like PLCs; level 2 monitoring and supervisory functions like SCADA servers and HMIs; level 3 for manufacturing operations management functions; and level 5 for business planning and logistics.

Incident data have been analyzed from a variety of sources within industrial networks. According to information compiled from ICS-CERT, the Repository for Industrial Security Incidents (RISI), and research from firms including Verizon, Symantec, McAfee, and others, trends begin to appear that impact the broader global market:

- Most attacks seem to be opportunistic. However, not *all* attacks are opportunistic (see the section titled "Hacktivism, Cybercrime, Cyberterrorism, and Cyberwar" in this chapter)
- Initial attacks tend to use simpler exploits; thwarted or discovered attacks lead to increasingly more sophisticated methods.
- The majority of cyberattacks are financially motivated. Espionage and sabotage have also been identified as motives.

- Malware, Hacking, and Social Engineering are the predominant methods of attack among those incidents classified as "espionage." Physical attacks, misuse, and environmental methods are common in financially motivated attacks but are almost completely absent in attacks motivated by espionage.
- New malware samples are increasing at an alarming rate. New samples have slowed somewhat in late 2013, but there are still upward of 20 million new samples being discovered each quarter.[10]
- The majority of attacks originate externally and leverage weak or stolen credentials. The pivoting that follows once the initial compromise occurs can be difficult to trace due to the masquerading of the "insider" that occurs from that point. This further corroborates a high incidence of social engineering attacks and highlights the need for cybersecurity training at all levels of an organization.
- The majority of incidents affecting industrial systems are unintentional in nature, with control and software bugs accounting for the majority of unintentional incidents.[11]
- New malware code samples are increasingly more sophisticated, with an increase in rootkits and digitally signed malware.
- The percentage of reported industrial cyber incidents is high (28%) but has been steadily declining (65% in the last 5 years).[12]
- AutoRun malware (typically deployed via USB flash drive or similar media) has also risen steadily. AutoRun malware is useful for bypassing network security perimeters and has been successfully used in several known industrial cyber security incidents.
- Malware and "hacking as a service" is increasingly available and has become more prevalent. This includes an increasing market of 0-day and other vulnerabilities "for sale."
- The number of incidents that are occurring via remote access methods has been steadily increasing over the past several years due to an increasing number of facilities that allow remote access to their industrial networks.[13]

The attacks themselves tend to remain fairly straightforward. The most common initial vectors used for industrial systems include **spear phishing**, **watering hole**, and **database injection** methods.[14] Highly targeted spear phishing (customized e-mails designed to trick readers into clicking on a link, opening an attachment, or otherwise triggering malware) is extremely effective when using Open Source Intelligence (OSINT) to facilitate social engineering. For example, spear phishing may utilize knowledge of the target corporation's organization structure (e.g., a mass email sender that masquerades as legitimate email from an executive within the company), or of the local habits of employees (e.g., a mass email promising discounted lunch coupons from a local eatery).[15] The phishing emails often contain malicious attachments, or direct their targets to malicious websites. The phished user is thereby infected and becomes the initial infection vector to a broader infiltration.[16]

The payloads (the malware itself) range from freely available kits such as Webattacker and torrents, to commercial malware such as Zeus (ZBOT), Ghostnet (Ghostrat), Mumba (Zeus v3), and Mariposa. Attackers prevent detection by antivirus and other detection mechanisms by obfuscating malware.[17] This accounts for the large rate at which new malware samples are discovered. Many new samples are code variants of existing malware, created as an evasion against common detection mechanisms such as antivirus and network intrusion protection systems. This is one reason that Conficker, a worm initially discovered in 2008, remained one of the top threats facing organizations infecting as many as 12 million computers until it began to decline in the first half of 2011.[18]

Once a network is infiltrated and a system infected, malware will attempt to propagate to other systems. When attacking industrial networks, this propagation will include techniques for pivoting to new systems with increasing levels of authorization, until a system is found with access to lower integration "levels." That is, a system in level 4 will attempt to find active connectivity to level 3; level 3 to level 2, and so on. Once connectivity is discovered between levels, the attacker will use the first infected system to attack and infiltrate the second system, burrowing deeper into the industrial areas of the network in what is called "pivoting." This is why strong defense-in-depth is important. A firewall may only allow traffic from system A to system B. Encryption between the systems may be used. However, if system A is compromised, the attacker will be able to communicate freely across the established and authorized flow. This method can be thought of as the "exploitation of trust" and requires additional security measures to protect against such attack vectors.

APTs and weaponized malware

More sophisticated cyberattacks against an industrial system will most likely take steps to remain hidden because a good degree of propagation may be needed to reach the intended target. Malware attempts to operate covertly and may try to deactivate or circumvent antimalware software, install persistent rootkits, delete trace files, and perform other means to stay undetected prior to establishing backdoor channels for remote access, open holes in firewalls, or otherwise spread through the target network.[19] Stuxnet, for example, attempted to avoid discovery by bypassing host intrusion detection (using 0-day exploits that are not detectable by traditional IDS/IPS prior to its discovery, and by using various autorun and network-based vectors), disguised itself as legitimate software (through the use of stolen digital certificates), and then covered its tracks by removing trace files from systems if they are no longer needed or if they are resident on systems that are incompatible with its payload.[20] As an extra precautionary measure, and to further elude the ability to detect the presence of the malware, Stuxnet would automatically remove itself from a host if it were not the intended target once it had infected other hosts a specific number of times.[21]

By definition, Stuxnet and many other modern malware samples are considered "advanced persistent threats" (APT). One aspect of an APT is that the malware utilized is often difficult to detect and has measures to establish persistence, so that it can continue to operate even if it is detected and removed or the system is rebooted. The term APT also describes cyber campaigns where the attacker is actively infiltrating systems and exfiltrating data from one or more targets. The attacker could be using persistent malware or other methods of persistence, such as the reinfection of systems and use of multiple parallel infiltration vectors and methods, to ensure broad and consistent success. Examples of other APTs and persistent campaigns against industrial networks include Duqu,[22] Night Dragon,[23] Flame,[24] oil and natural gas pipeline intrusion campaigns,[25] Dragonfly,[26] Industroyer, Pipedream,[27] and more.

Malware can be considered "weaponized" when it obtains a certain degree of sophistication and shows a clear motive and intent. The qualities of APTs and weaponized malware differ, as does the information that the malware targets, as can be seen in Tables 3.1 and 3.2. While many APTs will use simple methods, weaponized malware (also referred to as military-grade malware) trend toward more sophisticated delivery mechanisms and payloads. Stuxnet is, again, a useful example of weaponized malware. It is highly sophisticated—the most sophisticated malware by far when it was first discovered—and also extremely targeted. It had a clear purpose: to discover, infiltrate, and sabotage a specific target system. Stuxnet utilized multiple 0-day exploits for infection. The development of one 0-day requires considerable resources in terms of either the financial resources to purchase commercial malware or the intellectual resources with which to develop new malware. Stuxnet raised a high degree of speculation about its source and its intent at least partly due to the level of resources required to deliver the worm through so many 0-days. Stuxnet also used "insider intelligence" to focus on its target control system, which again implied that the creators of Stuxnet had significant resources and that they either had access to an ICS with which to develop and test their malware, or they had enough knowledge about how such a control system was built that they were able to develop it in a simulated environment.

Table 3.1 Distinctions Between Common Advanced Persistent Threats (APT) and Weaponized Malware.

APT qualities	Weaponized malware qualities
Often uses simple exploits for initial infection	Uses more sophisticated vectors for initial infection
Designed to avoid detection over long periods of time	Designed to avoid detection over long periods of time
Designed to communicate information back to the attacker using covert command and control	Designed to operate in isolation, not dependent upon remote command and control
Mechanisms for persistent operation even if detected	Mechanisms for persistent operation or reinfection if detected
Not intended to impact or disrupt network operations	Possible intentions include network disruption

Table 3.2 Information Targets of Advanced Persistent Threats (APT) and Cyberwar

APT targets	Weaponized industrial malware targets
Intellectual property	
Application code	Certificates and authority
Application design	Control protocols
Protocols	Functional diagrams
Patents	PCS command codes
Industrial designs	
Product schematics	Control system designs and schematics
Engineering designs and drawings	Safety controls
Research	PCS weaknesses
Chemicals and Formulas	
Pharmaceutical formulas	Pharmaceutical formulas
Chemical equations	Pharmaceutical safety and allergy information
Chemical compounds	Chemical hazards and controls

The developers of Stuxnet could have used stolen intellectual property—which is the primary target of the APT—to develop a more weaponized piece of malware. In other words, a cyberattack that is initially classified as "information theft" may seem relatively benign, but it may also be the logical precursor to weaponized code. Some other recent examples of weaponized malware include Shamoon, as well as previously mentioned Duqu and Flame campaigns.

Details surrounding the Duqu and Pipeline Intrusion campaigns remain restricted at this time and are not appropriate for this book. A great deal can be learned from Night Dragon and Stuxnet, as they both have components that specifically relate to industrial systems.

Industroyer

Industroyer and industroyer 2 represent the only malware at the time of this writing that is designed to target energy distribution and trigger blackouts. Industroyer was first utilized in 2016 and resulted in sustained blackouts in Ukraine, while industroyer 2 was utilized against Ukraine more recently during the 2022 Russian invasion of Ukraine. Industroyer is unique in its ability to communicate using industrial protocols to interact with electrical substation ICS hardware, including circuit breakers and protective relays.[28]

Night dragon

In February 2011, McAfee announced the discovery of a series of coordinated attacks against oil, energy, and petrochemical companies. The attacks, which originated primarily in China, were believed to have commenced in 2009, operating continuously and covertly for the purpose of information extraction, as is indicative of an APT.

Night Dragon is further evidence of how an outside attacker can (and will) infiltrate critical systems once it can successfully masquerade as an insider. It began with SQL database injections against corporate, Internet-facing web servers. This initial compromise was used as a pivot to gain further access to internal, intranet servers. Using standard tools, attackers gained additional credentials in the form of usernames and passwords to enable further infiltration to internal desktop and server computers. Night Dragon established **command and control** (C2) servers as well as **remote administration toolkits** (RATs), primarily to extract email archives from executive accounts.[29] Although the attack did not result in sabotage, as was the case with Stuxnet, it did involve the theft of sensitive information, including operational oil and gas field production systems (including ICSs) and financial documents related to field exploration and bidding of oil and gas assets[30]. The intended use of this information is unknown at this time. The information that was stolen could be used for almost anything, and for a variety of motives. None of the ICSs of the target companies were affected; however, certain cases involved the exfiltration of data collected from operational control systems[31]—all of which could be used in a later, more targeted attack. As with any APT, Night Dragon is surrounded with uncertainty and supposition. After all, APT is an act of cyber espionage—one that may or may not develop into a more targeted cyber war.

Stuxnet

It is impossible to discuss industrial network security without talking about Stuxnet. Stuxnet is largely considered as a "game changer" in the industry because it was the first targeted, weaponized cyberattack against an ICS. Prior to Stuxnet, it was still widely believed that industrial systems were either immune to cyberattack (due to the obscurity and isolation of the systems) and were not being targeted by hackers or other cyberthreats. Proof-of-concept cyberattacks, such as the Aurora project, were met with skepticism prior to Stuxnet. The "threat" pre-Stuxnet was largely considered to be limited to accidental infection of computing systems or the result of an insider threat. It is understandable, then, why Stuxnet was so widely publicized, and why it is still talked about today. Stuxnet proved many assumptions of industrial cyberthreats to be wrong and did so using malware that was far more sophisticated than anything seen before.

Today, it is obvious that ICSs are of interest to malicious actors, and that the systems are both accessible and vulnerable. Perhaps the most important lesson that Stuxnet taught us is that a cyberattack is not limited to PCs and servers. While Stuxnet used many methods to exploit and penetrate Windows-based systems, it also proved that malware could alter an automation process by infecting systems within the ICS, over-writing process logic inside a controller, and hiding its activity from monitoring systems. Stuxnet is discussed in detail in Chapter 7, "Hacking Industrial Control Systems."

TRISIS

TRISIS represents another fundamental shift in the industrial cybersecurity threat landscape. Until TRISIS, cybersecurity risk had been tempered by the resiliency of the control environment, which includes the use of Safety Instrumented Systems (SIS), a subset of a control system that is designed to maintain safe conditions in response to various types of failures. TRISIS is the first malware to directly target SIS, and by doing so, it has the potential to prevent the SIS from maintaining safe conditions. Consequences of a successful TRISYS attack could include an uncontrolled plant shut-down or even the purposeful introduction of a physically unsafe state.[32]

Advanced persistent threats and cyber warfare

One can make two important inferences when comparing APT and cyberwarfare. The first is that cyberwarfare is higher in sophistication and in consequence, mostly due to available resources of the attacker and the ultimate goal of destruction versus profit. The second is that in many industrial networks, there is less profit available to a cyberattacker than from others, and so it requires a different motive for attack (i.e., socio-political). If the industrial network you are defending is largely responsible for commercial manufacturing, signs of an APT are likely evidence of attempts at intellectual theft. If the industrial network you are defending is critical and could potentially impact lives, signs of an APT could mean something larger, and extra caution should be taken when investigating and mitigating these attacks.

Still to come

Infection mechanisms, attack vectors, and malware payloads continue to evolve. Greater sophistication of the individual exploits and bots is expected, as well as more sophisticated blends of these components. Because advanced malware is expensive to develop (or acquire), it is reasonable to expect new variations or evolutions of existing threats in the short term, rather than additional "Stuxnet-level" revolutions. Understanding how existing exploits might be fuzzed or enhanced to avoid detection can help plan a strong defense strategy. It is important to realize the wealth of information available in the open-source community. Tools like the Metasploit Framework by Rapid7 offer the ability to alter exploits and payloads to avoid detection, as well as transport this code between different mechanisms (DLL, VBS, OCX, etc.).

What can be assumed is that threats will continue to grow in size, sophistication, and complexity.[33] New 0-day vulnerabilities will likely be used for one or more stages of an attack (infection, propagation, and execution). The attacks will become more focused, attempting to avoid detection through minimized exposure. Stuxnet spread easily through many systems and only fully activated its entire payload within certain environments. If a similar attack was less promiscuous and more tactically inserted into the target environment, it would be much more difficult to detect.

In early 2011, additional vulnerabilities and exploits that specifically target ICS were developed and released publicly, including the broadly publicized exploits developed by two separate researchers in Italy and Russia. The "Luigi Vulnerabilities," identified by Italian researcher Luigi Auriemma included 34 total vulnerabilities against systems from Siemens (FactoryLink), Iconics (Genesis), 7-Technologies (IGSS), and DATAC (RealWin).[34] Additional vulnerabilities and exploit code, including nine 0-days, were released at that time by the Russian firm Gleg as part of the Agora + SCADA exploit pack (now called the SCADA + pack) for the Immunity CANVAS toolkit.[35] Today, Gleg consistently offers regular updates to the SCADA + exploit pack often including ICS-specific 0 days.[36]

Unfortunately, since then, the threat has evolved, and the industry now faces multiple nation-state level industrial attack frameworks (more information about using these and other tools will be covered further in Chapter 7 "Hacking Industrial Systems").

Luckily, many tools are already available to defend against these newer and more sophisticated attacks. Most network-based threat detection platforms, including Snort, Suricate, and Zeek (formerly named "Bro") are able to decode and analyze common industrial protocols by default, and/or have optional packages that can be added to provide this functionality (see Chapter 11, "Implementing Security and Access Controls"). The results can be very positive when they are used appropriately in a blended, sophisticated defense based upon "Advanced Persistent Diligence."[37]

Defending against modern cyber threats

As mentioned in Chapter 2, "About Industrial Networks," the security practices that are recommended in this book are aimed high because the threat environment in industrial networks has already shifted to these types of advanced cyberattacks, if not outright cyberwar. These recommendations are built around the concept of "advanced persistent diligence" and a much higher than normal level of situational awareness because the APT is evolving specifically to avoid detection by known security measures.[38]

Advanced persistent diligence requires a strong **defense-in-depth** (DiD) approach, both in order to reduce the available attack surface exposed to an attacker and in order to provide a broader perspective of threat activity for use in incident response, analysis, remediation, restoration, and investigation. The APT is evolving to avoid detection even through advanced event analysis, making it necessary to examine more data about network activity and behavior from more contexts within the network.[39]

The application of traditional security recommendations is not enough because the active network defense systems such as stateful firewalls are no longer capable of blocking the same threats that carry with them the highest consequences. APT threats can easily slide through these legacy cyber defenses and is why new technologies like **next-generation firewalls** (NGFW), **unified threat management** (UTM) appliances, and ICS protocol aware intrusion protection systems (IPS) can be deployed to perform deeper inspection into the content that actual comprises the network communications.

Having situational awareness of what is attempting to connect to the system, as well as what is going on within the system is the only way to start to regain control of the network and the systems connected to it. This includes information about systems and assets, network communication flows and behavior patterns, organizational groups, user roles, and policies. Ideally, this level analysis will be automated and will provide an active feedback loop in order to allow information technology (IT) and **operational technology** (OT) security professionals to successfully mitigate a detected APT.

The insider

One of the most common pitfalls within manufacturing organizations is the deployment of a cybersecurity program in the absence of a thorough risk assessment process. This often leads to the commissioning of security controls that do not adequately represent the unique risks that face a particular organization, including the origin of their most probable threats: the insider. It is essential to have a clear definition of exactly what is meant when someone is called an "insider." A commonly used definition of an insider is an individual who has "approved access, privilege, or knowledge of information systems, information services, and missions"[40]. This definition can be expanded to the unique operational aspects of ICS to include a wide range of individuals[41]:

- Employees with direct access to ICS components for operation
- Employees with highly privileged access for administration and configuration
- Employees with indirection access to ICS data
- Subcontractors with access to specific ICS components or subsystems for operation
- Services providers with access to specific ICS components or subsystems for support

It is easy to realize that there are many viable pathways into a secure industrial network through what could be thought of as "trusted connections" or trusted relationships that are not commonly identified on system architecture and network topology diagrams. Each one of these trusted insiders has the ability to introduce unauthorized content into the ICS while masquerading as a legitimate, authorized, and often time's privileged user. The security controls deployed in these cases are typically not designed to detect and prevent these inside attacks but are focused more heavily on preventing traditional attacks that are expected to originate on external, untrusted networks. A common symptom of this approach is the deployment of firewalls between the business and industrial networks where the deployed rules are designed to only aggressively block and log "inbound" traffic from the business network with little or no monitoring of "outbound" traffic from the industrial networks.

The Repository of Industrial Security Incidents (RISI) tracks and updates a database of ICS cyberevents and publishes an annual report that includes a yearly summary along with cumulative findings. The 2013 report showed that of the incidents analyzed, only 35% originated from outsiders.[42] If the primary defenses are based on protecting from external threats, then it can be expected to only mitigate one-third of the potentials threats facing the ICS!

Many organizations find it difficult to accept the fact that their industrial security program needs to include controls to protect the system from the actual users and administrators. The reason is not that they do not understand the risk, but that they do not understand or accept that an employee could intentionally cause harm to the system or the plant under their control. In most cases, the event is the result of an "unintentional" or "accidental" action that is no longer directed at any particular employee, but rather on the overall security policies deployed within the architecture. According to RISI, 80% of the analyzed cyber events in ICS architectures were classified as "unintentional" in nature.[43]

This should in no manner diminish the importance of maintaining diligence with trusted individuals with granted access to industrial networks who could in fact initiate intentional attacks. Even fully vetted insiders could be pressured to initiate an attack through bribery or blackmail. The widespread deployment of remote access techniques has increased the need for heightened awareness and appropriate controls resulting from more individuals allowed access to industrial networks from potentially insecure locations and assets. Remote access is a leading point of entry for cyberevents, with approximately one-third of the events originating via remote connections.[44] An example of this occurred in 2003 when a contractor's Slammer-infected computer connected via a virtual private network (VPN) connection to his company's network that had a corresponding secure site-to-site connection to a nuclear power generating station's business network. The worm was able to traverse the two VPNs and eventually penetrate the firewall protecting the industrial network and a safety monitoring system that was disabled by the worm. The plant engineers responsible for the system, which was targeted did not realize that a patch for the bug was available 6-months earlier.[45]

Hacktivism, cybercrime, cyberterrorism, and cyberwar

The risk against industrial networks, especially those that support critical infrastructures (local, regional, or national), has increased steadily in the past years. This can be attributed in part to an increase in cyber-security research of ICSs resulting from the global awareness of ICS security following the disclosure of Stuxnet, as well as the easy availability of tools such as ICS-specific exploit packages within both open-source and commercial penetration testing tools such as Metasploit and CANVAS. Figure 3.3 depicts the year-over-year disclosure counts as logged in the Open Source Vulnerability Database (OSVDB)[46] and shows a significant increase in disclosures beginning in 2010 (Figure 3.7). To remotely breach an industrial network and execute a targeted cyberattack, the attacker still requires a certain degree of specialized knowledge that may not be as readily available. Unfortunately, this logic—while valid—is too often used to downplay the risk of a targeted cyberattack. Of the more than 700 SCADA vulnerabilities listed in the OSVDB, most involve vulnerabilities of devices that are *not* typically used in highly critical systems. On the other hand, over 40% of those vulnerabilities have a Common Vulnerability Scoring System (CVSS) score of 9.0 or higher.

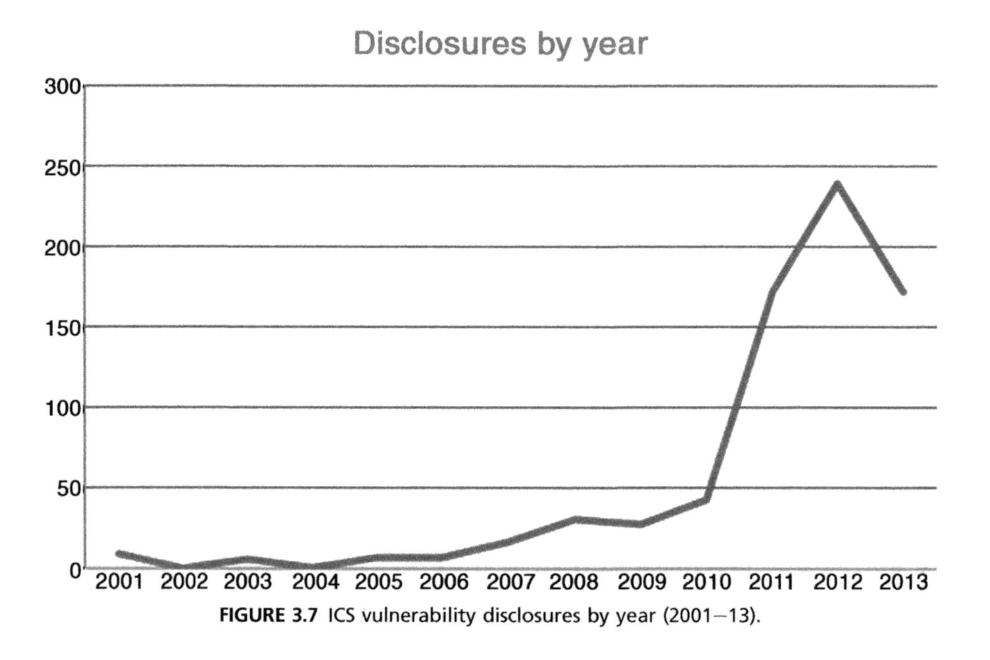

FIGURE 3.7 ICS vulnerability disclosures by year (2001–13).

What it comes down to is simple: There are vulnerable industrial systems, and because these systems are vulnerable, anyone willing to perform some research, download some freely available tools, and put forth, some effort can launch an attack. With a minimal amount of system- and industry-specific training, the likelihood of a successful attack with moderate consequences is significantly increased. The real question is one of motive and resources. While the average citizen may not be motivated enough to plan and execute an attack on critical infrastructures, there are hacktivist groups who are highly motivated. While the average citizen may not have the resources to craft a targeted payload, develop a 0-day exploit to penetrate network defenses, steal digital certificates, or execute targeted spear-phishing campaigns, all of these services are available for hire—anonymously. In a report by McAfee Labs, the use of digital currencies to anonymously buy and sell illegal products and services is becoming more prevalent, fostering an enormous digital black market. Cybercrime and cyberterrorism are no longer isolated to organized syndicates and terrorist groups but are now services available for hire. A fully weaponized attack against critical infrastructures at any level no longer needs to be military because it can be mercenary—bought as a service, online.

Taking into consideration the possibility of "hacking as a service" from potentially very large and capable anonymous entities, the known vulnerability data (which is compelling on its own) becomes an almost moot argument. The real attacks are far more likely to involve the unknown, using 0-day exploits and highly sophisticated techniques.

Summary

Industrial networks are both vital and vulnerable—there are potentially devastating consequences in the event of a successful cyber incident. Examples of real cyber incidents have grown progressively more severe over time, highlighting the evolving nature of threats against industrial systems. The attacks are evolving as well, to the point where modern cyberthreats are intelligent and adaptable, difficult to detect and highly persistent. The intentions have also evolved, from information theft to industrial sabotage and the actual disruption of critical infrastructures. Combined with a rise of criminal cyberservices that are becoming increasingly available via anonymous systems and that are paid for with anonymous digital currencies, this trend is worrisome, and should send a clear message to owners and operators of critical infrastructures to improve cybersecurity wherever and whenever possible.

Securing industrial networks requires a reassessment of your security practices, realigning them to a better understanding of how industrial protocols and networks operate (see Chapter 4, "Introduction to Industrial Control Systems and Operations" and Chapter 5, "Industrial Network Design and Architecture"), as well as a better understanding of the vulnerabilities and threats that exist (see Chapter 8, "Risk and Vulnerability Assessments").

Endnotes

1. Galloway, Brendan; Hancke, Gerhard P. (2012). "Introduction to Industrial Control Networks". IEEE Communications Surveys and Tutorials. 15 (2).
2. https://www.automation.com/en-us/articles/2011-1/hannover-messe-2011-more-than-230000-visitors
3. J. Pollet, Red Tiger, Electricity for free? The dirty underbelly of SCADA and smart meters, in: Proc. 2010 BlackHat Technical Conference, Las Vegas, NV, July 2010.
4. Ibid.
5. The Open-Source Vulnerability Database (OSVDB) Project, ID Nos. 79,399/79,400. http://osvdb.org (cited: December 20, 2013).
6. 2013 Data Breach Investigations Report. 2013. Verizon.
7. Microsoft. KB 103,884 "The OSI Model's Seven Layers Defined and Functions Explained", http://support.microsoft.com/kb/103884 (cited: December 21, 2013).
8. International Society of Automation (ISA). Standards & Practices 95. http://www.isa-95.com/subpages/technology/isa-95.php (cited: December 21, 2013).
9. Purdue Enterprise Reference Architecture (PERA), "Purdue Reference Model for CIM". http://www.pera.net/Pera/PurdueReferenceModel/ReferenceModel.html (cited: December 21, 2013).
10. McAfee Labs. McAfee Labs Threat Report: Third Quarter 2013. McAfee. 2013.
11. Repository of Industrial Security Incidents (RISI). 2013 Report on Cyber Security Incidents and Trends Affecting ICSs, June 15, 2013.
12. Ibid.
13. Ibid.
14. Ibid.
15. J. Pollet, Red Tiger, Understanding the advanced persistent threat, in: Proc. 2010 SANS European SCADA and Process Control Security Summit, Stockholm, Sweden, October 2010.
16. J. Pollet, Red Tiger, Understanding the advanced persistent threat, in: Proc. 2010 SANS European SCADA and Process Control Security Summit, Stockholm, Sweden, October 2010.
17. Ibid.

18. Threat Post. Move Over Conficker, Web Threats are Top Enterprise Risk. http://threatpost.com/move-over-conficker-web-threats-are-top-enterprise-risk/99762 (cited: December 20, 2013).
19. J. Pollet, Red Tiger, Understanding the advanced persistent threat, in: Proc. 2010 SANS European SCADA and Process Control Security Summit, Stockholm, Sweden, October 2010. J. Pollet.
20. N. Falliere, L.O. Murchu, E. Chien, Symantec. W32.Stuxnet Dossier, Version 1.1, October 2010
21. Ibid.
22. Symantec. W32.Duqu: The precursor to the next Stuxnet, v1.4. November 23, 2011.
23. McAfee. Global Energy Cyberattacks: "Night Dragon". February 10, 2011.
24. Symantec. Flamer: Highly Sophisticated and Discreet Threat Targets the Middle East, May 28, 2012. http://www.symantec.com/connect/blogs/flamer-highly-sophisticated-and-discreet-threat-targets-middle-east (cited: December 20, 2013).
25. ICS-CERT, U.S. Dept. of Homeland Security. ICSA-12-136-01P, Gas Pipeline Intrusion Campaign Indicators and Mitigations, May 15, 2012.
26. "Dragonfly: Cyberespionage Attacks Against Energy Suppliers," Symantec Security Response v.1.21, July 7, 2014 (v1.0 first published June 30, 2014).
27. PIPEDREAM: CHERNOVITE'S EMERGING MALWARE TARGETING INDUSTRIAL CONTROL SYSTEMS. Dragos, Inc. April 2022.
28. Robert Lipovsky, Anton Cherepanov. "Industroyer2: Sandworm's Cyberwarfare Targets Ukraine's Power Grid Again". BlackHat USA, August, 2022.
29. Ibid.
30. Ibid.
31. Ibid.
32. TRISIS Malware Analysis of Safety System Targeted Malware. Dragos, Inc. December 2017.
33. Ibid.
34. D. Peterson, Italian researcher publishes 34 ICS vulnerabilities. Digital Bond. http://www.digitalbond.com/2011/03/21/italian-researcher-publishes-34-ics-vulnerabilities/, March 21, 2011 (cited: April 4, 2011).
35. J. Langill, SCADAhacker.com. Agora + SCADA Exploit Pack for CANVAS http://scadahacker.blogspot.com/2011/03/agora-scada-exploit-pack-for-canvas.html, March 17, 2011. (cited: December 20, 2013).
36. J. Langill, SCADAhacker.com. Gleg releases Ver 1.28 of the SCADA + Exploit Pack for Immunity Canvas, October 8, 2013. (cited: October 8, 2013).
37. D. Peterson, Friday News and Notes. http://www.digitalbond.com/2011/03/25/friday-news-and-notes-127, March 25, 2011 (cited: April 4, 2011).
38. Ibid.
39. US Department of Homeland Security, US-CERT, Recommended Practice: Improving Industrial Control Systems Cybersecurity with Defense-In-Depth Strategies, Washington, DC, October 2009.
40. M. Maybury, "How to Protect Digital Assets from Malicious Insiders", Institute for Information Infrastructure Protection.
41. M. Luallen, "Managing Insiders in Utility Control Systems", SANS SCADA Summit 2011, March 2011.
42. Repository of Industrial Security Incidents (RISI), "2013 Report on Cyber Security Incidents and Trends Affecting Industrial Control Systems", June 15, 2013.
43. Ibid.
44. Ibid.
45. Security Focus, "Slammer worm crashed Ohio nuke plant network", August 19, 2003, http://www.securityfocus.com/news/6767, (cited: January 6, 2014).
46. Open-Source Vulnerability Database (OSVDB) Project. http://osvdb.org/search?search[vuln_title]=scada (cited: January 1, 2013).

4 Introduction to Industrial Control Systems and Operations

Information in this chapter

- System assets
- System operations
- Process management
- Safety instrumented systems
- Smart grid operations
- Network architectures

It is also necessary to have a basic understanding of how commonly used industrial control system (ICS) components interact within an industrial network in addition to knowledge of how industrial network protocols operate. This information may seem overly basic for operators of ICSs. It is also important to remember that "how control systems *are* connected" and "how they *should be* connected" are not always the same. One can quickly assess whether there are any basic security flaws in an industrial network design by taking a short step back to the basics. This requires an understanding of the specific assets, architectures, and operations of a typical industrial network.

System assets

The first step is to understand the components used within industrial networks and the roles that they play. These devices discussed in this chapter, include field components such as sensors, actuators, motor drives, gauges, indicators, and control system components such as programmable logic controllers (PLCs), remote terminal units (RTUs), intelligent electronic devices (IEDs), human–machine interfaces (HMIs), engineering workstations, application servers, data historians, and other business information consoles or dashboards.

Programmable logic controller

A programmable logic controller is a specialized industrial computer used to automate functions within manufacturing facilities. Unlike desktop computers, PLCs are typically physically hardened (making them suitable for deployment in a production environment) and may be specialized for specific industrial uses with multiple specialized inputs and outputs. PLCs do not typically use a commercially available operating system

Industrial Network Security. https://doi.org/10.1016/B978-0-443-13737-2.00011-7

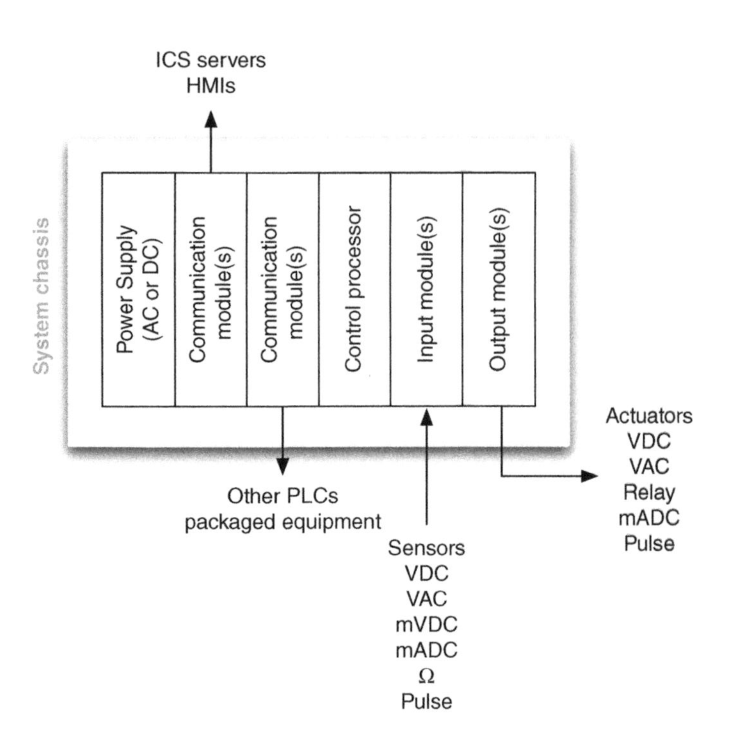

FIGURE 4.1 Components of a programmable logic controller.

(OS). They instead rely on specific application programs that allow the PLC to function automatically generating output actions (e.g., to pump motors) in response to specific inputs (e.g., from sensors) with as little overhead as possible. PLCs were originally designed to replace electromechanical relays. Very simple PLCs may be referred to as programmable logic relays (PLRs). Figure 4.1 illustrates the typical structure of a PLC.

PLCs typically control real-time processes, and so they are designed for simple efficiency. For example, in plastic manufacturing, a catalyst may need to be injected into a vat when the temperature reaches a very specific value. If processing overhead or other latency introduces delay in the execution of the PLC's logic, it would be very difficult to precisely time the injections, which could result in quality issues. For this reason, the logic used in PLCs is typically very simple and is programmed according to an international standard set of languages as defined by IEC-61131-3.

Ladder diagrams

PLCs can use "ladder logic" or "ladder diagrams (LD)," which is a simplistic programming language included within the IEC-61131-3 standard that is well suited for industrial applications. Ladder logic gets its name from the legacy method of implementing discrete logic via electromechanical relays and was initially referenced as "relay ladder logic." Ladder logic can be thought of as a set of connections between inputs (relay contacts) and outputs (relay coils). Ladder logic follows a relay function diagram, as

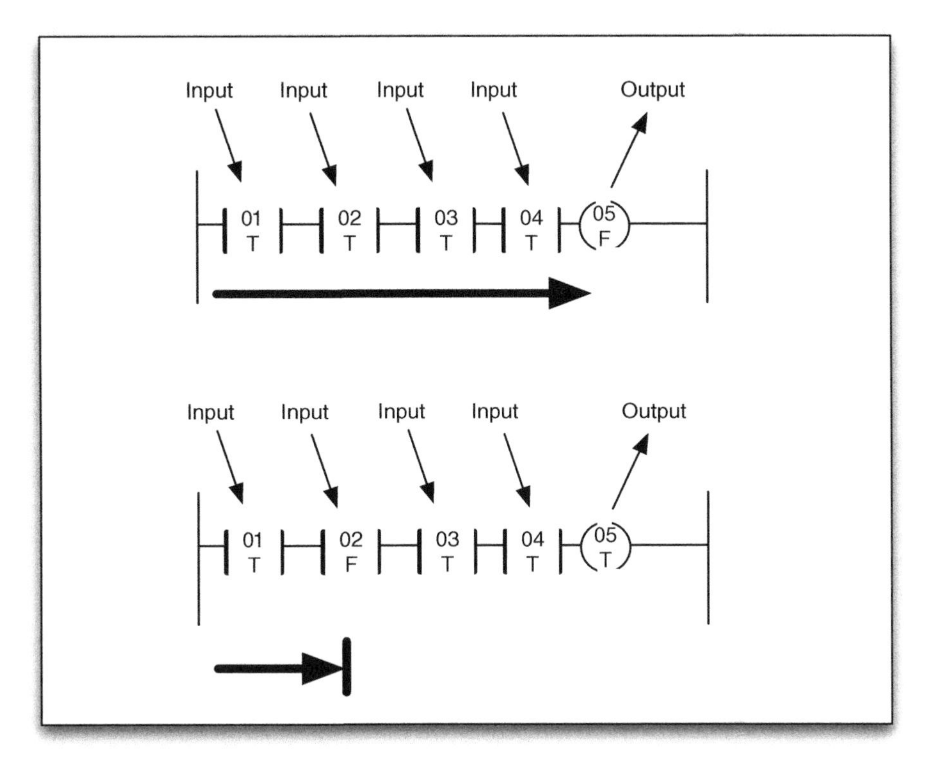

FIGURE 4.2 Example of simple ladder logic with both complete and incomplete conditions.

shown in Figure 4.2. A path is traced on the left side, across "rungs" consisting of various inputs. If an input relay is "true," the path continues, and if it is "false," it does not. If the path to the right side completes (there is a complete "true" path across the ladder), the ladder is complete and the output coil will be set to "true" or "energized." If no path can be traced, then the output remains "false," and the relay remains "de-energized."[1] This was implemented before PLCs, with a (+) bus on the left-hand side and a (–) bus on the right-hand side. The "path" just described represented electrical current flow through the logic.

The PLC applies this ladder logic by looking at inputs from discrete devices that are connected to the manufacturing equipment and performing a desired output function based on the "state" of these inputs. These outputs are also connected to manufacturing equipment, such as actuators, motor drives, or other mechanical equipment. PLCs can use a variety of digital and analog communications methods, but typically use a fieldbus protocol such as Modbus, ControlNet, Ethernet/IP, PROFIBUS, PROFINET, or similar (see Chapter 6, "industrial network protocols"). A switch is used to convert an analog or "continuous" value from a sensor to a "discrete" on or off value by comparing the input

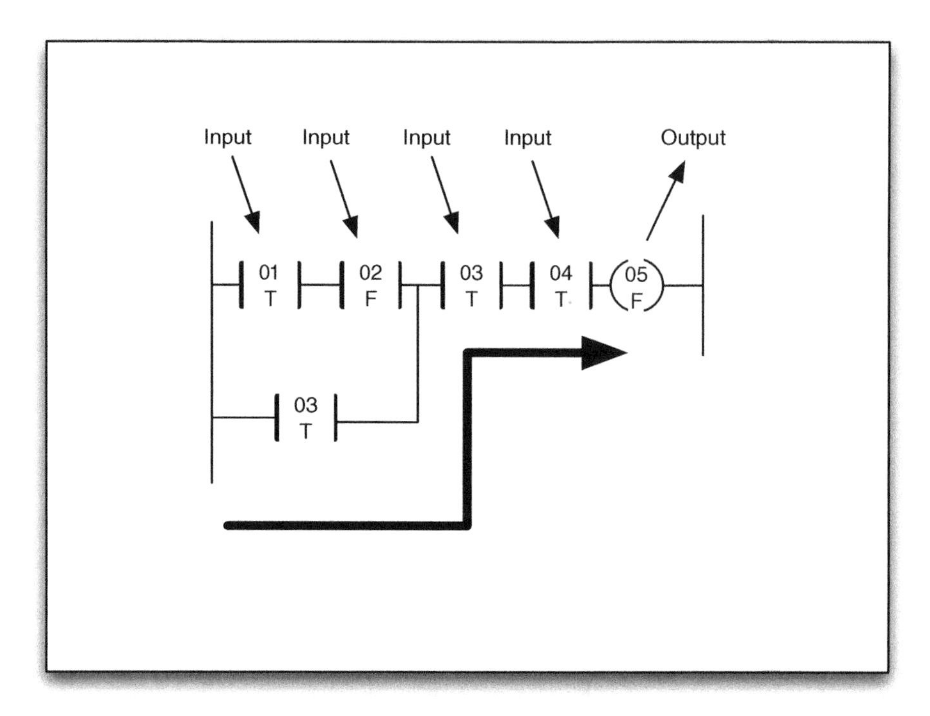

FIGURE 4.3 Example of simple ladder logic containing an "OR" condition.

to a setpoint. If a set point is satisfied, the input is considered "true," and if it is not it is considered "false." Processes defined by ladder logic can be simple or very complex. For example, an "or" condition can allow the rung to complete based on an alternate input condition, as shown in Figure 4.3.

When an output coil is finally reached it becomes "true," and the PLC activates the output. This allows the PLC to automate a function (e.g., turning a pump on or off) based on set point parameters (e.g., high and low water levels within a tank).[2]

Internal relays may also be used within a PLC—these relays, unlike input relays, do not use inputs from the physical plant, but rather are used by the ladder logic to lock an input on (true) or off (false) depending upon other conditions of the program. PLCs also use a variety of other function "blocks" including counters, timers, flip-flops, shift registers, comparators, mathematical expressions/functions, and many others allowing PLCs to act in defined cycles or pulses, as well as storage.[3]

Sequential function charts

Another programming language used by PLCs and defined within the IEC-61131-3 standard is "sequential logic" or "sequential function charts (SFC)." Sequential logic differs from ladder logic in that each step is executed in isolation and progresses to the next step only upon completion, as opposed to ladder logic where every step is tested in

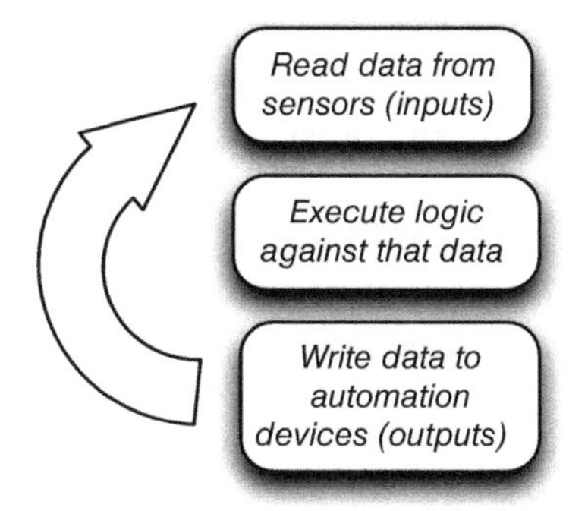

FIGURE 4.4 Programmable logic controller (PLC) operational flow diagram.

each scan. This type of sequential programming is very common in batch-oriented operations. Other common languages defined by IEC-61131-3 include "structured text (ST)," "function block diagram (FBD)," and "instruction list (IL)" methods. No matter what programming language is used with a particular PLC, the end goal is ultimately to automate the legacy electromechanical functions common in industrial systems by checking inputs, applying logic (the program), and adjusting outputs as appropriate,[4] as shown in Figure 4.4.

The logic used by the PLC is created using a software application typically installed on an engineering workstation that combines similar tools or may be combined with other system functions like the HMI. The program is compiled locally on the computer and then downloaded from the computer to the PLC by either direct serial (RS-232) or Ethernet connections, where the logic code is loaded onto the PLC. PLCs can support the ability to host both the source and compiled logic programs, meaning that anyone with the appropriate engineering software could potentially access the PLC and "upload" the logic.

Remote terminal unit

A remote terminal unit (RTU) typically resides in a substation, along a pipeline, or some other remote location. RTUs monitor field parameters and transmit that data back to a central monitoring station—typically either a master terminal unit (MTU) that is may be an ICS server, a centrally located PLC, or directly to an HMI. RTUs commonly include remote communications capabilities consisting of a modem, cellular data connection, radio, or other wide area communication technology. They are often installed in locations that may not have easy access to electricity and can be supplied with local solar power generation and storage facilities. It is common for RTUs to be placed outdoors, which means they are subjected to extreme environmental conditions (temperature, humidity, lightning, animals, etc.). Their communications bandwidth is generally limited, and in order to maximize the amount of information transmitted, they favor protocols that support "report by exception" or other "publish/subscribe" mechanisms to minimize unnecessary repetition or transmission of the data as described in Chapter 6, "Industrial Network Protocols."

RTUs and PLCs continue to overlap in capability and functionality, with many RTUs integrating programmable logic and control functions, to the point where an RTU can be thought of as a remote PLC that has been combined with integrated telecommunications equipment.

Intelligent electronic device

Each industry has unique physical and logical requirements, and for this reason, ICS equipment varies to some extent from industry to industry. A pipeline typically has pumping (liquids) or compressor (gases) stations distributed along the pipeline. The RTU is well suited for install installation in this application as was previously described. The electric utility sector has a similar requirement except that instead of pumping stations, their transmission lines consist of numerous electrical substations that are distributed throughout the grid to manage electrical loads and provide local isolation when needed. The IED was developed for these types of installations that require not only local direct control functionality and integrated telecommunications support but also can be installed in areas that involve high-voltage energy sources and the associated electrical "noise" that is typically present in these environments.

As with all technology, IEDs are growing more and more sophisticated over time, and an IED may perform other tasks, blurring the line between device types. To simplify things for the purposes of this book, an IED can be considered to support a *specific* function (i.e., substation automation) within the overall control system, whereas RTUs and PLCs are designed for *general* use (i.e., they can be programmed to control the speed of a motor, to engage a lock, to activate a pump, or rail crossing gate).

As technology evolves, the line blurs between the PLC, RTU, and IED, as can be seen in Emerson process management's ROC800L liquid hydrocarbon remote controller. This device performs measurement, diagnostics, remote control, and telecommunications in a single device that supports several programmable languages.

Human–machine interface

Human–machine interfaces are used as an operator's means to interact with PLCs, RTUs, and IEDs. HMIs replace manually activated switches, dials, and other electrical controls with graphical representations of the digital controls used to sense and influence that process. HMIs allow operators to start and stop cycles, adjust set points, and perform other functions required to adjust and interact with a control process. Because the HMI is software based, they replace physical wires and controls with software parameters, allowing them to be adapted and adjusted very easily. Figure 4.5 shows how the HMI integrates with the overall ICS architecture as explained so far.

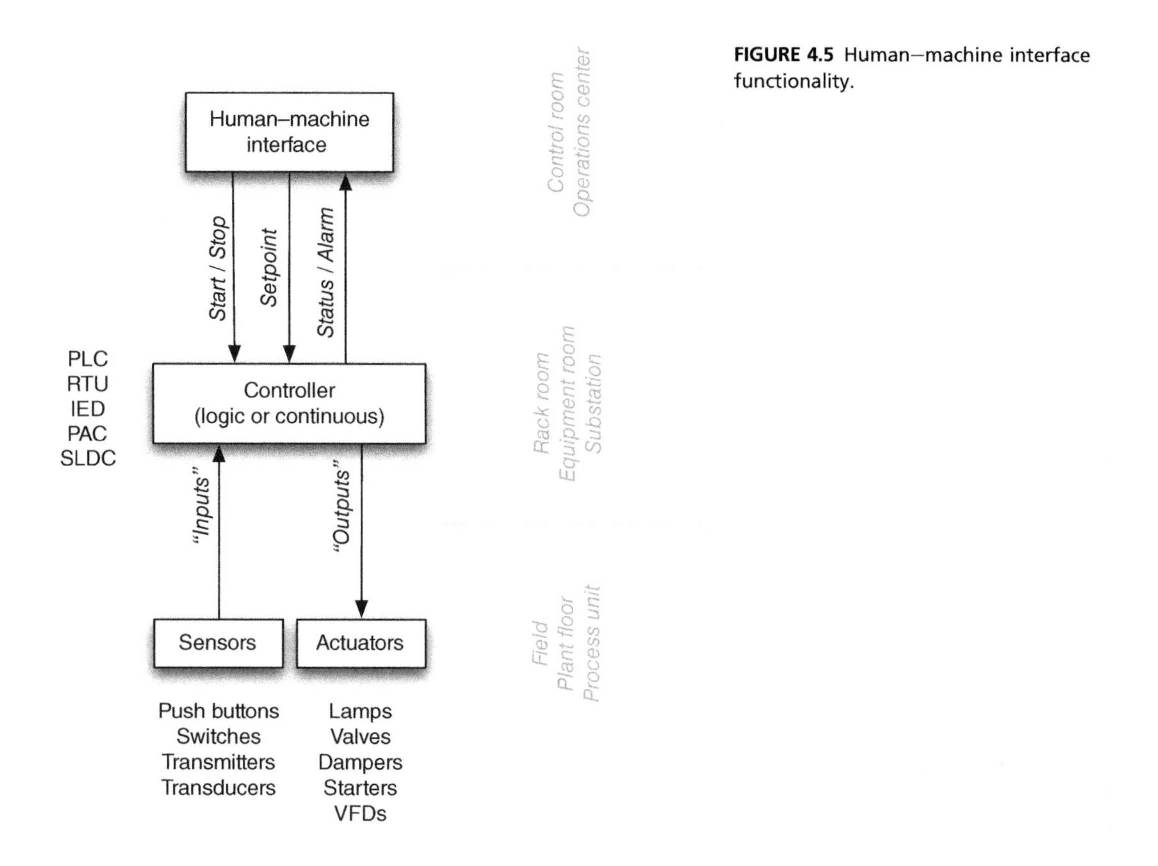

FIGURE 4.5 Human–machine interface functionality.

HMIs are modern software applications that come in two predominant form factors. The first runs on modern OSs like Windows 7 and are capable of performing a variety of functions. The other form combines an industrial hardened computer and local touch panel and is packaged to support door or direct panel mounting. These devices typically utilize an embedded OS like Windows Embedded (CE, XP, 7, 8, Compact) and are programmed with a separate computer and associated engineering software. They act as a bridge between the human operator and the complex logic of one or more PLCs, allowing the operator to focus on how the process is performing rather than on the underlying logic that controls many functions across distributed and potentially complex processes from a centralized location. To accomplish this, the user interface will graphically represent the process being controlled, including sensor values and other measurements, and visible representation of output states (which motors are on, which pumps are activated, etc.).

Humans interact with the HMI through a computer console but do not generally authenticate to the station with a password because during an abnormal event, a password lockout or any other mechanism that would block access to the HMI would be considered unsafe and violates the basic principal of guaranteed availability. At first this may seem insecure, but considering that these devices are typically installed in areas that possess strong physical security and are only operated by trained and authorized personnel, the resulting risk is tolerable. Because HMIs provide supervisory data (visual representation of a control process's current state and values) as well as control (i.e., set point changes), user access controls are usually part of the ICS allowing specific functions to be locked out to specific users. The HMI interacts either directly or indirectly through an ICS server with one or more controllers using industrial protocols such as OLE for process control (OPC) or fieldbus protocols such as Ethernet/IP or Modbus (see Chapter 6, "Industrial Network Protocols").

There are other more appropriate methods of securing HMIs from both unauthorized access by the intended user, as well as unauthorized access resulting from a cyberevent. Many vendors are aware of the importance of least privileges and now are providing local- and domain-based group policies that can be installed to restrict the authorization granted at the local workstation. Microsoft provides the ability to enforce these policies either by computer or user, making this well suited for workstations placed in common areas. These policies can not only restrict the execution of local applications and the functionality of the Windows GUI but also prevent unauthorized access to removable media and USB access ports. The security of the industrial process therefore relies heavily on access control and host security of the HMI and the underlying control system.

Supervisory workstations

A supervisory workstation collects information from assets used within a control system and presents that information for supervisory purposes. Unlike an HMI, a supervisory workstation is primarily read only. These workstations have no control element to interact directly with the process, only the presentation of information about that process. These workstations are typically authorized with the ability to change certain parameters that an operator is usually not allowed to manipulate. Examples may include alarm limits and, in some situations, process setpoints.

A supervisory workstation will consist of either an HMI system (with read-only or supervisory access restrictions) or a dashboard or workbook from a data historian (a device specifically designed to collect a running audit trail of control system operational data). Supervisory workstations can reside in a variety of locations throughout the industrial networks, as well as the ICS semi-trusted demilitarized zones (DMZ) or business networks, up to and including Internet-facing web portals and intranets (See "Control Processes" in this chapter).

Caution

When a supervisory system monitors a control system remotely, the connection between the workstation and the underlying ICS supervisory components must be carefully established, controlled, and monitored. Otherwise, the overall security of control systems' network could be weakened (because the supervisory system becomes an open attack vector to the ICS). For example, by placing a supervisory console in the business network, the console can be more easily accessed by an attacker and then utilized to communicate back to the ICS. If remote supervision can be provided via read-only data, a one-way communication path or some form of secure data replication should be used to prevent such an inbound attack. This is covered in detail in Chapter 9, "Establishing Zones and Conduits."

Data historian

A data historian is a specialized software system that collects point values, alarm events, batch records, and other information from industrial devices and systems and stores them in a purpose-built database. Most ICS vendors including ABB, Areva, Emerson, GE, Honeywell, Invensys, Rockwell, Schneider, Siemens, and others provide their own proprietary data historian systems. There are also third-party industrial data historian vendors, such as Aspen Technologies (www.aspentech.com), Canary Labs (www.canarylabs.com), Modiüs (www.modius.com), and OSIsoft (www.osisoft.com), which interoperate with ICS assets and even integrate with proprietary ICS historians in order to provide a common, centralized platform for data historization, analysis, and presentation.

Data that are historized and stored within a data historian is referred to as "tags" and can represent almost anything—the current speed of a motor or turbine, the rate of airflow through a heating, ventilation, and air-conditioning (HVAC) system, the total volume in a mixing tank, or the specific volumes of injected chemical catalysts in a tank. Tags can even represent human-generated values, such as production targets, acceptable loss margins, and manually collected data.

Information used by both industrial operations and business management is often replicated across industrial and business networks and stored in data historians. This can represent a security risk since a data historian in a less secure zone (i.e., the business network) could be used as a vector into more secure zones (i.e., the ICS network). Data historians should therefore be hardened to minimize vulnerabilities and utilize strict user and network access controls.

Note

The information collected by a data historian is stored centrally within a database. Depending upon the data historian used, this could be a commercial relational database management system (RDBMS), specialized columnar or time-series database system, or some other proprietary data storage system. Most data historians technologies deployed today depend on a hybrid approach that includes fast, proprietary data "collectors" that are deployed close to the production equipment and associated ICS components (to allow high-frequency data collection), and replication to central "shadow" server that relies more on standard RDBMS technologies like Microsoft SQL Server and Oracle. The type of database used is important for several reasons. The data historian will typically be responsible for collecting information from thousands or even millions of tags at very fast collection rates. In larger networks, the capabilities of the database in terms of data collection performance can impact the data historian's ability to collect operational information in real time. More importantly within the context of this book is that commercial RDBMSs may present specific vulnerabilities potentially leading to a cyberattack. The data historian and any auxiliary systems (database server, network storage, etc.) should be included in any vulnerability assessment, and care should be taken to isolate and secure these systems along with the data historian server.

OSIsoft holds a dominant position in the data historian market at the time of this writing, with 65% market penetration in global industrial automated systems.[5] The OSIsoft PI System integrates with many IT and OT systems including other data historians and is a premium target for attack. Applying the latest updates and patches can minimize vulnerabilities. Properly isolating and securing data historian components that connect with assets is less trusted networks within a semi-trusted DMZ significantly help to minimize accessibility. It is important to consider special component-level cyber security testing of assets such like data historians in order to ensure that they do not introduce vulnerabilities not common in the traditional public disclosure realm (e.g., Microsoft monthly security bulletins) to the ICS. For more information about the role of data historians within control system operations, see "Control Processes: Feedback Loops" and "Control Processes: Business Information Management" later in the chapter.

Business information consoles and dashboards

Business information consoles are extensions of supervisor workstations designed to deliver business intelligence to upper management. They typically consist of the same data obtained from HMI or data historian systems. A business information console in some cases may be a physical console such as a computer display connected to an HMI or historian within the ICS DMZ but physically located elsewhere (such as an executive office or administration building). The physical display in these cases is connected using a remote display or secure remote keyboard video mouse (KVM) switching system. Business information may also be obtained by replicating HMI or data historian systems within the business network or by publishing exported information from these systems using an intermediary system. An example of such an intermediary system may be exporting values from the data historian into a spreadsheet and then publishing that spreadsheet to a corporate information portal or intranet. This publishing model may be streamlined and automated depending upon the sophistication of the data historian. Many vendors have developed special platforms that allow the reuse of process-level

HMI graphics to be deployed and populated with real-time and historical data via replicated read-only servers placed on less-secure networks using web services (e.g., HTML, HTTPS, etc.) for the presentation of data to business network users. Any published data should be access controlled, and any open communication path from ICS systems to more openly accessible workstations or portals should be carefully controlled, isolated, and monitored.

Other assets

There are many other assets that may be connected to an industrial network other than PLCs, RTUs, HMIs, historians, and workstations. Devices such as printers and print servers may be connected to corporate networks, or they may connect directly to a control loop. Access control systems such as badge scanners and biometric readers may be used along with closed-circuit television (CCTV) systems all networked (probably over TCP/IP) together. There are also common infrastructure components like active directory and time servers that are deployed throughout an industrial network.

Although this book does not attempt to cover every aspect of every device that may be present within an industrial network, it is important to recognize that every device has an attack surface, and therefore a potential impact to security and should be assessed if:

1. It is connected to a network of any kind (including wireless networks originating from the device itself).
2. It is capable of transporting data or files, such as removable media (mobile devices).

Even the most seemingly harmless devices should be assessed for potential security weaknesses—either inherent to the device itself, or a result of configuration of the device. Check the documentation of devices to make sure that they do not have wireless capabilities and, if so, secure or disable those features. Many commercially produced devices contain multipurpose microprocessors, which may contain radio or Wi-Fi antennae receivers or transmitters *even if the device is not intended for wireless communication.* Many of today's Wi-Fi components include both wireless LAN (WLAN) and Bluetooth capability. This is because it is sometimes more cost-effective for a supplier to use a commercial, off-the-shelf (COTS) microprocessors with unneeded capabilities. The manufacturer may never enable those capabilities, but if the hardware exists malicious actors can use it as an attack vector.[6]

System operations

All of the industrial network protocols, devices, and topologies discussed up to this point are used to create and automate some industrial operation: refining crude oil, manufacturing a consumer product, purifying water, generating electricity, synthesizing and combining chemicals, etc. A typical industrial operation consists of several layers of programmed logic designed to manipulate mechanical controls in order to automate the operation. Each specific function is automated by what is commonly referred to as a control loop. Multiple control loops are typically combined or stacked together to automate larger processes.

Control loops

Industrial controllers are made up of many specific automated processes, called control loops. The term "loop" derives from the ladder logic that is widely used in these systems. A controller device such as a PLC is programmed with specific logic. The PLC cycles through its various inputs, applying the logic to adjust outputs, and then starts over scanning the inputs. This repetitive control action is necessary in order to perform a specific function. This cycle or "loop" automates that function.

In a closed loop, the output of the process affects the inputs, fully automating the process. For example, a water heater is programmed to heat water to a setpoint of 90°C. An electric heating coil is energized to heat the water, and the water temperature is measured and fed back as an input into the control process. When 90°C is reached, the heater turns off the heating coil and continues to monitor the temperature until it drops below the setpoint. In an open loop, the input from the process (temperature in this case) does not affect the outputs (the heating coil). Stated another way, closed loops provide automated control whereas open loops provide manual control.

Control loops can be simple, checking a single input, as illustrated in Figures 4.6 and 4.7. For example, a simple loop in an automated lighting process might check a single input

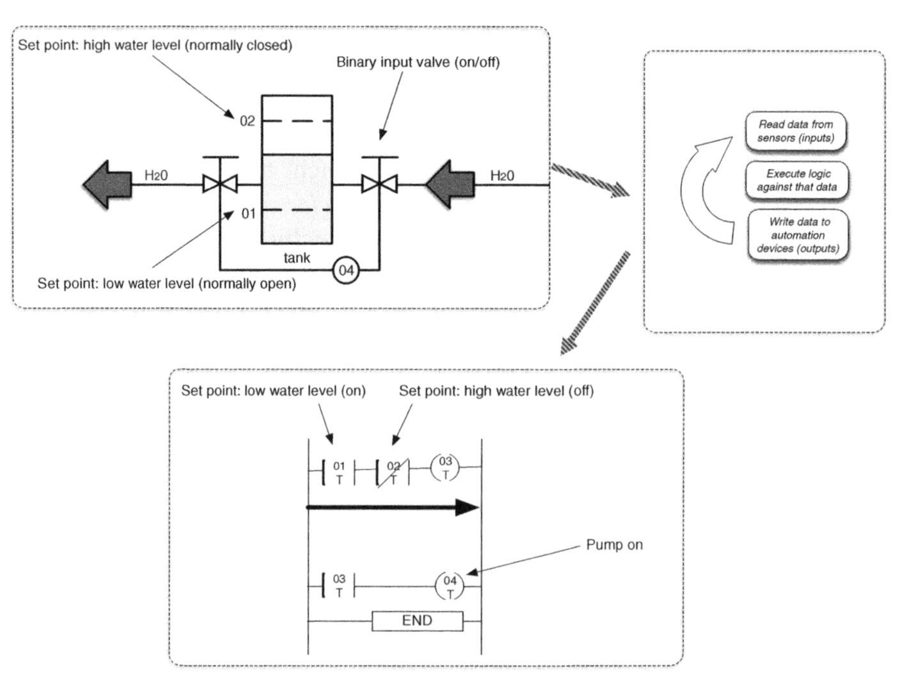

FIGURE 4.6 A simplified control loop in the "ON" state showing the applied ladder logic.

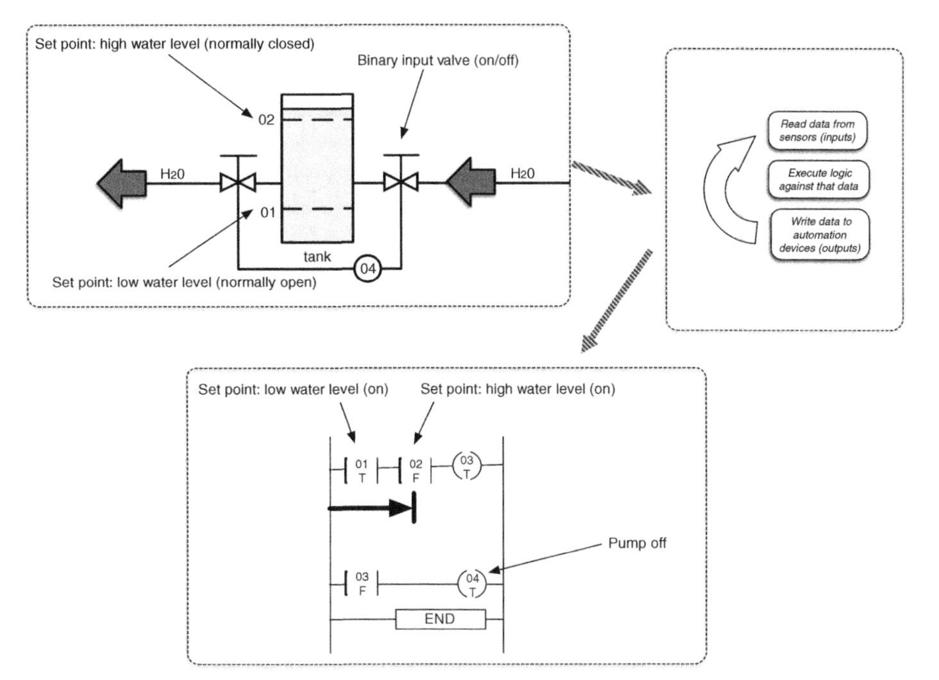

FIGURE 4.7 A simplified control loop in the "OFF" state showing the applied ladder logic.

(e.g., a light sensor to measure ambient light) and adjust a single output (e.g., the switch controlling flow of electricity to a lamp). Complex loops might use multiple inputs (e.g., pressure, volume, flow, and temperature sensors) and adjust multiple outputs (e.g., valves and pump motors) to perform a function that is inherently more complex. An example of such a complex loop might be controlling water level (input) in a boiler drum based on steam demand (input) and feedwater inlet flow (input/output) variations. There are actually multiple control loops in this case applied to perform a single control function. As control complexity increases, control loops may be distributed across multiple controllers requiring critical "peer-to-peer" communications across the network.

Control loops can also be complex, as shown in Figure 4.8. This particular example illustrates several common aspects of process control, including improved variable accuracy through compensation techniques, and stable performance through feed-forward and cascade control strategies. Figure 4.8 shows how increasing or decreasing make-up water into the drum is controlled to account for fluctuations in steam demand. Feed-forward techniques are used to account for the lag time associated with heating water into steam.

Control processes

A "control process" is a general term used to define larger automated processes within an industrial operation. Many control processes may be required to manufacture a product or to generate electricity, and each control process may consist of one or many

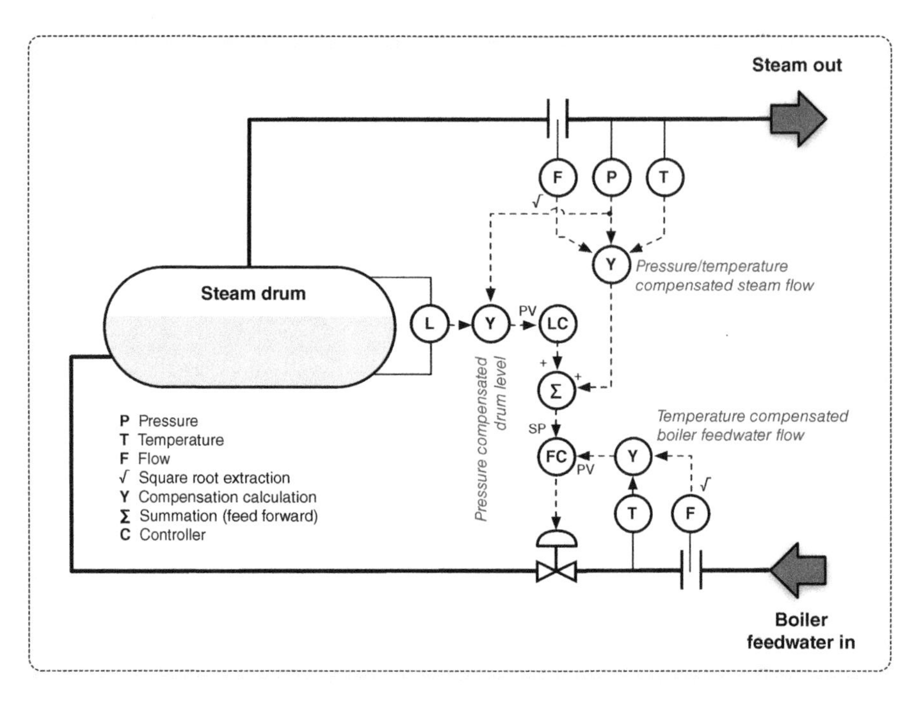

FIGURE 4.8 A more "complex" control loop typical in process control.

control loops. For example, one process might be to inject an ingredient into a mixer utilizing a control loop that opens a valve in response to volume measurements within the mixer, temperature, and other conditions. Several such processes or "steps" can automate the correct timing and combination of several ingredients, which in turn complete a larger process (to make a batter), which is known as a "phase." The mixed batter might then be transported to other entirely separate control processes for baking, packaging, and labeling—all additional "phases" each containing their own unique "steps" and control loops.

Each process is typically managed using an HMI, which is used to interact with the process. An HMI will provide relevant readings from one or more control loops in a graphical fashion, requiring communication to all subordinate systems, including controllers like PLCs and RTUs. HMIs include readouts of sensors and other feedback mechanisms or "alarms" used to inform the operator of an action that is required in response to a process condition. HMIs are also used to issue direct control operations and provide mechanisms to adjust the set points of the ongoing control process.

An HMI usually controls a process consisting of many control loops. This means that the HMI's network connectivity is typically heterogeneous, connecting to networks using routable protocols (TCP/IP) that include specialized ICS and fieldbus protocols, as well as other industrial network protocols to the various components that make up the ICS. HMIs are a common attack vector between the business and routable ICS networks.

Feedback loops

Every automated process relies on some degree of feedback both within a control loop and between a control loop or process and a human operator. Feedback is generally provided directly from the HMI used to control a specific process. A sample HMI graphical schematic of an automated process is shown in Figure 4.9. Feedback may also be centralized across multiple processes, through the collection, analysis, and display of information from many systems. For example, a refinery may have several crude oil and product storage tanks, each used in a replicated control process (e.g., local pump level and flow control). Information from each process can be collected and analyzed together to determine production averages, overages, and variations.

Production information management

The centralized information management of an ICS is typically performed by one or more data historian systems. The process of removing data from the real-time environment of an automated industrial process and storing it over time is called "historizing" the data. Once historized, the information can be further analyzed using tools such as statistical process control (SPC)/statistical quality control (SQC), either directly from within the data historian or by using an external analysis tool such as a spreadsheet. Historical data can be replayed at some point in the future to compare past and present plant operations.

Specific ICS components may use their own data historian system to historize data locally. For example, an ABB 800xA control system may use the 800xA information management historian, while an Emerson ovation control system may use the Ovation Process Historian. Industrial operations tend to be heterogeneous in nature and require data to be collected and historized from multiple systems. These operations involve different processes that may utilize assets manufactured by different vendors, yet all processes need to be evaluated holistically in order to manage and fine-tune overall production operations. There also may be value in collecting information from other devices and systems within the industrial network, such as HVAC systems, CCTV, and Access Control systems. The shift from process-specific data historization to operation-wide business intelligence has led to the development of specialized features and functionality within data historians.

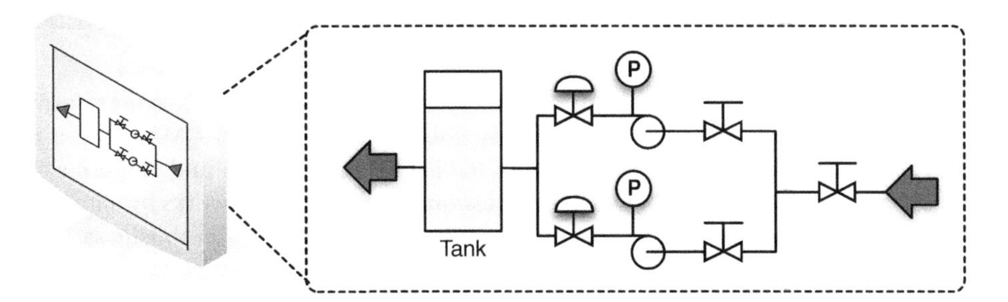

FIGURE 4.9 A human–machine interface (HMI) displaying current operational parameters.

Business information management

Operational monitoring and analysis provides valuable information that can be used by plant management to fine-tune operations, improve efficiencies, minimize costs, and maximize profits. This drives a need for replication of operational data into the business network.

Supervisory data can be accessed using an HMI or a data historian client, with each presenting their own security challenges. HMIs provide supervisory and control capabilities, meaning that an HMI user with the proper authorization can adjust parameters of control process (see "Process Management"). By placing an HMI outside of the ICS DMZ, any firewalls, IDS/IPS, and other security monitoring devices that are in place need to be configured to allow the communication of the HMI into and out of the ICS DMZ. This effectively reduces the strength of the security perimeter between the industrial and business networks to user authentication only. If not properly deployed, a user account that is compromised on the business HMI system can be used to directly manipulate control process(es), without further validation from perimeter security devices. This can be mitigated to some extent by leveraging more of the ICS system "authorization" capabilities that can restrict what a particular is used to do on the system irrespective of any prior user authentication that has occurred. This can be used to restrict business network HMI users from any "write" or "change" operations that impact the process.

The use of a data historian for business intelligence management presents a similar concern. The security perimeter must be configured to allow communication between the data historian in the business network and the various systems within the ICS DMZ that need to be monitored. Best practices recommend that in this case, the only component in the DMZ connected to the historian on the business network is a historian. This allows for replication of historical data out of the DMZ via well-defined communication ports using a one-to-one relationship while maintaining strict access control between the supervisory ICS components and the historian in the DMZ. Unlike an HMI, a data historian generally does not explicitly allow control of the process (however, some historians do support read and write capabilities to the ICS). The data historian instead provides a visual dashboard that can be configured to mimic the informational qualities and graphical representation of an HMI so that information about a process can be viewed in a familiar format.

Tip

Because the replication of data historian systems into the business network is for information purposes only, these systems can be effectively connected to the ICS DMZ using a **unidirectional gateway** or data diode (see Chapter 9, "Establishing Zones and Conduits"). This preserves the security perimeter between business and supervisory networks by allowing only outbound data communications. Data outbound (from the DMZ to the business network) should also be secured, if possible, using one or more security devices such as a firewall, IDS/IPS, or **application monitor**.

Data are collected by a historian through a variety of methods including direct communication via industrial network protocols such as Modbus, PROFIBUS, DNP3, and OPC (see Chapter 6, "Industrial Network Protocols"); history-oriented industrial protocols like OPC Historical Data Access (OPC-HDA); direct insertions in the data historian's database using Object Linking and Embedding Database (OLEDB), Open Database Connectivity (ODBC), Java Database Connectivity (JDBC), etc. Most data historians support multiple methods of data collection to support a variety of industrial applications. Once the information has been collected, it is stored within a database schema along with relevant metadata that helps to apply additional context to the data, such as batch numbers, shifts, and more depending upon the data historian's available features, functionality, and licensing.

Data historians also provide access to long-term data using many of the same methods mentioned above. Dashboards utilizing technologies like Microsoft SharePoint are becoming common allowing historical information to be retrieved and presented via web services for display on clients using standard Internet browser capabilities (HTTP/HTTPS). Custom applications can be created to access historical data via direct SQL queries and can be presented in almost any format, including binary files, XML, CSV, etc.

Historized data can also be accessed directly via the data historian's client application, as well as integrated at almost any level into supplementary Business Information Management Systems (BIMS). The data historian may in some case be integrated with Security Information and Event Management systems (SIEMs), Network Management Systems (NMSs), and other network and/or security monitoring systems.

Tip

Unnecessary ports and services are a security concern on data historians, just as they are on any other ICS cyber asset. Reference the data historian vendor's documentation for guidance disabling unused data interfaces and other hardening techniques that can be used to minimize the available attack surface of the data historian.

Process management

A control process is initially established through the programming of a controller and the building a control loop. In a fully automated loop, the process is controlled entirely through the comparison of established set points against various inputs. In a water heater, a set point might be used to establish the high-temperature range of 90°C, and an input would take temperature measurements from a sensor within the water tank. The controller's logic would then compare the input to the set point to determine whether the condition has been met (it is "true") or not (it is "false"). The output or heating element would then be energized or de-energized.

An HMI is used by an operator to obtain real-time information about the state of the process to determine whether manual intervention is required to manage the control process by adjusting an output (open loop) or modifying established set points (closed loop). The HMI facilitates both, by providing software controls to adjust the various set points of a control loop while also providing controls to directly affect the loop.

In the case of set point adjustments, the HMI software is used to write new set points in the programmable logic of the loop controller. This might translate to Function Code 6 ("Write Single Register") in a Modbus system, although the specific protocol function is typically hidden from the operator, and performed as part of the HMI's functionality. The HMI translates the function into human-readable controls presented within a graphical user interface (GUI), as represented in Figure 4.10.

In contrast, the HMI could also be used to override a specific process and force an output, for example, using Function Code 5 ("Write Single Coil") to write a single output to either the on ("true") or the off ("false") state.[7] The specific function code used to write the output state is hidden from the operator.

Note

The specific function codes used vary among industrial network protocols, and many protocols support vendor-proprietary codes. Although these protocols are discussed in Chapter 6, "Industrial Network Protocols," this book does not document protocol function codes. External resources are readily available describing many common industrial protocols.[8]

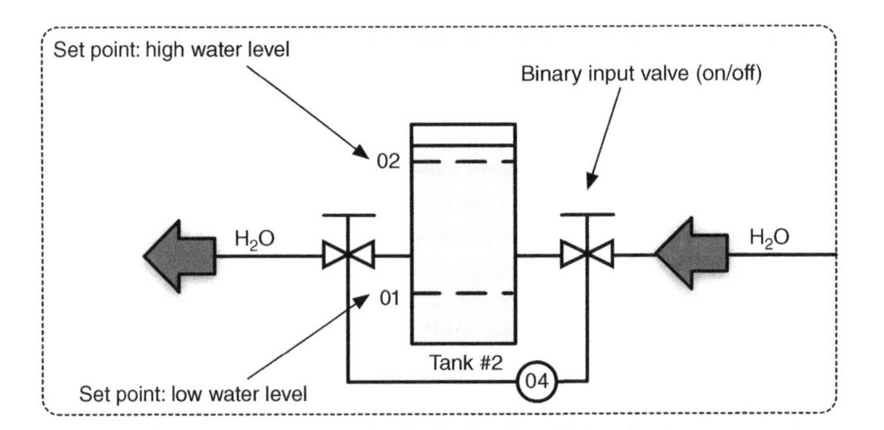

FIGURE 4.10 A human–machine interface (HMI)'s graphical user interface (GUI) representation of a control loop.

This represents a significant security concern. If an attacker is able to successfully compromise the HMI, fully automated systems can be permanently altered through the manipulation of set points. For example, by changing the high-temperature set point to 100°C, the water in a tank could boil, potentially increasing the pressure enough to rupture the tank. An attacker can also force direct changes to a process loop's output controls. In this example, the attacker could energize the water heater's coil manually. In the case of Stuxnet, malware inserted into a PLC listened to PROFIBUS-DP communication looking for an indication of a specific frequency converter manufacturer and the device operating at a specific frequency range. If those conditions were found, multiple commands were sent to the controller, alternating the operating frequency and essentially sabotaging the process.[9] It is important to understand that in both the water heater and Stuxnet examples described above, an attacker must have significant knowledge of the specific process and operational procedures in order to convert an HMI breach into an attack against the manufacturing process. Put another way, the attacker must know the exact register to change in order to alter the setpoint of the water heater from 90 to 100°C. This makes a "casual" cyberattack of this type much less probable but should not be considered a defense against a targeted cyberattack. It has been proven that sophisticated threat actors can and will obtain the knowledge necessary to launch a targeted attack of this type, and that "security by obscurity" cannot be considered a valid defensive strategy.

Note

This book does not claim to discuss all aspects of control theory, as this is not really necessary in order to understand ICS fundamentals necessary to deploy appropriate network security controls. It is worth mentioning, however, in the heater example that there are many more aspects that complicate what appears to be a rather simple process. All control loop examples thus far have been based on a simple "on-off" logic, which means the heating element (output) is either on or off based on the status of the temperature (input). This typically results in poor closed loop control, because if the corresponding setpoint to turn the output off is the same as that which turned it on, the output would basically "bounce" between on and off - something very undesirable in process control. High and low limits are established creating an effective "deadband" of control. So if the high limit was set to 92°C and the low limit 88°C, the output would energize when the input dropped below the low limit and de-energize when reaching the high limit. An obvious malicious action could be to change the limits.

To eliminate this swing in the measured variable (temperature), control loops implement "PID" or Proportional + Integral + Derivative loops that simply solve a first-order differential equation resulting in an output that can be held very close to the desired setpoint. This requires a modulating output such as a burner adjustment on a gas-fired heater that can be adjusted to control the amount of heat applied to the tank. A new attack vector could now be to change the constants associated with the P–I-D components making the control loop unstable, and possibly unsafe.

What if the output needed to be de-energized to apply heat to the tank? This is referred to as "control action" and represents whether a "true" input should generate a "true" output. Many industrial processes use indirect action that means a "true" input generates a "false" output. A simple parameter change on control action could obviously cause process instability.

What if the temperature in the water tank was at 90°C and someone began to use hot water decreasing the level in the tank resulting in cold water to be added to the tank to maintain level and the tank temperature to fall? All of the previous examples used what is called "feedback" control. In this case, as the water level drops and cold water is added, the heating element is energized in anticipation that the water temperature is going to drop as well. This is referred to as "feed-forward" control. There is a "gain" associated with feed-forward control that a threat actor could modify causing adverse process response.

These topics will be important in understanding the scope of exploiting not only vulnerabilities, but also capabilities in Chapter 7, "Hacking Industrial Systems."

Safety instrumented systems

Safety instrumented systems (SISs) are deployed as part of a comprehensive risk management strategy utilizing layers of protection to prevent a manufacturing environment from reaching an unsafe operating condition. The basic process control system (BPCS) is responsible for discrete and continuous control necessary to operate a process within normal operational boundaries. In the event that an abnormal situation occurs, which place the processing outside of these normal limits, the SIS is provided as an automated control environment that can detect and respond to the process event and maintain or migrate it to a "safe" state—typically resulting in equipment and plant shutdowns. As a final layer of protection, manufacturing facilities utilize significant physical protective devices including relief valves, rupture disks, flare systems, governors, etc. to act as a final level of safety prior to the plant entering dangerous operating limits. These events and corresponding actions are shown in Figure 4.11 below.

The risks that originate within the SIS relating to cyberincidents are twofold. First, since the system is responsible for bringing a plant to a safe condition once it is determined to be outside normal operational limits, the prevention of the SIS from properly performing its control functions can allow the plant to transition into a dangerous state that could result in operational disruptions, environmental impact, occupational safety, and mechanical damage. In other words, simple denial-of-service (DoS) attacks can translate into significant risk from a cyberevent.

On the other side, since the SIS operationally overrides the BPCS and its ability to control the plant, the SIS can also be used malicious to cause unintentional equipment or plant shutdowns, which can also result in similar consequences to a service denial attack. In other words, an attacker that gains control of an SIS can effectively control the final operation of the facility.

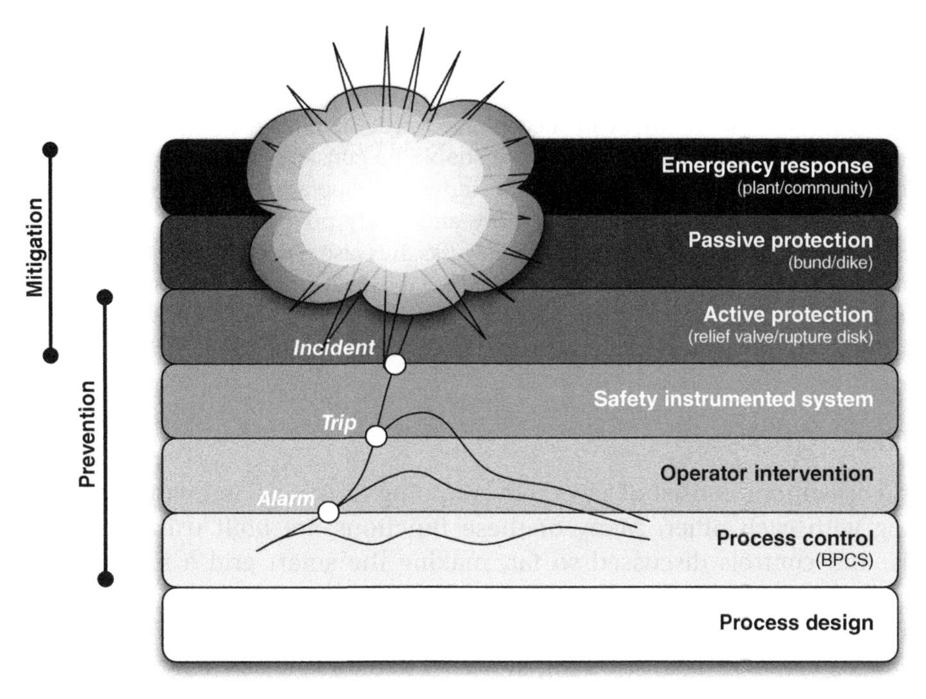

FIGURE 4.11 Layers of protection in plant safety design.

In both cases, the need to isolate the SIS to the greatest extent possible from other basic control assets, as well as eliminate as many potential threat vectors as possible is a reasonable approach to improving cyber security resilience. SIS programs, though performing in a similar manner to controller programming previously discussed, is not typically allowed in operational mode. This means that highly authorized applications like SIS programming tools and SIS engineering workstations can be removed from ICS networks until they are required. SIS systems must be tested on a periodic basis to guarantee their operation. This provides a good time to also perform basic cyber security assessments, including patching and access control reviews in order to make sure that the safety AND security of the SIS remains at the original design levels.

Note

With the introduction of the TRISIS malware in 2017, the discipline of process control safety and industrial cybersecurity began to merge. TRISIS was engineered to target Schneider Electric Triconex SIS and proved for the first time that a coordinated cyber campaign could include components designed to maximize damage by manipulating safety controls. While TRISIS (at the time of this writing) targeted Triconex SIS, the TRISIS framework could potentially be modified to target any SIS.

The presence of malware that targets SIS is concerning but does not necessarily mean that malware can compromise the safety of an ICS. ICS safety is a discipline of its own, and is complex and nuanced.

The presence of malware that can successfully target SIS does mean that cyberthreats can potentially impact plant safety and therefore should be considered as a part of all ICS safety planning. It is highly recommended that industrial operators implement programs to encourage cooperation between cybersecurity and safety professionals in order to develop more comprehensive safety and security frameworks. It is also recommended that more mature industrial operators, whom already have well-established cybersecurity programs, should consider a broader HAZOP-based approach to cybersecurity risk management (see Chapter 8, "Risk and Vulnerability Assessments: Thinking of Cybersecurity in terms of Safety").

The smart grid

Smart grid operations consist of several overlapping functions, intercommunicating and interacting with each other. Many of these functions are built using the ICS assets, protocols, and controls discussed so far, making the smart grid a nexus of many industrial networks. This can be problematic, because the smart grid is complex and highly interconnected. It is not the convergence of a few systems, but of many including: customer information systems, billing systems, demand response systems, meter data management systems, and distribution management systems, distribution SCADA and transmission SCADA, protection systems, substation automation systems, distributed measurement (synchrophasors), and many more. Most of these systems interconnect and intercommunicate with many others. For example, customer information systems communicate with distribution management systems, load management systems, customer service systems, and the advanced metering infrastructure (AMI).

The AMI Headend in turn feeds local distribution and metering, as shown in Figure 4.12. The AMI Headend will typically connect to large numbers of smart meters, serving a neighborhood or urban district, which, in turn, connect to home or business networks, and often to home energy management systems (HEMS), which provide end-user monitoring and control of energy usage.

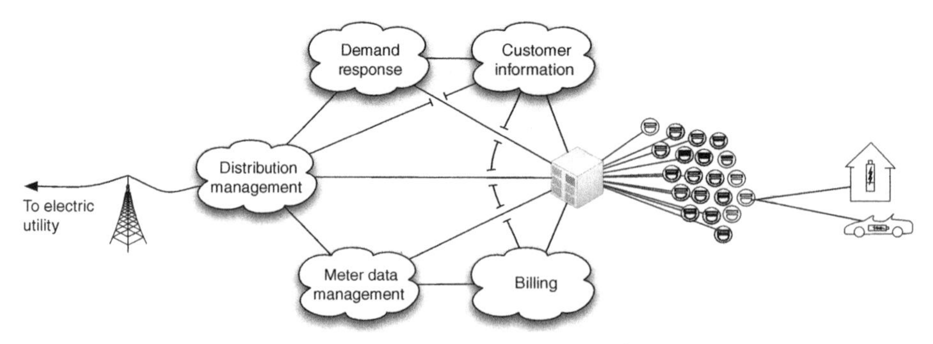

FIGURE 4.12 Components of a typical smart grid deployment.

Each system in a smart grid serves specific functions that map to different stakeholders, including: bulk energy generation; service providers; operations; customers; transmission; and distribution. For example, the customer information system is an operations system that supports the business relationship between the utility and the customer, and may connect to both the customer premise (via customer service portals) as well as the utility back-end systems (e.g., corporate CRM). Meter data management systems store data, including usage statistics, energy generation fed back into the grid, smart meter device logs, and other meter information, from the smart meter. Demand response systems connect to distribution management systems and customer information systems as well as the AMI Headend to manage system load based on consumer demand and other factors.[10]

Smart grid deployments are broad and widely distributed, consisting of remote generation facilities and micro-grids, multiple transmission substations, and so on, all the way to the end user. In metering alone, multiple AMI Headends may be deployed, each of which may interconnect via a mesh network (where all Headends connect to all other Headends) or hierarchical network (where multiple Headends aggregate back to a common Headend), and may support hundreds of thousands or even millions of meters. All of this represents a very large and distributed network of intelligent end nodes (smart meters) that ultimately connect back to energy transmission and distribution,[11] as well as to automation and SCADA systems used for transmission and distribution. The benefits of this allow for intelligent command and control of energy usage, distribution, and billing.[12] The disadvantage of such a system is that the same end-to-end command and control pathways could be exploited to attack one, any, or all of the connected systems.

There are many threat vectors and threat targets in the smart grid—in fact any one of the many systems touched on could be a target. Almost any target can also be thought of as a vector to an additional target or targets because of the interconnectedness of the smart grid. For example, considering the advanced metering infrastructure, some specific threats include:

- Bill manipulation/energy theft—An attack initiated by an energy consumer with the goal of manipulating billing information to obtain free energy.[13]
- Unauthorized access from customer end point—Use of an intelligent AMI end node (a smart meter or other connected device) to gain unauthorized access to the AMI communications network.[14]
- Interference with utility telecommunications—Use of unauthorized access to exploit AMI system interconnections in order to penetrate the bulk electric generation, transmission, and distribution system.[15]
- Mass load manipulation—The use of mass command and control to manipulate bulk power use, with the goal of adversely affecting the bulk electric grid.[16]

- Denial of service—Using intelligent nodes to communicate to other nodes in a storm condition, with the goal of saturating communications channels and preventing the AMI from functioning as designed.

The AMI is a good example of a probable threat target due to its accessibility with meters accessible from the home, often with wireless or infrared interfaces that can be boosted, allowing for covert access. The AMI is also used by many smart grid systems. Almost all end nodes, business systems, operational systems, and distributed control systems connect to (or through) the Headend, or utilize information provided by the Headend. Compromise of the AMI Headend would therefore provide a vector of attack to many systems. If any other connected system were compromised, the next hop would likely be to the Headend. All inbound and outbound communications at the Headend should be carefully monitored and controlled (see Chapter 9, "Establishing Zones and Conduits").

This is a very high-level overview of the smart grid. If more detail is required, please refer to "Applied Cyber Security and the Smart Grid," by Raj Samani and Eric Knapp.

Network architectures

The ICS systems and operations discussed so far are typically limited to specific areas of a larger network design, which at a very high level consist of business networks, production networks, and control networks, as shown in Figure 4.13.

Nothing is simple—in reality, industrial networks consist of multiple networks, and they are rarely so easily and neatly organized as in Figure 4.13. This is discussed in detail in Chapter 5, "Industrial Network Design and Architecture." It is enough to know for now that the ICS systems and operations being discussed represent a unique network, with unique design requirements and capabilities.

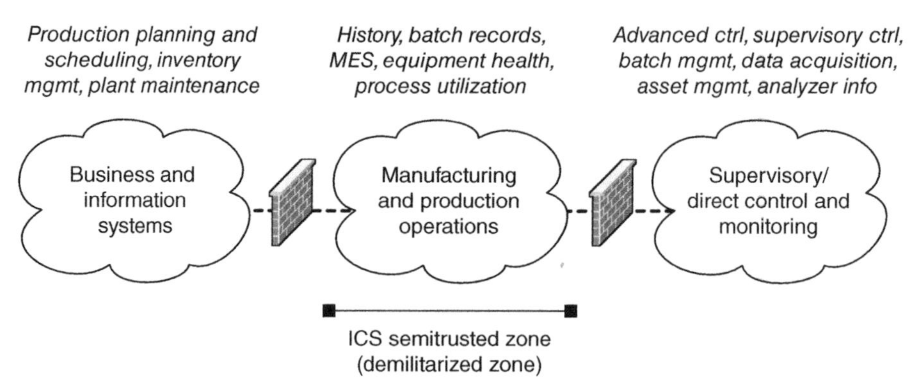

FIGURE 4.13 Functional demarcation of industrial networks.

Summary

Industrial networks operate differently from business networks and use specialized devices including PLCs, RTUs, IEDs, HMIs, application servers, engineering workstations, supervisory management workstations, data historians, and business information consoles or dashboards. These devices utilize specialized protocols to provide the automation of control loops, which, in turn, make up larger industrial control processes. These automated control processes are managed and supervised by operators and managers within both ICS and business network areas, which requires the sharing of information between two disparate systems with different security requirements.

This is exemplified in the smart grid, which shares information between multiple disparate systems, again across different networks each of which has its own security requirements. Unlike traditional industrial network systems, the smart grid represents a massive network with potentially hundreds of millions of intelligent nodes, all of which communicate back to energy providers, and residences, businesses, and industrial facilities all consuming power from the grid.

By understanding the assets, architectures, topologies, processes, and operations of industrial systems and smart grids, it is possible to examine them and perform a security assessment in order to identify prevalent threat vectors, or paths of entry that a malicious actor could use to exploit the industrial network and the manufacturing process under its control.

Endnotes

1. PLCTutor.com, Ladder logic, October 19, 2000 (cited: November 29, 2010).
2. P. Melore, PLC operations. http://www.plcs.net/chapters/howworks4.htm, (cited: November 29, 2010).
3. P. Melore, The guts inside. http://www.plcs.net/chapters/parts3.htm, (cited: November 29, 2010).
4. PLCTutor.com, PLC operations, October 19, 2000 (cited: November 29, 2010).
5. OSIsoft, OSIsoft company overview. http://www.osisoft.com/company/company_overview.aspx, 2010 (cited: November 29, 2010).
6. J. Larson, Idaho National Laboratories, Control systems at risk: sophisticated penetration testers show how to get through the defenses, in: Proc. 2009 SANS European SCADA and Process Control Security Summit, October 2009.
7. The Modbus Organization, Modbus application protocol specification V1.1b, Modbus Organization, Inc. Hopkinton, MA, December 2006.
8. "List of Automation Protocols", Wikipedia, http://en.wikipedia.org/wiki/List_of_automation_protocols (cited: January 6, 2014).
9. E. Chien, Symantec. Stuxnet: a breakthrough. http://www.symantec.com/connect/blogs/stuxnet-breakthrough, November 2010 (cited: November 16, 2010).
10. G. Locke, US Department of Commerce and Patrick D. Gallagher, National Institute of Standards and Technology, Smart Grid Cyber Security Strategy and Recommendations, Draft NISTIR 7628, NIST Computer Security Resource Center, Gaithersburg, MD, February 2010.
11. UCA International Users Group, AMI-SEC Task Force, AMI system security requirements, UCA, Raleigh, NC, Dec 17, 2008.

12. Ibid.
13. Raymond C. Parks, SANDIA Report SAND2007-7327, Advanced Metering Infrastructure Security Considerations, Sandia National Laboratories, Albuquerque, New Mexico and Livermore, California, November 2007.
14. Ibid.
15. Ibid.
16. Ibid.

5

Industrial Network Design and Architecture

Information in this chapter

- Introduction to Industrial Networking
- Common Topologies
- Network Segmentation
- Network Services
- Wireless Networks
- Remote Access
- Performance Considerations
- Safety Instrumented Systems
- Special Considerations

It is important to understand the similarities and differences of typical enterprise or business networks before we get too involved in securing industrial networks. This requires an understanding of how industrial control systems (ICSs) work, as explained previously in Chapter 4, "Introduction to Industrial Control Systems and Operations," because portions of these networks have been designed around specific criteria relating to how an ICS must operate. This includes not only host-to-host network communications utilizing familiar IT technologies like remote procedure calls (RPC) but also support for legacy fieldbus protocols and vendor-specific protocols that are unlike those seen on business networks. Chapter 6, "Industrial Network Protocols" provides a closer look at these technologies and how many have evolved from original serial-based point-to-point communications to today's high-speed switched and routed network methods. There are many functions to be served in an industrial network in addition to the control system, along with consideration for many distinct network areas. For example, each controller, and each process that is subordinate to it, is a network consisting of control devices, human–machine interfaces (HMIs) and possibly I/O modules. The supervisory components that oversee these basic control systems are interconnected via a network of specialized embedded systems, workstations, and various types of servers. Many supervisory networks may constitute a larger plant network. In addition, the business network cannot be forsaken here. While not an industrial network, per se, the business network contains systems that indirectly impact industrial systems.

Each area, depending upon its function, capacity, system vendor, and owner/operator, will have its own topologies, performance considerations, remote access requirements, and network services. These must all be taken into account when

Industrial Network Security. https://doi.org/10.1016/B978-0-443-13737-2.00003-8

considering one of the most important security design considerations—network segmentation. Network segmentation helps make each network area more manageable and secure and is a simple but effective weapon in the cyber security arsenal.

Note

As often is the case when dealing with industrial networking, terms that originated in IT may conflict or overlap with similar terms that were adopted by and are often used in OT. The term "segmentation" is one example where the same word has subtly different meanings depending on the context that it is used. Without a clear understanding of these various meanings, designing a modern, robust, and reliable industrial network that is also secure will prove very difficult.

From an IT infrastructure design perspective, segmentation is most often used and referred to in terms of *network segmentation*, referring to the division of a larger network into smaller networks, by imposing appropriate network controls at a given layer of the open systems interconnection (OSI) model.

From an ICS perspective, the term segmentation is most often used in terms of *zone segmentation*. Zone segmentation refers to the division of industrial systems into grouped subsystems, for the primary purpose of reducing the attack surface of a given system, as well as minimizing attack vectors into and out of that system. This is accomplished by "limit[ing] the unnecessary flow of data" between zones.[1] This will be covered in depth in Chapter 9, "Establishing Zones and Conduits." Chapter 9 will also introduce the concept of a "security zone" with respect to ICS system-level security design. It is important to understand early in the book that this concept is not the same as a "network segment" as a security zone is focused on the grouping of assets based purely on security requirements. For example, assets that may not be able to be patched due to specific vendor requirements may be placed in a separate security zone, yet be part of a network segment that comprises assets from other security zones.

It is also important to understand that, while the similarity of the two terms often causes confusion, both uses of "segmentation" are correct. Also, while network segmentation is primarily concerned with improving network uptime, and zone segmentation is primarily concerned with improving security, the two will often map easily to each other within a common infrastructure design. This is because the act of network segmentation will, by its nature, isolate any networked assets from communicating openly between the segmented networks. If each zone is given a dedicated and protected network segment, zone segmentation and network segmentation are very closely aligned and nearly identical. However, this is not always the case. In some cases, zone segmentation may be required within a single network segment, while in others, a single zone may consist of multiple network segments.

Last, and certainly not least, areas of the ICS may require zone separation where Ethernet and IP networking is not used at all. As mentioned at the start of this chapter, each controller, and each process that is subordinate to it, is a network consisting of control devices, HMIs, and I/O modules connected via legacy serial or point-to-point connections. These scenarios will occur more frequently deeper within the industrial network hierarchy, where it may be necessary to perform *zone segmentation* where *network segmentation* is not applicable at all.

That said, it is extremely difficult to avoid using the general term "segmentation" interchangeably, and so every attempt has been made in this book to denote *network* versus *zone* segmentation to avoid confusion. Both network segmentation and zone segmentation are strong security controls because, by limiting the scope of a network or system, they can minimize the impact of a cyberattack or incident.

What are your thoughts on network and zone segmentation? Continue the discussion at @smartgridbook

Introduction to industrial networking

In this book, an "industrial network" is any network that supports the interconnectivity of and communication between devices that make up or support an ICS. These types of ICS networks may be local area switched networks as common with distributed control system (DCS) architectures or wide area routed networks more typical of supervisory control and data acquisition (SCADA) architectures. Everyone should be familiar with networking to some degree (if not, this book should probably not be read before reading several others on basic network technology and design). The vast majority of information on the subject is relevant to business networks—primarily Ethernet and IP-based networks using the TCP transport that are designed (with some departmental separation and access control) primarily around information sharing and collaborative workflow. The business network is highly interconnected, with ubiquitous wireless connectivity options, and is extremely dynamic in nature due to an abundance of host-, server-, and cloud-based applications and services, all of which are being used by a large number of staff, supporting a diversified number of business functions. There is typically a network interface in every cubicle (or access to a wireless infrastructure), and often high degrees of remote access via virtual private networks (VPN), collaboration with both internal and external parties, and Internet-facing web, email, and business-to-business (B2B) services. Internet connectivity from a business network is a necessity, as is serving information from the business to the Internet. In terms of cybersecurity, the business network is concerned with protecting the confidentiality, integrity, and availability (in that order) of information as it is transmitted from source generation to central storage and back to destination usage.

An industrial network is not much different technologically—most are Ethernet and IP based, and consist of both wired and wireless connectivity (there are certainly still areas of legacy serial connectivity using RS-232/422/485 as well). The similarities end there. In an industrial network, the availability of data is often prioritized over data integrity and confidentiality. As a result, there is a greater use of real-time protocols, UDP transport, and fault-tolerant networks interconnecting endpoints and servers. Bandwidth and latency in industrial networks are extremely important because the applications and protocols in use support real-time operations that depend on deterministic communication often with precise timing requirements. Unfortunately, as more industrial systems migrate to Ethernet and IP, ubiquitous connectivity can become an unwanted side effect that introduces significant security risk unless proper design considerations are taken.

Table 5.1 addresses some of the many differences between typical business and industrial networks.

Note that these differences dictate network design in many cases. The requirement for high reliability and resiliency dictates the use of ring or mesh network topologies, while the need for real-time operation and low latency requires a design that minimizes switching and routing hops or may dictate purpose-built network appliances. Both of

Table 5.1 Differences in Industrial Network Architectures by Function

Function	Industrial network (control and process areas)	Industrial network (supervisory areas)	Business network
Real-time operation	Critical	High	Best effort
Reliability/Resiliency	Critical	High	Best effort
Bandwidth	Low	Medium	High
Sessions	Few, explicitly defined	Few	Many
Latency	Low, consistent	Low, consistent	N/A, retransmissions are acceptable
Network	Serial, Ethernet	Ethernet	Ethernet
Protocols	Real-time, proprietary	Near real-time, open	Non-real-time, open

these requirements may result in a vendor requiring the use of specific networking equipment to support the necessary configuration and customization necessary to accomplish the required functionality. The use of specific protocols also drives design, where systems dependent solely upon a given protocol must support that protocol (e.g., serial network buses).

The network shown in Figure 5.1 illustrates how the needs of a control system can influence design (redundancy will not be shown on most drawings for simplicity and clarity). While on the surface, the connectivity seems straightforward (many devices connected to layer two or layer three Ethernet devices, in a star topology), when taking into account the five primary communication flows that are required, represented as TCP Session one through five in Figure 5.1, it becomes obvious how logical information flow maps to physical design. In Figure 5.2, we see how these five sessions require a total

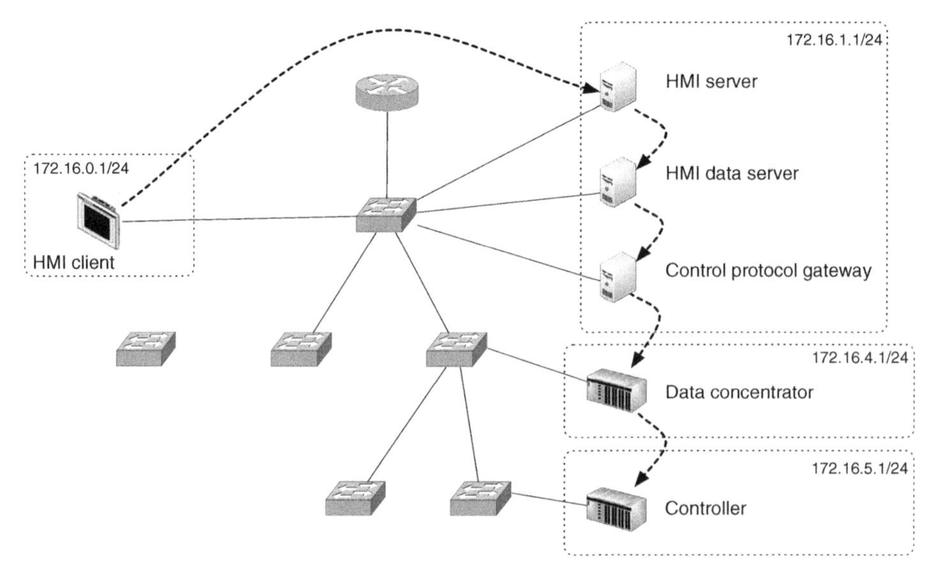

FIGURE 5.1 Communication flow represented as sessions.

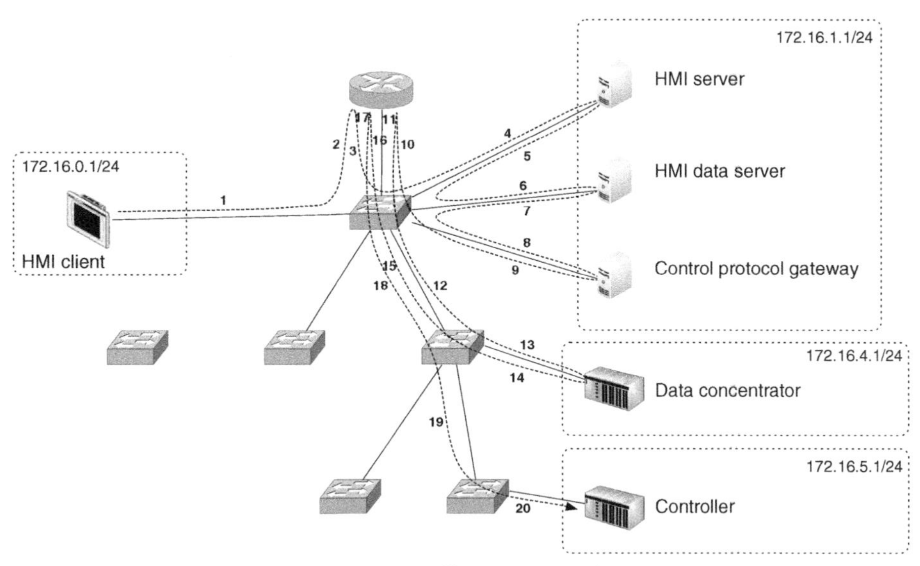

FIGURE 5.2 Communication flow represented as connections.

of 20 paths that must be traversed. It is therefore necessary to minimize latency wherever possible to maintain real-time and deterministic communication. This means that Ethernet "switching" should be used where possible, reserving Ethernet "routing" for instances where the communication must traverse a functional boundary. This concept, represented in Figures 5.1 and 5.2 as subnets, is important when thinking about network segmentation and the establishment of security zones (see Chapter 9, "Establishing Zones and Conduits). It becomes even more obvious that the selection of Ethernet "firewalls" deployed low in the architectural hierarchy must be designed for industrial networks in order to not impact network performance. One common method of accomplishing this is through the use of "transparent" or "bridged" mode configurations that do not require any IP routing to occur as the data traverses the firewall.

Figures 5.1 and 5.2 illustrate a common design utilizing Ethernet switches for low-latency connectivity of real-time systems, such as data concentrators and controllers, and a separate router (typically implemented as a layer 3 switch) to provide connectivity between the multiple subnets. Note that in this design, the total end-to-end latency from the HMI client to the controller would be relatively high—consisting of 11 total switch hops and 3 router hops. An optimized design, represented in Figure 5.3, would replace the router with a layer 3 switch (an Ethernet switch capable of performing routing functions[2]). Layer 3 switches provide significantly improved performance, and by replacing separate Layer 2 and Layer 3 devices with a single device, several hops are eliminated.

In Figure 5.4, a design typical of one vendor's systems has been provided. Redundancy is provided here by connecting systems to two separate Ethernet

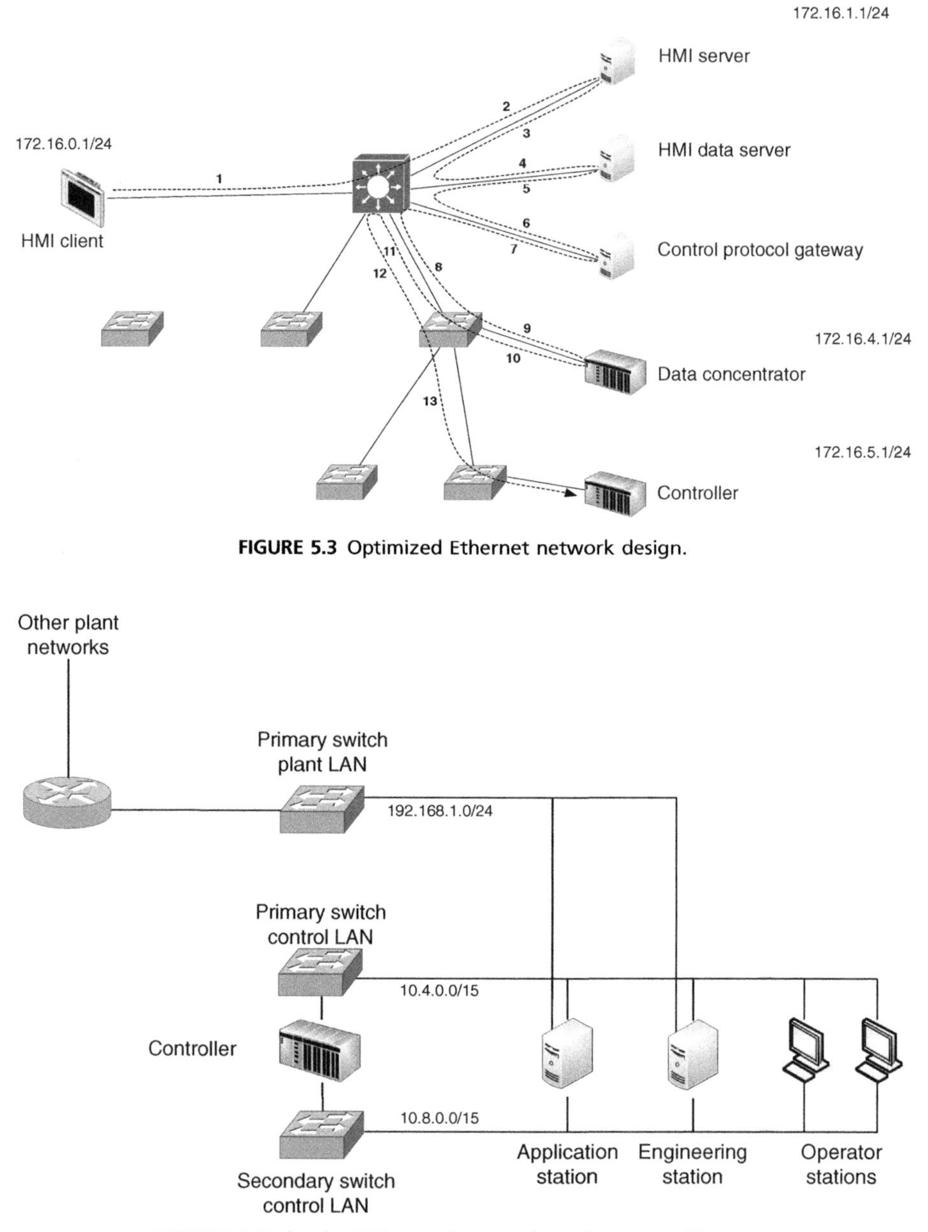

FIGURE 5.3 Optimized Ethernet network design.

FIGURE 5.4 Redundant Ethernet in a vendor reference architecture.

connections. While Figure 5.4 shows a very simple redundant network, more sophisticated networks can be deployed in this manner as well. The use of spanning tree protocol will eliminate loops (in a switched environment) and dynamic routing protocols will enable multipath designs in a routed environment. In more sophisticated designs, redundant switching and routing protocols, such as VSRP and VRRP, enable the use of multiple switches in high-availability, redundant configurations.

As we get lower into the control environment, functionality becomes more specialized, utilizing a variety of open and/or proprietary protocols, in either their native form or adapted to operate over Ethernet. Figure 5.5 illustrates a common fieldbus network based on FOUNDATION Fieldbus using serial two-wire connectivity and reliant upon taps (known as couplers) and bus terminations. Many fieldbus networks are similar, including PROFIBUS-PA, ControlNet, and DeviceNet.

It should be evident by now that specific areas of an industrial network have unique design requirements and utilize specific topologies. It may be helpful at this point to fully understand some of the topologies that are used before looking at how this affects network segmentation.

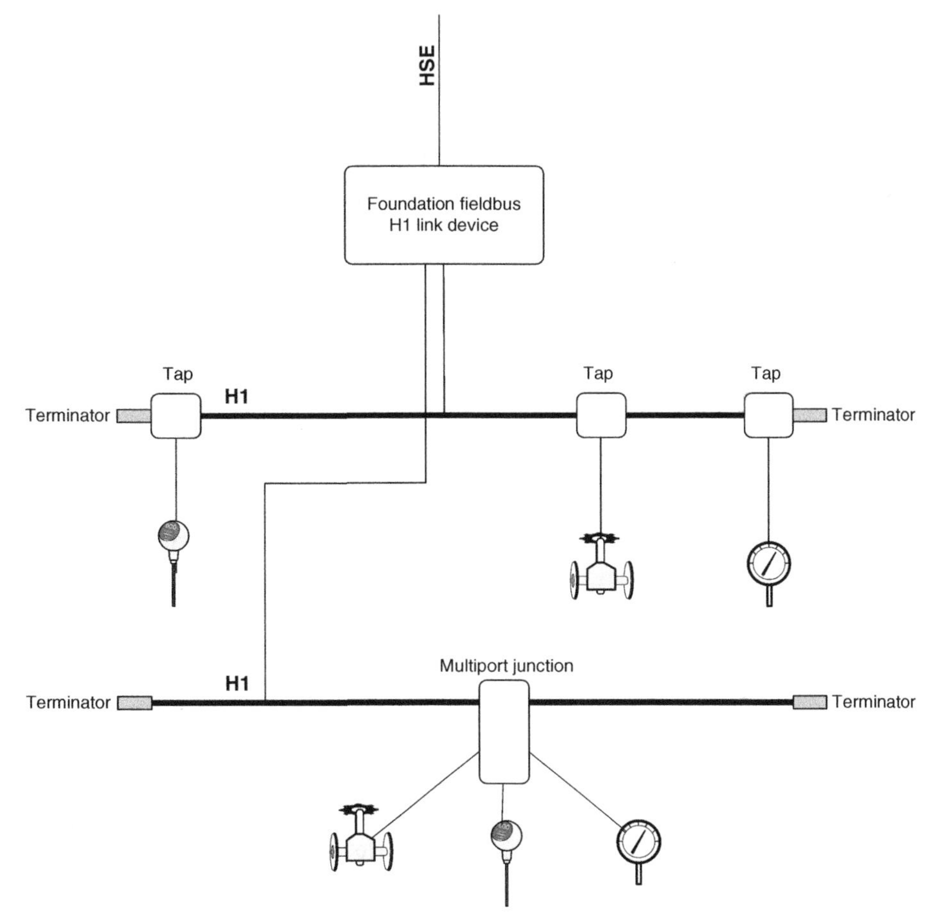

FIGURE 5.5 FOUNDATION fieldbus H1 network topology.

Common topologies

Industrial networks are typically distributed in nature and vary considerably in all aspects, including the link layer characteristics, network protocols used, and topology. In business environments, Ethernet and IP networks are ubiquitous and may be implemented in any number of topologies—including star, tree, and even full-mesh topologies (though mesh technologies tend to be only for the uplinks between network devices and not between endpoints and their network access devices). Like in a business, ICS networks may utilize various topologies as well. Unlike business network topologies, those deployed to support industrial systems are also likely to use bus and ring topologies in addition to star, tree, and mesh topologies. This is because, while these topologies have fallen out of favor in business (due to cost, performance, and other considerations), they are often necessary within ICS.

Topologies, such as rings, easily support the necessary redundancy commonly required in industrial networks. A bus topology represents a shared message transmission domain, where many nodes are competing for a finite amount of bandwidth and relying on traffic coordination or synchronous communication to provide best-effort connectivity. Many ICS architectures are based on underlying technologies like publish-subscribe and token-rings encapsulated in UDP packets well suited for bus technologies. In modern business networks, however, this is impractical—switched Ethernet provides a dedicated Ethernet segment with associated guaranteed "first-hop" bandwidth to every node and has become a commodity, making star topologies extremely common. Likewise, ring topologies (which promise redundant paths for greater reliability) have fallen out of favor with enterprises because full mesh topologies are relatively inexpensive and highly effective (essentially, each node is given two dedicated Ethernet connections to each other node, typically between core network infrastructure devices and/or business-critical servers). In industrial networks, it is more common for the access switches to be connected in a ring configuration while a star topology is used to connect to end devices.

There is still a strong need for both bus and ring topologies in industrial networks depending upon the specific type of control process that is in operation and the specific protocols that are used, as shown in Figure 5.6. In industrial environments that depend on wired communication for reliability, it can be cost prohibitive to implement mesh topologies over traditional bus and ring configurations. Mesh networks have become the de facto standard for wireless industrial networks. For example, an automated control process to sanitize water may use a bus topology with the PROFIBUS-PA protocol, while another control process may use Modbus/TCP in a ring topology to control pumping or filtration systems. As we move farther away ("up the architecture") from the process and closer to the business network, "typical" IT designs become more prevalent, to the point where many plant networks are designed similarly to corporate data centers, with meshed core switches and routers supporting switched access to smaller workgroup switches.

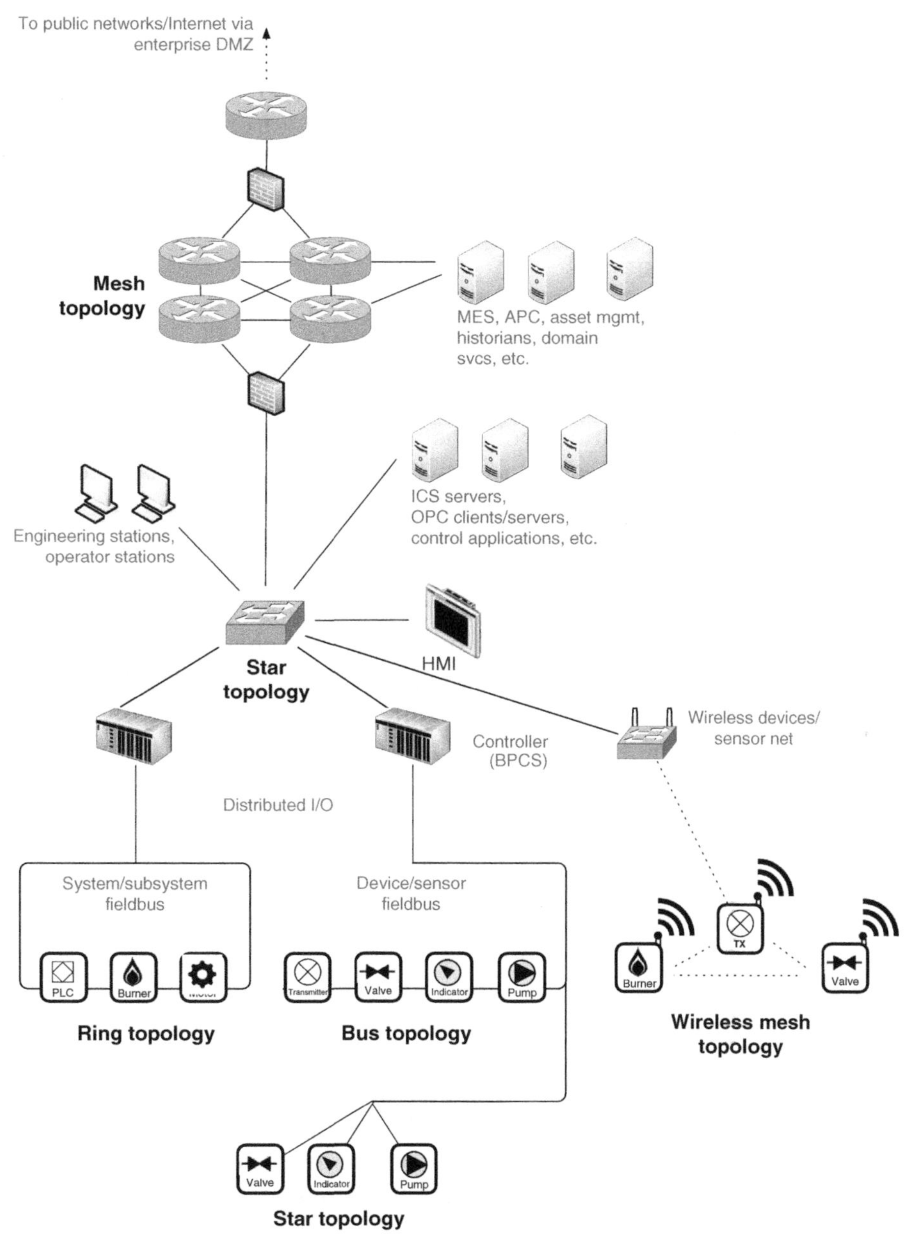

FIGURE 5.6 Common network topologies as used in industrial networks.

- **Bus topologies** are linear and often used to support either serially connected devices or multiple devices connected to a common bus via taps. Bus topologies often require that the bus network be terminated at either end by a terminator used to prevent signal reflections. In a bus topology, the resources of the network are shared among all of the connected nodes, making bus networks inexpensive but also limited in performance and reliability. The number of devices connected to a single bus segment is relatively small for this reason.

- **Mesh topologies** are common for the connectivity of critical devices that require maximum performance and uptime, such as core Ethernet network devices like switches and routers, or critical servers. Because many paths exist, the loss of one connection—or even the failure of a device—does not (necessarily) degrade the performance of the network.
- **Wireless mesh** topologies are logically similar to wired mesh topologies, only using wireless signaling to interconnect compatible devices with all other compatible devices. Unlike wired meshes where the physical cabling dictates the available network paths, wireless meshes rely on provisioning to control information flow.
- **Star topologies** are point-to-multipoint networks where a centralized network resource supports many nodes or devices. This is most easily illustrated with a standard Ethernet switch that provides individual connections to endpoints or other switches that can also be connected to additional endpoints.
- **Branch or tree topologies** are hierarchically connected topologies where a single topology (typically a bus, representing the "trunk") supports additional topologies (typically bus or star topologies, representing the "branches"). One practical example of this is the "chicken foot" topology used in FOUNDATION Fieldbus H1 deployments where a bus is used to interconnect several junction boxes or "couplers," which then allows a star connection to multiple field devices.
- **Ring topologies** are, as the name implies, circular, with each node connected serially, but with the final node connected back to the first node, rather than terminating the network at either end. This topology can cover endpoints, but is more commonly used to interconnect network access switches.
- **Multihoming or dual-homing** describes the connection of a single node to two or more networks. Dual homing can be used for redundancy (as illustrated in Figure 5.4), to essentially provide two networks over which a single device can communicate. Dual-homing has also been used as a method of making resources assessable to multiple zones (as illustrated in Figure 5.7), but this is not recommended. In the case of a dual-homed connection between a plant zone and a business zone, any successful break of the dual-homed server would provide a bridge between the two zones, fully exposing the plant zone to the outside world.
- **Virtual LANs, virtual networks, software-defined networking and microsegmentation,** while not technically topologies, are technologies that allow flexible control over communications flows—often despite the underlying physical topology. It is important to remember that any routable networks such as TCP/IP offer logical communication paths in addition to physical paths. Advancements such as **software-defined networking** and **microsegmentation** offer increased flexibility and control, which is important when superimposing logical security zones and conduits atop an incompatible physical topology (See Chapter 9, "Establishing Zones and Conduits" for more information).

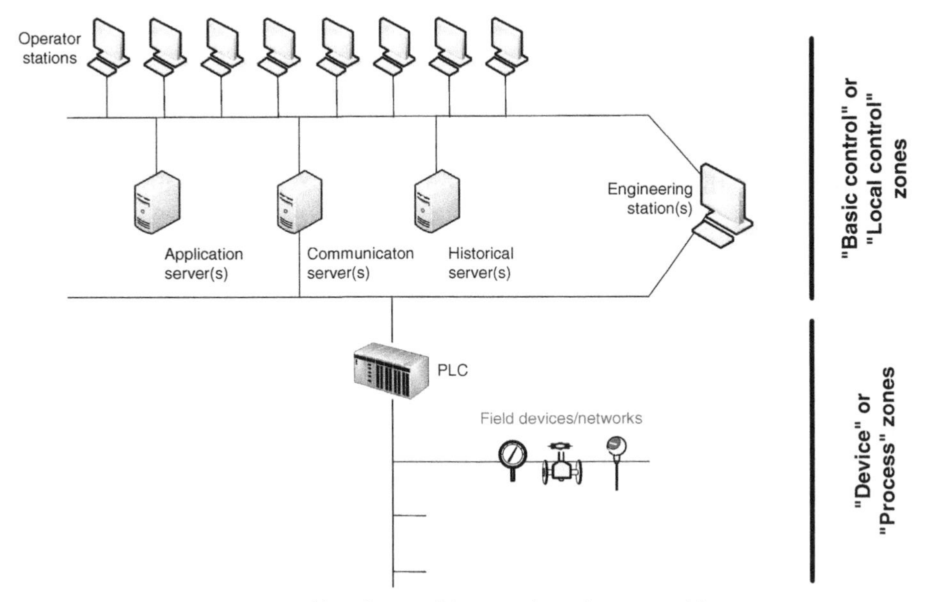

FIGURE 5.7 Dual-homing used in a vendor reference architecture.

Tip

If dual-homed systems are currently being used where a single device requires access to resources from two networks, consider an alternative method with fewer negative security implications. The shared resource could be placed within a semi-trusted DMZ, or data could be transferred out of the more secure network into the less secure network using a read-only mechanism, such as a data diode or unidirectional gateway.

The specific topology and network design can have a significant impact on the security and reliability of a particular network. Network topology will also impact your ability to effectively segment the network and to control network traffic flow—ultimately impacting your ability to define security zones and to enforce security communication channels via conduits (see Chapter 9, "Establishing Zones and Conduits"). Implementing router access control lists (ACLs), intrusion prevention systems, and application firewalls between two zones can add significant security. If there are dual-homed devices between these two zones, it is possible for an attacker to bypass these security controls altogether, eliminating their value. It is therefore necessary to understand topologies and network designs from the perspective of network segmentation.

Network segmentation

Segmentation is important for many reasons, including network performance considerations and cybersecurity, and so on. The concept of network segmentation was originally developed as a means to limit the broadcast domain of an Ethernet network that was designed at that time around 10 MB connections typically using either a "hub" (10BaseT) or a shared "trunk" (10Base2) as an access medium. Segmentation typically occurs at layer 3 (the network layer) by a network device providing routing functions (i.e., traditional routers, layer 3 switches, firewalls, etc.). Among other functions, the router blocks broadcasts, enabling a large flat Ethernet network to be broken up into discrete Ethernet segments; each segment having fewer nodes, and therefore fewer broadcasts and less contention. Networks became larger as switched Ethernet technology became commoditized, and the capabilities of network processing increased, providing an alternative method for segmentation. This relatively new development allowed broadcasts to be contained at layer 2 using virtual LANs (VLANs), which utilize a tag in the Ethernet header to establish a VLAN ID (802.1Q). VLANs enable compatible Ethernet switches to forward or deny traffic (including broadcasts) based upon either the 802.1Q tag or the port's VLAN ID (PVID). To communicate between VLANs, traffic would need to be explicitly routed between VLANs at layer 3, using a routing device. Essentially, each VLAN behaved as if it were connected to a dedicated sub-interface on the router, only the segmentation occurred at layer 2, separating the function from the main physical router interface. This meant that VLANs could segment traffic much more flexibly, and much more cost effectively as it minimized the amount of routers that needed to be deployed.

Note

It is important to note that VLANs are implemented at OSI layer 2. What this effectively means is that if two devices connected to the same switch share the same IP address space (for example, both are in the subnet 192.168.1.0/24) but have different VLAN IDs, they are logically segregated and will not be able to communicate with each other. This configuration, though allowed, is against best practices—it is recommended to have unique subnet ranges for each VLAN ID. VLANs can also support segmentation of non-IP-based traffic, which is sometimes used in industrial networks.

Today, there are layer three switches that combine the benefits of a VLAN switch with the added control of a layer 3 router, making VLANs much easier to implement and maintain. This book will not go into the specifics of VLAN design since there are numerous resources available on this subject if further detail is needed. In this book, it is enough to know the basics of what VLANs are and how they function for the purposes of industrial network design and security. VLANs are an important tool, and it is highly recommended that the reader pursue the topic further and become expert in VLAN behavior, design, and implementation.

How does segmentation apply to industrial networks and to industrial cybersecurity? As with all networks, industrial networks vary considerably. It has already been discussed how there are many obvious and clearly delineated functions—for example, "business systems" and "plant systems"—as well as specific network topologies, system functions, protocols used, and other considerations that will dictate where a network must be segmented and/or segregated.

Note

Further confusion arises between the use of the terms "segmentation" and "segregation."

"Segmentation" pertains to the division of networks (network segmentation) or zones (zone segmentation) into smaller units. Segmented networks still must intercommunicate over a common infrastructure—while this intercommunication may be controlled using additional mechanisms, it is inherently allowed. The term "segregation" pertains to the elimination of communication or data flow, either within or between the networks and/or zones, in order to fully isolate systems. For example, two networks that lack any physical connections are physically segregated. Examples include the "air gap," which is typically only found in myths, legends, on fully analog systems, and on the *Battlestar Galactica*. For clarity, segregation denotes an absolute separation in a black and white manner. Segmentation indicates tighter, more granular levels of controls while allowing authorized communications, and is much more of a "gray area" in terms of implementation.

Segregation, like segmentation, can occur at any layer of the OSI model, provided that the segregated environments do not share hardware or protocol implementations. These segregation methodologies are physical, network, and application.

Two VLANs on the same switch are not segregated because of the sharing of common hardware (the switch). If there is a network-based attack that affects the operation of the switch, both VLANs can be negatively affected; hence, the environments are not fully segregated. Conversely, if two, stand-alone, nontrunked VLANs exist on two different switches, and those switches are uplinked to a layer 3 device, those VLANs can be considered layer 2 segregated from themselves, but not the native VLAN that exists on both switches. This is an example of both physical and layer 2 network segregation.

If the same environment does trunk the uplinks to the router and its configuration prevents inter-VLAN communication, the VLANs are effectively segregated at layer 3 from each other but again not the other layer 2 implementations in the same environment. This is an example of layer 3 network segregation. Segregation, therefore, is a possible byproduct of segmentation but not all segments are necessarily segregated. If all network segments were fully segregated from all other segments, full scope, cross-network communications over the infrastructure would be impossible due to the lack of a direct or transitional communication pathway.

In the context of security, (logical) segregation between security zones will be enforced mainly through security controls implemented on the communication channels and conduits that exist between zones. This will be discussed in more detail in Chapter 9, "Establishing Zones and Conduits."

Segmentation and segregation are useful security controls in that they are vital in mitigating the propagation or lateral movement (i.e., "pivoting") of an attack once a network intrusion has occurred. This will be discussed further in Chapter 9, "Establishing Zones and Conduits."

Network segmentation allows us to enforce these demarcations by taking larger networks and splitting them up into smaller, more manageable networks, and then utilizing additional security controls to prevent unauthorized communications between these networks. Another way to think of this is as the division of endpoints across distinct networks. For example, ICS servers, controllers, and process-connected devices belong in an "industrial" network and the corporate web server and enterprise resource planning (ERP) systems in the "business" network. Segmentation, therefore, provides an inherent degree of access control at each demarcation point.

Network segmentation should be used to support zone segmentation whenever possible (see Note at the start of this chapter on network and zone segmentation). Some of the network areas that are candidates for segmentation in support of security zones include the following:

- Public networks like the Internet
- Business networks
- Operations networks
- Plant control networks
- Supervisory control networks (ICS servers, engineering workstations, and HMIs)
- Basic or local control networks (controllers, programmable logic controllers [PLCs], remote terminal units [RTUs], field devices, intelligent electronic devices [IEDs], and subsystems)
- Process networks (device networks, analyzer networks, equipment monitoring networks, and automation systems)
- Safety networks (safety instrumented systems [SISs] and devices).

Network segmentation results in hierarchical networks, such that communication between two networks might require traversal of several networks. Using Figure 5.8 as an example, to get from process network B1a to process network B2a, traffic would need to communicate through control network B1, supervisory control network B, and control network B2. This has only been shown for illustrative purposes, as it is unlikely there would be any traffic flow between process networks (in the form of peer-to-peer communications), which is why they were segmented in the first place. Note that we have specifically omitted the devices between networks that would form the basis of this segmentation as this will be covered later. Also note that just because a segmented network architecture *supports* communication flows between segments, it does not mean that this traffic should be *allowed* between segments. In the previous example, traffic flow should not be allowed between process networks.

Depending upon how the network infrastructure is configured, the division of the network can be absolute, conditional, bidirectional, or unidirectional, as shown in Table 5.2.

Higher layer segmentation

While network segmentation is traditionally enforced at layer 2 (VLANs) or layer 3 (subnets), the concepts of segmentation—the containment of certain network

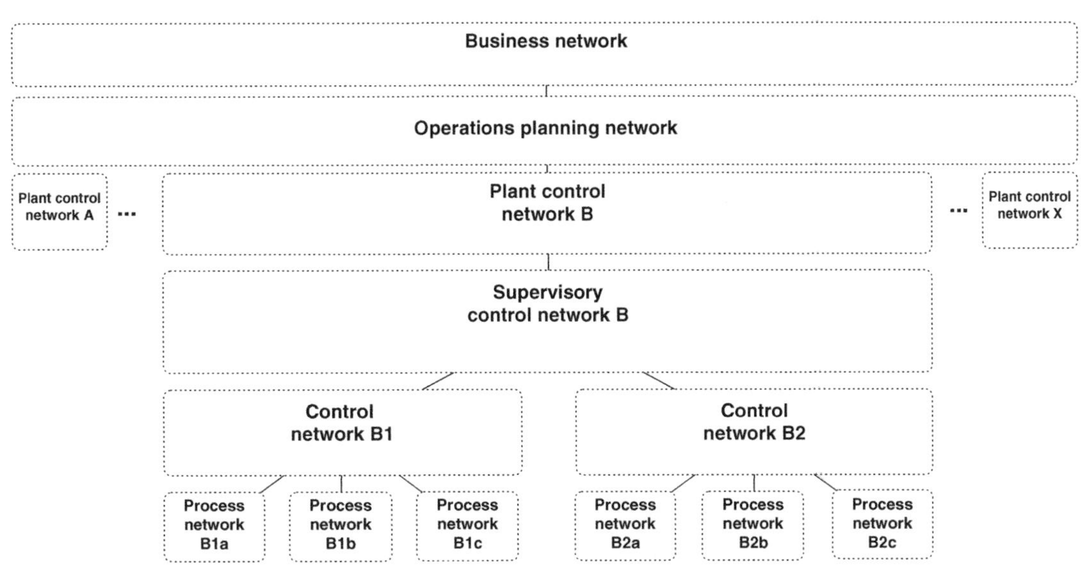

FIGURE 5.8 A conceptual representation of network segmentation in industrial systems.

Table 5.2 Types of Communication Flow Control

Absolute	No. communication is allowed (i.e., all traffic is blocked in both directions).
Conditional	Only explicitly defined traffic is allowed (e.g., via access control lists, filters, etc.).
Bidirectional	Traffic is allowed in both directions. Conditions may be enforced in both directions.
Unidirectional	Traffic is only allowed in one direction (e.g., via a data diode or unidirectional gateway).

activities—can be implemented at essentially any layer of the OSI model, often to great effect. For example, by limiting sessions and applications at OSI Layers four to seven instead of layers two to three, it becomes possible to isolate certain communications between carefully defined groups of devices, while allowing other communications to operate more freely. This is defined in Table 5.3.

Note

This concept is often referred to as "protocol filtering" or "network whitelisting" because it defines the network behaviors that are allowed, and filters the rest—essentially limiting the network to specific protocol, session, and application use. This can be enforced generally (only PROFINET is allowed) or very granular (PROFINET is allowed, only between these specific devices, using only explicitly defined commands). This level of control usually requires the use of a network-based IPS or a "next-generation" firewall (NGFW) that is able to inspect and filter traffic up to the application layer.

Table 5.3 Types of Segmentation

Method	Description	Security considerations
Physical layer segmentation	Refers to separation of two networks at the physical layer, meaning that there is a change or disruption in the physical transmission medium that prevents data from traversing from one network to another. An example could be as simple as a disconnected phone cable to a modem or a data diode to block wired transmission, a faraday cage or jammer to isolate wireless signals, etc. The mythical "air gap" is a physical layer segmentation method. Note that the term "physical layer segmentation" should not be confused with "physical segmentation," as defined below under "physical versus Logical segmentation."	Can be physically bypassed, via "sneaker net" attacks. In many cases, the excessively restrictive nature of the control motivates end users to bypass security by carrying data on thumb drives or other portable removable media, introducing new attack vectors that may not have controls in place.
Data-link layer segmentation	Occurs at layer 2, and as discussed earlier, it is typically performed using virtual local area networks, or VLANs. Network switches are used to separate systems, and VLANs are used to limit their broadcast domains. VLANs therefore cannot communicate with other VLANs without traversing at least one layer 3 hop to do so (when trunks are used), or by physically connecting VLAN access ports (when untagged access ports are used). The use of VLANs provides easy and efficient segmentation. If inter-VLAN communication is only allowed via a layer 3 device, VLANs can also enforce some security by implementing segregation via access control lists (ACLs) on the intermediary router(s). Newer layer 2 switches provide the capability to implement ACLs at the port level as traffic enters the switch, allowing options to help improve VLAN security since this ACL is applied to all VLANs on a given port.	Because VLANs are easy to implement, they are commonly used for network segmentation, which in turn will minimize the impact of many Ethernet issues and attacks, such as floods and storms. However, VLANs are also the least secure method of segmentation. Improperly configured networks are susceptible to VLAN Hopping attacks, easily allowing an attacker to move between VLANs. See "VLAN vulnerabilities," in this chapter.
Network layer segmentation	Occurs at layer 3, and is performed by a network router, a network switch with layer 3 capabilities, or a firewall. For any protocols utilizing the internet protocol (IP)—including industrial protocols that are encapsulated over TCP/IP or UDP/IP—routing provides good network layer segmentation as well as strong security through the use of router ACLs, IGMP for multicast control, etc. However, IP routing requires careful IP addressing. The network must be appropriately separated into address subnets, with each device and gateway interface appropriately configured. Network firewalls can also filter traffic at the network layer to enforce network segregation.	Most layer 3 switches and routers support access control lists (ACLs) that can further strengthen access controls between networks. Layer 3 network segmentation will help to minimize the attack surface of network-layer attacks. In order to protect against higher-layer attacks such as session hijacking, application attacks, etc. "extended" ACLs must be deployed that can restrict on communication port and IP addresses. This reduces the attack surface to only those allowed applications when configured using a "least privilege" philosophy.

Table 5.3 Types of Segmentation—cont'd

Method	Description	Security considerations
Layer 4–7 segmentation	Occurs at layers 4–7 and includes means of controlling network traffic carried over IP (i.e., above the network layer). This is important because most industrial protocols have evolved for use over IP, but are often still largely self-contained—meaning that functions such as device identity and session validation occur within the IP packet payload. For example, two devices with the IP addresses of 10.1.1.10/24 and 10.1.1.20/24 are in the same network, and should be able to communicate over that network according to the rules of TCP/IP. However, if both are slave or client devices in an ICS, they should never communicate directly to each other. By "segregating" the network based on information contained within the application payload rather than solely on the IP headers, these two devices can be prevented from communicating. This can be performed using variable-length subnet masking (VLSM) or "classless" addressing techniques.	This is a powerful method of segmentation because it offers granular control over network traffic. In the context of industrial network security, application layer "content filtering" is able to enforce segregation based upon specific industrial protocol use cases. Application layer segregation is typically performed by a "next generation firewall" or "application aware IPS," both of which are terms for a device that performs deep packet inspection (DPI) to examine and filter upon the full contents of a packet's application payload. Filtering can be very broad, limiting certain protocol traffic from one IP address to another over a given port, or very granular, limiting certain protocols to performing specific functions between predefined devices—for example, only allowing a specific controller to write values that are within a certain range to specific, explicitly defined outputs.
Microsegmentation	Provided as a feature by network infrastructure devices, typical modern switches, routers and/or firewalls. Some vendors provide microsegmentation via an in-line "bump in the wire" type of device. Microsegmentation defines and enforces user- and application-communication policy using a "zero trust" model: If a policy is not defined to allow a communication, that communication will be blocked. Microsegmentation is not limited to network interface or even hosts: Different communication policies can be defined and enforced for specific users and applications on a single host.	The zero-trust model of microsegmentation provides excellent security. Because no network communication can occur except for explicitly defined source- and destination- users and applications, the available paths open to attackers are extremely limited. An attacker would need to compromise a specific user/application on a specific host in order to reach the target system via the network.
Cryptographic microsegmentation	Like microsegmentation as described above, with the addition of network encryption using unique key pairs for each authorized communication.	Because each communication flow is encrypted separately, cryptographic microsegmentation adds additional protection. This effectively prevents network capture/replay attacks by making each allowed session "invisible" outside of the microsegment.

One point worth mentioning is that the more security that you can deploy at the various layers of the OSI model, the more resilient your architecture will be to attack. The attack surface within the communication stack typically decreases as you move "down" the stack. This is one reason why data diodes and unidirectional gateways provide one of the highest levels of segregation control because they are implemented at the physical layer. Another example is that by implementing static MAC address tables within the layer 2 switches, communication between devices can be restricted irrespective of any IP addressing (layer 3) or application (layers 4—7) vulnerabilities that may compromise the network. MAC addresses and IP addresses can both be discovered and spoofed, and application traffic can be captured, altered, and replayed. So at what layer should security be implemented? Risk and vulnerability assessments should help answer this dilemma. The first step is to focus on protecting areas that represent the greatest risk first, which is usually determined by those areas that possess the greatest impact and not necessarily those that contain the most vulnerabilities. Subsequent assessments will then indicate if additional layers of security are required to provide additional layers of protection and offer greater resilience to other cyber weaknesses.

VLAN segmentation is common on networks where performance is critical as it imposes minimal performance overhead and is relatively easy to manage. It should be noted that VLANs are not a security control. VLANs can be circumvented, and can allow an attacker to pivot between network segments (see "VLAN vulnerabilities," in this chapter). More sophisticated controls should be considered in areas where security is more important than network performance.

The relative benefits of various network segmentation methods are summarized in Table 5.4.

In order to realize the benefits of security from an application layer solution shown in Table 5.4, and it must be able to recognize and support those applications and protocols used with ICS architectures. At the time of publishing, there are still relatively few devices that provide this support, and the number of applications and protocols included is very small in relation to that observed in a variety of ICS installations. Consideration must always be given to any restrictions in place regarding the installation of third-party or "unqualified" software and controls on ICS components by the ICS vendors. ICS components are subjected to rigorous stability and regression testing to help ensure high levels of performance and availability, and for this reason, ICS vendor recommendations and guidelines should always be given due consideration.

VLANs are susceptible to a variety of layer two attacks. This includes flood attacks, which are designed to cripple Ethernet switches by filling up their MAC address table, spanning tree attacks, ARP poisoning, and many more.

Some attacks are specific to VLANs, such as VLAN hopping, which works by sending and receiving traffic to and from different VLANs. This can be very dangerous if VLAN switches are trunked to a Layer 3 router or other device in order to establish inter-VLAN access controls, as it essentially invalidates the benefits of the VLAN. VLAN Hopping can be performed by spoofing a switch, or by the manipulation of the 802.1Q header.

Table 5.4 Characteristics of Segmentation

Segmentation/ Segregation	Provided by	Management	Performance	Network security	ICS protocol support	OT Applicability
Physical layer	Air GapData diode	None	Good	Absolute	N/A	High
DataLink layer	VLAN	Moderate	Good	Very broad	High	High
Network layer	Layer 2 switch (via VLAN interfaces only) layer 3 SwitchRouter	Low	Moderate	Broad	High	High
Session layer	FirewallIPSProtocol anomaly detection	Moderate	Low	Specific	Moderate	Moderate
Application layer	Application Proxy/ IPS"Next generation" Firewall/IPSContent filter	High	Poor	Very specific	Low	Low
Microsegmentation	Compatible switch/ router and/or in-line device	High	Good	Absolute, zero trust	N/A	High
Cryptographic microsegmentation	Compatible switch/ router and/or in-line device	High	Good	Absolute, zero-trust, Stealth'	N/A	High

Switch spoofing occurs when an attacker configures a system to imitate a switch by mimicking certain aspects of 802.1Q. VLAN trunks allow all traffic from VLANs to flow, so that by exploiting the Dynamic Trunking Protocol (DTP), the attacker has access to all VLANs.

Manipulation of the VLAN headers provides a more direct approach to communicating between VLANs. It is normal behavior for a VLAN trunk to strip the tag of its native VLAN. This behavior can be exploited by double tagging an Ethernet frame with both the trunk's native VLAN and that target network's VLAN. The result is that the trunk accepts the frame and strips the first header (the trunk's native VLAN ID), leaving the frame tagged with the target network VLAN.

VLAN hopping can be countered by restricting the available VLANs that are allowed on the trunk or, when possible, disabling VLAN trunking on certain links. VLAN trunks allow multiple VLANs to be aggregated into a single physical communication interface (i.e., switch port) for distribution to another switch or router via an uplink. Without VLAN trunking, each VLAN resident in a switch that needs to be distributed would require a separate uplink.

Firewalls can operate at many layers and have evolved considerably over the years. As the firewall is able to inspect traffic "higher up" in the layers of the OSI model, they are also able to make filtering and forwarding decisions with greater precision. For example, session-aware firewalls are able to consider the validity of a session and can therefore protect against more sophisticated attacks. Application layer firewalls are application aware, meaning that they can inspect traffic to the application layers (OSI layers 5–7), examining and making decisions on the application's contents. For example, a firewall may allow traffic through to "read" values from a PLC, while blocking all traffic that wants to "write" values back to the PLC.

Similarly, the degree to which a network should be segmented requires both consideration and compromise. A highly segmented network (one with more explicitly defined networks and fewer nodes per network) will benefit in terms of performance and manageability.

Tip

Implementing IP address changes to accommodate routing or address translation may be difficult or even impossible in many existing industrial control environments. While many firewalls provide routing and/or network address translation features, firewalls that can operate in "transparent mode" or "bridge mode" are often easier to deploy.

Physical versus logical segmentation

It is important to understand the difference between physical and logical segmentation and is why this has been used in a variety of scenarios throughout this chapter. In the lexicon of network design, physical segmentation refers to the use of two separate physical network devices (both passive and active components) to perform the isolation between networks. For example, Switch 1 would support Network 1, and Switch 2 would support Network 2 with a router managing traffic between the two. In contrast, logical segmentation refers to the use of logical functions within a single network device to achieve essentially the same result. In this example, two different VLANs are used in a single switch and a trunk connection to a layer 3 switch or router is used to control access between the networks.

Physical *separation* of systems ("air gap" separation) is still widely used in industrial networks when talking about the coexistence of basic process control and safety systems overseeing the same process. Physical layer controls are still popular in highly critical areas (such as between safety- and non-safety-related levels in a nuclear power generating station) via the use of data diodes and unidirectional gateways. This has led to some confusion between the terms *physical segmentation* (multiple physical network devices) and the concept of *physical-layer separation* (isolation at the physical layer).

As shown in Figure 5.8, network segmentation often results in a hierarchical or tiered design. Because of this, it will take more hops to reach some networks (e.g., process networks) than others (e.g., plant networks). This facilitates the use of increasingly stricter access controls when a network is designed properly, defense-in-depth strategies can (and should) add additional layers of security controls as one navigates deeper into the network hierarchy.

Proper network segmentation is important for both process and control networks that often utilize UDP multicasts to communicate between process devices with the least amount of latency. Layer 2 network segmentation within a common process may be impossible because it would break up the required multicast domain. The lack of segmentation between unrelated processes could also cause issues because multicasts would then be transmitted between disparate processes, causing unnecessary contention as well as potential security risks. Process networks often segment broadcast domains using VLANs when segmentation is possible, supporting multiple processes from a single Ethernet switch. Each process should utilize a unique VLAN unless open communication between processes is required, and/or communication between services should be limited or disabled at the switch. Communication between control networks and process networks are handled at a higher tier of the overall architecture using layer 3 switching or routing.

The implementation of additional security controls within a process network can be difficult for the same reason as just explained. This may be of some concern because VLAN segmentation can be bypassed. In larger process networks, or in broadly distributed process networks (where geographically distributed devices make physical network access more difficult to prevent), this can introduce an unacceptable level of risk. This concept is discussed within ISA 62443-3-3 in terms of a relative "security level" assigned to each segment or zone. Logical segmentation is only allowed between those segments/zones that require minimal security against cyberthreats.

To address this risk

- Implement defense-in-depth security controls at the demarcation points where networks can be segmented. Example: Deploy a network-based security control in the process network, using a transparent firewall or IPS that can monitor and enforce traffic without blocking multicasts or other expected process control traffic. Implement network security controls immediately upstream of the process network VLAN switch where this is not possible.
- Monitor process network activity. If network controls are deployed, these controls can provide security event logging and alerting to provide security analysts with the needed visibility to the process network. If they are not (or cannot) be deployed, consider deploying IDS devices on mirrored or spanned switch interfaces, so that the same degree of monitoring can occur out of band.

Attention must be given to physical and environmental conditions that exist within a production environment before any decision is made on a particular security control deployed within an industrial network. Devices must typically be able to operate over extended temperature ranges and even hazardous environments—requirements not typically of standard security technologies deployed in business networks. It is not acceptable to increase security at the price of decreased availability and loss of production when securing industrial networks and systems.

Microsegmentation

Like VLANs, microsegmentation is a method of managing network communications at a more granular level than a physical network interface or host. Microsegmentation typically controls data flows by application, users, virtual hosts or containers, or combinations thereof. Unlike VLANs and other traditional network segmentation methods, which do not explicitly imply the presence of additional security controls, microsegmentation is built upon the principle of zero-trust access.[3] That is, regardless of network reachability, any active network communication between any two resources can be allowed or denied based on an explicitly defined policy. Like a properly configured firewall, the zero trust principle will "deny all", such that unknown or unplanned communications requests will be denied, which greatly reduces the attack surface of the microsegmented network. This is sometimes referred to as an "east-west firewall" or "application segmentation" because it can effectively control traffic even within a particular VLAN or in some cases even within a particular host.[4]

In Figure 5.9, we can see several hosts connected via a network switch. Because all of these hosts are in a common VLAN they can communicate freely with the traffic needing to traverse off of the switch. Broadcast and multicast packets will reach all hosts within VLAN 1. This is layer 2 or data link layer segmentation. The host that is in VLAN 2 because it is in a

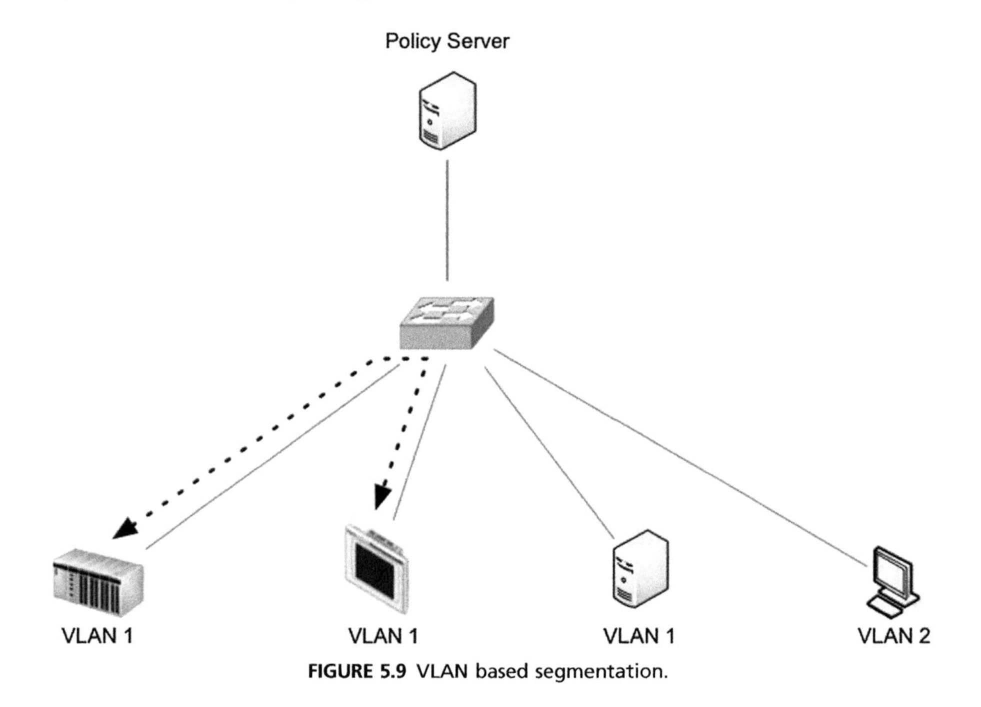

FIGURE 5.9 VLAN based segmentation.

different VLAN, cannot communicate directly to the others and broadcast and multicast packets will not reach those hosts; this host would need to be routed first before communication could occur. This is layer 3 or network layer segmentation.

While VLANs do effectively segment the network, they lack inherent security controls. Between VLANs, the security control is typically limited to simple ACLs and VLAN-to-VLAN policies enforced by the switch or router.

Assuming that the VLANs represent distinct zones, it is easy enough to disallow intercommunication. But what if a specific user of a specific application on a host in VLAN 1 has a legitimate need to communicate to a specific application on a host in VLAN 2? With micro-segmentation, this specific communication could be allowed, while all other traffic between these two hosts can be denied. This is typically enforced in one of two ways: the first is by verifying all traffic requests through a policy server, and if the request is granted, the network infrastructure allows the direct communication, as shown in Figure 5.10; the second is by using in-line devices that provide policies directly to the relevant hosts, as shown in Figure 5.11. In the first option, the explicit policies are managed enforced at the policy server. The policy server acts as an advanced firewall. In the second the policies are managed and enforced at each in-line device.

Regardless of the method used, microsegmentation creates a very well-defined and controlled communication conduit between two otherwise distinct networks. Because of this, microsegmentation is well suited to establishing zones and conduits within industrial networks (see Chapter 9, "Establishing Zones & Conduits").

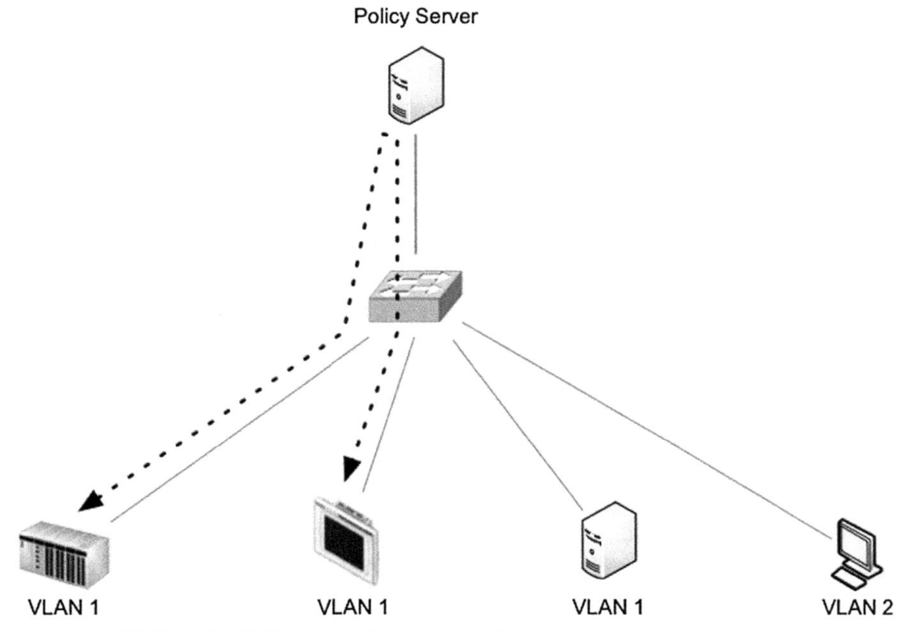

FIGURE 5.10 VLAN based microsegmentation using a central policy server.

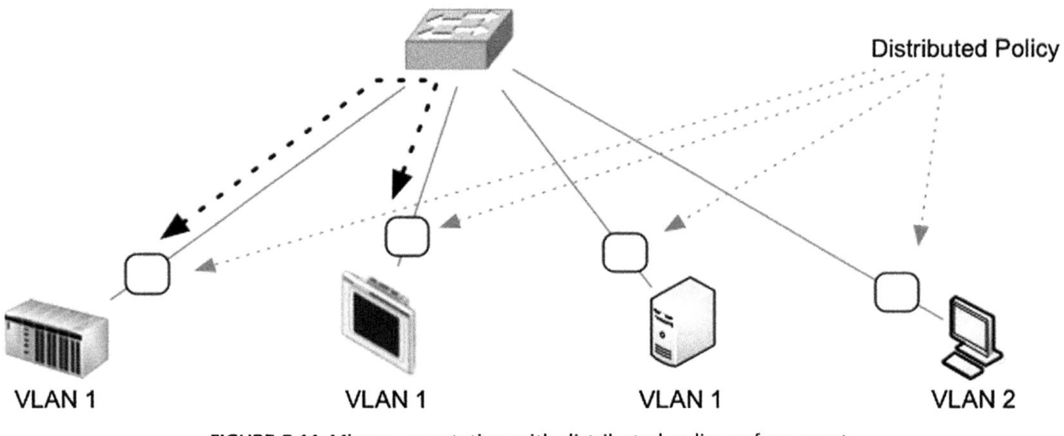

FIGURE 5.11 Microsegmentation with distributed policy enforcement.

Note that the security provided by microsegmentation varies by product and vendor. In most microsegmentation products, each dataflow is controlled by checking all traffic through an integrated firewall at the switch, which allows granular rules to limit communications. However, there is no standard for microsegmentation and no guaranteed interoperability between vendors or products.

Cryptographic microsegmentation

In some solutions, cryptographic microsegmentation is used. Each microsegment is encrypted with unique certificates making any networked device outside of a given zone effectively invisible to those within and vice-versa. Microsegmentation is often referred to as "stealth" or "cloaked" networking for this reason. As a security control, it can prevent many attacks outright because attackers will have no visibility to the segment they are attempting to infiltrate. There is no ability to capture traffic, inject traffic, replay commands, etc., making many attack methods ineffective (see Chapter 7, "Hacking Industrial Control Systems" for examples of specific attack methods).

Unfortunately, microsegmentation has also become an effective buzz-word in the industry, and so some products that claim to provide microsegmentation are doing nothing at all beyond simple layer 3 VLAN switching. This confusion, coupled with mostly proprietary solution, introduces some risk; be diligent when selecting vendors for segmentation projects to ensure that they meet your specific segmentation needs.

Network services

Network services, such as identity and access management (IAM), directory services, domain services, and others are required to ensure that all industrial zones have a baseline of access control in place. While these systems are most likely already in place within the business network, utilizing them within industrial networks can introduce risk.

Domain servers and other identity and access control systems should be maintained separately for the industrial network. This is counter-intuitive to most IT security professionals who recognize the value of centralized network services. However, the risk that a domain controller in the business zone could be compromised is much higher than the risk to a domain controller that is isolated within the plant zone. The user credentials of OT managers should therefore not be managed by IAM systems that have been deployed within the business zone. Rather, they should be managed exclusively from within the plant zone. Note that an authoritative source of identity information (e.g., human resource systems) still has value to an industrial system—it is only that the authoritative source needs to reside within that system. Any federation of information into the plant zone from centralized IT services should be very carefully controlled, and no supporting authentication and authorization systems should be allowed to serve both zones. In this way, if servers in the business domain are breached, valid credentials of OT users cannot be compromised because they reside only within OT-located systems.

As a general rule, when providing for network services in industrial systems, abide by the principle of least route, which states that in purpose-built networks, such as those used for industrial automation, a node should only be given the connectivity necessary to perform its function.[5] Any required connectivity should be provided as directly as possible to a given system (see the callout "The Principle of Least Route," in this chapter). If a critical system needs a specific network service, provide that source locally, and do not share the resource to other systems in unrelated networks (see also, Chapter 9, Establishing Zones and Conduits).

Much like the *Principle of Least Privilege/Use*, which states that a user or service must only possess the minimum privilege required to satisfy its job function, the *Principle of Least Route* follows a similar concept. The Principle of Least Route states that a node must only possess the minimum level of network access that is required for its individual function. In the past, the argument has been made that Least Route "is essentially the Least Privilege or Least Use," yet only in network form. While on the surface and with the most basic of fundamental viewpoint, this notion is correct, it is only correct in the same way that a Chevrolet Silverado 2500 Pickup truck and a Fiat 500 are both automobiles.

In order to fully understand the practical application of the Principle of Least Route, one must understand the concept of the "purpose-built network." A purpose-built network is a specialty network designed to fulfill a single, well-established purpose. There are many examples of purpose-built networks in modern life, which include broadcast networks, Internet-facing and general-purpose DMZ networks, storage area networks, voice and video networks, as well as industrial networks. With these special purpose environments in mind, the network engineering supporting these architectures requires an additional level of due care and attention to specific use in their creation. In the original explosive proliferation of TCP/IP over Ethernet networks during the 1990s, the general-purpose network philosophy included the basic idea of treating the network as a utility. In other words, an entity that was pervasive in its existence as well as reliable as the light switch on the wall. The purpose was to serve as a ubiquitous and seamless medium providing end-to-end communication to every node on the network.

Purpose-built networks that follow the Principle of Least Route are the antithesis of the modern, open, general-purpose networks of today.

In ICS environments today, a properly engineered and secured IP network environment will have considered the due care and specific use requirements in their creation. A basic example of this can be seen in the subnet and VLAN elements (implemented as organizational constructs and not security controls) that can be deployed in an ICS environment to further reduce the variables with a specific application. In a basic production line arrangement, this could mean that "line 1" to "line 2" communication is either blocked by ACLs or is null routed, provided that there is no control, functional, or business reason for "line 1" to "line 2" communication to exist.

Wireless networks

Wireless networks might be required at almost any point within an industrial network, including plant networks, supervisory networks, process control networks, and field device networks. Wireless networks are bound by the same design principles as wired networks; however, they are more difficult to *physically* contain because they are bound by the range of the radio wave propagation from an access point rather than by physical cables and network interfaces. This means that any device that is equipped with an appropriate receiver and is within the range of a wireless access point can physically receive wireless signals. Similarly, any device equipped with a suitable transmitter that is within range of an access point can physically transmit wireless signals.

There is no sure way to prevent this physical (wireless) access, as the effective range of the wireless network can easily be extended. While it is possible to block transmissions by using jammers or signal-absorbing materials (such as a Faraday containment), these measures are costly and rarely implemented. For this reason, industrial networks that implement outdoor wireless networks typically conduct thorough radiofrequency surveys in order to not only place antennas in optimal locations considering a location's unique physical obstructions but also prevent unnecessary transmission of signals into untrusted and unrestricted areas.

Some might argue that the inherent lack of physical containment makes wireless networking a poor fit for industrial networks, as it presents a very broad attack surface. However, as is often the case, there are legitimate use cases where wireless networking makes sense to the process. The existence of such use cases has spurred a rapid growth in wireless industrial networking, led by the use of WirelessHART and OneWireless. WirelessHART is a wireless implementation of the HART Communication Protocol using IEEE 802.15.4 radio and TDMA communication between nodes, while OneWireless is an implementation of ISA 100.11a wireless mesh networking based on IEEE 802.11 a/b/g/n standards and is used to transport common industrial protocols, such as Modbus, HART, OPC, General Client Interface (GCI), and other vendor-specific protocols.

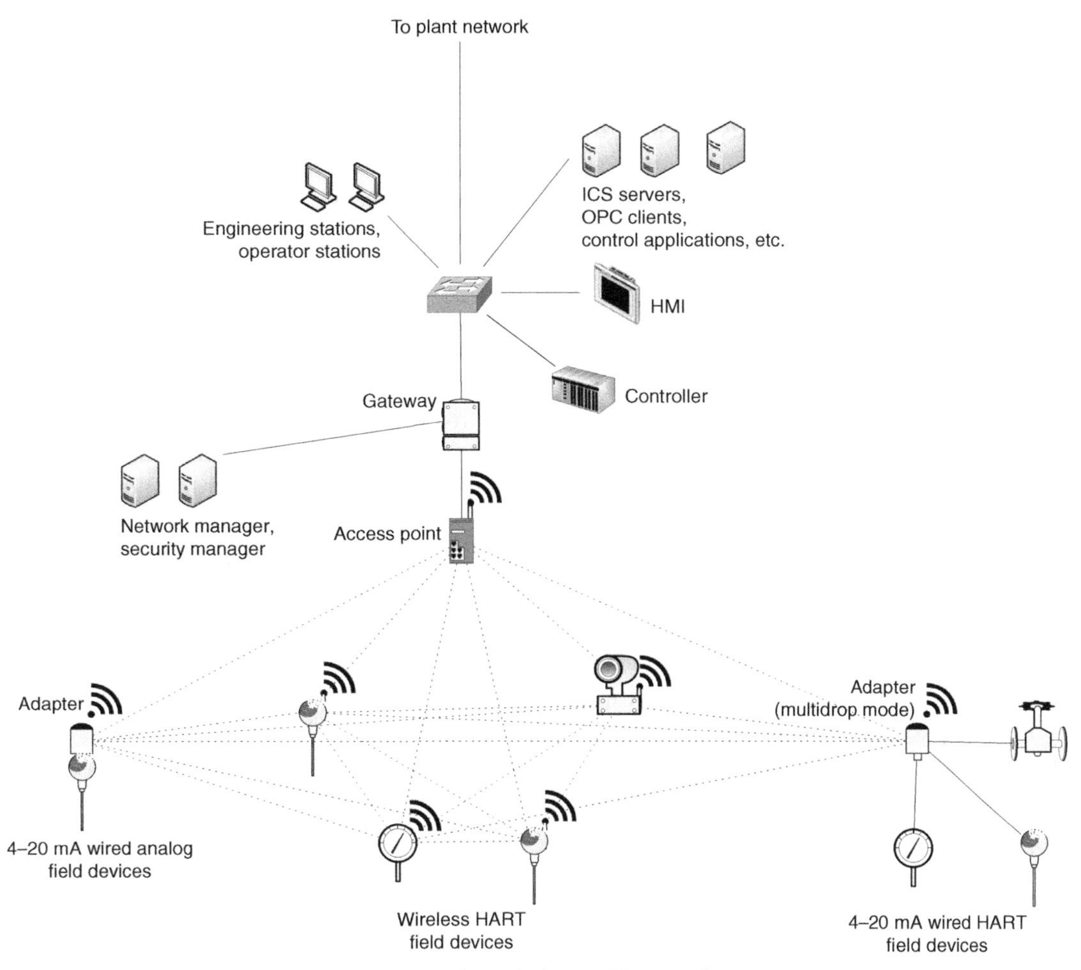

FIGURE 5.12 A wireless HART network.

Both systems support mesh networking and use two devices: one to manage connected nodes and communications between nodes and one to enforce access control and security. A common implementation of WirelessHART is shown in Figure 5.12 illustrating how the Network Manager and Security Manager are connected via wired Ethernet to the WirelessHART gateway. One or more access points also connect to the gateway with each wireless device acting as a router capable of forwarding traffic from other devices, building out the wireless network.

One important consideration in deploying wireless networks in ICS architectures is that they are commonly used to support remote, difficult, and/or costly connectivity between field devices and basic control components like PLCs and asset management systems. In areas where local power is unavailable, power can be extracted from the same line used for communications (e.g., power over Ethernet, or PoE), or utilize local batteries. This is an important consideration, as the availability of power directly impacts the availability of the process. In the case of battery power, battery life versus communication speed and update rate must be considered and typically limits the deployment of wireless field technologies in closed-loop control applications.

Remote access

Remote access is a necessary evil that must be considered when designing a secure industrial network. Remote access serves many needs of an organization. For example, an ICS commissioned in a manufacturing facility will typically include third-party contracts with explicitly defined service requirements, often requiring 24 × 7 response, with measured response times and guarantees around problem resolutions. The ICS vendor might staff support personnel in multiple time zones around the globe to meet strict service demands, while dictating that remote access be provided to allow technicians to connect to the ICS remotely for diagnostics and problem resolution. Distributed workforces within the company may also pose an issue. If engineers work remotely or from home offices, remote access to engineering systems must also be provided. In some cases (e.g., wind turbines, pipelines, oil and gas production fields) devices may be physically difficult to access, making remote access a functional necessity.

Remote access can introduce multiple attack vectors at the same time. Even if secure remote access methods are used, such as virtual private networks, two-factor authentication, and so on, a node can be compromised remotely because the underlying infrastructure used with remote access is connected to public, untrusted networks like the Internet.

To address the risks of remote access, all access points should be considered an open attack vector and should only be used when necessary. Strict security controls should be used, including the following:

- Minimize attack vectors. Only provide one path over which remote access may occur when implementing a remote access solution. This allows the single path into and out of the network to be carefully monitored and controlled. If multiple paths are allowed, it is more likely that security controls might be eliminated (due to the added cost of securing multiple paths) or that a specific security control might be overlooked or misconfigured.[6]
- Follow the principle of "least privilege," allowing users to only access those systems or devices with which they have a specific need or authority.[7]
- This means that if a user only needs to view data, they should not be provided mechanisms to download and change data.
- To enforce "least privilege," the network may require further segmentation and segregation to isolate systems that allow remote access from other systems not accessed remotely. Ideally, third parties, such as subcontractors and vendors, should be restricted access to only their devices, which may impact network segregation design, and only allowed to perform those functions they are authorized to

perform remotely (e.g., view configuration vs. download new configuration and software to devices). This will be explained in greater detail in Chapter 9, "Establishing Zones and Conduits."[8]

- Application control may also be required to further limit remote users to only those applications with which they are authorized. Requiring remote users to authenticate directly to a secure application server rather than just using a remote access server (RAS) limits the remote access session to a specific application rather than to the network on which the server resides.[9]
- Prevent direct access to any system that is considered critical or where the risk to a system outweighs the benefit of remote access. Force remote access through a secure semi-trusted or demilitarized zone (DMZ) or proxy so that additional security controls and monitoring can be implemented if remote access is required for these systems.[10]
- The security policy deployed for an endpoint connecting via remote access should be equal to or better than that of the hosts directly connected to the trusted industrial network. This can be very difficult to enforce, especially with third parties, and is why the preferred approach may be to create a "jump station" that is always used to provide a landing point for the remote user before accessing the final trusted industrial network-connected device. This physically separates the remote user's local computer and associated resources (removable media, file system, clipboard, etc.) from that computer accessing the industrial network.
- Avoid storing credentials on the remote end of the connection (e.g., the vendor support personnel) that are transmitted and utilized on the most trusted industrial network, even if they are transmitted within encrypted tunnels.
- Procedures should be established and tested that allow for site personnel to terminate and disconnect remote access mechanisms locally in the event of a cyberincident.
- Log everything. Remote access, by its nature, represents an attack vector where only one end of the connection is 100% known and controlled. All remote access attempts, successful or not, should be logged, and all activity performed by remote users during their entire session should be logged. This provides a valuable audit trail for investigators during incident response and disaster recovery efforts. In addition, if security analytics—such as advanced security information and event management systems (SIEMs) or anomaly detection systems—are used, these logs can provide proactive indicators of an attack, and can greatly reduce incident response times, which in turn will minimize losses in the event of an attack.

Performance considerations

When talking about network performance, it is necessary to consider four components: bandwidth, throughput, latency, and jitter.

Latency and jitter

Latency is the amount of time it takes for a packet to traverse a network from its source to destination host. This number is typically represented as a "round-trip" time that includes the initial packet transfer plus the associated acknowledgment or confirmation from the destination once the packet has been received.

Networks consist of a hierarchy of switches, routers, and firewalls interconnected both "horizontally" and "vertically" making it necessary for a packet to "hop" between appliances as it traverses from host to destination (see Figures 5.1 and 5.2). Each network hop will add latency. The deeper into a packet the device reads to make its decision, the more latency will be accrued at each hop. A layer 2 switch will add less latency than a layer 3 router, which will add less latency than an application layer firewall. This is a good rule of thumb but is not always accurate. The adage "you get what you pay for" is true in many cases, and network device performance is one of them. A very complex and sophisticated application layer device can outperform a poorly defined software-based network switch built on underpowered hardware if built with enough CPU and NPU horsepower or custom-designed high-performance ASICs.

Jitter on the other hand is the "variability" in latency over time as large amounts of data are transmitted across the network. A network introduces zero jitter if the time required transferring data remain consistent over time from packet-to-packet or session-to-session. Jitter can often be more disruptive to real-time communications than latency alone. This is because, if there is a tolerable but consistent delay, the traffic may be buffered in device memory and delivered accurately and with accurate timing—albeit somewhat delayed. This translates into deterministic performance, meaning that the output is consistent for a given input—a desirable feature in real-time ICS architectures. Latency variation means that each packet suffers a different degree of delay. If this variation is severe enough, timing will be lost—an unacceptable condition when transporting data from precision sensors to controls within a precisely tuned automation system.

Bandwidth and throughput

Bandwidth refers to the total amount of data that can be carried from one point to another in a given period of time, typically measured in megabits per second (Mbps) or gigabits per second (Gbps). Contention refers to competition between active nodes in a network segment for the use of available bandwidth. Bandwidth is not usually a concern in industrial networks, as most ICS devices require very little bandwidth to operate (often much less than 100 Mbps, across the entire ICS during normal operation), while most Ethernet switches provided 100 Mbps or 1000 Mbps per switch interface. (It is not uncommon for embedded ICS devices like PLCs and RTUs to contain 10 Mbps network interfaces that may require special configuration at the switch level to prevent undesirable network traffic from impacting communication performance.) Industrial network designs must accommodate bursts of event-related data (often in the form of multicast traffic) that can be seen during upsets or disturbances to the manufacturing process. Contention for available bandwidth can still be an issue on heavily populated networks, large flat (layer 2) networks, or "noisy" networks. Areas to watch out for include links between large VLAN-segmented networks and a centralized switch or router that connects these to upstream networks (e.g., the supervisor control network shown in Figure 5.8 may need to process traffic from all subordinate networks including the individual process networks).

Throughput refers to the volume of data that can flow through a network. Network throughput is impacted by a variety of physical, MAC, network, and application layer factors—including the cabling (or wireless) medium, the presence of interference, the capabilities of network devices, the protocols used, and so on. Throughput is commonly measured in packets per second (pps). The correlation between bandwidth and throughput is dependent on the size of the packet. A device that can transfer data at the full capability of the network interface is considered to support *line rate* throughput. Some networking hardware may not be able to move packets through the device at line rate even though the rated speed of a fast Ethernet connection might be 100 Mbps. Throughput is an important measurement when real-time networking is a requirement. If the network traffic generated in real-time networks (such as in process and control networks) exceeds the rated throughput of the network infrastructure, packets will be dropped. This will cause added delay in TCP/IP communications since lost packets are retransmitted. In UDP/IP communications (common with broadcast and multicast traffic), lost packets are not immediately transmitted per the UDP standard, but rather retransmitted based on error correction in the application layer. Depending on the applications and protocols used, this could result in communications errors (see Chapter 6, "Industrial Network Protocols").

Note

The monitoring of networks communications and collection of cybersecurity event data is an important part of strong cybersecurity program. However, data must be somehow be obtained from throughout the infrastructure and aggregated centrally. This can create unplanned load. Inline network monitoring can create additional latency. Enabling port mirroring or network span ports in order to monitor traffic out-of-band can impact overall performance of the network switch or router, which can also impact bandwidth and latency. Refer to Chapter 11, "Implementing Security and Access Controls," for tips on when, where, and how to implement cybersecurity controls.

Type of service, class of service, and quality of service

Quality of service (QoS) refers to the ability to differentiate and prioritize some traffic over other traffic. For example, prioritizing real-time communications between a PLC and an HMI over less critical communications. Type of service (ToS) and class of service (CoS) provide the mechanisms for identifying the different types of traffic. CoS is identified at layer 2 using the 802.1p protocol—a subset of the 802.1Q protocol used for VLAN tagging. 802.1p provides a field in the Ethernet frame header that is used to differentiate the service class of the packet, which is then used by supporting network devices to prioritize the transmission of some traffic over other traffic.

Type of service is similar to CoS, in that it identifies traffic in order to apply a quality of service. However, ToS is identified at layer 3 using the 6-bit ToS field in the IPv4 header.

Both ToS and CoS values are used by QoS mechanisms to shape the overall network traffic. In many network devices, these levels will map to dedicated packet queues, meaning that higher priority traffic will be processed first, which typically means lower latency and less latency variation. Note that QoS will not improve the performance of a network above its baseline capabilities. QoS can ensure that the most important traffic is successfully transmitted in conditions where there is a resource constraint that might prevent the transmission of some traffic in a timely manner (or at all).

Network hops

Every network device that traffic encounters must process that packet, creating varying degrees of latency. Most modern network devices are very high performance, and do not add much, if any, measurable latency. Routers and some security devices that operate at layers 4–7 may incur measurable amounts of latency. Even low amounts of latency will eventually add up in network designs that use many hops. For example, in Figure 5.2, there are 20 total hops, with three (3) of these processed by a router. In the optimized design, which replaces the router with a layer 3 switch, there are only 13 hops, and all of them are done at high speed.[11] The network design should be optimized wherever possible because industrial networks are time critical and deterministic in nature.

Note

Consideration must be given to each ICS vendor's unique network design requirements when deploying or modifying an industrial network. System performance and reliability can be negatively impacted by unnecessary network latency, and for this reason, vendors may have specific limits on the number of network appliances that can be "stacked" in a given segment or broadcast domain.

Network security controls

Network security controls also introduce latency, typically to a greater degree than network switches and routers. This is because, as in switches and routers, every frame of network traffic must be read and parsed to a certain depth, in order to make decisions based upon the information available in Ethernet frame headers, IP packets headers, and payloads. The same rule applies as before—the deeper the inspection, the greater the imposed latency.

The degree of processing required for the analysis of network traffic must also be considered. Typically, when performing deep packet inspection (a technique used in many firewalls and IDS/IPS products), more processing and memory is required. This will increase relative to the depth of the inspection and to the breadth of the analysis, meaning the more sophisticated the inspection, the higher the performance overhead. This is typically not a problem for hardware inspection appliances, as the vendor will typically ensure that this overhead is accommodated by the hardware. However, if a network security appliance is being asked to do more than it has been rated for in its specifications, this could result in errors, such as increased latency, false negatives, or even dropped traffic. Examples include monitoring higher bandwidth than it is rated for, utilizing excessive numbers of active signatures, and monitoring traffic for which preprocessors are not available. This is one reason why the deployment of traditional IT controls like IDS/IPS in OT environments must be carefully reviewed and "tuned" to contain only the signatures necessary to support the network traffic present (this will also help to reduce false positives). If an industrial network does not have Internet access, then signatures relating to Internet sites (i.e., gaming websites or other business-inappropriate sites) could easily be removed or disabled.

Safety instrumented systems

A safety instrumented system consists of many of the same types of devices as a "regular" ICS—controllers, sensors, actuators, and so on. Functionally, the SIS is intended to detect a potentially hazardous state of operation, and place the system into a "safe state" before that hazardous state can occur. SISs are designed for maximum reliability (even by the already-high standards of automation) and often include redundancy and self-diagnostics to ensure that the SIS is fully functional should a safety event occur. The idea is that the SIS must be available when called upon to perform its safety function. This requirement is measured as a statistical value called the average probability of failure on demand (PFD). This probability is stated as a safety integrity level (SIL) ranging from 1 to 4 (SIL1 has a PDF of $<10^{-1}$, SIL2 $<10^{-2}$, SIL3 $<10^{-3}$, and SIL4 $<10^{-4}$.)

Note

There is a great deal of correlation between industrial security and functional safety, and for this reason, ISA has leveraged the activities of the SP85 committee on safety with the SP99 committee on security. The premise of the SIL is to allow a quantitative value to be calculated that presents the integrity "capability" of a component or the integrity "assurance" of a deployed system in relation to ensuring health, safety, and environmental (HSE) protection in the event of a component failure. A corresponding criterion called the security level (SL)[12] has been established to provide a mechanism to qualitatively represent a security zone's (or conduit's) "capability" (SL-C) based on selected components against a particular design "target" (SL-T) and "achieved" (SL-A) levels of security assurance. The idea behind the development of the SL was to shift thinking regarding security from an individual device or standalone system basis to a more integrated zone-based approach that more accurately represents the integrated, heterogeneous nature of deployed ICSs.

Ideally, safety systems are built using dedicated controllers known as "logic solvers" to support a specific process. The SIS can either be "interfaced" to the basic process control system (BPCS) components via hardwired connections, or "integrated" via higher-level connectivity that may include a common or shared network. More recent standards and trends allow safety devices to coexist and interoperate with standard BPCS devices in the process network (example: Emerson DeltaV SIS[13] and Honeywell Safety Manager[14]). Some SIS solutions are also available that allow process and safety functions to exist within the same device (example: ABB AC 800M HI[15] and Siemens S7-400FH, S7-300F and ET-200[16]). Some industrial protocols allow safety and basic control messaging to share a common messaging and control infrastructure. This trend introduces new security concerns[17] While SIS cannot protect against cyberattacks directly, they should be able to prevent catastrophe from being caused by a cyberattack against an industrial process by putting the system into a secure state before the catastrophe can occur.

Entire books have been written solely on the topic of securing SIS. In this book, the advice will be limited and general:

- SIS exists to prevent unsafe conditions. When implementing an SIS, do so in a way that a malicious actor who successfully compromises control and process zones will not be able to also compromise the SIS. Preference should be to keeping the SIS completely segregated from upstream networks (including supervisory networks), and when integration or interfacing is necessary, direct point-to-point connections are recommended.
- Comply with the Principle of Least Privilege when implementing an SIS to minimize the potential vectors that an attacker might take to access the safety systems.
- Consider failures and unsafe states when implementing an SIS that may be the result of a manipulation of the controller, process, protocols, and systems of the industrial network by an attacker.

Special considerations

Industrial control systems are used for a variety of purposes across many industries, and because of this, there will always be special circumstances that need to be considered when designing the industrial networks. The use of specialized wide area networks will grow as businesses become increasingly global. As systems are tuned to specific purposes—such as the advanced metering requirements for the smart grid—specialized networks, such as the advanced metering infrastructure (AMI), will evolve to accommodate them. It is important to give specialized systems their due consideration while continuing to apply the fundamental principles of secure network design.

Wide area connectivity

Long-range, wide area connectivity requirements are common when interconnecting central control rooms to remote plants, microgrids, pipelines, offshore oil platforms, remote wind farms, and other far-reaching locations. Wide area connectivity can be provided by private infrastructure or by leased connectivity from public carriers. The technologies vary widely, as do the transport mediums, which may include satellite, microwave, radio, fiber optic, cellular, and others.

Wide area connectivity should be given the same consideration as any other network connection when designing a secure network. By its nature, the WAN infrastructure is physically accessible to unknown users who could potentially be threat actors, especially at unmanned sites with network connectivity. Access can also be provided through the use of appropriate wireless transmitters and receivers or by physically splicing or taping cables and wires. These connections should therefore be considered higher risk, and extra measures should be taken to ensure the confidentiality, integrity, and availability of any wide area connection.

When performing risk and vulnerability assessments, make sure that specialized wide area overlay networks are not overlooked. In smart grid applications, distributed phase measurement devices called synchrophasors require precisely synchronized timing and utilize GPS network timing. The GPS network is a globally accessible network, and researchers have proven that GPS spoofing can result in real-world impact. A study by the University of Texas and Northrup Grumman showed how GPS spoofing was able to manipulate synchrophasor readings and cause a plant to trip.[18] In another study by the University of Texas, GPS spoofing was used to alter GPS coordinates to a cruise ship, enabling the researchers to steer the ship off of its intended course.[19]

As GPS, cellular and similar technologies become increasingly popular for the interconnection of highly distributed remote devices; they will continue to introduce new threat vectors to systems that utilize them.

Smart grid network considerations

One area that deserves special consideration is the smart grid. As mentioned in Chapter 4, "Introduction to Industrial Control Systems and Operations," the smart grid is an extensive network providing advanced metering and communications capabilities to energy generation, transmission, and distribution. It may be specific to the energy industry, yet is also a concern for any other industrial sector that may connect to the smart grid as a client of the electric utility industry.

The smart grid varies widely by deployment, and the topologies and protocols used vary accordingly. There is one primary quality that is consistent across any smart grid deployment and that is its scale and accessibility. As a distribution system designed to deliver power ubiquitously to industrial facilities, residences, offices, storefronts, and all aspects of urban infrastructure, even small smart grid deployments create large numbers of nodes and network interconnections. These networks can exceed hundreds of thousands to even millions of interconnected devices. The scale of a smart grid requires the use of some mechanism to "tier" or hierarchically distribute the nodes.

Represented in terms of an addressable attack surface, smart grids provide broad and easy access to a network that ultimately interconnects the electric utility transmission and distribution infrastructure to many homes and businesses. Figure 5.13 illustrates the attack surface as being exponentially larger as one radiates outward from core electric power generation through long-distance transmission to regional distribution and the outer reaches of the smart grid.

Scalability also plays a role in the development of smart grid devices, putting significant cost pressure on the end-node devices (smart meters). Any device deployed at such a large

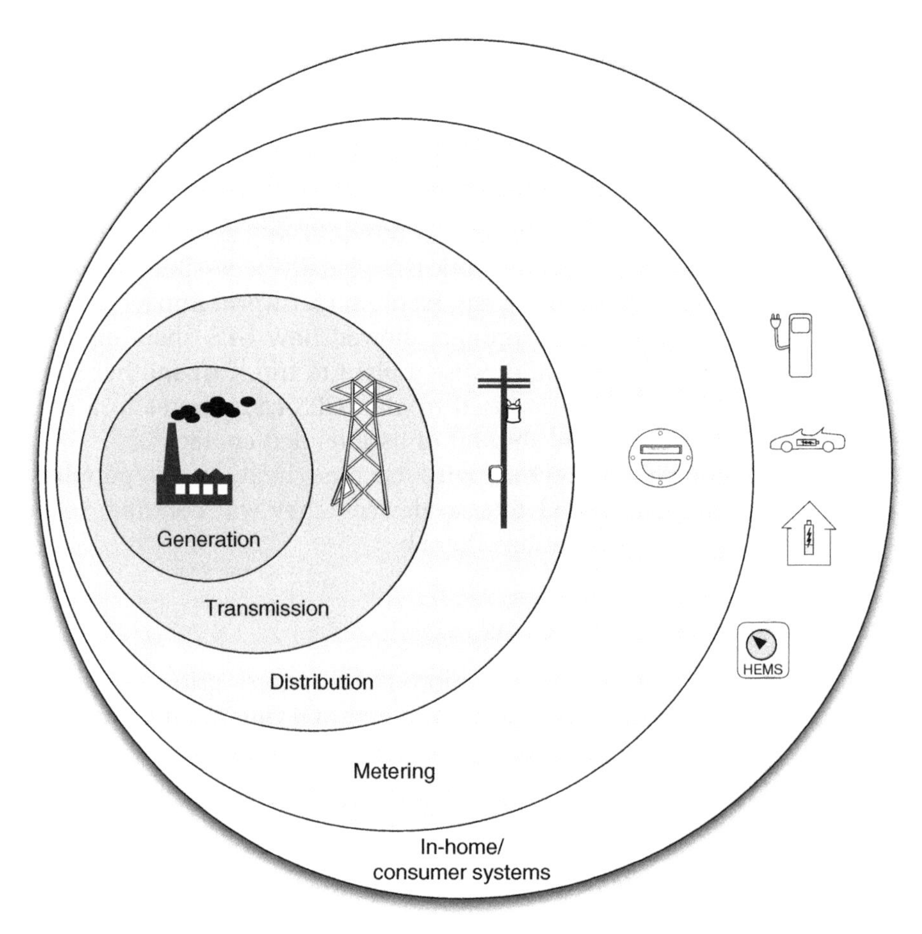

FIGURE 5.13 The expanding attack surfaces within a smart grid.

scale needs to be as efficient to build, deploy, operate, and maintain as possible. This business driver is a real concern because of the costs and complexity of providing security assurance and testing throughout the supply, design, and manufacturing stages of smart meter development. As pressures force costs down, there is an increased chance that some physical or network-based vulnerability will find its way into production, and therefore into one of the most easily reachable networks ever built.

Advanced metering infrastructure

Advanced metering infrastructure systems are utilized by electric, water, and gas utilities. AMI is a good example of a specialized industrial network—it has unique characteristics in that it is highly distributed, massively scalable to millions of nodes, uses specialized systems and protocols, and presents a number of new security and privacy considerations. It also operates very similar to many industrial networks in that it is built of operator-owned devices that function in a (theoretically) closed system. Unlike many industrial networks, which are isolated behind physical security controls, and protected behind multiple layers of network defenses, the metering infrastructure is extremely accessible.

Advanced metering infrastructure architecture consists of smart meters, a communication network, and an AMI server or headend. The smart meter is a digital device consisting of a solid state measuring component for real-time data collection, a microprocessor and local memory to store and transmit measurements and at least one network interface to communicate to the headend. The headend will typically consist of an AMI server, which is primarily responsible for collection of meter data, and a meter data management system (MDMS), which manages that data and shares it with demand response systems, historians, billing systems, and other business applications. The headend maintains communications with the meters to read data (to measure consumption), push data (to transmit rate information for demand-response systems), and to establish control (for remote disconnects). The headend also intercommunicates with many other systems in the smart grid—transmission and distribution ICS servers, demand response servers, energy management systems (EMS), in home networks, and many others (for more detail on smart grid architecture, please refer to "Applied Cyber Security and the Smart Grid").

Some common issues that have already been discussed with regard to other industrial networks become obvious. The specialized devices are essentially computing platforms—they have microprocessors, memory, and storage and can execute code. This means that the system can be exploited, data can be manipulated, and an attack can easily propagate to other interconnected systems. In the United States alone, nearly 65 million smart meters will have been deployed by 2015[20], with a global estimate of 602.7 million smart meters deployed by 2016.[21] This rapid deployment makes AMI a highly scalable communication network and in turn a vast attack surface that is comparable to the Internet itself. To further complicate matters, a variety of less common network technologies are used in AMI systems, including Broadband over Power Line (BPL), Power Line Communications (PLC), radio networks (VHF/UHF), and telecommunications (landline, cellular, paging, etc.) networks.

Summary

By understanding how ICSs and automation processes function, and by adhering to the basic principles of secure network design, it is possible to accommodate ICSs on modern Ethernet networks. This becomes especially important when considering how industrial protocols operate, which is covered in Chapter 6, "Industrial Network Protocols."

Endnotes

1. 1International Society of Automation (ISA), 62443-3-1, "Security for industrial automation and control systems: System security requirements and security levels," December 2012.
2. Cisco. "Layer two and Layer three Switch Evolution." (cited: December 21, 2013).
3. Cisco. "What Is Micro-Segmentation?" Accessed Feb 20, 2023, https://www.cisco.com/c/en/us/products/security/what-is-microsegmentation.html
4. Ibid
5. Brad Hegrat. Industrial Infrastructure Design for Safety and Security. ISA Safety & Security Symposium, Houston. 2008.
6. Paul Didier, Fernando Macias, James Harstad, Rick Antholine, Scott A. Johnston, Sabina Piyevsky, Mark Schillace, Gregory Wilcox, Dan Zaniewski, Steve Zuponcic. Converged Plantwide Ethernet (CPwE) Design and Implementation Guide. Cisco Systems, Inc. and Rockwell Automation, Inc. Sep. 9, 2011.
7. Ibid
8. Ibid
9. Ibid
10. Ibid
11. Cisco, "Design Best Practices for Latency Optimization," December 2007.
12. International Society of Automation (ISA), "Security for industrial automation and control systems: System security requirements and security levels," ISA 62443-3-1:2013.
13. Emerson Process Management, "DeltaV SIS for Process Safety Systems: A Modern Safety System - for the Life of Your Plant," September 2013.
14. Honeywell Process Solutions, "Safety Manager - Product Information Note," PN-12—25-ENG, March 2013.
15. ABB, "800xA High Integrity Emergency Shutdown Solution," 2009.
16. Siemens, "Safety Integrated for Automation - Reliable, Flexible, Easy," April 2008.
17. ABB, "The rocky relationship between safety and security - Best practices for avoiding common cause failure and preventing cyber security attacks in Safety Systems."
18. Shepard Daniel P, Humphreys Todd E, Fansler Aaron A. Evaluation of the vulnerability of phasor measurement units to GPS spoofing attacks, In *Sixth annual IFIP WG 11.10 international conference on critical infrastructure protection.* Washington, DC; March 19—21, 2012.
19. University of Texas at Austin. UT Austin Researchers Successfully Spoof an $80 million Yacht at Sea. July 29, 2013. Article on Internet. http://www.utexas.edu/news/2013/07/29/ut-austin-researchers-successfully-spoof-an-80-million-yacht-at-sea/
20. The Edison Foundation, "Utility-Scale Smart Meter Deployments, Plans, and Proposals," IEE Report, May 2012.
21. K. Rowland, "602.7 million installed smart meters globall by 2016," (cited: December 23, 2013).

6 Industrial Network Protocols

Information in this chapter

- Overview of Industrial Network Protocols
- Fieldbus Protocols
- Backend Protocols
- AMI and the Smart Grid

Understanding how industrial networks operate requires a basic understanding of the underlying communications protocols that are used, where they are used, and why. There are many highly specialized protocols used for industrial automation and control, most of which are designed for efficiency and reliability to support the economic and operational requirements of large industrial control system (ICS) architectures. Industrial protocols are designed for real-time operation to support precision operations involving deterministic communication of both monitoring and control data.

This means that most industrial protocols forgo any feature or function that is not absolutely necessary for the sake of efficiency. More unfortunate is that this often includes the absence of even basic security features such as authentication or encryption, both of which require additional overhead. To further complicate matters, many of these protocols have been modified to run over Ethernet and Internet Protocol (IP) networks as suppliers moved away from proprietary networks and networking hardware and leveraged commercial off-the-shelf technologies. This, however, has now left these "fragile" protocols potentially vulnerable to cyberattack.

Overview of industrial network protocols

Industrial network protocols are deployed throughout a typical ICS network architecture spanning wide-area networks, business networks, plant networks, supervisory networks, and fieldbus networks. Most of the protocols discussed have the ability to perform several functions across multiple network zones, and so will be referred to here more generically as industrial protocols.

Industrial protocols are real-time communications protocols, developed to interconnect the systems, interfaces, and instruments that make up an industrial control system. Many were designed initially to communicate serially over RS-232/485 physical connections at low speeds (typ. 9.6–38.4 kbps), but have since evolved to operate over Ethernet networks using routable protocols such as Transmission Control Protocol (TCP)/IP and User Datagram Protocol (UDP)/IP.

Industrial Network Security. https://doi.org/10.1016/B978-0-443-13737-2.00006-3

Industrial protocols for the purposes of this book will be divided into two common categories: fieldbus and backend protocols. Fieldbus is used to represent a broad category of protocols that are commonly found in process and control (see Chapter 5, "Industrial Network Design and Architecture"). Beginning in the early 1980s, there was a push from ICS vendors and end users to establish a global fieldbus standard. This effort continued for over 20 years and resulted in the creation of a wide range of standards devoted to industrial protocols. The International Electrotechnical Commission (IEC) 61158 standard was one of the early documents that established a base of eight different protocol sets called "types." Some of the major protocols at that time (HART and CIP to name two) were missing from this list. The IEC 61784 standard was introduced in the early 2000s to amend the list originally contained in the IEC 61158 standard, and includes a total of nine protocol "profiles": FOUNDATION Fieldbus, CIP, PROFIBUS/PROFINET, P-NET, WorldFIP, INTERBUS, CC-Link, HART, and Serial Real-time Communications System (SERCOS).[1] Fieldbus protocols in this book are commonly deployed to connect process-connected devices (e.g., sensors) to basic control devices (e.g., Programmable Logic Controller [PLC]), and control devices to supervisory systems (e.g., ICS server, Human-Machine Interface [HMI], historian).

Backend protocols are those protocols that are commonly deployed on or above supervisory networks and are used to provide efficient system-to-system communication, as opposed to data access. Examples of backend protocols include connecting a historian to an ICS server, connecting an ICS from one supplier to another supplier's systems, or connecting two ICS operation control centers.

Four common industrial network protocols will be discussed in some depth, others will be touched upon more briefly, and many will not be covered here. There are literally dozens of industrial protocols, many developed by manufacturers for their specific purposes. The two fieldbus protocols analyzed include the Modicon Communication Bus (Modbus) and the Distributed Network Protocol (DNP3). Two backend protocols will also be discussed in detail; Object Linking and Embedding for Process Control (OPC) and the Inter-Control Center Protocol (ICCP, also referenced by standard IEC 60870–3 TASE.2 or Telecontrol Application Service Element). These particular protocols have been selected for more in-depth discussion because they are all widely deployed and they represent several unique qualities that are important to understand within the context of security. These unique qualities include the following:

- Each is used in different (though sometimes overlapping) areas within an industrial network.
- Each provides different methods of verifying data integrity and/or security.
- The specialized requirements of industrial protocols (e.g., real-time, synchronous communication) often make them highly susceptible to disruption.

It should be possible to assess the risks of other industrial network protocols that are not covered here directly by understanding the basic principles of how to secure these protocols.

Fieldbus protocols

Modicon communication bus (Modbus)

The PLC dates as far back as 1968 when General Motors set out to find a new technology to replace their hard-wired electromechanical relay system with an electronic device. The first PLC was developed by Bedford Associates and designated 084 (representing the Bedford's 84th project), and released by the product name Modicon or MOdular DIgital CONtroller.[2] This Modbus protocol was designed in 1979 to enable process controllers to communicate with real-time computers (e.g., MODCOMP FLIC, DEC PDP-11), and remains one of the most popular protocols used in ICS architectures. Modbus has been widely adopted as a de facto standard and has been enhanced over the years into several distinct variants.

Modbus' success stems from its relative ease of use by communicating raw messages without restrictions of authentication or excessive overhead. It is also an open standard, is freely distributed, and is widely supported by members of the Modbus Organization, which still operates today.

What it does

Modbus is an application layer messaging protocol, meaning that it operates at layer 7 of the OSI model. It allows for efficient communications between interconnected assets based on a "request/reply" methodology. Extremely simple devices such as sensors or motors use Modbus to communicate with more complex computers, which can read measurements and perform analysis and control. To support a communications protocol on a simple device requires that the message generation, transmission, and receipt all require very little processing overhead. This same quality also makes Modbus suitable for use by PLCs and Remote Terminal Units (RTUs) to communicate supervisory data to an ICS system.

Because Modbus is a layer 7 protocol, it operates independently of underlying network protocols residing at layer 3, allowing it to be easily adapted to both serial and routable network architectures. This is shown in Figure 6.1.[3]

How it works

Modbus is a request/response protocol using three distinct Protocol Data Units (PDUs): Modbus Request, Modbus Response, and Modbus Exception Response, as illustrated in Figures 6.2 and 6.3.[4]

Modbus can be implemented on either an RS-232C (point-to-point) or RS-485 (multidrop) physical layer. Up to 32 devices could be implemented on a single RS-485 serial link, requiring each device communicating via Modbus be assigned a unique address. A command is addressed to a specific Modbus address, and while other devices may receive the message, only the addressed device will respond. Implementations using RS-232C were relatively simple to commission; however, do to the many variations in the way RS-485 could be implementing (2-wire, 4-wire, grounding, etc.), it was sometimes very challenging to commission a multidrop topology when using devices from many different vendors.

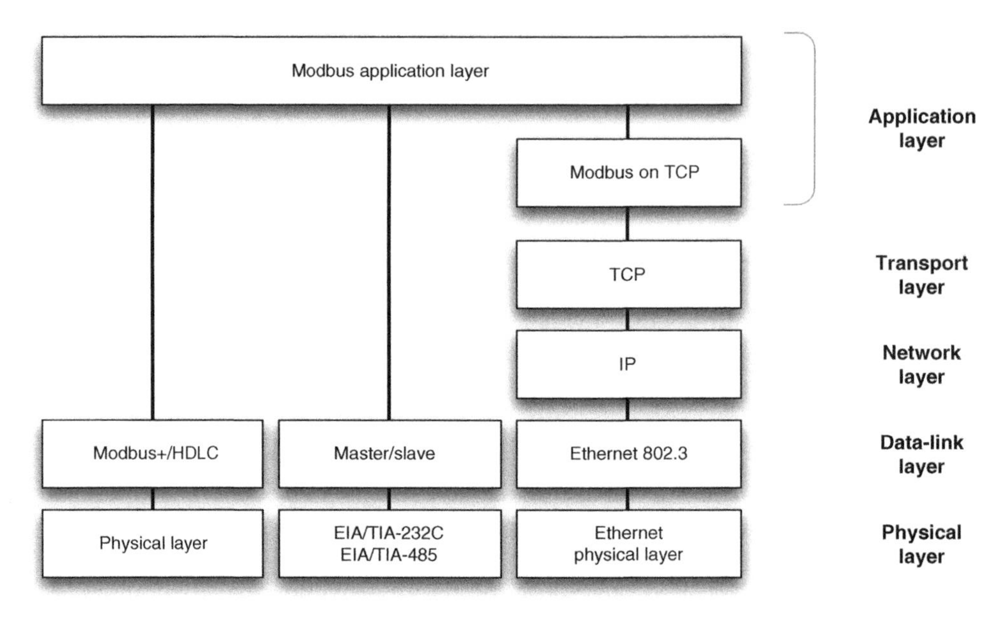

FIGURE 6.1 Modbus alignment with OSI 7-layer model.

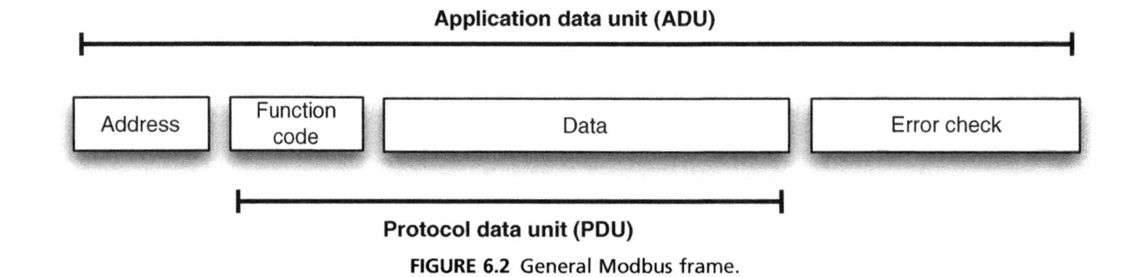

FIGURE 6.2 General Modbus frame.

A "transaction" begins with the transmission of an initial Function Code and a Data Request within a Request PDU. The receiving device responds in one of two ways. If there are no errors, it will respond with a Function Code and Data Response within a Response PDU. If there are errors, the device will respond with an Exception Function Code and Exception Code within a Modbus Exception Response.

Data are represented in Modbus using four primary tables as shown in Table 6.1. The method of handling each of these tables is device specific, as some may offer a single data table for all types, while others over unique tables. Careful review of the device documentation is needed in order to understand the device's data model, because the original Modbus definitions provided for only addresses in the range 0–9999. The specification has since been appended to allow up to 65,536 addresses across all four data tables. Another caveat within the standard is the original definition provided for the first digit of the register to identify the data table.

Function Codes used in Modbus are divided into three categories and provide the device vendor with some flexibility in how they implement the protocol within the

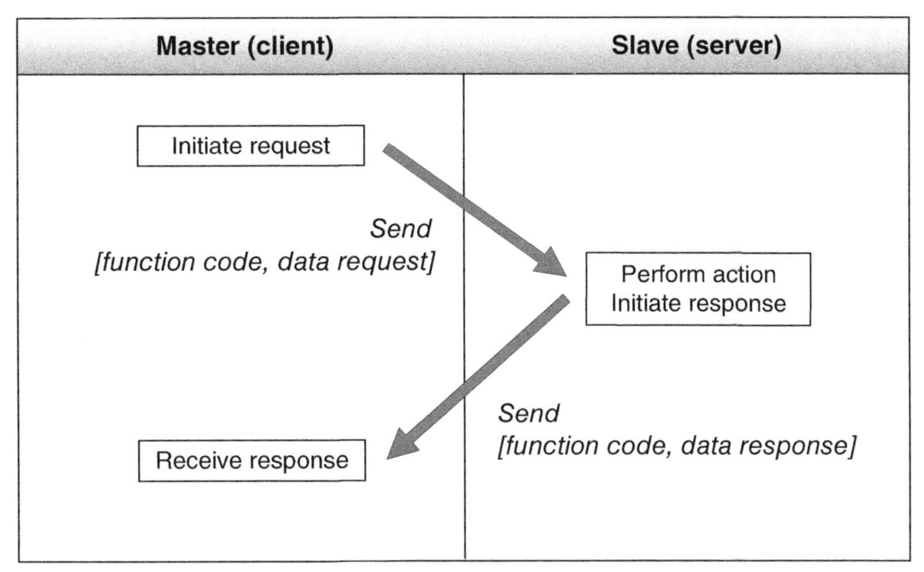

FIGURE 6.3 Modbus protocol transaction (error-free).

device. Function codes in the range of 01–64, 73–99 and 111–127 are defined as "Public" and are validated by the Modbus-IDA community and are guaranteed unique. This range is not entirely implemented, allowing codes to be defined in the future. "User-Defined" function codes in the range 65–72 and 100–110 are provided to allow a particular vendor to implement functionality to suit their particular device and application. These codes are not guaranteed to be unique and are not supported by the standard. The final category of codes represents "Reserved" functions that are used by some companies for legacy products, but are not available for general public use. These reserved codes include 8, 9, 10, 13, 14, 41, 42, 90, 91, 125, 126, and 127.

Function Codes and Data Requests can be used to perform a wide range of commands. Some examples of Modbus commands include the following:

- Read the value of a single register.
- Write a value to a single register.
- Read a block of values from a group of registers.
- Write a block of values to a group of registers.
- Read files.

Table 6.1 Modbus Data Tables

Data table	Object type	Access	Data provided by	Register range (0–9999)	Register range (0–65535)
Discrete input	Single Bit	Read-only	Physical I/O	00,001–09,999	000,001–065,535
Coil	Single Bit	Read-write	Application	10,001–19,999	100,001–165,535
Input register	16-bit Word	Read-only	Physical I/O	30,001–39,999	300,001–365,535
Holding register	16-bit Word	Read-only	Read-write	40,001–49,999	400,001–465,535

- Write files.
- Obtain device diagnostic data.

Variants

The popularity of Modbus has led to the development of several variations to suit particular needs. These include **Modbus RTU** and **Modbus ASCII**, which support binary and ASCII transmissions over serial buses, respectively. Modbus TCP is a variant of Modbus developed to operate on modern networks using the IP. **Modbus Plus** is a variant designed to extend the reach of Modbus via interconnected busses using token passing techniques.[5]

Modbus RTU and Modbus ASCII

These similar variants of Modbus are used in asynchronous serial communications, and they are the simplest of the variants based on the original specification. Modbus RTU (Figure 6.4) uses binary data representation, whereas Modbus ASCII (Figure 6.5) uses ASCII characters to represent data when transmitting over the serial link. Modbus RTU is the more common version and provides a very compact frame over Modbus ASCII. Modbus ASCII represents data as a hexadecimal value coded as ASCII, with 2 characters required for each byte of data (ASCII PDU is twice the size of RTU PDU). Each uses a simple message format carried within an Application Data Unit (ADU) (see Figure 6.2), consisting of an address, function code, a payload of data, and a checksum, to ensure the message was received correctly.

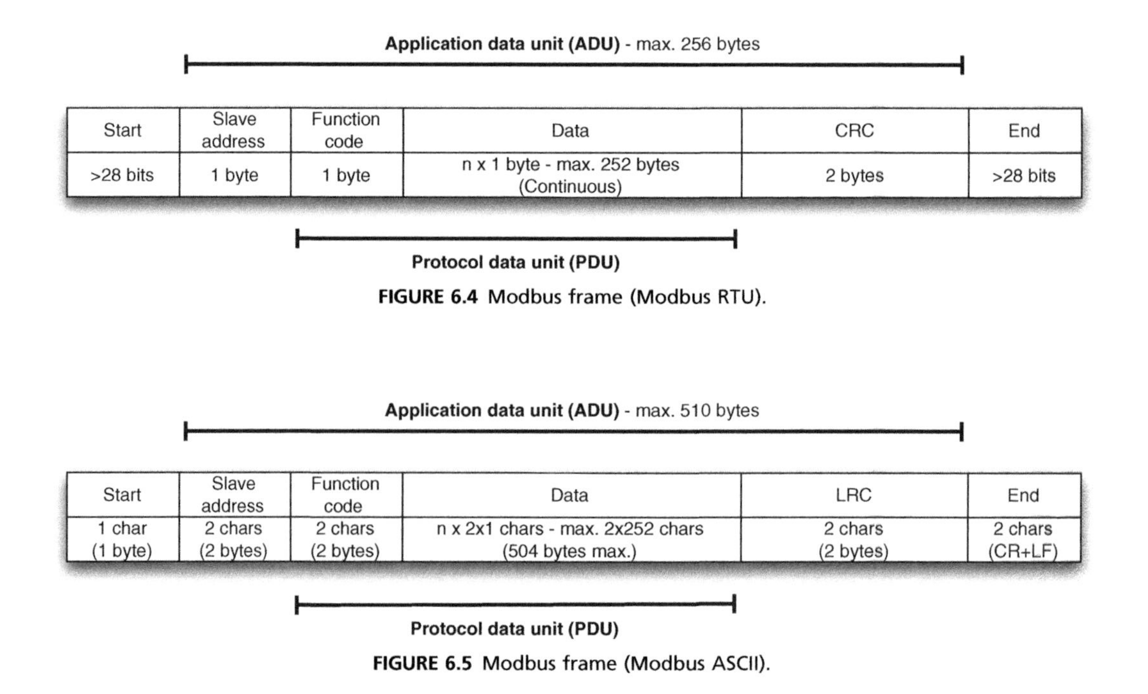

FIGURE 6.4 Modbus frame (Modbus RTU).

FIGURE 6.5 Modbus frame (Modbus ASCII).

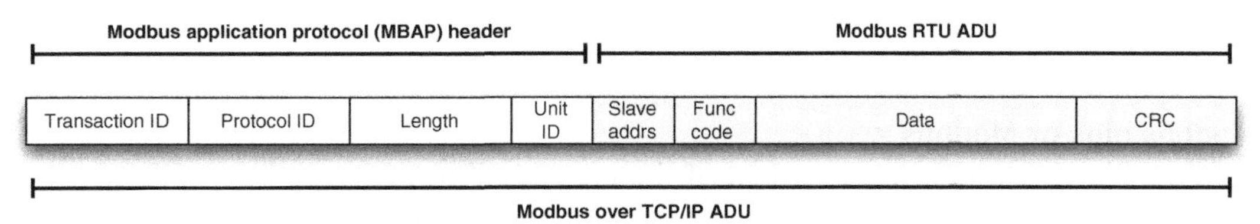

FIGURE 6.6 Modbus frame (Modbus over TCP/IP).

Modbus TCP

Modbus can also be transported over Ethernet using TCP in two forms. The basic form takes the original Modbus RTU ADU (as shown in Figure 6.4) and applies a Modbus Application Protocol (MBAP) header to create a new frame (Figure 6.6) that is passed down through the remaining layers of the communication stack adding appropriate headers (Figure 6.7) before being placed on the Ethernet network. This new frame includes all of the original error checking and addressing information. This protocol behavior is very common with older, legacy devices that contain a Modbus RTU serial interface: the devices connect to a "device server" that converts serial data so that it can be transmitted on an industrial network; and is received by a similar "device server" that converts it back to serial RTU form.

Modbus TCP is the more common form and uses TCP as a transport over IP to issue commands and messages over modern routable networks. Modbus/TCP removes the legacy address and error checking, and places only the Modbus PDU together with an MBAP header into a new frame (see Figure 6.8). The "Unit ID" acts as the new network

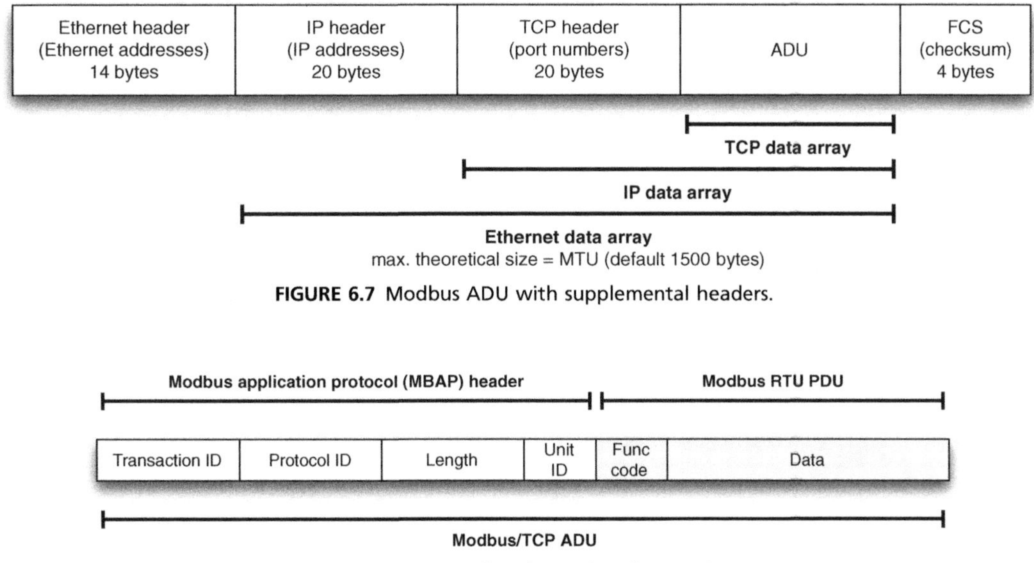

FIGURE 6.7 Modbus ADU with supplemental headers.

FIGURE 6.8 Modbus frame (Modbus/TCP).

device address and is part of the MBAP. Error checking is performed as part of the composite Ethernet frame.

Modbus plus or Modbus+

Modbus Plus is actually not a variant of the base Modbus protocol, but a different one that utilizes token passing mechanisms to send embedded Modbus messages over an RS-485 serial communications link with transmissions rates up to 1Mbps using single (nonredundant) and dual-cable (redundant) topologies. The network supports the ability to broadcast data to all nodes and allows "bridges" to be added to network creating segmented Modbus networks that each can contain up to 64 addressable nodes. This allows for very large Modbus networks to be created. Modbus + remains a proprietary protocol to Schneider-Electric.[6]

Where it is used

Modbus is typically deployed between PLCs (slave) and HMIs (master), or between a master PLC and several slave devices such as PLCs, drives, and sensors as shown in Figure 6.9. Modbus devices can act as a "master" to some, while acting at the same time as a "slave" to other devices. This function is common in a Master Terminal Unit that is

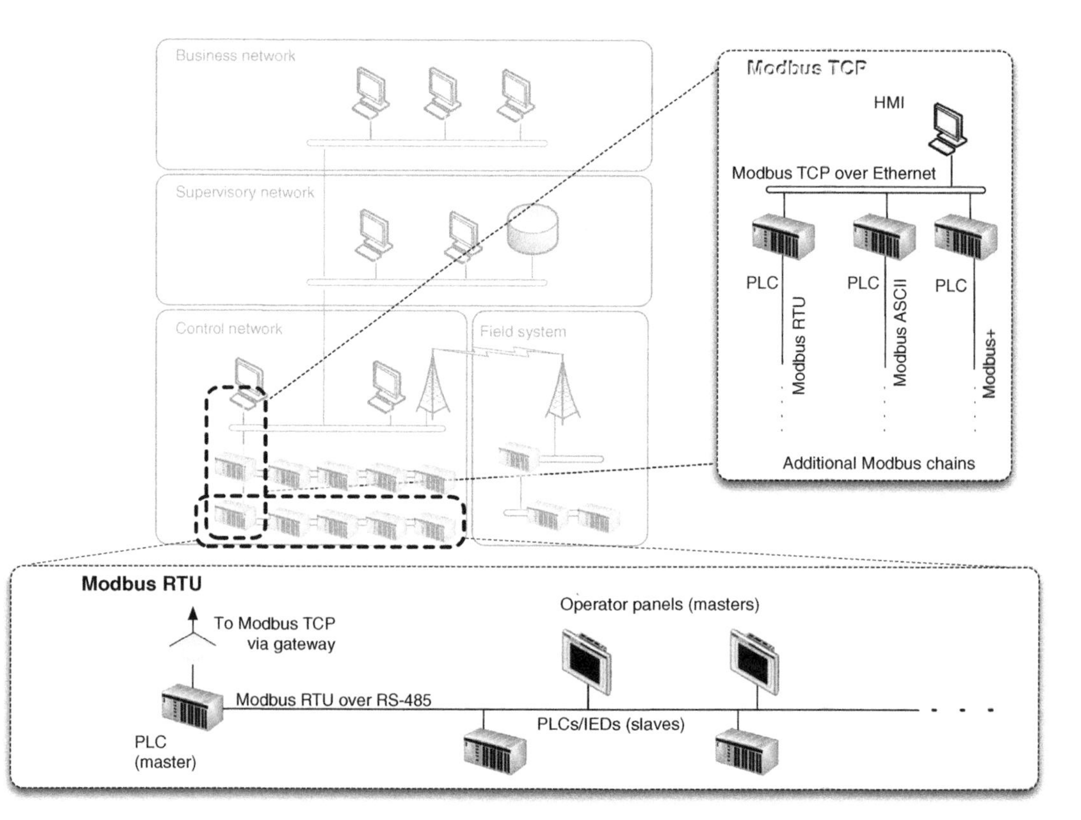

FIGURE 6.9 Typical Modbus use within the industrial network architecture.

polling data as a master from several slave PLCs and intelligent electronic devices (IEDs), while supporting requests for data as a slave to other master devices like ICS servers and HMIs.

Security concerns

Modbus represents several security concerns:

- Lack of authentication. Modbus sessions only require the use of a valid Modbus address, function code, and associated data. The data must contain the values of legitimate registers or coils contained in the slave device, or the message will be rejected. This requires additional information of the target in order to provide a valid message; however, this can be obtained from either analysis of network traffic or the configuration of the device. Modbus supports additional function codes that can be used without specific knowledge of the target (e.g., function code 43). There is no verification that the message originated from a legitimate device allowing for simple man-in-the-middle (MitM) and replay style attacks.
- Lack of encryption. Commands and addresses are transmitted in clear text and can therefore be easily captured and spoofed or replayed due to the lack of encryption. Network packet capturing of communications to/from a Modbus device can also disclose significant information pertaining to the configuration and use of the device.
- Lack of message checksum (Modbus/TCP only). A command can easily be spoofed by building up the Modbus/TCP ADU with the desired parameters, as the checksum is generated at the transmission layer, not the application layer.
- Lack of broadcast suppression (serial Modbus variants only used in a multidrop topology). All serially connected devices will receive all messages, meaning a broadcast of unknown addresses can be used for effective denial of service (DoS) to a chain of serially connected devices.

Security recommendations

Modbus, like many industrial control protocols, should only be used to communicate between sets of known devices, using expected function codes. In this way, it can be easily monitored by establishing clear network zones and by baselining acceptable behavior. This baseline behavior can then be used to establish access controls on the conduit into the zone via appliances that provide protocol inspecting and filtering capabilities (e.g., industrial firewall with deep-packet inspection capabilities). It is also possible at the network level to create fingerprints of normal behavior patterns that facilitate network **whitelists** that can be implemented on in-line and out-of-band devices. For more information about creating whitelists, this topic is discussed in detail in Chapter 11, "Anomaly and Threat Detection."

Some specific examples of Modbus messages that should be of concern include the following:

- Modbus TCP packets that are of wrong size or length.
- Function codes that force slave devices into a "listen only" mode.
- Function codes that restart communications.
- Function codes that clear, erase, or reset diagnostic information such as counters and diagnostic registers.
- Function codes that request information about Modbus servers, PLC configurations, or other device-specific, need-to-know information.
- Traffic on port 502/tcp that is not Modbus or is using Modbus over malformed protocol(s).
- Any message within an Exception PDU (i.e., any Exception Code).
- Modbus traffic from a server to many slaves (i.e., a potential DoS).
- Modbus requests for lists of defined points and their values (i.e., a configuration scan).
- Commands to list all available function codes (i.e., a function scan).

ICS-aware intrusion protection systems can be configured to monitor for these activities using Modbus signatures such as those developed and distributed by Digital Bond under the QuickDraw project. In more critical areas, an application-aware firewall, industrial protocol filter, or application data monitor may be required to validate Modbus sessions and ensure that Modbus has not been "hijacked" and used for covert communication, command, and control (i.e., the underlying TCP/IP session on port 502/tcp has not been altered to hide additional communications channels within otherwise normal-looking Modbus traffic). This device can also be used to limit function codes communicated into the zone to only those allowed for normal operation. This is discussed in detail in Chapter 9, "Establishing Zones and Conduits." Figure 6.10 illustrates configuration of an application-layer firewall on the conduit into an EIP zone separating four HMIs, one Engineering Workstation (EWS) and two PLCs.

Caution

Intrusion prevention systems are able to actively block suspect traffic by dropping packets or resetting TCP connections. However, intrusion prevention systems deployed on industrial networks should be only be configured to block traffic after careful consideration and tuning. Unless you are confident that a given signature will not inadvertently block a legitimate control command, the signature should be set to alert, rather than block (i.e., operate in "detection" mode rather than active "prevention" mode).

Distributed network protocol (DNP3)

DNP3 began as a serial protocol much like Modbus designed for use between "master stations" or "control stations" and slave devices called "outstations. It is also commonly

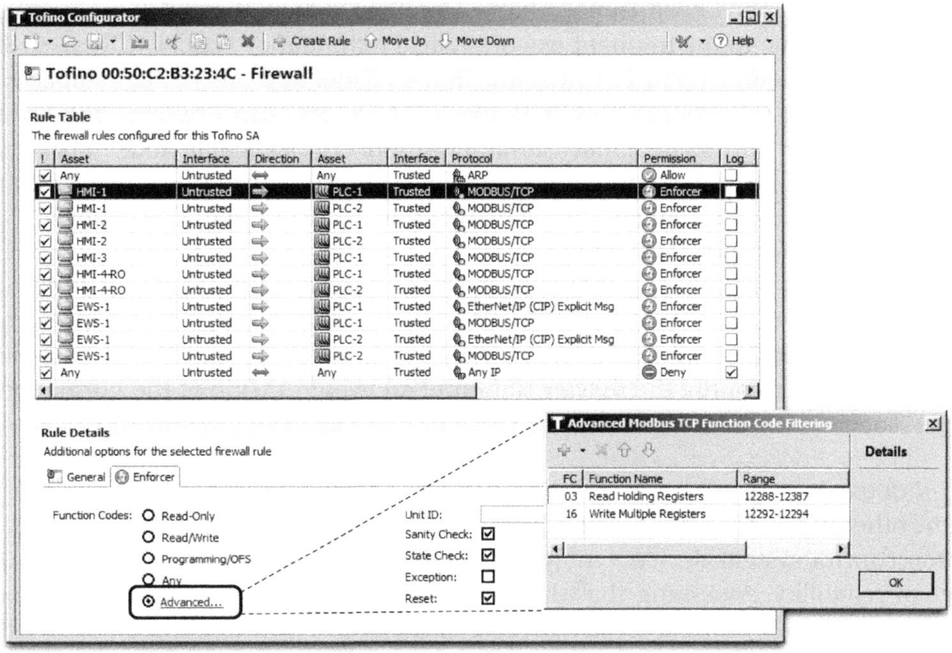

FIGURE 6.10 Application-layer firewall—Modbus/TCP zone protection. *Image courtesy of Tofino Security—A Belden Brand.*

used to connect RTUs configured as "master stations" to IED "outstations" in electric substations. The Inter-Control Center Communication Protocol (ICCP) discussed later in this chapter is commonly used for communication between master stations. DNP3 was initially introduced in 1990 by Westronic (now GE-Harris Canada) and was based on early drafts of the IEC 60870-5 standard. The primary motivation for this protocol was to provide reliable communications in environments common within the electric utility industry that include high level of electromagnetic interference (EFI) and poor transmissions media (at that time based on analog telephone lines). DNP3 was extended to work over IP via encapsulated in TCP or UDP packets in 1998 and is now widely used in not only electric utility, but also oil and gas,[7] water, and wastewater industries. One of the leading reasons for some industry migration from Modbus to DNP3 includes features that apply to these other industries including report by exception, data quality indicators, time-stamped data including sequence-of-events, and a two-pass "select before operate" procedure on outputs.[8] Other markets, including Europe, have adopted the IEC 60870-5 versus of the protocol as it was ratified. Though DNP3 was based on IEC 60870-5, differences do exist between the two.

One distinction of DNP3 is that it is very reliable, while remaining efficient and well suited for real-time data transfer. It also utilizes several standardized data formats and supports time-stamped (and time-synchronized) data, making real-time transmissions more efficient and thus even more reliable. Another reason that DNP3 is considered highly reliable is due to the frequent use of cyclical redundancy checks (CRCs)—a single DNP3 frame can include up to 17 CRCs: one in the header and one per data block within the payload (see the section "How it Works"). There are also optional link-layer acknowledgments for further reliability assurance, and—of particular note—variations of DNP3 that support link-layer authentication as well. Because all of this is done within the link-layer frame, it means that additional network-layer checks may also apply if DNP3 is encapsulated for transport over Ethernet.

Unlike Modbus and ICCP, DNP3 is both bidirectional (supporting communications from both Master to Slave and from Slave to Master) and it supports exception-based reporting. It is therefore possible for a DNP3 outstation to initiate an unsolicited response, in order to notify the master station of an event outside of the normal polling interval (such as an alarm condition).

What it does

Like the other industrial protocols, DNP3 is primarily used to send and receive messages between control system devices—only in the case of DNP3, it also does it with a high degree of reliability. Assuming that the various CRCs are all valid, the data payload is then processed. The payload is very flexible and can be used to simply transfer informational readings. It can also be used to send control functions, or even direct binary or analog data for direct interaction with devices such as RTUs and IEDs.

Both the link-layer frame (or LPDU) header and the data payload contain CRCs, and the data payload actually contains a pair of CRC octets for every 16 data octets. This provides a high degree of assurance that any communication errors will be detected. DNP3 will retransmit the faulty frames if any errors are detected. There are also physical layer integrity issues in addition to frame integrity. However, it still remains possible that a correctly formed and transmitted frame will not arrive at its destination. DNP3 uses an additional link layer confirmation to overcome this risk. When link layer confirmation is enabled, the DNP3 transmitter (source) of the frame requests that the receiver (destination) confirms the successful receipt of the frame. If a requested confirmation is not received, the link layer will retransmit the frame. This confirmation is optional because although it increases reliability, it adds overhead that directly impacts the efficiency of the protocol. In real-time environments, this added overhead might not be appropriate.[9]

Once a successful and (if requested) confirmed frame arrives, the frame is processed. Each frame consists of a multipart header and a data payload. The header is significant as it contains a well-defined function code, which can tell the recipient whether it should confirm, read, write, select a specific point, operate a point (initiate a change to a point), directly operate a point (both selecting and changing a point in one command), or directly operate a point without acknowledgment.[10]

These functions are especially powerful when considering that the data payload of the DNP3 frame supports analog data, binary data, files, counters, and other types of data objects. At a high level, DNP3 supports two kinds of data, referred to as class 0 or static data (data that represent a static value) and event data (data that represent a change such as an alarm condition). Event data are rated by priority from class 1 (highest) to class 3 (lowest). The differentiation of static and event data, as well as the classification of event data, allows DNP3 to operate more efficiently by allowing higher-priority information to be polled more frequently, for example, or to enable or disable unsolicited responses by data type. The data itself can be binary, analog input or output, or a specific control output.[11]

How it works

DNP3 provides a method to identify the remote device's parameters and then use message buffers corresponding to event data classes 1 through 3 in order to identify incoming messages and compare them to known point data. In this way, the master station is only required to retrieve new information resulting from a point change or change event on the outstation.

Initial communications are typically a class 0 request from the master station to an outstation, used to read all point values into the master station's database. Subsequent communications will typically either be direct poll requests for a specific data class from the master station; unsolicited responses for a specific data class from an outstation; control or configuration requests from the master station to an outstation; or subsequent periodic class 0 polls. When a change occurs on an outstation, a flag is set to the appropriate data class. The master station is then able to poll only those outstations where there is new information to be reported.

This is a major departure from constant data polling that directly results in improved responsiveness and more efficient data exchange. The departure from a real-time polling mechanism does require time synchronization because the time between a change event and a successful poll/request sequence is variable. This means that all responses are time-stamped so that the events between polls can be reconstructed in the correct order.

Communication is initiated by the master station to the outstation, or in the case of unsolicited responses (alarms) from the outstation to the master station, as shown in Figure 6.11. Because DNP3 operates bidirectionally and supports unsolicited responses, as shown in Figure 6.12, each frame requires both a source address and a destination address so that the recipient device knows which messages to process, and which device to return responses to. The addition of a source address does add some overhead. Remember that with purely master/slave protocols, there is no need for a source address as the originating device is always the master. This overhead provides a return benefit of dramatically increased scalability and functionality. As many as 65,520 individual addresses are available within DNP3, and any one of them can initiate communications. An address equals one device (every DNP3 device requires a unique address), although there are reserved DNP3 addresses, including one for broadcast messages (which will be received and processed by all connected DNP3 devices).[12]

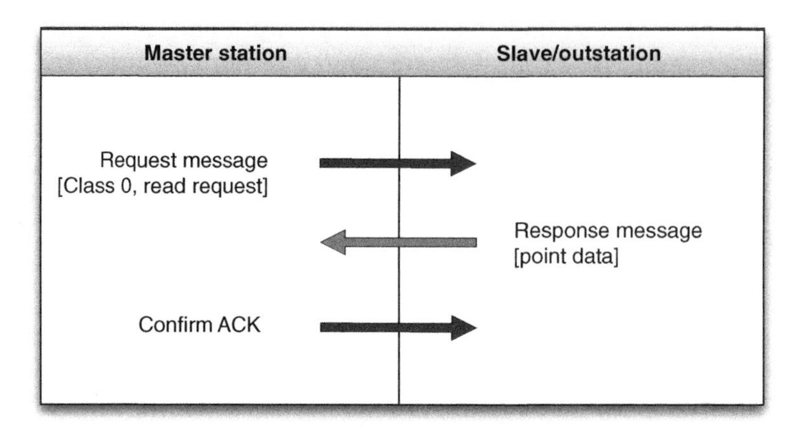

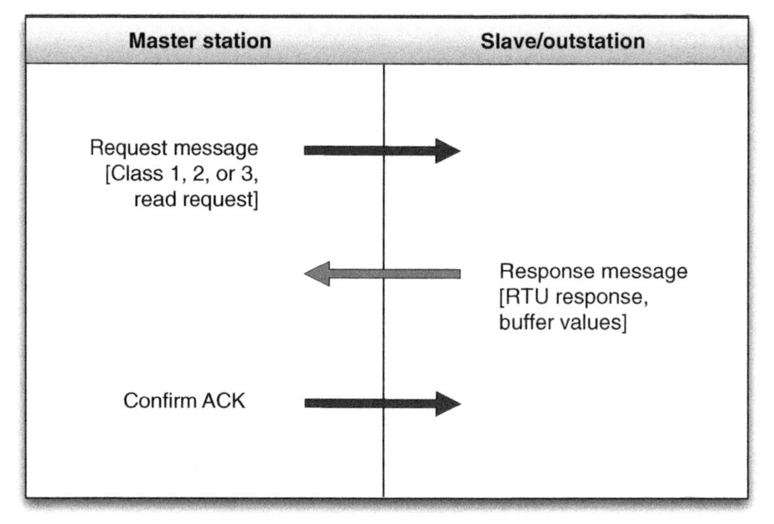

FIGURE 6.11 DNP3 protocol operation.

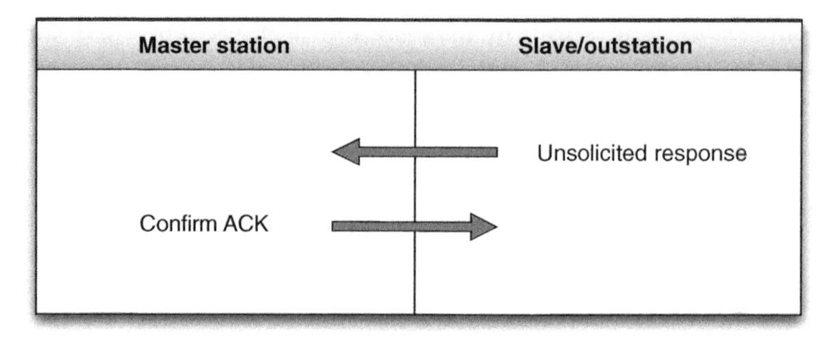

FIGURE 6.12 DNP3 protocol operation: unsolicited responses allow remote alarm generation.

Secure DNP3

Secure DNP3 is a DNP3 variant that adds authentication to the response/request process, as shown in Figure 6.13. Authentication is issued as a challenge by the receiving device. A challenge condition occurs upon session initiation (when a master station initiates a DNP3 session with an outstation), after a preset period of time (the default is 20 min), or upon a "critical" request such as writes, selects, operates, direct operates, starts, stops, and restarts. It is possible to know which requests are critical because the data types and functions of DNP3 are well defined.[13]

Authentication occurs using a unique session key that is hashed together with message data from the sender and from the challenger. The result is an authentication method that verifies authority (checksum against the secret key), integrity (checksum against the sending payload), and pairing (checksum against the challenge message) at the same time. In this way, it is very difficult to perform data manipulation or code injection, or to spoof or otherwise hijack the protocol.[14]

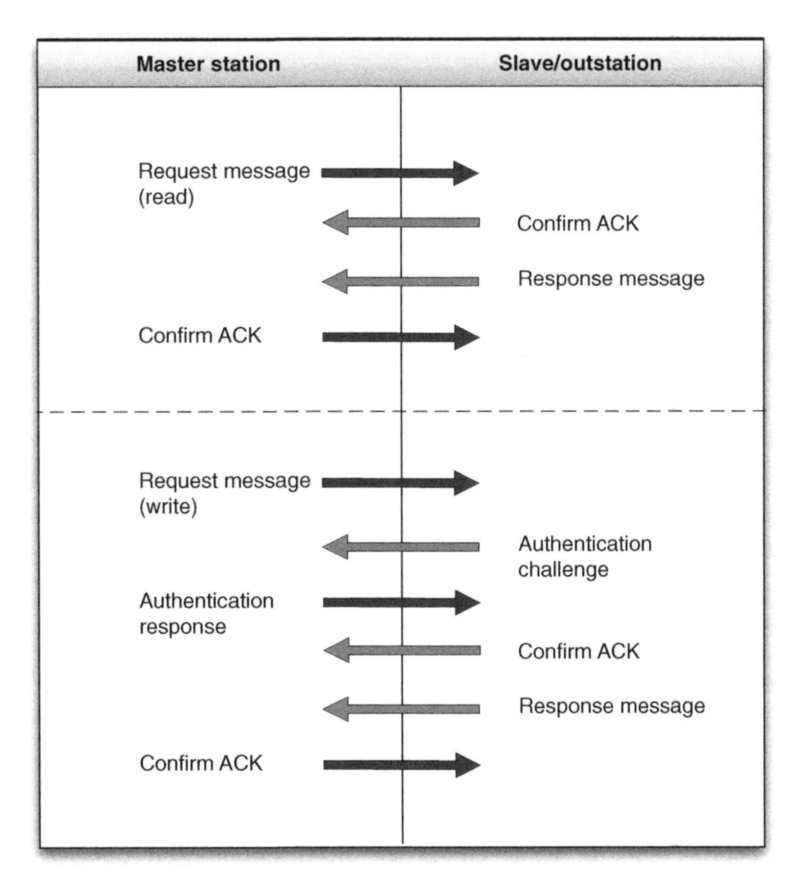

FIGURE 6.13 Message confirmation and secure DNP3 authentication operation.

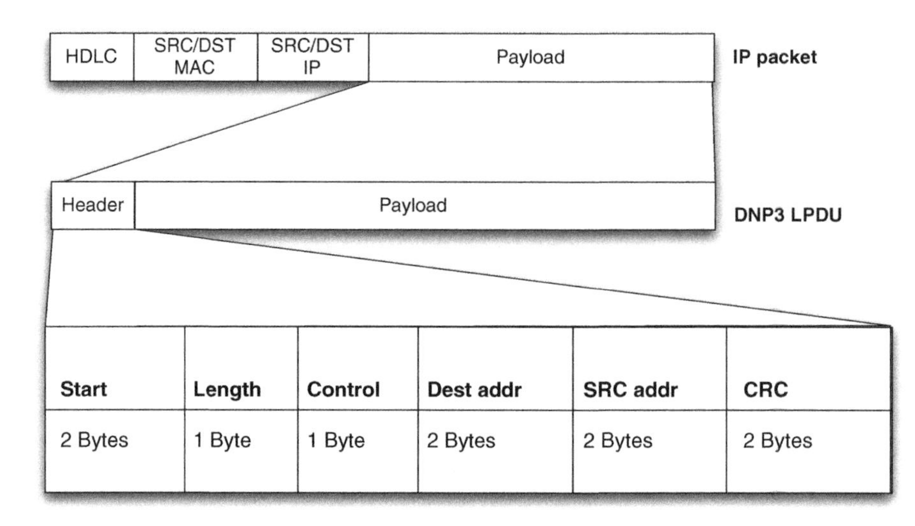

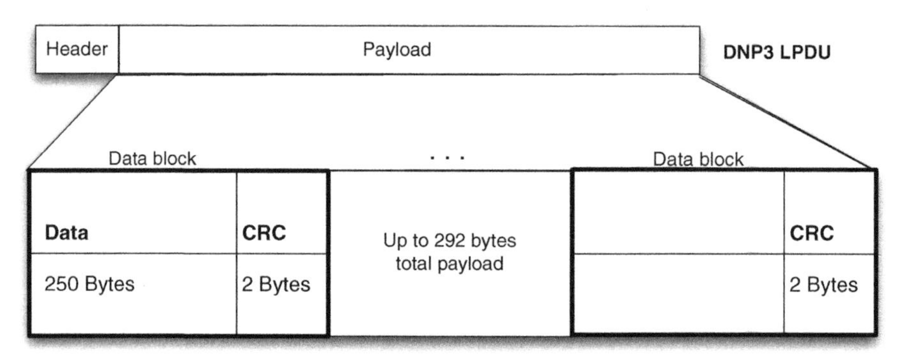

FIGURE 6.14 DNP3 protocol framing.

The DNP3 layer 2 frame provides the source, destination, control, and payload, and can operate over a variety of application layers including TCP and UDP transports over IP (defaults include 19,999/tcp when using Transport Layer Security (TLS) for confidentiality and 20,000/tcp or 20,000/udp when using application layer only secure authentication). The function codes are resident within the CNTRL bytes in the DNP3 frame header, as shown in Figure 6.14.

Where it is used

DNP3 is primarily used between a master control station and an RTU in a remote station as shown in Figure 6.15, Transmissions medium can include wireless, radio, and dial-up. DNP3 is also widely used between to interconnect RTUs and IEDs. It can be applied in many applications like the Modbus protocols through a typical ICS architecture. Unlike Modbus, however, DNP3 is well suited for hierarchical and aggregated point-to-

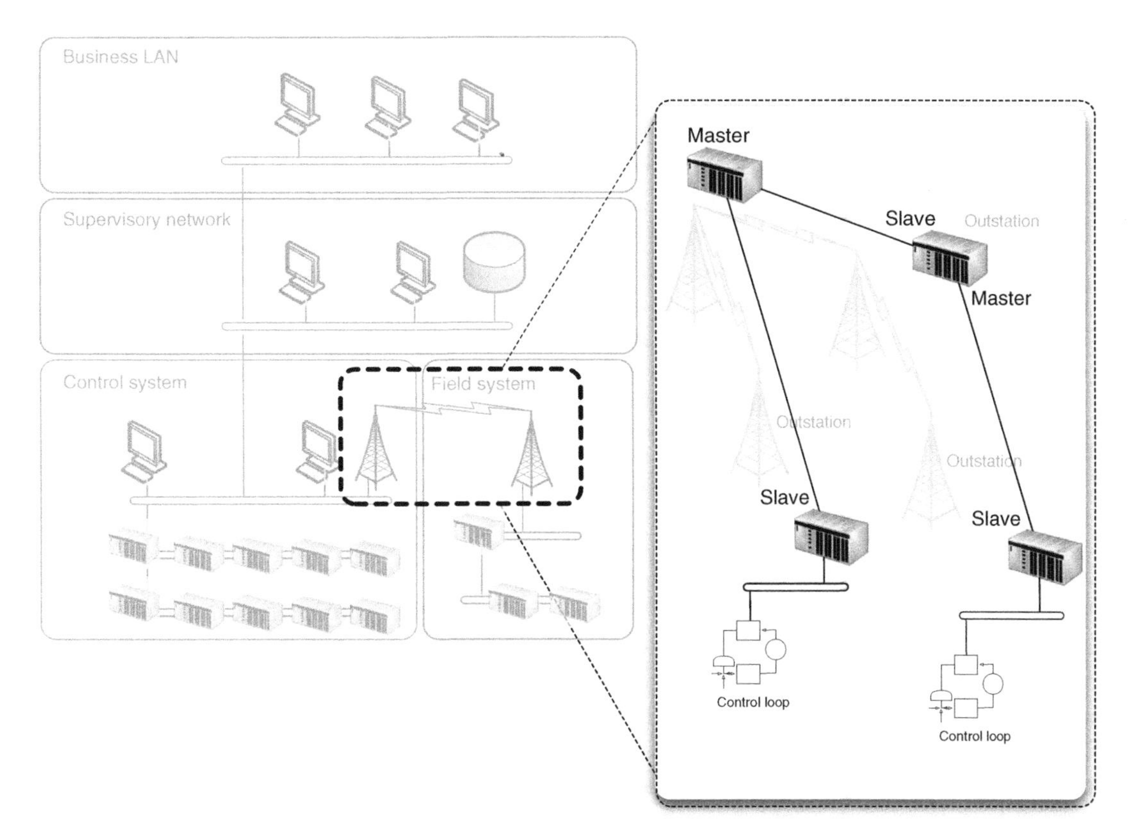

FIGURE 6.15 Typical DNP3 use within the industrial network architecture.

multipoint topologies in addition to the linear point-to-point and serial point-to-multipoint topologies that are supported by Modbus.[15]

Security concerns

While much attention is given to the integrity of the data frame, there is no authentication or encryption inherent within DNP3 (although there is within Secure DNP3). It then becomes relatively easy to manipulate a DNP3 session because of the well-defined nature of DNP3 function codes and data types in much the same way as it was the Modbus protocol.

DNP3 does include security measures; however, this added complexity of the protocol increases the chances of vulnerabilities. As of this writing, there are several known vulnerabilities with DNP3 that have been reported by the Industrial Control System Cyber Emergency Response Team (ICS-CERT). Proper system hardening, regular security assessments, and patching of DNP3 interconnections (both master stations and outstations) are recommended because there are known exploits in the wild and DNP3 is a heavily deployed protocol.

Some examples of realistic hacks against DNP3 include the use of MitM attacks to capture addresses, which can then be used to manipulate other system components. Examples of such manipulation include the following:

- Turning off unsolicited reporting to suppress alarms.[16]
- Spoofing unsolicited responses to the master station to falsify events and trick an operator into taking inappropriate actions.
- Performing a DoS attack through the injection of broadcasts, creating storm behavior within the full extent of the DNP3 system.
- Manipulating the time synchronization data, resulting in synchronization loss and subsequent communication errors.
- Manipulating or eliminating confirmation messages forcing a state of continuous retransmission.
- Issuing unauthorized stops, restarts, or other functions that could disrupt operations.

Security recommendations

Because a secure implementation of DNP3 is available, the primary recommendation is to implement only Secure DNP3. This can pose problems with legacy installations due to backwards compatibility, as Version 5 of the standard (adopted as IEEE-1815-2012) is not backwards compatible, and Version 2 (adopted as IEEE-1815-2010) is now deprecated and should be upgraded. It may not always be possible to implement Secure DNP3 due to varying vendor support and other factors. Secure use of the transport layer protocol is advised in these cases, such as the use of TLS. In other words, treat your encapsulated DNP3 traffic as highly sensitive information and use every TCP/IP security best practice to protect it.

DNP3 master stations and outstations should always be isolated into a unique zone consisting only of authorized devices (multiple zones can be defined for devices communicating to multiple clients, or for hierarchical Master/Slave pairs), and the zone(s) should be thoroughly secured using standard defense-in-depth practices, including an industrial firewall and/or intrusion protection system that enforces strict control over the type, source, and destination of traffic over the DNP3 link across conduits between zones. Preference should be given to security practices that are capable of deep-packet inspection of DNP3 traffic. Many of the recommendations described for Modbus are equally applicable for DNP3, including the creation of network baselines and deployment of network whitelists.

Many threats can be detected through monitoring of DNP3 sessions, and looking for specific function codes and behaviors, including the following:

- Use of any non-DNP3 communication on a DNP3 Port (19,999/tcp, 20,000/tcp, 20,000/udp).
- Use of configuration function code 23 (Disable Unsolicited Responses).

- Use of control function codes 4, 5, or 6 (Operate, Direct Operate, and Direct Operate without Acknowledgment).
- Use of application control function 18 (Stop Application).
- Multiple, unsolicited responses over time (Response Storm).
- Any unauthorized attempt to perform an action requiring authentication.
- Any authentication failures.
- Any DNP3 communication sourced from or destined to a device that is not explicitly identified as a DNP3 master station or outstation device.

As with other industrial protocols, ICS-aware intrusion protection systems can be configured to monitor for these activities using DNP3 signatures such as those developed and distributed by Digital Bond under the QuickDraw supervisory control and data acquisition (SCADA) IDS project. An application-aware firewall or application data monitor may be required to validate DNP3 sessions.

Caution

Intrusion prevention systems are able to actively block suspect traffic by dropping packets or resetting TCP connections. However, intrusion prevention systems deployed on industrial networks should be only be configured to block traffic after careful consideration and tuning. Unless you are confident that a given signature will not inadvertently block a legitimate control command, the signature should be set to alert, rather than block (i.e., operate in "detection" mode rather than active "prevention" mode).

Process fieldbus (PROFIBUS)

PROFIBUS (PROcess FIeldBUS) is a fieldbus protocol that was originally developed in the late 1980s in Germany by a group of 21 companies and institutions known as the Central Association for the Electrical Industry (ZVEI). ZVEI published their first protocol specification known as PROFIBUS FMS (Fieldbus Message Specification) designed primarily to allow PLCs to communicate with host computers. This protocol was found to be too complex to implement in process control applications, so in 1993 the PROFIBUS DP (Decentralized Periphery) specification was released providing easier configuration and faster messaging. In 1989, the PROFIBUS User Organization (PROFIBUS Nutzerorganization or PNO) was established to maintain the specifications and ensure device compliance and certification. A larger user community was established in 1995 called PROFIBUS International to continue the advancement of PROFIBUS on a global level.

Several specialized variants of PROFIBUS exist, including PROFIBUS PA (for instrumentation used for process automation (PA)), PROFIsafe (for safety applications), and PROFIdrive (for high-speed drive applications). The most widely deployed variant is PROFIBUS DP, which itself has three variants: PROFIBUS DP-V0, DP-V1, and DP-V2, each of which represents a minor evolution of capabilities within the protocol. There are also three profiles for PROFIBUS communication: asynchronous, synchronous, and via

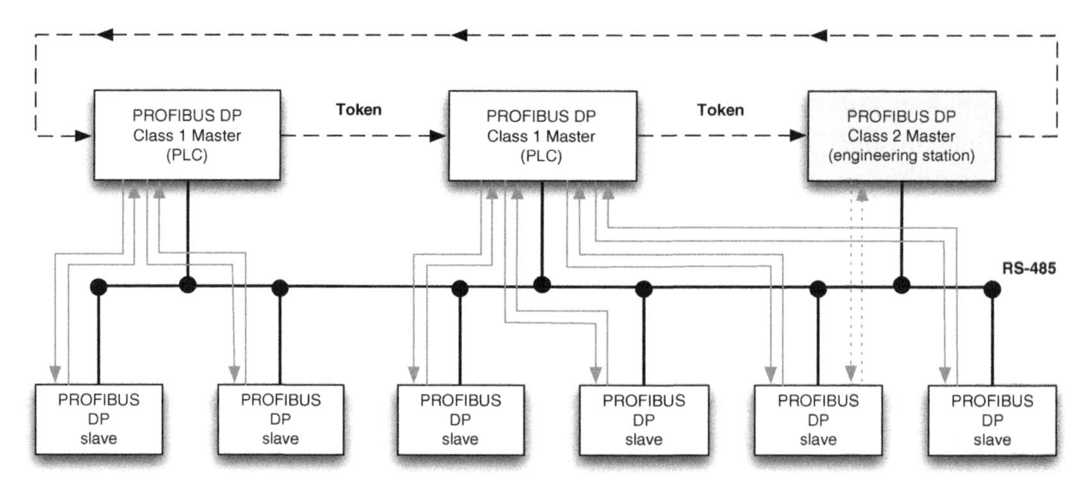

FIGURE 6.16 PROFIBUS DP communications.

Ethernet using ethertype 0x8892. PROFIBUS over Ethernet is also called PROFINET[17] and will be discussed separately as part of a category of protocols referred to as "Industrial Ethernet."

PROFIBUS is a master-slave protocol that supports multiple master nodes through the use of token sharing: when a master has control of the token, it can communicate with its slaves (each slave is configured to respond to a single master). Figure 6.16 illustrates how this token-based, master-slave topology operates. In PROFIBUS DP-V2, slaves can initiate communications to the master or to other slaves under certain conditions. A master PROFIBUS node is typically a PLC or RTU, and a slave is sensor, motor, or some other control system device.

PROFIBUS DP supports several different physical layer deployments with RS-485 as the most common. The existing RS-485 specification was extended to allow PROFIBUS to operate at speeds up to 12Mbps using two wires. The PA specification was developed to address the unique needs of field instrumentation in a manner similar to FOUNDATION Fieldbus. These installations must support wiring and communication with devices that are commonly installed in hazardous areas where explosive vapors and dusts are common. A concept known as "intrinsic safety" is used to limit the amount of available power on these communication lines to levels below that necessary to ignite the dust or vapor. The Manchester-encoded, bus-powered, intrinsically safe physical layer is used in these cases to address this requirement providing both limited levels of device power and communication on a single pair of wires.

Security concerns

PROFIBUS lacks authentication inherent to many of its functions, allowing a spoofed node to impersonate a master node, which in turn provides control over all configured slaves. A compromised master node or a spoofed master node could also be used to capture the token, inject false tokens, or otherwise disrupt the protocol functions, causing a DoS. A rogue master node could alter clock synchronization to slave devices, snoop query responses (across all masters), or even inject code into a slave node. It is important to remember that PROFIBUS DP utilizes a serial connection between the master and slave devices, so the security concerns mentioned require physical access to connect to the DP network. This means that a DP network is not generally susceptible to industrial network-based attacks. However, the master device is typically connected to an Ethernet network and is therefore no less susceptible to attack from authorized network access than any other Ethernet-connected device. PROFINET is a real-time Ethernet protocol, and as such it is susceptible to any of the vulnerabilities of Ethernet. When used over the IP, it is also susceptible to any vulnerabilities of IP.

Note

Stuxnet (see Chapter 3, "Industrial Cyber Security, History and Trends") is an example of PROFIBUS exploitation. Stuxnet compromised PLCs (PROFINET devices acting as PROFIBUS DP master nodes) via an initial network attack on an EWS or HMI. It then monitored the PROFIBUS DP network and looked for specific behaviors associated with frequency controllers (PROFIBUS DP slave nodes). Once the sought-after conditions were detected, Stuxnet then issued commands to the relevant slave nodes to sabotage the mechanical equipment (centrifuges used to enrich Uranium) by altering their operating parameters (speed of the centrifuges).

Security recommendations

PROFIBUS DP is a naturally segmented serial network utilizing a topology that is generally contained within a small geographical area such as a section of a plant or manufacturing process. The network and connected devices are very susceptible to attack if unauthorized physical access is obtained. For the purposes of this book, physical security must always be provided, since the threat events that can be performed via local access are relatively easy and can provide significant disruption to the operation of the ICS. This is outside the scope of this book.

Industrial ethernet protocols

Industrial Ethernet is a term used to reference the adaptation of the IEEE 802.3 Ethernet standard to real-time industrial automation applications. One of the primary objectives of these extensions is the move toward more "synchronous" mechanisms of communication in order to prevent data collisions and minimize jitter inherent with "asynchronous" communications like standard Ethernet. This will allow the technology to be deployed in critical time-dependent applications like safety and industrial motion control. This concept may seem abstract in a time when 1Gbps switched networks are readily available;

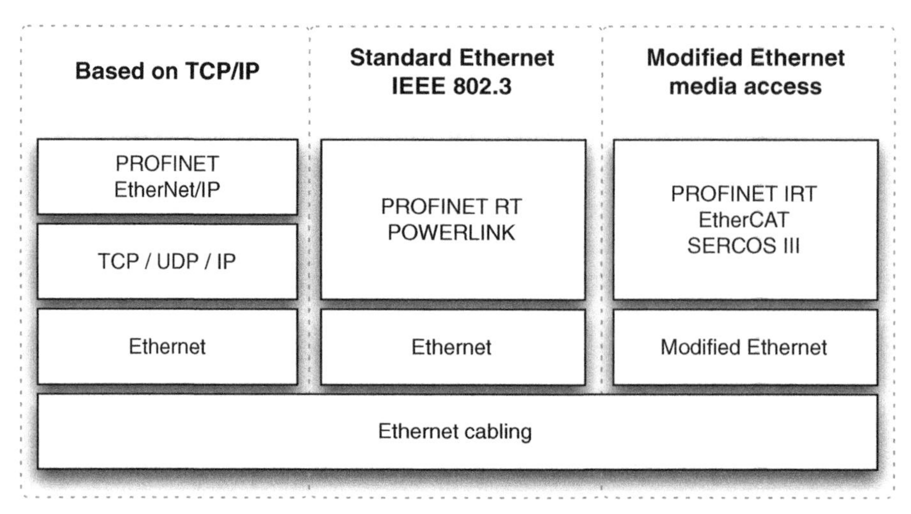

FIGURE 6.17 Methods for real-time Ethernet implementation.

however, as one moves into the industrial sector, the applications must be applicable to not only "lightweight" and simple devices that may not have the capacity for these modern IT networks, but also the deployment of network topologies on the factory floor that be more suited for bused or trunked style topologies (e.g., automobile networks).

Industrial Ethernet also provides physical enhancements to "harden" the office-grade nature of standard Ethernet technologies with ruggedized wiring, connectors, and hardware designed to meet the environment of industrial applications. Conditions that are addressed with Industrial Ethernet include electrical noise and interference, vibration, extended temperatures and humidity (high and low), power requirements, and extensions to support real-time performance (low latency, low jitter, minimal packet loss).[18]

There are some 30 different varieties of Industrial Ethernet[19]; however, for the purposes of this book, attention will be given to five as they are not only widely accepted and deployed in industry global (e.g., market leaders), but they introduce new concepts and concerns regarding industrial network security. These include Ethernet/IP, PROFINET, EtherCAT, Ethernet POWERLINK, and SERCOS III. Studies conducted by Institute of Management Services and ARC Advisory Group show that approximately 75% of all Ethernet installation in industrial environments use EtherNet/IP, PROFINET, or Modbus/TCP (already discussed), with the next two leading technologies based on POWERLINK and EtherCAT.[20] Figure 6.17 provides an illustration of how these various technologies compare.

Ethernet industrial protocol (EtherNet/IP)

It is important to understand the Common Industrial Protocol (CIP) in order to appreciate its versatility and application to the Ethernet/IP implementation. CIP, originally known as "Control and Information Protocol," is a publicly available protocol managed through the Open DeviceNet Vendors Association (ODVA). CIP is an application layer protocol that provides a consistent set of messages and services that can be

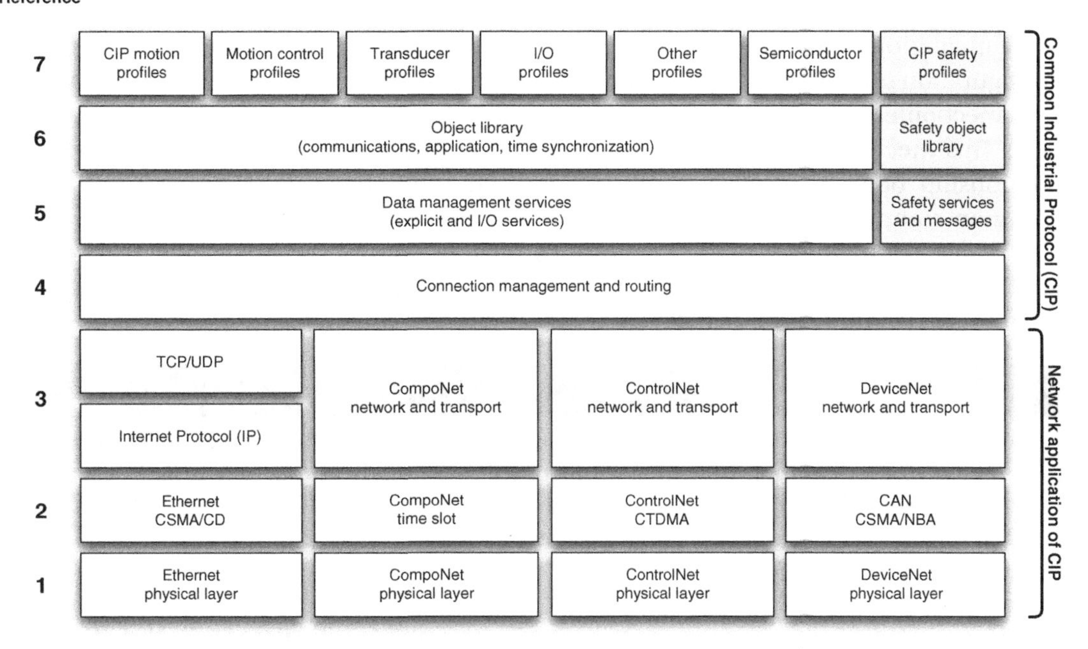

FIGURE 6.18 Overview of common industrial protocol.[22,52]

implemented in a variety of ways using different network and link layer techniques, all supporting interoperability. These variations include EtherNet/IP (CIP on Ethernet), DeviceNet (CIP on CAN), CompoNet, and ControlNet (CIP on Concurrent Time Domain Multiple Access [CTDMA]) with extensions that include safety (CIP Safety), motion control (CIP Motion), and synchronization (CIP sync). Figure 6.18 illustrates the deployment model for CIP against the OSI layers.[21]

Note

The Controller Area Network (CAN) is a bus developed in 1985 by Bosch and adopted as international standard ISO 11898 in 1993 originally used for vehicle networks. It is a low-cost network utilizing a trunk-drop technology while suppling by power and signal to interconnect simple devices.

Note

CTDMA provides the enhancements over traditional Carrier Sense Multiple Access/Collision Domain found in Ethernet to support deterministic, high-speed communication of time-critical I/O and control data. The design allows for all addresses to have access to the network through the implementation of a time slice algorithm that provides both "scheduled' and "unscheduled" data transfers.

EtherNet/IP (EIP) or CIP on Ethernet uses standard Ethernet frames (ethertype 0x80E1) in conjunction with the CIP suite to communicate with nodes. As with all CIP implementations, EIP supports integration of I/O, control, data collection, and device configuration on a single network. For real-time I/O and control-related data, EIP utilizes a connectionless multicast UDP transport called "implicit messaging" using port 2222/udp. This mechanism optimizes performance by establishing a "producer-consumer" relationship between devices sending data and those devices requiring the data—a common communications model within ICS architectures. A unicast TCP transport is also available to transmit larger quantities of data commonly associated with device configuration, diagnostics, and event information using an "explicit messaging" service commonly found on port 44,818/tcp.

Note

The "IP" in Ethernet/IP derives from "Industrial Protocol" and not "Internet Protocol," because of the use of the Common Industrial Protocol (CIP). Similarly, the acronym "CIP" meaning "Common Industrial Protocol" should not be confused with "Critical Infrastructure Protection" of NERC CIP.

CIP uses object models to define the various qualities of a device. Each CIP object possesses attributes (data), services (commands), connections, and behaviors (relationships between attribute values and services). There are three types of objects:

- Required Objects: define attributes such as device identifiers such as the manufacturer, serial number, date of manufacture, etc. (Identity Object), routing identifiers for object-to-object messaging (Message Router Object), and physical connection data (Network Object)
- Application Objects: define input and output profiles for devices
- Vendor-specific Objects: enable vendors to add proprietary objects to a device

Objects (other than vendor-specific objects) are standardized by device type and function, to facilitate interoperability. If one brand of pump is exchanged for another brand, for example, the Application Objects will remain compatible, eliminating the need to build custom drivers. The wide adoption and standardization of CIP has resulted in an extensive library of device models, which can facilitate interoperability but can also aid in control network scanning and enumeration (see Chapter 8, "Risk and Vulnerability Assessments").

While the Required Objects provide a common and complete set of identifying values, the Application Objects contain a common and complete suite of services for control, configuration, and data collection that includes both implicit (control) and explicit (information) messaging.[23]

Security concerns

EIP is a real-time Ethernet protocol, and as such it is susceptible to any of the vulnerabilities of Ethernet. EIP implicit messaging over UDP is transaction-less and so there is no inherent network-layer mechanism for reliability, ordering, or data integrity checks. CIP also introduces some specific security concerns, due to its well-defined object model.

The following concerns are specific to Ethernet/IP:

- The CIP does not define any explicit or implicit mechanisms for security.
- The use of common Required Objects for device identification can facilitate device identification and enumeration, facilitating a targeted attack.
- The use of common Application Objects for device information exchange and control can enable broader industrial attacks, able to manipulate a broad range of industrial devices.
- Ethernet/IP's use of UDP and multicast traffic—both of which lack transmission control—for real-time transmissions facilitate the injection of spoofed traffic or (in the case of multicast traffic) the manipulation of the transmission path using injected Internet Group Management Protocol controls.

Security recommendations

EIP is a real-time Ethernet protocol using TCP and UDP transports making it necessary to provide Ethernet and IP-based security at the perimeter of any EIP network. Consideration should be given to placing EIP devices in dedicated zones that include either an application layer appliance capable of performing inspection in EIP packets and only allowing required functions within the zone. A stateful, packet-filtering firewalls that can be used to limit unnecessary inbound traffic (such as device configuration) to the zone. Figure 6.19 illustrates configuration of an application-layer firewall on the conduit into an EIP zone separating four HMIs, one EWS and two PLCs.

It is also recommended that passive network monitoring be used to ensure the integrity of the EIP network, ensuring that the EIP protocol is only being used by explicitly identify devices, and that no EIP traffic is originating from an unauthorized, outside source. This can be accomplished using an ICS-aware intrusion prevention system or other network monitoring devices capable of detecting and interpreting the EIP. Additional guidance can be obtained through ODVA.[24]

PROFINET

PROFINET is an open standard Industrial Ethernet developed by the PROFIBUS User Organization (PNO) and Siemens and is included as part of the IEC 61158 and IEC 61784 international standards for fieldbus communications. PROFINET was designed for scalability, and can be deployed at varying degrees of determinism and network performance. The first version of PROFINET utilized standard Ethernet and TCP/IP packets without modification for nonreal time automation applications and generation integration. The software-based real-time technology included in version 2 added support for time-critical communications with cycle times of 5–10 ms incorporating an optimized protocol stack bypassing OSI layers 3 and 4, limiting communications to a single

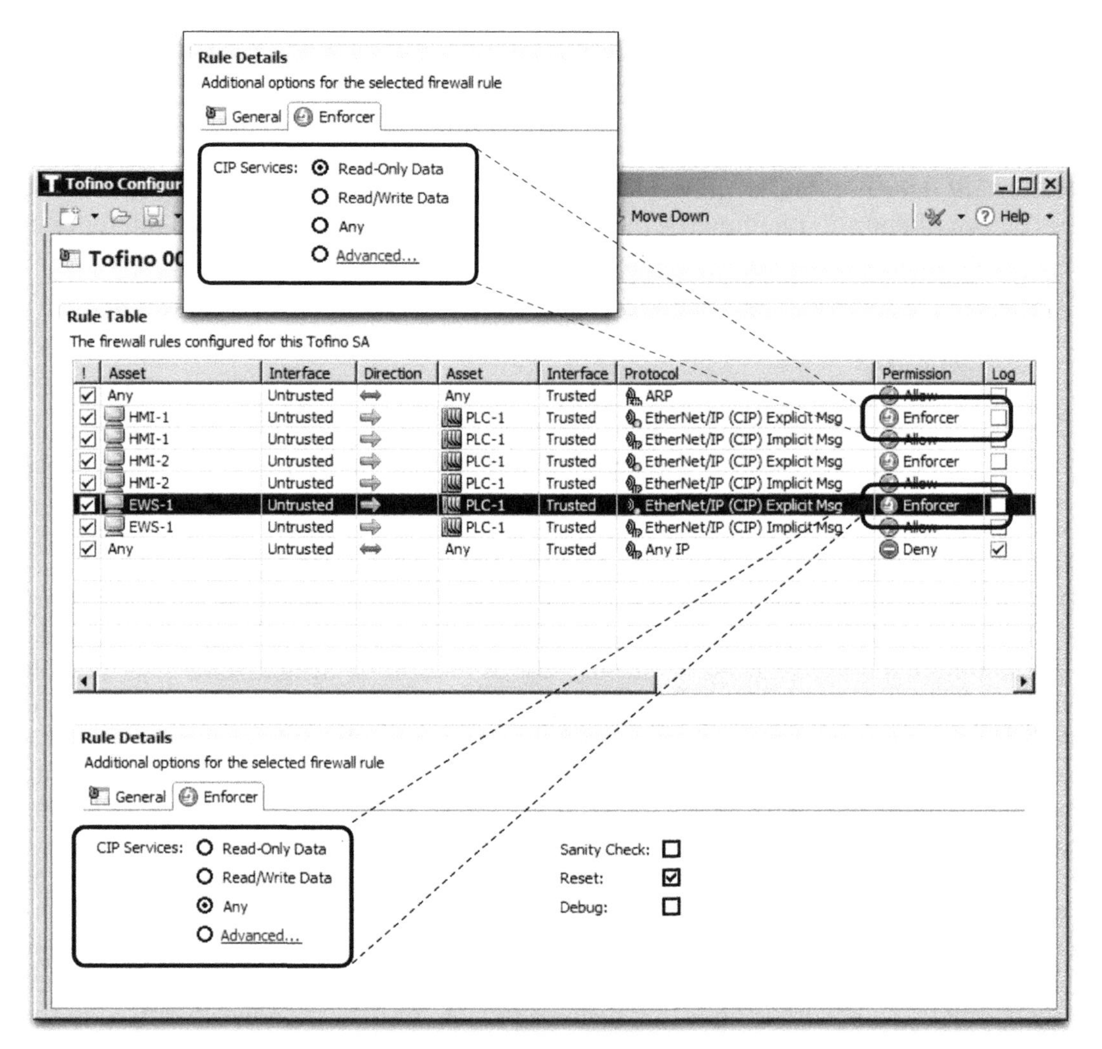

FIGURE 6.19 Application-layer firewall—EtherNet/IP zone protection. *Image courtesy of Tofino Security—A Belden Brand.*

broadcast domain with no routing capability. PROFIBUS Isochronous Real Time (IRT) was introduced in version 3 of the standard, and provides cycle times of less than 1 ms with jitter less than 1 μs common in high-speed motion control applications. PROFIBUS IRT is a hardware-based solution that incorporates extensions to the Ethernet stack (OSI layer 2) requiring special application-specific integrated circuits at the device level and IRT-compatible network switches designed to minimize jitter. IRT is a layer 2 technology, so there is no routing capability possible with these data packets. Figure 6.20 illustrates the different classes of PROFINET.

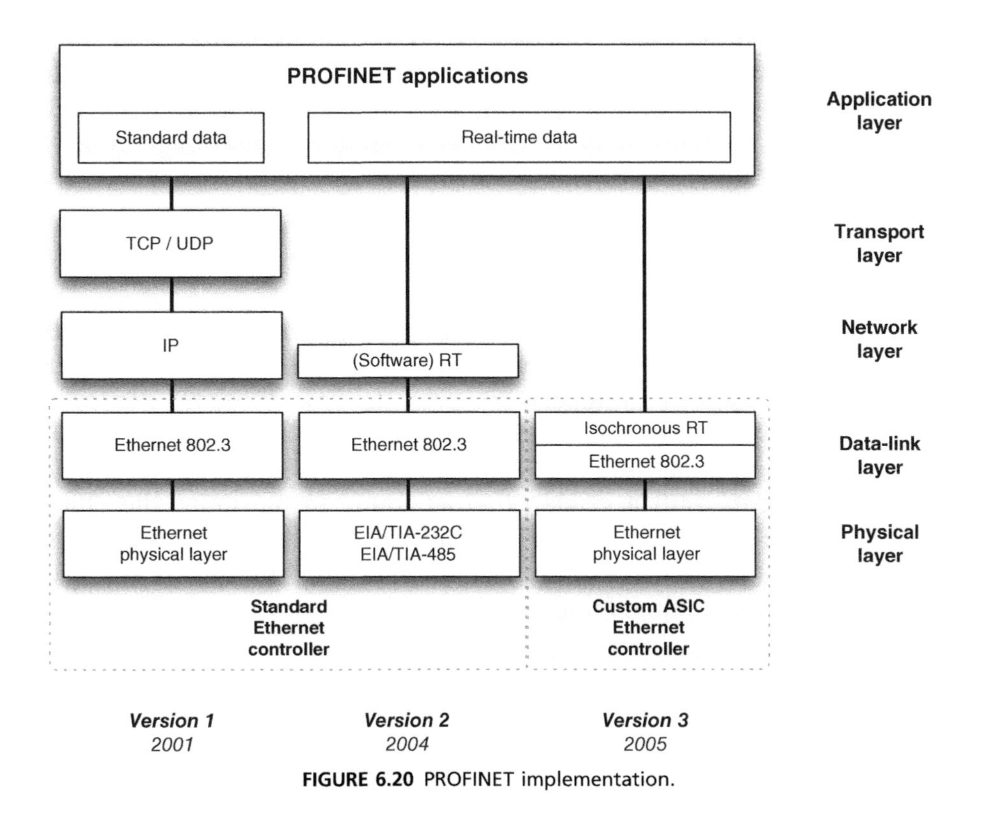

FIGURE 6.20 PROFINET implementation.

Security concerns

PROFINET is a real-time Ethernet protocol, and as such it is susceptible to any of the vulnerabilities of Ethernet. The extend of the risk is highly dependent on the technology deployed, since newer devices can utilize proprietary hardware making unauthorized network access more challenging than the general-purpose TCP/IP implementation. When used over the IP, it is also susceptible to any vulnerabilities of IP; however, the real-time implementations of PROFINET also employ nonroutable network communications offering some protection against remote or adjacent network vectors.

Security recommendations

As with many fieldbus protocols, the inherent lack of authentication and vulnerability of the protocol requires strong isolation of the bus. PROFINET TCP/IP represents the greatest risk as it can be transmitted over standard business and industrial networks. It should be tightly controlled, and when used within less-trusted business networks, it should be limited to authenticated and encrypted networks. It is not possible to segment PROFINET networks that contain devices that must communicate with each other (e.g., virtual local area networks (LANs) are not supported between PROFINET devices for logical segmentation); therefore, careful consideration in the deployment of zones and conduits should be taken (see Chapter 9, "Establishing Zones and Conduits"). Monitoring of Ethernet networks for unauthorized or suspicious use of PROFINET should be implemented including monitoring of all conduits into PROFINET zones. Firewalls and ICS-aware Intrusion Prevention Systems should be configured to explicitly deny PROFINET traffic outside of well-defined areas. Additional guidance can be obtained through PNO.[25]

EtherCAT

EtherCAT is another real-time Ethernet-based fieldbus protocol classified as "Industrial Ethernet" (see PROFINET for more information), which uses a defined ethertype (0x88A4) to transport ICS communications over standard Ethernet networks. These messages can either be transported directly in an Ethernet frame or encapsulated as a UDP payload using port 34980/udp (0x88A4). EtherCAT communicates large amounts of distributed process data with a single Ethernet frame to maximize the efficiency of distributed process data communications requiring only a few bytes per cycle over Ethernet frames that may vary in size from 46 to 1500 bytes. This means that only one or two Ethernet frames are required for a complete cycle allowing for very short cycle times with low jitter easily allowing network synchronization tasks to occur as required by the IEEE 1588 Precision Time Protocol (PTP) standard. EtherCAT is able to meet the requirements of PTP without any additional hardware (not the case with other industrial protocols discussed). Slaves pass the frame(s) to other slaves in sequence, appending its appropriate response, until the last slave returns the completed response frame back.[26]

Security concerns

EtherCAT is a real-time Industrial Ethernet protocol, and as such it is susceptible to any of the vulnerabilities of standard Ethernet. EtherCAT over UDP is transaction-less and so there is no inherent network-layer mechanism for reliability, ordering or data integrity checks.

EtherCAT is sensitive and highly susceptible to DoS attacks as with many real-time Ethernet protocols. EtherCAT is easily disrupted via the insertion of rogue Ethernet frames into the network to interfere with time synchronization and is subject to spoofing and MitM attacks due to the lack of bus authentication, requiring the separation of EtherCAT from other Ethernet systems.

Security recommendations

EtherCAT is a real-time Industrial Ethernet protocol making it is necessary to provide Ethernet-based security at the perimeter of any EtherCAT network. It is also recommended that passive network monitoring be used to ensure the integrity of the EtherCAT network, and that the EtherCAT protocol is only being used by explicitly identified devices. No EtherCAT traffic should be allowed that is originating from an unauthorized, outside source. This can be accomplished using an ICS-aware intrusion prevention system or other network monitoring device capable of detecting and interpreting the EtherCAT protocol via UDP/IP. Static Ethernet address tables (MAC address) can be deployed to further protect real-time EtherCAT devices from external attack. Many switches provide features to provide MAC address control as well as tables to further restrict communications between EtherCAT devices. A network monitoring product or probe can also be used to detect Ethernet packets using EtherCAT's specific ethertype.

Ethernet POWERLINK

Ethernet POWERLINK is also an "Industrial Ethernet" technology that uses Fast Ethernet as the basis for real-time transmission of control messages via the direct encapsulation of Ethernet frames without a master node that is used to initiate and synchronize cyclic polling of slave devices. Communication is divided into three time periods, with the first being the transmission of a master "Start of Cycle" frame that provides a basis for the network synchronization. The master then polls each station. The second time period is devoted to synchronous communication allowing the slaves to respond only if they receive a poll request frame, ensuring that all master/slave communications occur in sequence. Slave responses are broadcast, eliminating source address resolution. Asynchronous communication occurs in the third period where larger, nontime-critical data are transmitted. POWERLINK is best used homogeneously because collisions are avoided solely via the carefully controlled request/response cycles. The introduction of other Ethernet-based systems could disrupt synchronization and cause a failure.[27]

POWERLINK is often used in conjunction with CANopen, an application layer protocol based on CAN. CANopen enables the communication between devices of different manufacturers, and the protocol stacks are widely available including open-source distribution for both Windows and Linux platforms. The open nature of CANopen makes POWERLINK/CANopen a desirable combination for industrial networks requiring inexpensive solutions in Linux environments.[28]

Security concerns

POWERLINK is a real-time Industrial Ethernet protocol, and as such it is susceptible to any of the vulnerabilities of other forms of Ethernet communication.

As with many real-time Ethernet protocols, POWERLINK is sensitive and highly susceptible to DoS attacks. POWERLINK is easily disrupted via the insertion of rogue Ethernet frames into the network, requiring the separation of POWERLINK from other Ethernet systems. The protocol itself is sensitive and highly susceptible to DoS attacks.

Security recommendations

POWERLINK implementations will most likely have a clear demarcation from other networks because sensitivity of the cyclic polling mechanism requires separation from other non–POWERLINK Ethernet services. This demarcation can be leveraged to further isolate the industrial protocol, through the establishment of appropriate security zones and the definition of strong perimeter defenses at these boundaries. Static Ethernet address tables (MAC address) can be deployed to further protect real-time POWERLINK devices from external attack, since these are pure Ethernet-based messages and typically represent the most critical communications. Many switches provide features to provide MAC address control as well as tables to further restrict communications between EtherCAT devices.

SERCOS III

SERCOS is a standardized open digital interface for communication between industrial controls, motion devices, and I/O devices. Versions I and II of the interface were based on a fiber-optic ring to establish interdevice communication. Version III of the interface is an "Industrial Ethernet"–based implementation of the SERCOS interface that supports deterministic real-time control of motion and I/O applications. Like EtherCAT and POWERLINK, SERCOS III has the ability to directly place Ethernet frames on the network in order to obtain high-speed communications with very low jitter.[29] Networks can support up to 511 slave devices in either straight or ring topologies.

SERCOS III is a master-slave protocol that operates cyclically, using a mechanism in which a single Master Synchronization Telegram is used to communicate to slaves, and the slave nodes are given a predetermined time (again synchronized by the master node) during which they can place their data on the bus. All messages for all nodes are packaged into a Master Data Telegram, and each node knows which portion of the MDT it should read based upon a predetermined byte allocation.[30]

SERCOS III dedicates the use of the bus for synchronized real-time traffic during normal cycles; however, like other Industrial Ethernet protocols discussed, it allows unallocated time within a cycle to be freed up for other network protocols such as TCP and UDP data using IP. This "IP Channel" allows the use of broader network applications from the same device—for example, a web-based management interface that would be accessible to "office and wide area networks."[31]

Security concerns

SERCOS III is a real-time Industrial Ethernet protocol, and as such it is susceptible to any of the vulnerabilities of other forms of Ethernet communication. SERCOS III introduces new security concerns through the option to support embedded, open TCP/IP, and UDP/IP communications. With this option enabled, a compromised RTU or PLC using SERCOS III could be used to launch an in-bound attack into other corporate communications systems, including industrial and business networks.

Security recommendations

As with other Industrial Ethernet–based protocols, static Ethernet address tables (MAC address) can be deployed to further protect real-time SERCOS III devices from external attack, since these are pure Ethernet-based messages and typically represent the most critical communications. Many switches provide features to provide MAC address control as well as tables to further restrict communications between SERCOS III devices. SERCOS III should be isolated to control loops that require the protocol, and the use of IP channels should be restricted and avoided if possible. If IP channels are used, the extent and reach of the IP channel should be enclosed within an explicitly defined zone consisting of the SERCOS III master node and only those TCP/IP network devices that are absolutely required. Strong perimeter defenses should be installed in-band for all conduits into this zone using least privilege principles. Active monitoring of security device logs on the perimeter should be enabled due to the heightened risk from pivoting through networks using SERCOS III.

Backend protocols

Object linking and embedding for process control

OPC is not actually an industrial protocol, but "a series of standards specifications"[32] designed to simplify integration of various forms of data on systems from different vendors. In order to appreciate the impact OPC had on industrial automation, a brief history of OPC is warranted.

The original standard released in 1996 provided a mechanism for a standardized way for systems to exchange data across an Ethernet network using a core set of Microsoft technologies including: Object Linking and Embedded (OLE), Component Object Model (COM), and Distributed Component Object Model (DCOM). The specification included standard sets of "objects," "interfaces," and "methods" to support this interoperability in industrial applications. The underlying mechanism to support this communication was based on interprocess communications using the Remote Procedure Call (RPC) protocol. The original set of standards that utilized the COM/DCOM infrastructure is today commonly references as "OPC Classic."

OPC has evolved significantly since its introduction nearly 20 years ago, and for that reason, the OPC Foundation (the organization that oversees the standards) has introduced new meaning to the dated acronym including "Open Platform Communications" and "Open Productivity and Connectivity." The "classic" set of standards originally focused on real-time data access (OPC-DA released 1996), historical data access (OPC-HDA released 2001), and alarms and events data (OPC-AE released 1999). This set was expected to include data access via web services using extensible markup language (OPC-XMLDA released 2003), server-to-server and machine-to-machine communications (OPC-DX released 2003), and batch applications (OPC Batch released 2000). Since OPC relied on the DCOM infrastructure, users encountered significant problems in trying to manage OPC communication across security zones that were protected with firewalls including the lack of network address translation support and session callbacks.

Technology was moving away from the DCOM infrastructure and toward the .NET Framework. Using Windows Communication Foundation OPC.NET (formerly known as

OPC-Xi or eXpress Interface) incorporates the functionality of DA, HDA, and AE on a simplified data model. This new technology provided users with significant security improvements to how OPC.NET traffic was managed on industrial networks across zones. The downside was that there was little vendor support for this enhanced standard resulting in a relative small number of "gateway" type products.[33]

All standards up to this point depended on some form of underlying Microsoft technology—COM, DCOM, or .NET. This significantly limited the deployment within ICS architectures much below the supervisory networks due to the fact that most of the embedded devices (Controllers, PLCs, RTUs, etc.) were not based on a Windows operating system that would support these classic standards. The idea was to move the communications model from COM/DCOM to a cross-platform service-oriented architecture to support broader deployment to non-Windows devices, and ... better security! The OPC Unified Architecture (OPC-UA) specification was first released in 2006, and offers numerous improvements to the "classic" specifications while still supporting the underlying data integration requirements.

OPC Data Access "classic" is still one of the most widely deployed OPC specifications, and for the purposes of this book, is the one that will be discussed in more detail.

What it does

OPC is one of the major "backend" protocols because it is designed to provide a higher level of integration between systems and subsystems, versus a fieldbus protocol that generally provides low-level data access and configuration.

OPC was originated motivated by the needs of end "users" and not system "vendors" to provide a common communications interface between diverse ICS components. The idea was to create a process industries technology that mirrored what Microsoft had done with device drivers in there newer Windows object-oriented operating systems. To digress briefly, many remember the days of Windows 3.11 and the requirement for every application to possess drivers necessary to utilize a dot-matrix printer. Microsoft solved that problem when they released Windows 95. The manufacturing community was no different—significant time and effort was spent in the 1980s and 1990s simply providing basic integration between the various systems that now are common components with in integrated ICS architecture.

This was accomplished by leveraging Microsoft's DCOM communications API, reducing the need for device-specific drivers. In place of specific communications drivers for each device, simple device drivers could be written to interface with OPC. The use of OPC therefore minimized driver development and allowed for better optimization of core OPC interfaces.[34]

OPC's strengths and weaknesses come from its foundation, which is based upon Microsoft's OLE technology. OLE is used extensively in office document generation allowing the presentation of data to be separated from the application that generated it. A Word document can either "link" to a value calculated by a local or remote spreadsheet, or "embed" the spreadsheet inside the document. This not only allows

OPC-connected devices to communicate and interact with minimal operator feedback (as in the case of the Office documents). The concept of cyber security did not really exist in 1996, which meant that there were significant security challenges that lie ahead to those implementing OPC.[35]

How it works

OPC works in a client/server manner, where a client application calls a local process, but instead of executing the process using local code, the process is executed on a remote server. The remote process is linked to the client application and is responsible for providing the necessary parameters and functions to the server, utilizing an RPC.

In other words, the stub process is linked to the client, but when a function is performed, the process is performed remotely, on the server. The server RPC functions then transmit the requested data back to the client computer. The client process then receives the data over the network, provides it to the requesting application, and closes the session, as shown in Figure 6.21.

In Windows systems, the requesting application typically loads RPC libraries at run-time, using a Windows dynamic link library.[36]

OPC is more complex than previous client/server industrial protocols because of this interaction with the calling application and the underlying DCOM architecture. It interacts with various aspects of the host operating system, tying it closely to other host processes and exposing the protocol to a very broad attack surface. OPC also inherently supports remote operations that allow OPC to perform common control system functions.[37]

One aspect that makes OPC and DCOM very challenging when characterizing industrial networks and the communications that occur across these networks and through various conduits is how DCOM begins the session on one port and then transfers to another. Figure 6.22 illustrates a typical OPC session that does not incorporate server "callbacks."

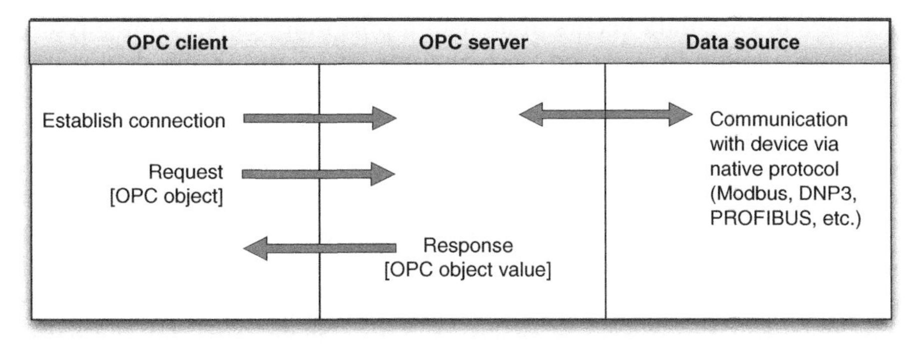

FIGURE 6.21 Typical OPC protocol operation.

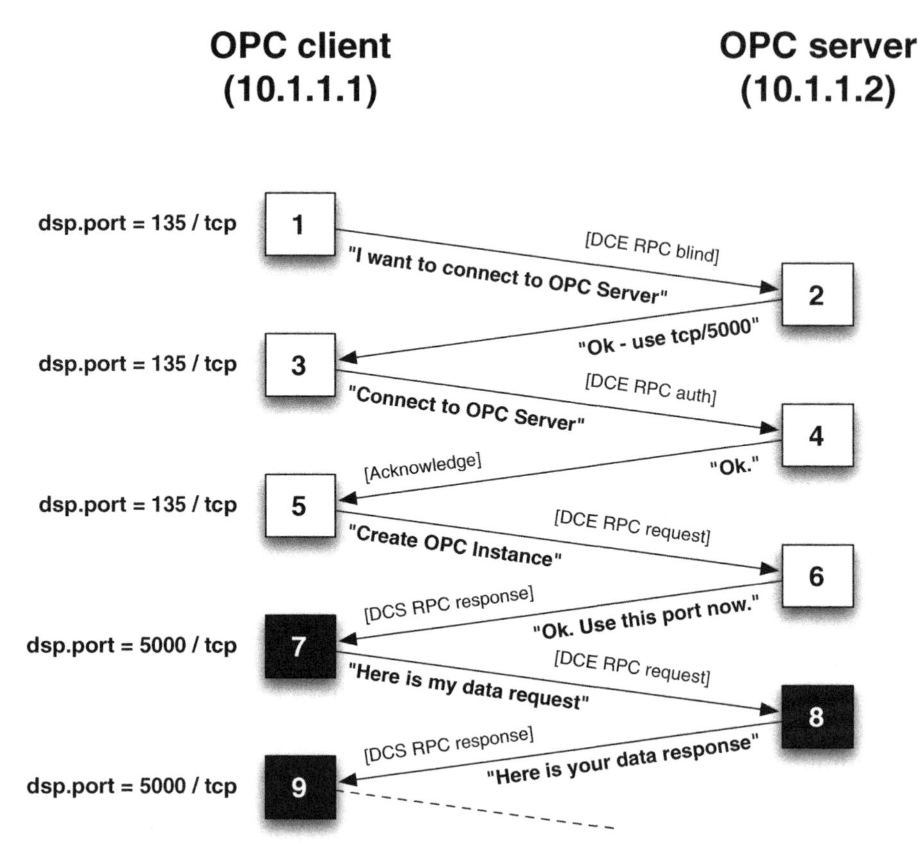

FIGURE 6.22 OPC client-server communications.

This figure shows how an initial request from an OPC Client to a corresponding OPC Server begins using a DCE BIND request to the Endpoint Mapper service listing on 135/tcp of the Server. Once the Client is authenticated against the Server and an OPC Instance created on the Server, the sessions shifts to a different connection, where the actual exchange of OPC data occurs. If a custom port range is not configured, this new port can be any randomly assigned port between 1024 and 65535 depending on the operating system. If Server callbacks are used, the original session actually disconnects after the OPC Instance is created, and the OPC Server initiates a new session with the OPC Client. In other words, the OPC Server is now the network "source address" and the OPC Client is now the "destination address." A "tunneler" application can be installed to address this problem by allowing a point-to-point tunnel be created using a single predefined port where all RPC traffic (135 and the subsequent session port) are directed. The tunneler must be installed on both OPC Client and Server hosts, and should be qualified by the respective vendor to ensure that there is no impact to the performance of the other applications and services.

Where it is used

As the name implies, OLE for "Process" Control is primarily used within industrial networks (i.e., not a common business network technology), including data transfer to data historians, data collection within HMIs, connectivity between serial fieldbus protocols like Modbus and DNP3 and ICS servers, and other supervisory controls, as shown in Figure 6.23. The deployment of OPC servers within ICS architectures can greatly simplify the data integration in the core ICS servers allowing all proprietary protocols and interfaces to be managed via local, distributed OPC Servers that contain the appropriate physical and application connectivity to a particular subsystem or device. This Server is then connected to various ICS servers and components using a single, consistent mechanism. OPC is a Windows interconnection, so all communications occur either between Windows-based devices, or via OPC gateways that translate the RPC to the native fieldbus format. Because of the common use of RPC protocols within OPC, this opens the ICS environment to a very broad attack surface.

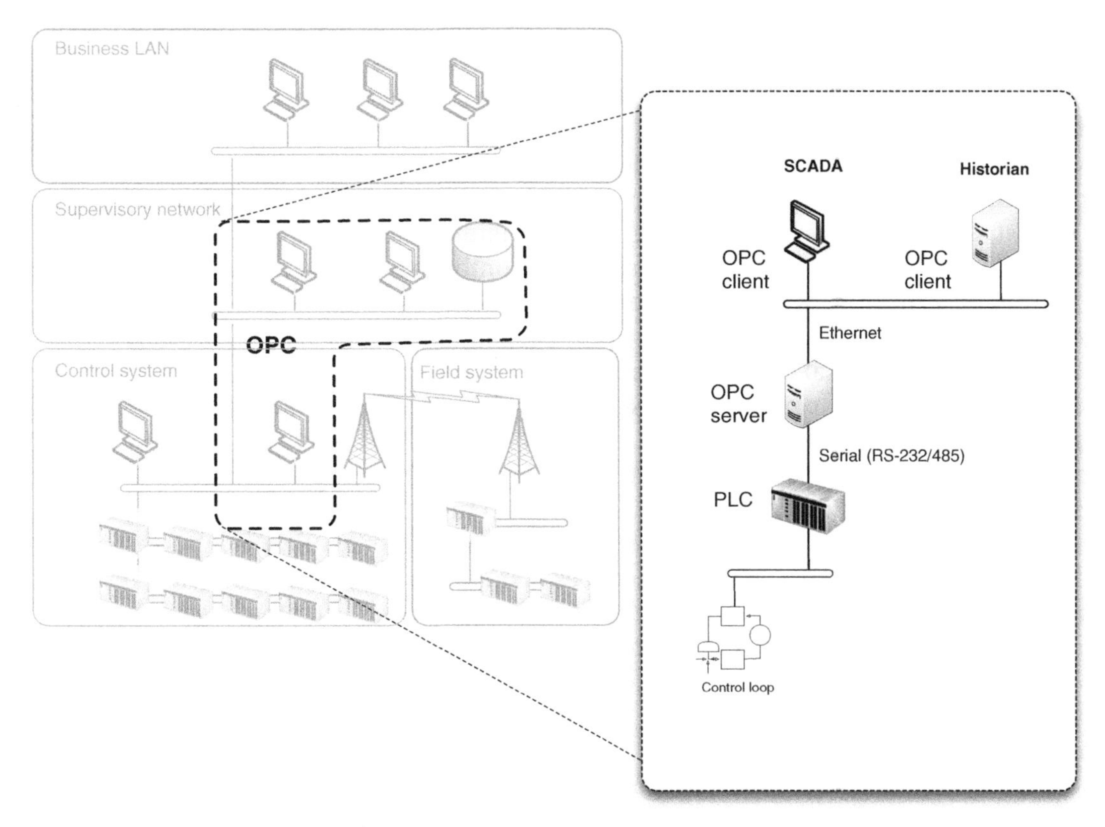

FIGURE 6.23 Typical OPC use within the industrial network architecture.

Security concerns

Modular attack frameworks including Trisis, Industroyer, and Incontroller are able to directly manipulate OPC-DA and OPC-UA, both for reconnaissance and enumeration of industrial systems, and also to read and write specific parameters for manipulation of control (see Chapter 7, "Hacking Industrial Control Systems" for details on these attack frameworks).

OPC's use of DCOM and RPC makes it highly vulnerable to attack using multiple vectors, as it is subject to the same vulnerabilities as the more ubiquitously used OLE.[38] Classic OPC is rooted in the Windows operating system and is therefore susceptible to attack through exploitation of any vulnerability inherent to the OS.[39] Support for the Windows XP with Service Pack 3 ended on April 2014 (XP-SP2 ended July 2010), meaning that OPC applications hosted on unsupported OS can introduce significant risk to the integrity of manufacturing operations and potential health, safety, and environment impact.

OPC and related ICS vulnerabilities can be tracked via a variety of sources including the U.S. Department of Homeland Security ICS-CERT and the Open Source Vulnerability Database. Many OLE and RPC vulnerabilities exist and are well known, including exploit modules for a variety of open-source and fee-based security frameworks like Metasploit and Canvas (see Chapter 7, "Hacking Industrial Systems"). It is difficult to patch production systems within an industrial network (see Chapter 8, "Risk and Vulnerability Assessments" and Chapter 10, "Implementing Security Controls") so many of these vulnerabilities may still be in place, even if there is an available patch from Microsoft. The SQL Slammer worm actually caused global damage despite the fact that Microsoft released a patch to correct the vulnerability 6 months prior to the release of the worm.

Many basic host security concerns apply because OPC is supported on Windows. RPC requires local authentication to occur on both client and server hosts. This requires the creation of either a local or domain-based account that can be used by RPC for the OPC sessions. This account can introduce significant risk if it is not properly secured using a least privilege approach for just the essential OPC/DCOM services. This account is common to all hosts utilizing OPC, and if not properly protected and managed can lead to a widespread compromise in large ICS architectures. Many OPC hosts utilize weak authentication, and passwords are often weak when authentication is enforced. Many systems support additional Windows services that are irrelevant to ICS systems, resulting in unnecessary processes, which often correspond to open "listening" communication ports accessible via the network. Inadequate or nonexistent logging exacerbates these potential weaknesses by providing insufficient forensic detail should a breach occur, as Windows 2000/XP auditing settings do not record DCOM connection requests by default.[40]

Unlike the simple and single-purpose fieldbus protocols discussed earlier, OPC must be treated as an overall system integration framework, and implemented and maintained according to modern OS and network security practices.

Other security concerns of OPC include the following:

- Legacy authentication services—systems within industrial networks are difficult to upgrade (due to limited maintenance windows, compatibility and interoperability concerns, and other factors); insecure authentication mechanisms remain in use. For example, Windows 2000 LAN Manager (LM) and NT LAN Manager (NTLM) authentication mechanisms are still used by default in many systems (enabled by default up to and included Windows XP and 2003 Server). These and other legacy authentication mechanisms may be vulnerable and susceptible to exploitation.[41]
- RPC vulnerabilities—OPC uses RPC making it susceptible to all RPC-related vulnerabilities, including several vulnerabilities that are exposed prior to authentication. Exploitation of underlying RPC vulnerabilities could result in arbitrary code execution, or DoS.[42]
- Unnecessary ports and services—OPC supports network protocols other than TCP/IP, including NetBIOS Extended User Interface (NetBEUI), Connection Oriented NetBIOS over InterNetwork packet Exchange (IPX), and Hyper Text Transport Protocol (HTTP) Internet services.[43]
- OPC Server Integrity—it is possible to create a rogue OPC server and to use that server for disruption of service, DoS, information theft through bus snooping, or the injection of malicious code.[44]

Security recommendations

The newer Unified Architecture (OPC-UA) specification was designed for security and should be used where possible in place of OPC-DA.

Regardless of the OPC specification used (Classic or Unified Architecture), all unnecessary ports and services should be removed or disabled from the OPC server. This includes any and all irrelevant applications, and all unused network protocols. All unused services may introduce vulnerabilities to the system that could result in a compromise of the Windows host, and therefore the OPC network.[45]

OPC servers should be isolated into a unique zones consisting only of authorized devices, and the zones(s) should be thoroughly secured using standard defense-in-depth practices, including a firewall and/or intrusion protection system that enforces strict control over the type, source, and destination of traffic to and from the OPC zone. Consideration should be given to application-aware firewalls that are capable of following the RPC session from the initial request (via 135/tcp) to response (a different port) and possible server "callbacks."

Because OPC is primarily used in a supervisory capacity, intrusion "prevention" systems can be considered in place of "detection" only, understanding that an IPS may block legitimate ICS traffic and result in a lack of visibility into control system operations potentially causing a Loss of View (LoV) or Loss of Control (LoC) situation. If information loss will be damaging to the control process or detrimental to business operations, use only an IDS.

Many threats can be detected through monitoring OPC networks and/or OPC servers (server activity can be monitored through the collection and analysis of Windows logs), and looking for specific behaviors, including the following:

- The use of non-OPC ports and services initiated from the OPC server (requires DCOM services to be configured to use specific port range to eliminate wide range of "randomly" generated response ports).
- The presence of known OPC (including underlying OLE RPC and DCOM) exploits.
- OPC services originating from unknown OPC servers (indicating the presence of a rogue server).
- Failed authentication attempts or other authentication anomalies on the OPC server.
- Successful authentication attempts on the OPC server from unknown or unauthorized users.

Most commercially available IDS and IPS devices support a wide range of detection signatures for OLE and RPC and therefore can also detect many of the underlying vulnerabilities of OPC. Most open-source and commercial log analysis and threat detection tools are capable of collecting and assessing Windows logs.

Tip

OPC vulnerabilities may require the use of an ICS-aware intrusion protection system rather than an enterprise equivalent. Enterprise devices typically detect exploits via inspection of OLE, RPC, and DCOM but may not be able to detect all threats targeting OPC. In some cases, enterprise IDS/IPS devices may be able to detect a wider range of OPC threats, using industrial protocol preprocessors and detection signatures.

Intercontrol center communications protocol (ICCP/IEC 60870-6 TASE.2)

The Inter-Control Center Communications Protocol (also known as TASE.2 or IEC60870-6, but more commonly referred to as simply ICCP) is a protocol designed for communication between control centers within the electric utility industry. Unlike fieldbus protocols like Modbus and DNP3, ICCP is classified as a "backend" protocol like OPC because of the fact it was designed for bidirectional Wide Area Network (WAN) communication between a utility control center and other control centers, power plants, substations, and even other utilities.

Much like the fundamental driver in the process industries developing OPC, electric utilities were also faced with ICS vendors and equipment suppliers utilizing many custom and proprietary protocols. A common protocol was needed to allow for reliable and standardized data exchange between utility control centers—especially when these control centers are operated by different owners, produce different products, or perform different operations. Standardization became necessary to support the unique business and operational requirements of the electrical utilities that require careful load balancing within a bulk system operated by many disparate facilities. In North America, the division of utilities among several responsible regional entities requires a means of sharing information between utilities as well as the regional entity. National and global energy markets require real-time information exchange for load distribution and trading that spans the boundaries of individual utilities.

A working group was formed in 1991 to develop and test a standardized protocol and to submit the specification to the IEC for ratification and approval. The initial protocol was called ELCOM-90, or Telecontrol Application Service Element-1 (TASE.1). TASE.1 evolved into TASE.2, which is the most commonly used form of ICCP.[46]

What it does

ICCP is used to perform a number of communication functions between control centers, including the following:

- Establishing a connection.
- Accessing information (read requests).
- Information transmission (such as email messages or energy market information).
- Notifications of changes, alarms, or other exception conditions.
- Configuration of remote devices.
- Control of remote devices.
- Control of operating programs.

How it works

The ICCP protocol defines communication between two control centers using a client-server model. One control center (the server) contains application data and defined functions. Another control center (the client) issues requests to read from the server with appropriate server responses. Communications over ICCP occur using a common format in order to ensure interoperability.

ICCP support is typically integrated either directly into an ICS, provided via a gateway product, or provided as software that can then be installed to perform gateway functions.

ICCP is primarily a unidirectional client-server protocol; however, most modern implementations support both functions, allowing a single ICCP device to function as both a client and a server, supporting bidirectional communication over a single connection.

ICCP can operate over essentially any network protocol, including TCP/IP; however, it is commonly implemented using the ISO transport on port 102/tcp, as defined in RFC 1006. ICCP is effectively a point-to-point protocol due to the use of a "bilateral table" that explicitly defines an agreement between two control centers connected with an ICCP link, as shown in Figure 6.24. The bilateral table acts as an access control list that identifies which data elements a client can access. The permissions defined within the bilateral tables in the server and the client are the authoritative control over what is accessible to each control center. The entries in the bilateral tables must also match on

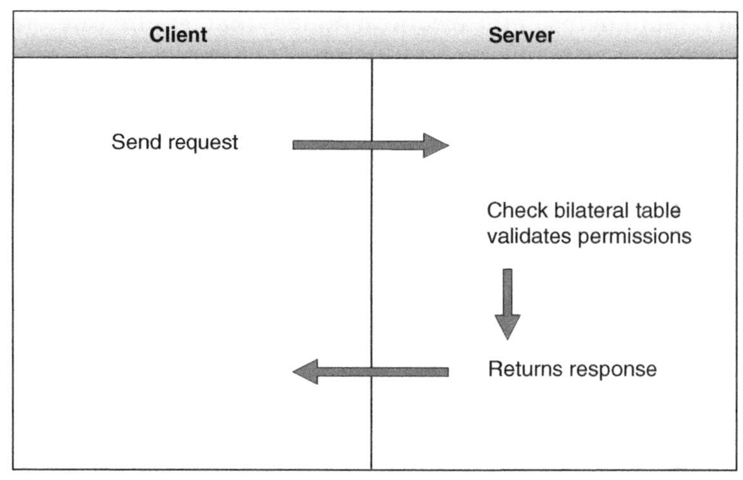

FIGURE 6.24 ICCP protocol operation.

both the client and the server, ensuring that the permissions are agreed upon by both centers (remembering that ICCP is used to interconnect to other organizations in addition to internal WAN links to substations).[47]

Where it is used

ICCP is widely used between control system zones and between distinct control centers, as shown in Figure 6.25. It is also commonly deployed between two electric utilities, between two control systems within a single electric utility, and between a main control center and a number of substations.

Security concerns

ICCP represents several security concerns much like most of the other fieldbus and backend protocols discussed. ICCP is susceptible to spoofing, session hijacking, and any number of attacks made possible because of:

- Lack of authentication and encryption—ICCP does not mandate authentication or encryption, most often deferring these services to lower protocol layers. Although "Secure ICCP"[48] does exist, it is not ubiquitously deployed.
- Explicitly defined trust relationships—the exploitation of bilateral tables could directly compromise security of ICCP servers and clients.
- Accessibility—ICCP is a Wide Area Protocol making it highly accessible and susceptible to many attacks including DoS attacks from being exposed to public and/or shared networks versus traditional closed or private industrial networks within a plant environment.

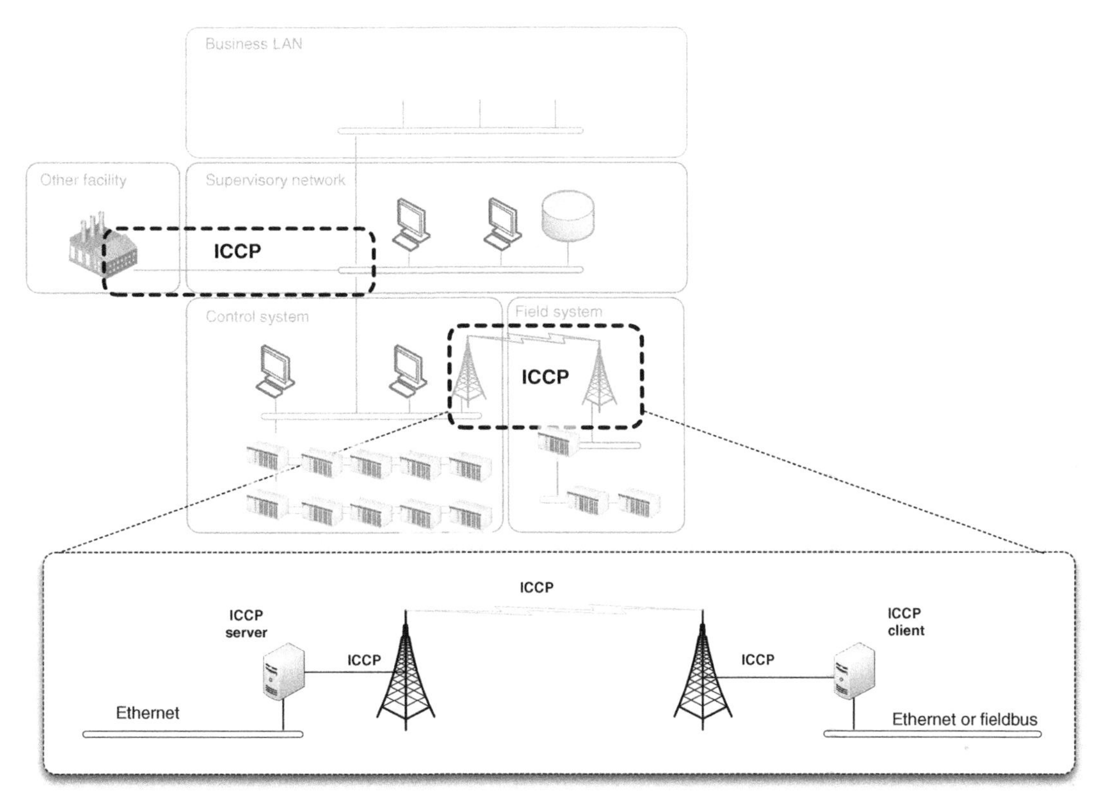

FIGURE 6.25 Typical ICCP use within the industrial network architecture.

The limited security mechanisms within ICCP are configured on the ICCP server, meaning that the successful breach of the server through an MitM or other attack opens the entire communication session up to manipulation.

Security improvements over Modbus

ICCP offers several improvements over more basic fieldbus protocols such as Modbus and DNP3, including the following:

- ICCP's use of bilateral tables provides basic control over the communication path by explicitly defining which ICCP clients and servers can communicate.
- A secure version of ICCP exists that incorporates digital certificate authentication and encryption.

Security recommendations

Secure ICCP variants should be used wherever possible and supported by the current vendors installed within a particular site. There are several known vulnerabilities with ICCP that have been reported by ICS-CERT. Proper system hardening and regular system assessments and patching of ICCP servers and clients are recommended because there are known exploits in the wild and ICCP is a WAN protocol.

Extreme care should be taken in the definition of the bilateral table. The bilateral table is the primary enforcement of policy and permissions between control centers. Malicious commands issued via ICCP could directly alter or otherwise impact control center operations.

ICCP clients and servers should also be isolated into a unique zone consisting only of authorized client-server pairs (multiple zones can be defined for devices communicating to multiple clients), and the zones(s) should be thoroughly secured using standard defense-in-depth practices, including a firewall (industrial grade if installed in production environments) and/or intrusion protection system that enforces strict control over the type, source, and destination of traffic over the ICCP link. As with other industrial protocols, preference should be given to security practices that are capable of deep-packet inspection of ICCP traffic, if available. Many of the recommendations described for other industrial protocols are equally applicable for ICCP, including the creation of network baselines and deployment of network whitelists.

Many malicious behaviors can be detected through monitoring of the ICCP link, including the following:

- Intruders gaining unauthorized access to the control center network, via overlooked access points such as dial-up or remote access connections to partner or vendor networks with weak access control mechanisms.
- Insider threats, including unauthorized information access and transmission, alteration of secure configurations, or other malicious actions can be the result of a physical security breach within a control center, or of a disgruntled employee.
- A DoS attack resulting from repeated information requests ("spamming") that utilize the server's available resources and prevent legitimate operation of the ICCP link.
- Malware infecting the ICCP server or other devices on the network could be used to exfiltrate sensitive information for purposes of sabotage (e.g., theft of command function codes), financial disruption (e.g., alteration of energy metrics used in trading), or various other malicious intents.
- Interception and modification of ICCP messages (i.e., MitM) attacks.

  Monitoring of ICCP protocol functions can also detect suspicious or malicious behavior, such as

- Function "read" codes that could be used to exfiltrate protected information.
- Function "write" codes that could be used to manipulate client or server operations.
- Traffic on port 102/tcp that is not ICCP or other authorized protocol (PROFINET utilizes 102/tcp ISO-TSAP for its industrial Ethernet communications).
- ICCP traffic that is not sourced by and destined to defined ICCP servers or clients.

An ICS-aware intrusion protection system can be configured to monitor for these activities using ICCP signatures such as those developed and distributed by Digital Bond under the QuickDraw SCADA ICS project. An application-aware firewall, industrial protocol filter, or application data monitor may be required to validate ICCP sessions and ensure that ICCP or the underlying RFC-1006 connection have not been "hijacked" and that messages have not been manipulated or falsified.

Caution

Intrusion prevention systems are able to actively block suspect traffic by dropping packets or resetting TCP connections. However, Intrusion prevention systems deployed on industrial networks should be only be configured to block traffic after careful consideration and tuning. Unless you are confident that a given signature will not inadvertently block a legitimate control command, the signature should be set to alert, rather than block (i.e., operate in "detection" mode rather than active "prevention" mode).

IEC 61850, 60870-5-101, and 60870-5-104

IEC 61850 and IEC 60870-5 are communication protocols used primarily in electric power systems. While there are numerous standards of this nature, IEC 61850, IEC 60870-5-101, and IEC 60870-5-104 are mentioned here specifically because there is known malware capable of manipulating these protocols (see Chapter 7, "Hacking Industrial Control Systems" for details on the Industroyer and Industroyer 2 attack frameworks).

IEC 60870 is a communication protocol used in SCADA systems and other applications for monitoring and controlling electrical power systems. It is designed to provide reliable and interoperable communication used in power plants, substations, and control centers. It is a part of the IEC 60870-5 series of standards, which define telecontrol protocols for SCADA systems. IEC 60870-5-101 is designed for serial communications, while 60870-5-104 is designed for communication over TCP/IP networks. [Gordon R. Clarke et al, Practical modern SCADA protocols: DNP3, 60870.5 and related systems, Newnes, 2004 ISBN 0-7506-5799-5]

IEC 60870-5-101 and IEC 60870-5-104, often referred to more simply as IEC-101 and IEC-104 for convenience, both function in a client—server model. The control center acts as the client, and the remote device act as servers. The control center sends requests for data to devices, and the devices respond.

IEC 61850 is a communication protocol specifically designed for electrical substations, providing a standardized method for communication between IEDs like circuit breakers, transformers, and protection relays. The protocol helps improve interoperability, reliability, and flexibility in the management and control of power systems.

How they work

60870-5-101 and 60870-5-104

Requests and responses are organized into Application Service Data Units (ASDUs), which include a 1 byte Type Identification field (TI), a 1 byte Variable Structure Qualifier, a variable sized Cause of Transmission (COT) field, a variable Common Address field.

The remainder of the ASDU consists of an information object: the actual data elements contained within the ASDU. This object could consist of a data value such as a measurement, a command, or a status response. [International Electrotechnical Commission (IEC). International Standard IEC 60870-5-101. Second Edition. IEC. Geneva 20, Switzerland. 2006.]

ASDUs, encapsulated with Application Protocol Data Units (APDUs), are transmitted over either serial (IEC-101) or TCP/IP (IEC-104).

Communication typically consists of the control center sending a request for data or a control command to one or more devices. The devices then respond with the requested information (in the case of a data request) or a command confirmation (in the case of a control command). Data requests could consist of alarms, measurements, or status information. Once the exchange is finished, either the client or the server can close the connection.

IEC 61850

IEC 61850 uses an object-oriented approach for modeling devices and their data. Each device is represented as a hierarchy of logical nodes, data objects, and data attributes. These models can be used to provide highly interoperable communications between devices, over a variety of communication protocols.

In substation applications, data is exchanged primarily via Generic Object-Oriented Substation Events (GOOSE) and Sample Values (SVs) messages. GOOSE messages are used for event-driven communication, like alarms or interlocking, while SV messages are used for transmitting high-speed sampled analog data, such as current and voltage measurements. Messages are exchanged using high-speed Ethernet networks and TCP/IP.

A single GOOSE message sent from one device can be received and used by several other devices, using standard broadcast and multicast mechanisms within IP, and a publish/subscribe data exchange model. One device (the sender) publishes information, while one or more other devices (the subscribers) receive that data act upon it. The action performed by each receiver upon receiving data depends on its configuration and functionality, as defined by its IEC 61850 device model. This enables automation of substation control, including protection. [Bill Lydon. IEC 61850 Power Industry Communications Standard. Automation.com. Feb 07, 2009. https://www.automation.com/en-us/articles/2003-1/iec-61850-power-industry-communications-standard].

Security concerns

As with most industrial protocols, the increased connectivity and reliance on digital communication also introduces new cybersecurity risks and challenges. Some of the key cybersecurity implications of these protocols are:

- These protocols are vulnerable to known cyber-physical attack frameworks, including Industroyer and Industroyer 2 (see Chapter 7, "Hacking Industrial Control Systems" for details on these attack frameworks).

- These protocols have been successfully manipulated to disrupt electrical distribution systems.
- The ability to monitor devices remotely using these protocols can be exploited by adversaries to obtain valuable information about the target systems, which can be used to develop targeted attacks.
- The ability to control devices remotely using these protocols can be exploited by adversaries to manipulate a process, potentially damaging physical equipment and/or introducing hazardous conditions.
- The ability to intercept and manipulate data can be used to hide malicious activities from operators.
- Network-based DoS attacks can disrupt communication between devices, leading to LoC, reduced situational awareness, and potential process failures.

Security recommendations

Devices communicating using these IEC protocols should be isolated into a unique zones consisting only of authorized devices, and the zones(s) should be thoroughly secured using standard defense-in-depth practices, including a firewall and/or intrusion detection or protection system that enforces strict control over the type, source, and destination of IEC traffic within this zone.

Because these protocols are often used to communicate between highly distributed systems, e.g., power substations, the integrity of communications should not be assumed. Monitoring relevant devices and the communications between those devices for abnormal behavior is recommended.

AMI and the smart grid

The smart grid is a term encompassing many aspects of modern power generation, transmission, and distribution. Although smart grid technology might seem irrelevant to many industrial network systems outside of the electric utility industry, it is discussed briefly here because of its broad reach and vulnerable attack surface. The smart grid is a widely distributed communication network that touches both power generation and transmission systems, along with many end user networks. The smart grid represents an easily accessible network that contains many vectors to many possible targets. Once compromised, an attacker could use the network to attack the power utility's network, or to attack the networks of connected home and businesses.

The term "smart grid" is widely used and generally refers to a new era of energy distribution built around an Advanced Metering Infrastructure (AMI). AMI promises many new features designed to increase the efficiency and reduce the costs of energy distribution. Common AMI features include remote meter reading, remote billing, demand/response energy delivery, remote connect/disconnect, and remote payment and prepayment.[49]

At a high level, the smart grid requires coordination among the following systems:

- Bulk Electric Generation Systems
- Electric Transmission Systems
- Electric Distribution Systems
- Customer Information and Management Systems
- Usage and Meter Management Systems
- Billing Systems
- Interconnected network systems, including neighborhood area networks (often using wireless mesh technologies); metropolitan area networks; home area networks (HANs); and business area networks (BANs)

The smart grid is essentially a large, end-to-end communications system interconnecting power suppliers to power consumers (see Figure 6.26). It is made of highly diverse systems, using diverse protocols and network topologies. Smart grids even introduce new protocols. To support home- and business-based service portals, smart metering introduces HAN and BAN protocols, such as Zigbee and HomePNA, as well as power line protocols such as IEC 61334, Control Network Power Line (PL) Channel Specification, and Broadband over Powerline (BPL). The data link and application protocols are too numerous to discuss in detail, though it is widely accepted that TCP/IP will be leveraged for network-layer communications.[50]

These specific protocols will not be discussed within this book, but it is still important to recognize that the disparate nature of these systems requires that several distinct operational models and network architectures combine to form a single end-to-end communications path, as illustrated in Figure 6.23. This means that while many distinct smart grid protocols may be used, the smart grid as a whole should be considered as a single, readily accessible communications network that is vastly interconnected.

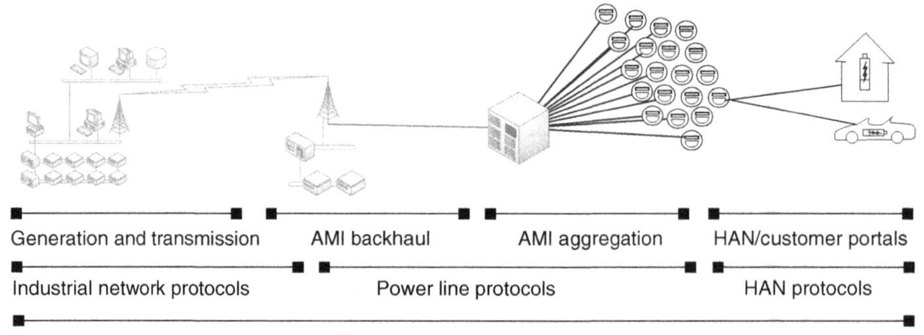

FIGURE 6.26 Smart grid operational areas and protocols.

Security concerns

The security concerns of the smart grid are numerous. AMI represents an extremely large network that touches many other private networks and is designed with command and control capabilities in order to support remote disconnect, demand/response billing, and other features.[51] Combined with a lack of industry-accepted security standards, the smart grid represents significant risk to connected systems that are not adequately isolated. Specific security concerns include the following:

- Smart meters are readily accessible and therefore require board- and chip-level security in addition to network security.
- Smart grid protocols vary widely in their inherent security and vulnerabilities.
- Neighborhood, home, and business LANs can be used both as an ingress to the AMI, and as a target from the AMI.
- Smart grids are ultimately interconnected with critical power generation and distribution systems.
- Smart grids represent a target to private hackers (for financial gain or service theft) as well as to more sophisticated and serious attackers (for sociopolitical gain or cyber warfare).

Security recommendations

The best recommendation for smart grid security at this point is for electric utilities to carefully assess smart grid deployments and to perform risk and threat analysis early in the planning stages. A similar assessment of the system should be performed for end users who are connected to the smart grid who could become a potential threat vector into the business (or home) networks.

Clear delineation, separation of services, and the establishment of strong defense-in-depth at the perimeters will help to mitigate the risk from threats associated with the smart grid. This could represent a challenge (especially in terms of security monitoring) for smart grid operators, due to the broad scale of smart grid deployments, which could contain hundreds of thousands or even millions of intelligent nodes. It may be necessary then to carve out smart grid deployments into multiple, smaller and more manageable security zones.

Industrial protocol simulators

One way to learn and understand how an industrial protocol operates is to purchase the appropriate hardware (i.e., PLC) and software (SCADA). This can be both expensive and time consuming. Another more practical approach is through the deployment of client and server simulators capable of mimicking the protocol within a physical or virtualized computing environment.

Simulators are readily available for royalty-free protocols like Modbus/TCP, but can be limited for the licensed protocols. In the latter cases, one alternative approach is the use of "trial" or "demonstration" software packages. The products below were available at the time of publishing, and are provided for illustrative purposes only.

Modbus/TCP

There are a range of Modbus simulators that will support both Modbus RTU and ASCII formats using both serial and Ethernet communication. The ModbusPal package available on SourceForge is particularly interesting because it is based on Java allowing it to be easily transported between different platforms (Windows, Mac, Linux). It also features an "automation" capability allowing it to vary inputs and outputs providing the ability to change data at the source. ModbusPal supports "user-defined" commands using function codes 65–72 and 100–110.

Triangle Microworks Communication Protocol Test Harness provides not only protocol simulation, but actual simulation of a variety of devices as well, allowing this to be a tool used by ICS software developers as part of protocol compliance testing. The Test Harness supports a range of protocols including Modbus/TCP, DNP3, and IEC 60870-5, and is available as a paid download or a 21-day evaluation version.

Modsak is a software package from Wingpath Software Development that supports either master or client modes. A 3-day trial version is available that offers a range of features, including support for Modbus "user-defined" functions.

DNP3

The Axon Group offers a free simulation package for both DNP3 and IEC 60870-5. The Communication Test Harness from Triangle Microworks also supports DNP3 and can operate as both the master station and outstation. More advanced options are available through a variety sources that provide DNP3 protocol libraries for custom application development.

OPC

Matrikon and Kepware are two leading suppliers of OPC products to a variety of ICS industry segments, both offering demonstration versions of their OPC applications. Matrikon offers a set of free OPC test tools that support the creation of OPC clients and servers, as well as trial versions of most of their applications including various system interface servers, protocol tunnelers, and more. Kepware offers similar trial licenses for their OPC server, as well as a linking package that can be used to connect two OPC servers.

ICCP/TASE.2

Triangle Microworks IEC 60870-6 (TASE.2/ICCP) Test Tool is available as a paid license or a 21-day evaluation version with support for both client and server roles. The package supports ICCP blocks 1, 2, and 5 with full support of writes, reads, controls, dynamic data sets, and dataset transfer sets. It also allows for models to be created via .csv files and .xml files.

Physical hardware

Investing in physical hardware to support a training and test laboratory does not have to be overly expensive. Many suppliers including ABB, Allen—Bradley, Schneider Electric, Siemens, and Wago offer affordable, compact programmable devices that can support multiple protocols within a single device. Nearly all products will offer support for Modbus/TCP due to its widespread use, but can also be supplied with Ethernet/IP, PROFINET, and EtherCAT capabilities. Another very economical method of obtaining physical hardware is through reseller or auction websites like eBay.

Summary

Industrial networks use a variety of specialized protocols at multiple layers in the network to accomplish specific tasks, often with careful attention to synchronization and real-time operation. Each protocol has varying degrees of inherent security and reliability, and these qualities should be considered when attempting to secure these protocols. All of these protocols are susceptible to cyberattack using relatively simple MitM mechanisms because industrial network protocols, in general, lack sufficient authentication or encryption. These attacks can be used to disrupt normal protocol operations or potentially alter or otherwise manipulate protocol messages to steal information, commit fraud, or potentially cause a failure of the control process itself including mechanical equipment sabotage (e.g., Stuxnet).

These protocols can be reasonably secured by understanding them and isolating each into its own carefully defined security zone with related conduits (see Chapter 9, "Establishing Zones and Conduits"). The creation of zones based purely on physical devices is possible and relatively simple because each protocol has specific uses within a control system. Since industrial network protocols are used more widely over Ethernet and TCP/IP—UDP/IP, the creation of clean zone boundaries becomes more difficult, as these boundaries begin to overlap. The use of "business" network protocols to transport fieldbus protocols should be avoided unless absolutely necessary for this reason, and be especially scrutinized and tested where they are necessary.

Endnotes

1. IEC 61784-1:2010 "Industrial communication networks - Profiles - Part 1: Fieldbus profiles", published June 1, 2011.
2. "Schneider Electric Modicon History", http://www.plcdev.com/schneider_electric_modicon_history (cited: January 7, 2014).
3. Modbus Organizations, "Modbus Application Protocol Specification", Version 1.1b, Published December 28, 2066.
4. Ibid.
5. Ibid.
6. AEG Schneider Autotmation, "Modicon Modbus Plus Nework Planning and Installation Guide", 890-USE-100.00 Version 3.0, April 1996.

7. Triangle MicroWorks, “Using DNP3 & IEC 60870-5 Communication Protocols in the Oil & Gas Industry”, Revision 1, published March 26, 2001.
8. Triangle MicroWorks, “Modbus and DNP3 Communication Protocols”, http://trianglemicroworks.com/docs/default-source/referenced-documents/Modbus_and_DNP_Comparison.pdf (cited: January 8, 2014).
9. The DNP Users Group, DNP3 Primer, Revision A. http://www.dnp.org/About/DNP3%20Primer%20Rev%20A.pdf, March 2005 (cited: November 24, 2010).
10. R. Clarke, Deon Reynders Practical Modern SCADA Protocols: DNP3, 60870.5 and Related Systems, Newnes, Oxford, UK and Burlington MA, 2004.
11. The DNP Users Group, DNP3 Primer, Revision A. http://www.dnp.org/About/DNP3%20Primer%20Rev%20A.pdf, March 2005 (cited: November 24, 2010).
12. Ibid.
13. Digitalbond SCADAPEDIA, Secure DNP3. http://www.digitalbond.com/wiki/index.php/Secure_DNP3, August 2008 (cited: November 24, 2010).
14. Ibid.
15. The DNP Users Group, DNP3 Primer, Revision A. http://www.dnp.org/About/DNP3%20Primer%20Rev%20A.pdf, March 2005 (cited: November 24, 2010).
16. A.B.M. Omar Faruk, Testing & Exploring Vulnerabilities of the Applications Implementing DNP3 Protocol, KTH Electrical Engineering, Stockholm, Sweden, June 2008.
17. V.M. Igure, Security assessment of SCADA protocols: a taxonomy based methodology for the identification of security vulnerabilities in SCADA protocols, VDM Verlag Dr. Müller Aktiengesellschaft & Co. KG, 2008.
18. “Industrial Ethernet: A Control Engineer's Guide”, Cisco, April 2010.
19. Prof. Dr.-Ing. J. Schwager, “Information about Real-Time Ethernet in Industry Automation”, Reutlinger University, http://www.pdv.reutlingen-university.de/rte/, (cited: January 10, 2014).
20. Industrial Ethernet Facts, “System Comparison: The five Major Technologies”, Ethernet POWERLINK Standardization Group, Issue 2, February 2013.
21. Ibid.
22. Open Device Vendors Association (ODVA), “Common Industrial Protocol”, PUB00122R0-ENGLISH, 2006.
23. Ibid.
24. Open-Device Vendors Association, “Securing Ethernet/IP Networks”, PUB00269R1, 2011.
25. PROFIBUS Nutzerorganisation e.V., “PROFINET Security Guidelines: Guideline for PROFINET”, Version 2.0, November 2013.
26. The EtherCAT Technology Group, Technical introduction and overview: EtherCAT—the Ethernet Fieldbus. http://www.ethercat.org/en/technology.html#5, May 10, 2010 (cited: November 24, 2010).
27. P. Doyle, Introduction to Real-Time Ethernet II. The Extension: A Technical Supplement to Control Network, vol. 5, Issue 4, Contemporary Control Systems, Inc., Downers Grove, IL, July 2004.
28. Ethernet POWERLINK Standardization Group, CANopen. http://www.ethernet-powerlink.org/index.php?id=39, 2009 (cited: November 24, 2010).
29. SERCOS International, Technology: Introduction to SERCOS interface. http://www.sercos.com/technology/index.htm, 2010 (cited: November 24, 2010).
30. SERCOS International, Technology: Cyclic Operation. http://www.sercos.com/technology/cyclic_operation.htm, 2010 (cited: November 24, 2010).
31. SERCOS International, Technology: Service & IP Channels. http://www.sercos.com/technology/service_ip_channels.htm, 2010 (cited: November 24, 2010).
32. OPC Foundation, “What is OPC?”, http://www.opcfoundation.org/Default.aspx/01_about/01_whatis.asp?MID=AboutOPC, (cited: January 9, 2014).
33. OPC Foundation, “Certified Products”, http://www.opcfoundation.org/Products/Products.aspx, (cited: January 9, 2014).
34. Ibid.

35. Digital Bond, British Columbia Institute of Technology, and Byres Research. OPC Security White Paper #2: OPC Exposed (Version 1-3c), Byres Research, Lantzville, BC and Sunrise, FL, November 13, 2007.
36. Microsoft Corporation, RPC Protocol Operation. http://msdn.microsoft.com/en-us/library/ms818824.aspx (cited: November 4, 2010).
37. European Organization for Nuclear Research (CERN), A Brief Introduction to OPC Data Access. http://itcofe.web.cern.ch/itcofe/Services/OPC/GeneralInformation/Specifications/RelatedDocuments/DASummary/DataAccessOvw.html, November 11, 2000 (cited: November 29, 2010).
38. "OPC Security Whitepaper #3: Hardening Guidelines for OPC Hosts", DigitalBond, British Columbia Institute of Technology, Byres Research, November 13, 2007.
39. Digital Bond, British Columbia Institute of Technology, and Byres Research. OPC Security White Paper #2: OPC Exposed (Version 1-3c), Byres Research, Lantzville, BC and Sunrise, FL, November 13, 2007.
40. Ibid.
41. Ibid.
42. Ibid.
43. Ibid.
44. Ibid.
45. Ibid.
46. J.T. Michalski, A. Lanzone, J. Trent, S. Smith, SANDIA Report SAND2007-3345: Secure ICCP Integration Considerations and Recommendations, Sandia National Laboratories, Albuquerque, New Mexico and Livermore, California, June 2007.
47. Ibid.
48. J. Michalski, A. Lanzone, J. Trent, S. Smith, "Secure ICCP Integration: Considerations and Recommendations", Sandia Report SAND2007-3345, printed June 2007.
49. UCA International Users Group, AMI-SEC Task Force, AMI System Security Requirements, UCA, Raleigh, NC, December 17, 2008.
50. National Institute of Standards and Technology, NIST Special Publication 1108: NIST Framework and Roadmap for Smart Grid Interoperability Standards, Release 1.0, February 2010.
51. UCA International Users Group, AMI-SEC Task Force, AMI system security requirements, UCA, Raleigh, NC, December 17, 2008.
52. Open Device Vendors Association (ODVA), "Common Industrial Protocol", PUB00122R0-ENGLISH, 2006.

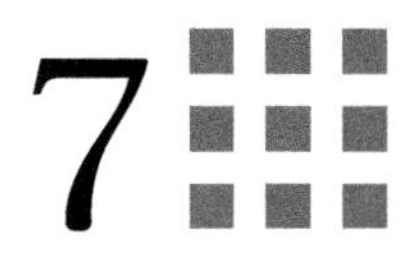

Hacking Industrial Control Systems

Information in this chapter

- Motives and Consequences
- Common Industrial Targets
- The Evolution of the Industrial Cyber Attack
- Common Attack Methods
- Examples of Advanced Industrial Cyber Threats
- Attack Trends
- Dealing with an Infection

Motives and consequences

Industrial networks are responsible for continuous and batch processing and other manufacturing operations of almost every scale, and as a result, the successful penetration of a control system network can be used to directly impact those operations. Consequences vary and can range from relatively benign disruptions, such as the interruption of the operation (taking a facility offline) and the alteration of an operational process (changing the formula of a chemical process or recipe), to deliberate acts of sabotage that are intended to cause harm. Manipulating the feedback loop of certain processes could, for example, cause pressure within a boiler to build beyond safe operating parameters. Cyber sabotage, on the other hand, can result in environmental damage (oil spill, fire, toxic release, etc.), injury or loss of life, the loss of critical services (blackouts, disruption in fuel supplies, unavailability of vaccines, etc.), or potentially catastrophic explosions.

Consequences of a successful cyberincident

A successful cyberattack on an ICS can have many undesirable consequences, including:

- Delay, block, or alter the intended process, that is, alter the amount of energy produced at an electric generation facility.
- Delay, block, or alter information related to a process, thereby preventing a bulk energy provider from obtaining production metrics that are used in energy trading or other business operations.
- Unauthorized changes to instructions or alarm thresholds that could damage, disable or shutdown mechanical equipment, such as generators or substations.

Industrial Network Security. https://doi.org/10.1016/B978-0-443-13737-2.00001-4

- Inaccurate information sent to operators could either be used to disguise unauthorized changes (see Stuxnet later in this chapter) or cause the operator to initiate inappropriate actions.

The end result could be anything from financial loss to physical safety liabilities, with impacts extending beyond the plant, to the local community, state, and even federal level (see Figure 7.1). Companies can incur penalties for regulatory noncompliance, or they may suffer financial impact from lost production hours due to misinformation or denial of service. An incident can impact the ICS in almost any way, from taking a facility offline, disabling or altering safeguards, to life-threatening incidents within the plant—up to and including the release or theft of hazardous materials or direct threats to national security.[1]

The possible damages resulting from a cyberincident vary depending upon the type of incident, as shown in Table 7.1.

Cybersecurity and safety

Most industrial networks employ automated safety systems to avoid catastrophic failures. However, many of these safety controls employ the same messaging and control protocols used by the industrial control network's operational processes, and in some cases, such as certain fieldbus implementations, the safety systems are supported directly within the same communications protocols as the operational controls on the same physical media (see Chapter 6, "Industrial Network Protocols," for details and security concerns of industrial control protocols).

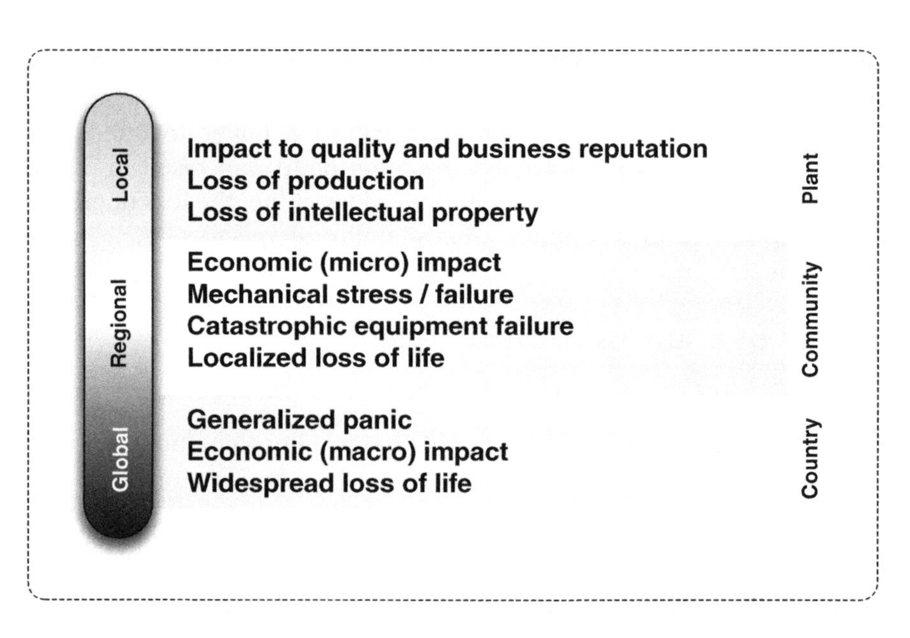

FIGURE 7.1 Consequences of a compromised industrial control system.

Table 7.1 The Potential Impact of Successful Cyberattacks

Incident type	Potential impact
Change in a system, operating system, or application configuration	Command and control channels introduced into otherwise secure systems. Suppression of alarms and reports to hide malicious activity. Alteration of expected behavior to produce unwanted and unpredictable results
Change in programmable logic in PLCs, RTUs, or other controllers	Damage to equipment and/or facilities. Malfunction of the process (shutdown). Disabling control over a process
Misinformation reported to operators	Inappropriate actions taken in response to misinformation that could result in a change to operational parameters. Hiding or obfuscating malicious activity, including the incident itself or injected code
Tampering with safety systems or other controls	Preventing expected operations, fail safes, and other safeguards with potentially damaging consequences
Malicious software (malware) infection	Initiation of additional incident scenarios. Production impact resulting from assets taken offline for forensic analysis, cleaning, and/or replacement. Assets susceptible to further attacks, information theft, alteration, or infection
Information theft	Leakage of sensitive information such as a recipe or chemical formula
Information alteration	Alteration of sensitive information such as a recipe or chemical formula in order to sabotage or otherwise adversely affect the manufactured product

Note

Critical, risk-based safety operations implemented within the ICS typically follow separate standards regarding the use of programmable logic solvers, field devices, and communication protocols (e.g., IEC 61508/61511, NFPA 85, ISA 84) and how these safety instrumented systems (SIS) can be interfaced and integrated with other ICS components. It is important to realize that not all "safety" controls and interlocks are implemented against these standards, and that it is possible for these systems to share infrastructure (including the controller platform itself) with other ICS systems and components. Regulatory requirements typically require standards-based SIS implementations for safety functions that represent significant unmitigated risk in terms of human health, safety, and environmental impact, and not on production uptime or reliability.

Although safety systems are extremely important, there is the perception that they have been used to downplay the need for heightened security of industrial networks. Research has shown that real consequences can occur in modeled systems. Simulations performed by the Sandia National Laboratories showed that simple man-in-the-middle (MitM) attacks could be used to change values in a control system and that a modest-scale attack on a larger bulk electric system using targeted malware (in this scenario, targeting specific ICS front-end processors) was able to cause significant loss of generation.[2]

The European research team VIKING (Vital Infrastructure, Networks, Information and Control Systems Management) is currently investigating threats of a different sort. The Automatic Generation Control (AGC) system within the electric power network is responsible for adjusting the output of multiple generators on the grid in response to changes in demand. It operates autonomously from human interaction—that is, output actions are based entirely on processing of input states with the logic of the AGC. Rather than breaching a control system through the manipulation of an HMI, VIKING's research attempts to investigate whether the manipulation of input data could alter the normal control loop functions, ultimately causing a disturbance.[3]

Tip

Think of security as separate from safety when establishing a cybersecurity plan. Do not assume that security leads to safety or that safety leads to security. If an automated safety control is compromised by a cyberattack (or otherwise disrupted), the necessity of having a strong digital defense against the manipulation of operations becomes even more important. Likewise, a successful safety policy should not rely on the security of the networks used. Both systems will be inherently more reliable by planning for safety and security controls that operate independently of one another. At the same time, safety systems are built around strong process assessments, to protect against identified physical risk conditions. These risk conditions may be the ultimate goal of a cyberattack, and so safety and security also need to work together within an organization to ensure that cyberdefenses are properly implemented.

With attack frameworks capable of manipulating SISs (see "TRISIS", and "Incontroller" below), it is more important than ever to consider safety controls from a cybersecurity perspective and to consider cybersecurity controls from a safety perspective. This concept is discussed in more detail in Chapter 8, "Risk and Vulnerability Assessments: Thinking of Cybersecurity in terms of Safety."

Common industrial targets

Industrial control systems may be comprised of similar components; however, each system is unique in terms of the exact composition, quantity, and criticality of these components. There are, however, some common targets within industrial networks despite these system differences. These include network services, such as active directory (directory services) and identity and access management (IAM) servers, which may be shared between business and industrial zones (though the best practice is to not share these services!); engineering workstations, which can be used to exfiltrate, alter, or overwrite process logic; operator consoles, which can be used to trick human operators into performing unintended tasks; and of course the industrial applications (SCADA server, historian, asset management, etc.), and protocols (Modbus, DNP3, EtherNet/IPI, etc.)

themselves, which can be used to alter, manipulate, blind, or destroy almost any aspect of an ICS. Table 7.2 highlights some of the common targets, how they are likely to be attacked, and what the consequences of such attacks might be.

The evolution of the industrial cyberattack

There was a time when cyberattacks were simpler. Malware was often self-contained, and there was a direct correlation between exposure to malware and a subsequent infection. Cybersecurity threats have evolved considerably in the past decade, and a cybersecurity threat is better thought of as a campaign: a series of tactics and techniques that might leverage multiple, disparate, purpose-built malwares used at various stages of a larger coordinated attack effort. In addition, most cyber-physical attacks targeting industrial control require two distinct phases, each of which has its own lifecycle: the first phase involves the various steps of a cyberattack required to reach industrial control systems; the second involves the various steps of a subsequent cyberattack to target those industrial control systems.

This is discussed in depth in Chapter 10, "Attack and Defense Lifecycles" but is mentioned here for context. In addition, some of the topics described below are components that could be used as part of a larger campaign (e.g., "Common Attack Methods"), while some refer to the larger campaigns that could be comprised of those methods (i.e., "Weaponized Industrial Cyber Threats").

Finally, the examples discussed below are obviously not a comprehensive list. There are billions of distinct malwares that have been discovered, and there are 14 categorized "tactics" as defined by MITRE, and hundreds of techniques utilized to achieve them.[4] The examples discussed here are intended to understand attacks that are most relevant to industrial network security, but they are not all inclusive. The threat landscape changes continuously; try to imagine the unimaginable and to expect the unexpected!

Common attack methods

There are many methods of attacking a target, once a target has been identified. MitM, denial-of-service (DoS), replay attacks, and countless more methods all remain very effective in industrial networks. The primary reason for this is a combination of insecure communication protocols, little device-to-device authentication, and delicate communication stacks in embedded devices. If an industrial network can be penetrated and malware deposited (on disk or in memory) anywhere on the network, tools such as Metasploit Meterpreter shell can be used to provide remote access to target systems, install keyloggers or keystroke injectors, enable local audio/video resources, manipulate control bits within industrial protocols, plus many other covert capabilities.

In some cases, the information that is available can be used as reconnaissance for further cyberattack capability. In many cases, systems can be attacked directly using disclosed exploits, with only basic system knowledge required. If an attack is successful,

Table 7.2 Attack Targets

Target	Possible Attack vectors	Possible Attack methods	Possible consequences
Access control system	- Identification cards - Closed-circuit television (CCTV) - Building management network - Software vendor support portal	- Exploitation of unpatched application (building management system) - RFID spoofing - Network access through unprotected access points - Network pivoting through unregulated network boundaries	- Unauthorized physical access - Lack of (video) detection capabilities - Unauthorized access to additional ICS assets (pivoting)
Analyzers/analyzer management system	- Subcontractor laptop - Maintenance remote access - Plant (analyzer) network	- Exploitation of unpatched application - Network access via insecure access points (analyzer shelters) - Remote access VPN via stolen or compromised subcontractor laptop - Remote access VPN via compromise of maintenance vendor site - Insecure implementation of OPC (communication protocol)	- Product quality—spoilage, loss of production, loss of revenue - Reputation—product recall, product reliability
Application servers	- Remote user access (interactive sessions) - Business application integration communication channel - Plant network - Software vendor support portal	- Exploitation of unpatched application - Installation of malware via unvalidated vendor software - Remote access via "interactive" accounts - Database injection - Insecure implementation of OPC (communication protocols)	- Plant upset/shutdown - Credential leakage (control) - Sensitive/confidential information leakage - Unauthorized access to additional ICS assets (pivoting)
Asset management system	- Plant maintenance software/ERP - Database integration functionality - Mobile devices used for device configuration - Wireless device network - Software vendor support portal	- Exploitation of unpatched application - Installation of malware via unvalidated vendor software - Remote access via "interactive" accounts - Database injection - Installation of malware via mobile devices - Access via insecure wireless infrastructure	- Calibration errors—product quality - Credential leakage (business) - Credential leakage (control) - Unauthorized access to additional business assets like plant maintenance/ERP (pivoting) - Unauthorized access to additional ICS assets (pivoting)
Condition monitoring system	- Subcontractor laptop - Maintenance remote access - Plant (maintenance) network - Software vendor support portal	- Exploitation of unpatched application - Installation of malware via unvalidated vendor software - Network access via unsecure access points (compressor/pump house) - Remote access VPN via stolen or compromised subcontractor laptop	- Equipment damage/sabotage - Plant upset/shutdown - Unauthorized access to additional ICS assets (pivoting)

Table 7.2 Attack Targets—cont'd

Target	Possible Attack vectors	Possible Attack methods	Possible consequences
		- Remote access VPN via compromise of maintenance vendor site - Remote access via "interactive" accounts - Database injection - Insecure implementation of OPC (communication protocols)	
Controller (PLC)	- Engineering workstation - Operator HMI - Standalone engineering tools - Rogue device in control zone - USB/removable media - Controller network - Controller (device) network	- Engineer/technician misuse - Network exploitation of industrial protocol—known vulnerability - Network exploitation of industrial protocol—known functionality - Network replay attack - Network DoS via communication buffer overload - Direct code/malware injection via USB - Direct access to device via rogue network (local/remote) PC with appropriate tools/software	- Manipulation of controlled process(es) - Controller fault condition - Manipulation/masking of input/output data to/from controller - Plant upset/shutdown - Command-and-control
Data historian	- Business network client - ERP data integration communication channel - Database integration communication channel - Remote user access (interactive session) - Plant network - Software vendor support portal	- Exploitation of unpatched application - Installation of malware via unvalidated vendor software - Remote access via "interactive" accounts - Database injection - Insecure implementation of required communication protocols - Exploitation of unnecessary/excessive openings on perimeter defense (firewall) due to insecure communication infrastructure between applications	- Manipulation of process/batch records - Credential leakage (business) - Credential leakage (control) - Unauthorized access to additional business assets like MES, ERP (pivoting) - Unauthorized access to additional ICS assets (pivoting)
Directory services	- Replication services - Print spooler services - File sharing services - Authentication services - Plant network - Software vendor support portal	- Exploitation of unpatched application(s) - Installation of malware via unvalidated vendor software - DNS spoofing - NTP reflection attack - Exploitation of unnecessary/excessive openings on perimeter defense (firewall) due to replication requirements between servers - Installation of malware on file shares	- Communication disruptions via DNS - Authentication disruptions via NTP - Authentication disruptions via LDAP/Kerberos - Credential leakage - Information leakage - file shares - Malware distribution - Unauthorized access to ALL domain-connected ICS assets (pivoting) - Unauthorized access to business assets (pivoting)

Continued

Table 7.2 Attack Targets—cont'd

Target	Possible Attack vectors	Possible Attack methods	Possible consequences
Engineering workstations	- Engineering tools and applications - Nonengineering client applications - USB/Removable media - Elevated privileges (engineer/administrator) - Control network - Software vendor support portal	- Exploitation of unpatched applications - Installation of malware via unvalidated vendor software - Installation of malware via removable media - Installation of malware via keyboard - Exploitation of trusted connections across security perimeters - Authorization to ICS applications without sufficient access control mechanisms	- Plant upset/shutdown - Delay plant startup - Mechanical damage/sabotage - Unauthorized manipulation of operator graphics—inappropriate response to process action - Unauthorized modification of ICS database(s) - Unauthorized modification of critical status/alarms - Unauthorized distribution of faulty firmware - Unauthorized startup/shutdown of ICS devices - Process/plant information leakage - ICS design/application credential leakage - Unauthorized modification of ICS access control mechanisms - Unauthorized access to most ICS assets (pivoting/own) - Unauthorized access to business assets (pivoting)
Environmental controls	- HVAC control - HVAC (building management) network - Software vendor support portal	- Exploitation of unpatched application (building management system) - Installation of malware via unvalidated vendor software - Network access through unprotected access points - Network pivoting through unregulated network boundaries	- Disruption of cooling/heating - Equipment failure/shutdown
Fire detection and suppression system	- Fire alarm/evaluation - Fire suppressant system - Building management network - Software vendor support portal	- Exploitation of unpatched application (building management system) - Installation of malware via unvalidated vendor software - Network access through unprotected access points - Network pivoting through unregulated network boundaries	- Unauthorized release of suppressant - Equipment failure/shutdown

Table 7.2 Attack Targets—cont'd

Target	Possible Attack vectors	Possible Attack methods	Possible consequences
Master and/or slave devices	- Unauthorized/unvalidated firmware - Weak communication problems - Insufficient authentication for "write" operations - Control network - Device network	- Distribution of malicious firmware - Exploitation of vulnerable industrial protocols via rogue PC on network (local/remote) - Exploitation of vulnerable industrial protocols via compromised PC on network (local) - Exploitation of industrial protocol functionality via rogue PC on network (local/remote) - Exploitation of industrial protocol functionality via compromised PC on network (local) - Communication buffer overflow via rogue PC on network (local/remote) - Communication buffer overflow via compromised PC on network (local)	- Plant upset/shutdown - Delay plant start - Mechanical damage/sabotage - Inappropriate response to control action - Suppression of critical status/alarms
Operator workstation (HMI)	- Operational applications (HMI) - Non-SCADA client applications - USB/Removable media - Elevated privileges (administrator) - Control network - Software vendor support portal	- Exploitation of unpatched applications - Installation of malware via unvalidated vendor software - Installation of malware via removable media - Installation of malware via keyboard - Authorization to ICS HMI functions without sufficient access control mechanisms	- Plant upset/shutdown - Suppression of critical status/alarms - Product quality - Plant/process efficiency - Credential leakage (control) - Plant/operational information leakage - Unauthorized access to ICS assets (pivoting) - Unauthorized access to ICS assets (communication protocols)
Patch management servers	- Software patches/hotfixes - Patch management software - Vendor software support portal - Business network - Plant network - Software vendor support portal	- Insufficient checking of patch "health" before deployment - Alternation of automatic deployment schedule - Installation of malicious software via trusted (supplier) media - Installation of malware via unvalidated vendor software	- Malware distribution server - Unauthorized modification of patch schedule - Credential leakage - Unauthorized access to ICS assets (pivoting)
Perimeter protection (firewall/IPS)	- Trusted connections (business-to-control) - Local user account database - Signature/rule updates	- Untested/unverified rules - Exploitation of unnecessary/excessive openings on perimeter defense (firewall)	- Unauthorized access to business network - Unauthorized access to DMZ network

Continued

Table 7.2 Attack Targets—cont'd

Target	Possible Attack vectors	Possible Attack methods	Possible consequences
		- Insecure office and industrial protocols allowed to cross security perimeter - Reuse of credentials across boundary	- Unauthorized access to control network - Local credential leakage - Unauthorized modification of rulesets/signatures - Communication disruption across perimeter/ boundary
SCADA servers	- Non-SCADA client applications - Application integration communication channels - Data historian - Engineering workstation - Control network - Software vendor support portal	- Exploitation of unpatched applications - Installation of malware via unvalidated vendor software - Remote access via "interactive" accounts - Installation of malware via removable media - Exploitation of trusted connections within control network - Authorization to ICS applications without sufficient access control mechanisms	- Plant upset/shutdown - Delay plant startup - Mechanical damage/ sabotage - Unauthorized manipulation of operator—inappropriate response to process action - Unauthorized modification of ICS database(s) - Unauthorized modification of critical status/ alarms - Unauthorized startup/ shutdown of ICS devices - Credential leakage (control) - Plant/operational information leakage - Unauthorized modification of ICS access control mechanisms - Unauthorized access to most ICS assets (pivoting/ own) - Unauthorized access to ICS assets (communication protocols) - Unauthorized access to business assets (pivoting)
Safety systems	- Safety engineering tools - Plant/emergency shutdown communication channels (DCS/SCADA) - Control (safety) network - Software vendor support portal	- Exploitation of unpatched applications - Installation of malware via unvalidated vendor software - Installation of malware via removable media - Installation of malware via keyboard - Authorization to ICS applications without sufficient access control mechanisms	- Plant shutdown - Equipment damage/ sabotage - Environmental impact - Loss of life - Product quality - Company reputation

Table 7.2 Attack Targets—cont'd

Target	Possible Attack vectors	Possible Attack methods	Possible consequences
Telecommunications systems	- Public key infrastructure - Internet visibility	- Disclosure of private key via external compromise - Exploitation of device "unknowingly" connected to public networks - Network access through unmonitored access points - Network pivoting through unregulated network boundaries	- Credential leakage (control) - Information leakage - Unauthorized remote access - Unauthorized access to ICS assets (pivoting) - Command and control
Uninterruptible power systems (UPS)	- Electrical management network - Vendor/subcontractor maintenance	- Exploitation of unpatched application (building management system) - Installation of malware via unvalidated vendor software - Network access through unprotected access points - Network pivoting through unregulated network boundaries	- Equipment failure/ shutdown - Plant upset/shutdown - Credential leakage - Unauthorized access to ICS assets (pivoting)
User – ICS engineer	- Social engineering—corporate assets - Social engineering - personal assets - E-mail attachments - File shares	- Introduction of malware through watering hole or spear-phishing attack on business PC - Introduction of malware via malicious email attachment on business PC from trusted source - Introduction of malware on control network via unauthorized/ foreign host - Introduction of malware on control network via shared virtual machines - Introduction of malware via inappropriate use of removable media between security zones (home - business - control) - Propagation of malware due to poor segmentation and "full visibility" from EWS - Establishment of C2 via inappropriate control-to-business (outbound) connections	- Process/plant information leakage - ICS design/application credential leakage - Unauthorized access to business assets (pivoting) - Unauthorized access to ICS assets (pivoting/own)
User – ICS technician	- Social engineering—corporate assets - Social engineering—personal assets - E-mail attachments - File shares	- Introduction of malware on control network via connection of unauthorized/foreign host - Introduction of malware on control network via shared virtual machines - Introduction of malware via inappropriate use of removable	- Plant upset/shutdown - Delay plant startup - Mechanical damage/ sabotage - Unauthorized manipulation of operator graphics—inappropriate response to process action

Continued

Table 7.2 Attack Targets—cont'd

Target	Possible Attack vectors	Possible Attack methods	Possible consequences
		media between security zones (home - business - control) - Exploitation of applications due to unnecessary use of administrative rights - Network disturbances resulting from connection to networks with poor segmentation	- Unauthorized modification of ICS database(s) - Unauthorized modification of status/alarms settings - Unauthorized download of faulty firmware - Unauthorized startup/ shutdown of ICS devices - Design information leakage - ICS application credential leakage - Unauthorized access to most ICS assets (pivoting/ own)
Users – plant operator	- Keyboard - Removable media—USB - Removable media—CD/ DVD	- Introduction of malware on control network via unauthorized/ foreign host - Introduction of malware via inappropriate use of removable media between security zones (home—business—control) - Exploitation of applications due to unnecessary use of administrative rights	- Plant upset/shutdown - Mechanical damage/ sabotage - Unauthorized startup/ shutdown of mechanical equipment - Process/plant operational information leakage - Credential leakage - Unauthorized access to ICS assets (pivoting) - Unauthorized access to ICS assets (communication protocols)

persistence can often be established, enabling an attacker to gather intelligence over time. In systems that make up a nexus between other systems (such as a control room SCADA server), a persistent presence can also be used to launch secondary attacks against other portions of the industrial network—such as basic control and process control zones that reside within the supervisory zone.

It is important to understand at this point the difference between *compromising* or "owning" a target and *attacking* a target. There is no formal definition that defines either, but for the purposes of this book, a compromise can be thought of as the ability to exploit a target and perform an *unknown* action (such as running a malicious payload). An attack, on the other hand, can be thought of as causing a target to perform an *undesirable* action. In this case, the device may be performing as designed, yet the ability to attack the device and cause it to perform an action that is not desired by the engineer may lead to negative consequences. Many ICS devices can therefore be attacked via the *exploitation of functionality* versus the *exploitation of vulnerabilities*. In other words, issuing a "shutdown" command to a control device does not represent any particular weakness in the device per se. However, if the lack of authentication enables a malicious user to inject a shutdown command (i.e., perform a replay attack), this is a major vulnerability.

Attack phases

Cyberattacks are complex and require multiple stages of development and execution if they are to be successful. This is an important consideration when planning cyber-defensive strategies and when responding to a cybersecurity incident. The attack that you are anticipating (from a defensive context) or are responding to (from an incident response context) might be part of a larger overall effort or attack campaign. Before an attack can even be developed, there is initial reconnassance that must be performed, and in many cases, initial cyberattack phases are needed to identify and enumerate target systems before further attack stages can be developed.

Understanding where a specific event falls within an overall attack strategy will help determine the best way to respond. To this end, frameworks such as the MITRE ATT&CK framework can be useful tools in understanding attack tactics and strategies. This will be covered in detail in Chapter 10, "OT Attack and Defense Lifecycles." However, when discussing cyberattacks against industrial systems, it is important to understand the basic approach of a cyber-physical attack, and it will be discussed here briefly.

For industrial networks, the target infrastructure is not directly accessible, with the exception of specifically allowed network connections (see Chapter 9, "Establishing Zones and Conduits"). Therefore, an attacker needs to divide their efforts into a minimum of two phases: an initial attack phase that is intended to gain access to the industrial network environment, followed by an additional attack phase once the industrial network has been breached.

Initial attack phases

Initial attack phases involve gaining access to the industrial network environment. Because industrial networks should be isolated from direct network connectivity, this means compromising an initial target to provide the attacker with a landing point from which to pivot into the industrial environment. This could be the penetration of nonindustrial networks within an organization, in an attempt to identify a method of reaching the target industrial network. This could be the discovery of legitimate communication paths in hopes of manipulating them, identifying architectural weaknesses such as a dual-homed server, etc.

The initial attack phases primarily concern nonindustrial networks and systems, and will typically require data exfiltration and some means of command and control, so that attackers can learn about their target environment and develop a more targeted attack to ultimately penetrate the industrial network.

Industrial attack phases

Once the attacker breaches the industrial network, additional attacks can be developed against the industrial environment. The various stages of the industrial attack phase differ because of the nature of industrial control systems and are the primary focus of this book. Industrial attack phases can include: additional stages of reconnaissance to learn as much as possible about the industrial control system and how it functions; implementation of remote access and command and control functions to allow an attacker to directly interact with the industrial environment; direct manipulation of the industrial control environment to develop and execute targeted cyber-physical outcomes; or perhaps something less sophisticated such as DoS disruptions, data destruction, or the encryption of highly critical systems for purposes of disruption and/or extortion (ransomware in particular represents a nonindustrial attack that can cause major disruption to industrial systems and will be discussed in more depth later in this chapter).

Cyber-physical attacks

Cyber-physical attacks refer to any attack that leverages cyber techniques (the manipulation of a computing or other digital system) in order to produce a physical outcome (typically via the manipulation of an industrial automation or control system). In properly segmented networks, cyber-physical attacks almost always consist of the above two attack phases: the first to reach the industrial network, which provides access to control system assets, and the second to manipulate those assets. However, cyber-physical attacks can take many forms, including:

- Insider attacks. An insider with direct access to industrial control assets could directly manipulate the process control system to achieve a specific physical outcome.
- Unintentional local access. Accessing industrial control assets directly, without the knowledge or consent of the user, using corrupt or infected physical media (see "Keylogging/Keystroke Injections/HID Attacks" below).
- Lateral attacks. Manipulating one part of a process control system in order to cause a specific physical outcome in another part of that process. For example, manipulating a controller that is responsible for valve within a pipeline in order to create bubbles that could cause a physical failure of downstream hardware components (known as a "cavitation attack").[5]

Rogue access devices

A rogue access device refers to any device capable of providing unwanted access to networks or computers. This could include anything from a misconfigured server to purpose-built covert hardware designed to provide a malicious backdoor to your network. Some examples include:

- A server that is unintentionally dual-homed to two networks.
- A server that is misconfigured to act as a WiFi access point, providing a wireless vector the physical network.
- A modem or remote access gateway that is installed by a service technician to facilitate troubleshooting, and then left operational.
- A server, such as a Raspberry Pi or other single-board computer, that is intentionally connected to the network to provide a covert backdoor to attackers.
- A server that is infected with remote access malware, allowing it to send unwanted communications over existing network paths.
- Networked devices such as smartphones that have broad communications capabilities, that are physically plugged into a computer (e.g., for charging).

Rogue access devices are not always malicious but present significant risk to industrial networks because they can be leveraged by an attacker to obtain information and establish command and control, which are necessary steps for many industrial cyberattacks (See Chapter 10, "OT Attack and Defense Lifecycles").

Rogue access devices are also not always obvious. There are numerous hardware attack platforms that are designed specifically to penetrate networks with a host of capabilities including full penetration testing software suites, that also present a variety of wired and wireless network connectivity options. When successfully implanted into an industrial network, these devices can potentially bypass the cybersecurity controls at the industrial network perimeter, providing attacks with direct access to industrial systems and assets.

Keylogging/keystroke injections/HID attacks

A common attack technique for industrial systems involves the manipulation of human interface devices (HIDs). Computers inherently trust input devices such as keyboards and mice; access controls apply to the user of the device, and it is therefore assumed that what the user is typing, pointing to, or clicking within the user interface is an intentional act of that user. This trust can be manipulated by attackers in several ways, including keylogging and keystroke injections.[6]

Keylogging refers to the monitoring and capture of keystrokes. The most obvious use case is for the capture of login credentials, but keylogging can be used to capture any data that is input by the user.

Keystroke injection refers to the process of emulating keyboard interaction, tricking the computer into believing that the user typed something that they did not. Because HID devices operate with the same privileges as their user, anything that is "typed" via a keystroke injection will function as if typed by the keyboard's user. For example, if a user is logged in as an administrator any command typed into the Windows console can be executed with administrator privileges as well.

These types of attacks are often referred to as HID attacks. They may be a function of malicious software (malware), although they are increasingly leveraging custom hardware, especially using the ubiquitous USB standard interface(s). Often referred to as a USB Attack Platforms, or UAPs, these devices are inexpensive and easily obtainable.[7] Often sold as penetration testing tools (e.g., the USB Rubber Ducky), they make keystroke injections extremely powerful, scriptable, and easy to deploy by disguising the UAP as a legitimate USB device (e.g., a USB thumb drive). While these tools have been available for many years, they are very popular and are still being actively developed and enhanced. In addition, new variants of UAPs continue to be introduced. In 2020, the system-on-chip hardware used to build such tools became sufficiently miniaturized to enable the development of highly covert UAPs that appear identical to USB charging cables. The power of these devices also continued to improve, and UAP cables are currently available that include built in exfiltration, communications channels, network command and control, keylogging, and keystroke injection.[8]

Keylogging and keystroke injection attacks are especially powerful when used together: first capturing information typed by a legitimate human user and then later injecting keystrokes as that same user. For example, capturing keystrokes in order to obtain the username and password of the user and then "typing" those credentials via keystroke injection after business hours to effectively log in to an unattended system.

When HID attacks are combined with network communications, these attacks can be developed and executed remotely, and if combined with rogue access devices, the network communications can occur via rogue networks that may bypass your organization's network security controls. That is, rather than communicating to a command and control server using your organization's network (which may be monitored), communication could occur via a dedicated Wi-Fi access point using an unmonitored wireless network, or communicate via a cellular network to provide reach.

While HID attacks are not limited to USB, the prevalence of USB devices and the ubiquitous support of USB devices by most computing platforms make USB a common vehicle for HID attacks. In addition, the USB standard's power delivery functionality allows UAP devices to power additional components (such as the aforementioned rogue access points). Please refer to Chapter 11, "Implementing Security and Access Controls" for specific guidance on how to protect against these types of attacks.

Man-in-the-middle attacks

A man-in-the-middle attack refers to an attack where the attacker goes between communicating devices and snoops the traffic between them. The attacker is actually connecting to both devices, and then relaying traffic between them so that it appears that they are communicating directly, even though they are really communicating through a third device that is eavesdropping on the interaction. To perform a MitM attack, the attacker must be able to intercept traffic between the two target systems and inject new traffic. If the connection lacks encryption and authentication—as is often the case with industrial protocol traffic—this is a very straightforward process. Where authentication or encryption are used, an MitM attack can still succeed by listening for key exchanges and passing the attacker's key in place of a legitimate key. This attack vector is somewhat complicated in industrial networks because devices can communicate via sessions that are established and remain intact for long periods of time. The attacker would have to first hijack an existing communication session. The biggest challenge to a successful MitM attack is successfully inserting oneself into the message stream, which requires establishing trust. In other words, the attacker needs to convince both sides of the connection that it is the intended recipient. This impersonation can be thwarted with appropriate authentication controls. Many industrial protocols unfortunately authenticate in clear text (if at all), facilitating MitM attacks within the various industrial control systems.

Denial-of-service attacks

Denial-of-service attacks occur when some malicious event attempts to make a resource unavailable. This is a very broad category of attacks and can include anything from loss of communications with the device, to inhibiting or crashing particular services within the device (storage, input/output processing, continuous logic processing, etc.). DoS attacks in traditional business systems do not typically result in significant negative consequences if resolved in a timely manner. Access to a web page may be slowed, or email delivery delayed until the problem is resolved. However, while there are rarely physical consequences associated with the interruption of services, a well-targeted DoS could bring very important systems off-line, and could even trigger a shutdown.

Automation systems are deployed to monitor or control a physical process. This process could be controlling the flow of crude oil in a pipeline, converting steam into electricity, or controlling ignition timing in an automobile engine. The inability of a controller such as an SIS to perform its action is commonly called "Loss of Control (LoC)" and typically results in the physical process being placed in a "safe" state—shutdown! This means that even simple disruptions of control functions can quickly translate into physical plant disturbances that can further lead to environmental releases, plant shutdowns, mechanical failure, or other catastrophic events. In the case of the HMI, it is not directly connected to the mechanical equipment; however, in many manufacturing industries, the inability of the HMI to perform its function can lead to "Loss of View (LoV)," which often requires the manufacturing process to be shut down if view of data cannot be restored in a timely manner. In the case of an automobile's ignition control system, if the controller stops performing, the engine stops running!

A hacker typically does not boast of a DoS attack on an Internet-facing website (unless you are part of a hacktivist group) but because a DoS can result in LOV or LOC, a similar DoS attack on an ICS can lead to far greater consequences: an oil spill, a plant fire and explosion, or spoiled batches of products. DoS in industrial environments is much more than an inconvenience but can lead to significant consequences if not managed accordingly.

Replay attacks

Initiating specific process commands into an industrial protocol stream requires an in-depth knowledge of industrial control system operations. It is possible to capture packets and simply replay them to inject a desired process command into the system because most industrial control traffic is transmitted in plain text. When capturing packets in a lab environment, a specific command can be initiated through a console, and the resulting network traffic captured. When these packets are replayed, they will perform the same command. When commands are in clear text, it is simple to find and replace a command from within captured traffic to create custom packets that are crafted to perform specific tasks. If traffic is captured from the field, authentication mechanisms (symmetric encryption, challenge-response, cleartext exchange, etc.) can be captured as well allowing an attacker to authenticate to a device via a replay attack, providing an authorized connection through which additional recorded traffic can be played back. This capability is actually part of many open-source and licensed industrial protocols and is why this can best be referred to as *exploitation of functionality*. If the device is a PLC or other process automation controller, such as the controller functions found in more advanced substation gateways, the behavior of an entire system could be altered. If the target is an IED, specific registers could be overwritten to inject false measurements or readings into a system.

Security researcher Dillon Beresford demonstrated a PLC replay attack at the 2011 Black Hat conference in Las Vegas, NV. The attack began by starting a Siemens SIMATIC STEP 7 engineering console and connecting to a PLC within a lab environment. Various commands were then initiated to the PLC via the STEP 7 console while traffic was being captured. This traffic included a valid STEP 7 to PLC session initiation, allowing the recorded traffic to be played back against any supported PLC to replay those same commands in the field.[9]

Replay attacks are useful because of the command-and-control nature of an ICS. A replay attack can easily render a target system helpless because commands exist to enable or disable security, alarms, and logging features. Industrial protocols also enable the transmission of new programmable code (for device firmware and control logic updates), allowing a replay attack to act as a "dropper" for malicious logic or malware. Researcher Ralph Langner described how simple it could be to write malicious ladder logic at the 2011 Applied Control Systems Cyber Security Conference. He was able to inject a time-bombed logic branch with just 16 bytes of code that was inserted at the front of existing control logic that will place the target PLC into an endless loop—preventing the remaining logic from executing and essentially "bricking" the PLC.[10]

For the subtle manipulation of industrial systems and automation processes, knowledge of specific ICS operations is required. Much of the information needed to attack a PLC can be obtained from the device itself. For example, in Beresford's example, packet replay was used to perform a PLC scan. Using SIMATIC requests to probe a device, Beresford was able to obtain the model, network address, time of day, password, logic files, tag names, data block names, and other details from the targeted PLC.[11]

If the goal is simply to sabotage a system, almost anything can be used to disrupt operations—a simple replay attack to flip the coils in a relay switch is enough to break most processes.[12] In fact, malware designed to flip specific bits could be installed within ICS assets to manipulate or sabotage a given process with little chance of detection. If only read values are manipulated, the device will report false values; if write commands are also manipulated, it would essentially render the protocol functionality useless for that device.

Compromising the human-machine interface

One of the easiest ways to obtain unauthorized command and control of an ICS is to leverage the capabilities of a human-machine interface (HMI) console. Whether an embedded HMI within a control zone, or the centralized command and control capability of DCS, SCADA, EMS or other systems, the most effective way to manipulate those controls is via their console interface. Rather than attacking via the industrial network using MitM or Replay attacks, a known device vulnerability is exploited to install remote access to the console leading to a host *compromise*. One example would be to use the Metasploit framework or similar penetration testing tool to exploit the target system and then using the Meterpreter shell to install a remote VNC server. Now, the HMI, SCADA, or EMS console is fully visible to and controllable by the attacker. This allows the hacker to directly monitor and control whatever that console is responsible for, remotely. There is no knowledge of industrial protocols needed, no specific experience in ladder logic, or control systems operations—only the ability to interpret a graphical user interface, click buttons, and change values within a console that is typically designed for ease of use.

Compromising the engineering workstation

The vectors used to compromise an Engineering Workstation (EWS) are not much different from those used previously with the HMI. The same vulnerabilities often apply because the system is managed consistently across all hosts. The same payloads (Meterpreter) can also be used to establish C2 functionality. What is important to consider in this case is the relative value of the logical assets contained on the EWS versus those on the HMI. The HMI does provide bidirectionality read/write capability with the process under control; however, many systems today incorporate role-based access control that may limit the extent of these functions in a distributed architecture consisting of multiple operators and multiple plant areas or units.

The EWS, on the other hand, is typically the single host that not only possesses the capability to configure such role-based access control mechanisms but also the specialized tools needed to directly communicate with, configure, and update the primary control equipment (PLC, BPCS, SIS, IED, etc.). It is also common for the EWS to contain significant amounts of sensitive documentation specific to the ICS design, configuration, and plant operation, making this target a much higher-valued asset than a typical HMI.

Blended attacks

Many attacks are more than single exploits against a single vulnerability on a single target. Sophisticated attacks commonly use a blended threat model. According to SearchSecurity, "a blended threat is an exploit that combines elements of multiple types of malware and usually employs multiple attack vectors to increase the severity of damage and the speed of contagion."[13]

In the past, blended attacks typically contained multiple types of malware that were used in succession—a spear phishing attack to access systems behind a firewall that would drop a Remote Access Trojan (RAT) and then obtain the credentials needed to access the trusted industrial networks, where targets may be compromised or exploited further.

Recently, blended threats have evolved to a much greater degree of complexity. This was first observed with Stuxnet where a single complex and mutating malware framework was deployed that was capable of behaving in multiple ways depending upon its environment. This concept has now been taken even further, with the discovery of Skywiper (also known as Flame) and other complex malware variants.

Weaponized industrial cyberthreats

Cyberattacks against industrial networks were, at one time, purely theoretical. We have now seen real cyberattacks targeting actual industrial systems. The first documented ICS cyber-attack "in the wild" was Stuxnet discovered in 2010, which was followed shortly by a string of incidents over the next few years. While many high-profile incidents occurred, often targeting the oil industry and countries of the Middle East, Stuxnet remains a strong example of what a modern, weaponized industrial cyberattack looks like. Stuxnet was very precise, sabotaging specific ICS devices to obtain a specific goal. Shortly after Stuxnet, Shamoon (also DistTrack) and Flame (also called Flamer or Skywiper) surfaced. Shamoon was widely publicized due to its highly destructive nature. Rather than performing a precision attack against target devices, like Stuxnet, Shamoon spread promiscuously and wiped systems clean, incurring huge impact to the computing infrastructure of infected companies. Flame showed signs of being a derivative of Stuxnet, with even greater sophistication. However, the intention of Flame seems to be espionage rather than sabotage or the direct destruction of target systems.

Stuxnet

Stuxnet is the poster child of industrial malware. When discovered, it was the first real example of weaponized computer malware, which began to infect ICSs as early as 2007.[14] Any speculation over the possibility of a targeted cyberattack against an industrial network has been overruled by this extremely complex and intelligent collection of malware. Stuxnet is a tactical nuclear missile in the cyberwar arsenal. It was not just a "shot across the bow," but rather it hits its mark and left behind the proof that extremely complex and sophisticated attacks can and do target industrial networks. The worst-case scenario has now been realized—industrial vulnerabilities have been targeted and exploited by a sophisticated threat actor more commonly called an advanced persistent threat (APT).

Although early versions of Stuxnet were released as early November 2007,[15] widespread discussions about it did not occur until the summer of 2010, after an Industrial Control Systems Cyber Emergency Response Team (ICS-CERT) advisory was issued.[16]

Stuxnet was armed with four 0-days in total at its disposal. Stuxnet was able to infect Windows-based computers covering four generations of kernels from Windows 2000 up to and including Windows 7/Server 2008R2. The primary target was a system comprising Siemens SIMATIC WinCC and PCS7 software along with specific models of S7 PLCs utilizing the PROFIBUS protocol to communicate with two specific vendors of variable frequency drives (VFD). These VFDs were used to control the centrifuges used in the process of enriching uranium.[17] (PROFIBUS is the industrial protocol used by Siemens and was covered in Chapter 6, "Industrial Network Protocols".) The subsequent steps taken by the malware depend on what software was installed on the infected host. If the host was not the intended target, the initial infection would load a rootkit that would automatically load the malware at boot and allow it to remain undetected. It then would deploy up to seven different propagation methods to infect other targets. For those methods using removable media, the malware would automatically remove itself after the media infected three new hosts. If the target contained Siemens SIMATIC software, methods existed to exploit default credentials in the SQL Server application allowing the malware to install itself in the WinCC database, or to copy itself into the STEP 7 project file used to program the S7 PLCs. It also had the ability to overwrite a critical driver used to communicate with the S7 PLCs effectively creating a MitM attack allowing the code running in the PLC to be altered without detection by the system users.

Although little was known at first, Siemens effectively responded to the issue, quickly issuing a security advisory, as well as a tool for the detection and removal of Stuxnet. Stuxnet drew the attention of the mass media through the fall of 2010 for being the first threat of its kind—a sophisticated and blended threat that actively targets ICS—and it immediately raised the industry's awareness of advanced threats by illustrating exactly why industrial networks need to dramatically improve their security measures.

Dissecting stuxnet

Stuxnet is very complex, as can be seen by the Infection Process shown in Figure 7.2. It was used to deliver a payload targeting not only a specific control system but also a specific configuration of the control system including unique model numbers of PLCs and vendors of field-connected equipment. It is the first rootkit targeting ICS. It can self-update even when cut off from the C2 servers (which is necessary should it find its way into a truly air-gapped system) by enumerating and remembering a complex peer-to-peer network necessary to allow external access. It is able to inject code into the PLCs,

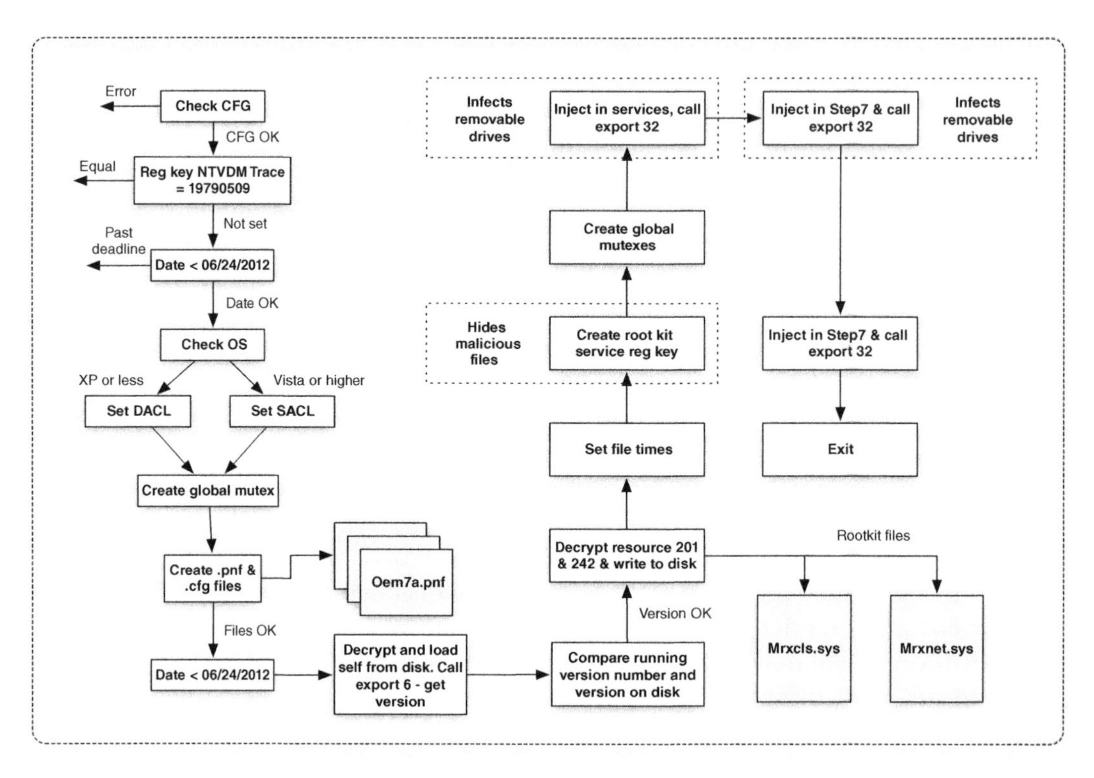

FIGURE 7.2 Stuxnet's infection processes.

and at that point alter the operations of the PLC as well as hide itself by reporting false information back to the HMI. It adapts to its environment. It uses system-level, hard-coded authentication credentials that were publicly disclosed as early as 2008[18] (indications exist that it was disclosed within the Siemens Support portal as early as 2006[19]). It was able to install malicious drivers undetected by Windows through the use of two different legitimate digital certificates manufactured using stolen keys. There is no doubt about it at this time—Stuxnet is an advanced new weapon in the cyber war.

What it does

The full extent of what Stuxnet is capable of doing remains uncertain at the time of this writing. What we do know is that Stuxnet does the following:[20]

- Infects Windows systems using a variety of 0-day exploits and stolen certificates and installing a Windows rootkit on compatible machines.
- Attempts to bypass behavior-blocking and host intrusion-protection-based technologies that monitor LoadLibrary calls by using special processes to load any required DLLs, including injection into preexisting trusted processes.

- Typically infects by injecting the entire DLL into another process and only exports additional DLLs as needed.
- Checks to make sure that its host is running a compatible version of Windows, whether or not it is already infected, and checks for installed **antivirus** before attempting to inject its initial payload.
- Spreads laterally through infected networks, using removable media, network connections, print services, WinCC databases, and/or Step 7 project files.
- Looks for target industrial systems (Siemens SIMATIC WinCC/PCS7). When found, it injects itself into an SQL database (WinCC) or project file (Step 7), and replaces a critical communication driver that will facilitate authorized and undetected access to target PLCs.
- Looks for target system configuration (S7-315-2/S7-417 PLC with specific PROFIBUS VFD). When found, it injects code blocks into the target PLCs that can interrupt processes, inject traffic on the Profibus-DP network, and modify the PLC output bits, effectively establishing itself as a hidden rootkit that can inject commands to the target PLCs.
- Uses infected PLCs to watch for specific behaviors by monitoring PROFIBUS.
- If certain frequency controller settings are found, Stuxnet will throttle the frequency settings sabotaging the centrifuge system by slowing down and then speeding up the motors to different rates at different times.
- It includes the capabilities to remove itself from incompatible systems, lay dormant, reinfect cleaned systems, and communicate peer to peer in order to self-update within infected networks.
- It includes a variety of stop execution dates to disable the malware from propagation and operation at predetermined future times.

What we do not know at this point is what the full extent of damage could be from the malicious code that is inserted within the PLC. Subtle changes in **set points** over time could go unnoticed that could cause failures down the line, use the PLC logic to extrude additional details of the control system (such as command lists), or just about anything. Another approach might be to perform man-in-the-middle attacks intercepting invalid process values received from the PLCs and forward to the WinCC HMI bogus values for display making the plant operator unaware of what is actually occurring in the plant. Because Stuxnet has exhibited the capability to hide itself and lie dormant, the end goal is still a mystery.

Lessons learned

Because Stuxnet is such a sophisticated piece of malware, there is a lot that we can learn from dissecting it and analyzing its behavior. A detailed white paper coauthored by one of the authors of this book has been developed that specifically analyzes Stuxnet in terms of its impact on industrial control systems, and how they are designed and deployed in actual operational environments.[21] How did we detect Stuxnet? It succeeded largely because it was so widespread and infected approximately 100,000 hosts searching for a single target. Had it been deployed more tactically, it might have gone unnoticed—altering PLC logic and then removing itself from the Siemens SIMATIC hosts that were used to inject those PLCs. How will we detect the next one? The truth is that we may not, and the reason is simple—our "barrier-based" methodologies do not work against cyberattacks that are this well researched and funded. Furthermore, since Stuxnet's propagation mechanisms were all LAN-based, the target host must be assumed on direct or adjacent networks to the initial infection. In other words, the attack originated from inside the targeted organization. They are delivered via 0-days, which means we do not detect them until they have been deployed, and they infect areas of the control system that are difficult to monitor.

So what do we do? We learn from Stuxnet and change our perception and attitude toward industrial network security (see Table 7.3). We adopt a new "need to know" mentality of control system communication. If something is not explicitly defined, approved, and allowed to execute and/or communicate, it is denied. This requires understanding how control system communications work, establishing that "need to know" and "need to use" in the form of well-defined security zones with equally defined perimeters, establishing policies and baselines around those zones, and then implementing cyber security controls and countermeasures to enforce those policies and minimize the risk of a successful cyber-attack.

It can be seen in Table 7.3 that additional security measures need to be considered in order to address new "Stuxnet-class" threats that go beyond the requirements of

Table 7.3 Lessons Learned From Stuxnet

Previous beliefs	Lessons learned from stuxnet
Control systems can be effectively isolated from other networks, eliminating risk of a cyberincident.	Control systems are still subject to human nature: a strong perimeter defense can be bypassed by a curious operator, a USB drive, and poor security awareness.
PLCs and RTUs that do not run modern operating systems lack the necessary attack surface to make them vulnerable.	PLCs can and have been targeted and infected by malware.
Highly specialized devices benefit from "security through obscurity." because industrial control systems are not readily available, it is impossible to effectively engineer an attack against them.	The motivation, intent, and resources are all available to successfully engineer a highly specialized attack against an industrial control system.
Firewalls and intrusion detection and prevention system (IDS/IPS) are sufficient to protect a control system network from attack.	The use of multiple 0-day vulnerabilities to deploy a targeted attack indicates that "blacklist" point defenses, which compare traffic to definitions that indicate "bad" code are no longer sufficient, and "whitelist" defenses should be considered as a catchall defense against unknown exploits.

compliance mandates and current best-practice recommendations. New measures include layer 7 application session monitoring to discover 0-day threats and to detect covert communications over allowed "overt" channels. They also include more clearly defined security policies to be used in the adoption of policy-based user, application, and **network whitelisting** to control behavior in and between zones (see Chapter 9, "Establishing Zones and Conduits").

Tip

The axiom "to stop a hacker, you need to think like a hacker" was often used before Stuxnet. This simply meant that in order to successfully defend against a cyberattack you need to think in terms of someone trying to penetrate your network. This philosophy still has merit, the only difference being that now the "hacker" can be thought of as having a much greater knowledge of deployed ICSs, an understanding of the manufacturing processes, and how the ICS is used to control this environment, along with significantly more resources and motivation. The ISA 62443 family of industry standards provides the ability to address each of these aspects in terms of a **security level**. In the post-Stuxnet world, imagine building a digital bunker in the cyber war, rather than simply defending a network, and aim for the best possible defenses against the worst possible attack. In other words, "think like an insider."

Shamoon/DistTrack

Shamoon, or W32.DistTrack (often shortened to "DistTrack"), possesses both information gathering and destructive capabilities. Shamoon will attempt to propagate to other systems once an initial infection occurs, exfiltrate data from the currently infected system, and then cover its tracks by overwriting files, including the system's master boot record (MBR). The system is then unusable, and overwritten data are not recoverable once the MBR is destroyed. The result, Shamoon left a path of inoperable systems in its wake.[22]

Shamoon accomplished this through three primary components:[23]

- Dropper—a modular component responsible for initial infection and network propagation (often through network shares)
- Wiper—a malware component responsible for system file and MBR destruction
- Reporter—a component designed to communicate stolen data and infection information back to the attacker.

Much of the details around Shamoon are protected from disclosure; however, Shamoon reportedly infected business systems of Saudi Aramco (an oil and gas company in the Kingdom of Saudi Arabia) and caused the destruction of at least 30,000 systems. Luckily, this destruction did not spread to industrial network areas and therefore did not directly impact oil production, refining, transportation, or safety operations.[24]

Flame/flamer/skywiper

Skywiper is an advanced persistent threat that spread actively, targeting Middle Eastern countries, with the majority of infections occurring in Iran. Like Stuxnet, Skywiper (Flame) redefined the complexity of malware in its time. Skywiper had been active for years prior to being discovered also like Stuxnet, mining sensitive data and returning them to a sophisticated C2 infrastructure consisting of over 80 domain names, and using servers that moved between multiple locations, including Hong Kong, Turkey, Germany, Poland, Malaysia, Latvia, the United Kingdom, and Switzerland.[25]

Over a dozen modules are present within Skywiper, including:[26]

- "Flame"—handles AutoRun infection routines (Skywiper is often referred to as Flame because of this package)
- "Gadget"—an update module that allows the malware to evolve and to accept new modules and payloads
- "Weasel" and "Jimmy"—handle disk and file parsing
- "Telemetry" and "Gator"—handle C2 routines
- "Suicide"—self-termination
- "Frog"—exploit payload to steal passwords
- "Viper"—exploit payload that captures screenshots
- "Munch"—exploit payload that captures network traffic

Skywiper seems to be focused on espionage rather than sabotage. No modules dedicated to manipulation or sabotage of industrial systems have been detected at the time of this writing. The modular nature of Skywiper would certainly allow the threat to include more damaging modules as needed, no doubt leveraging the "Gadget" update module to further evolve the malware into a directed cyber weapon.

Dragonfly

Dragonfly is an example of a nondestructive cyber campaign targeting industrial control systems. In 2014, the first reports of Dragonfly emerged from F-Secure and Symantec. The Dragonfly campaign is named after the Dragonfly group (also known as "Energetic Bear"). This group "initially targeted defense and aviation companies in the US and Canada before shifting its focus to US and European energy firms in early 2013," according to Symantec.[27]

The Dragonfly campaign used several different malicious payloads. However, the Havex RAT (also known as Backdoor.Oldrea) was the primary malware utilized in approximately 95% of Dragonfly payloads.[28] For this reason, Dragonfly is also referred to as Havex.

Dragonfly was initially believed to be a cyberespionage campaign targeting the energy sector,[29] although subsequent reports claimed that the actual target was the pharmaceutical industry.[30] Regardless of the intended target sector, Dragonfly is of interest largely because of the types of organizations that it was able to compromise and exfiltrate data from. In Symantec's analysis, "if they had used the sabotage capabilities open to them, could have caused damage or disruption to the energy supply in the affected countries."[31]

Dragonfly is also of interest because of its clever vectors for targeting industrial control systems. Initially, the Havex RAT was distributed via spear phishing campaigns, but the Dragonfly group shifted tactics to watering hole attacks using several energy-related websites. Legitimate websites were compromised and malicious frames were injected that would redirect to an additional compromised website, which in turn compromised the victims' computers by exploiting either Java or Internet Explorer.[32]

Finally, Dragonfly was able to compromise of a number of legitimate software packages from industrial control equipment vendors, inserting malware into software that the vendors had made available on their websites.[33]

While nondestructive, Dragonfly is considered to have been a successful reconnaissance campaign, and one can speculate that the knowledge obtained contributed to the development of future industrial control attack techniques.

BlackEnergy

Nearly a quarter of a million energy customers in Ukraine experienced a blackout on December 23, 2015, as the result of the first known real-world example of a successful, targeted cyberattack against the energy grid.[34] The attack—often called "Black Energy" in reference to the attack's utilization of a malware tool named "BlackEnergy 3"—illustrates the effectiveness of a true attack campaign versus more autonomous malware frameworks. The attackers combined targeted malware with "careful planning, discipline in execution, and capability in many of the discrete tasks exhibited" over almost an entire year prior to the final stage of the attack.[35]

The attack combined many techniques in a highly coordinated attack. This is perhaps best summarized in a DHS NCCIC in an Incident Alert issued in March 2016 that called for critical infrastructure owners to "implement enhanced cyber measures that reduce risks from the following types of adversary techniques:[36]

- Theft of legitimate user credentials to enable access masquerading as approved users
- Leveraging legitimate remote access pathways (VPNs)[37]
- The remote operation of human-machine interface (HMI) via company-installed remote access software (such as RDP, TeamViewer, or rlogin)[38]
- The use of destructive malware such as KillDisk to disable industrial control systems (ICSs) and corporate network systems[39]
- Firmware overwrites that disable/destroy field equipment[40]
- Unauthorized scheduled disconnects of uninterruptible power supplies (UPS) to devices to deny their availability[41]
- The delivery of malware via spear-phishing emails and the use of malicious Microsoft Office attachments[42]
- Use of Telephone Denial of Service (TDoS) to disrupt operations and restoration"[43]

After considerable research throughout the public and private sector, a more complete timeline of the attack became clear. In the report, "When The Lights Went Out: A Comprehensive Review Of The 2015 Attacks On Ukrainian Critical Infrastructure" the full extent of the campaign was described in 17 discrete steps[44] including:

- Initial reconnaissance, development of initial malware, and development of documents with which to deliver the initial infections.[45]
- The delivery of those documents via a phishing campaign and the subsequent installation of the BlackEnergy 3 RAT.[46]
- BlackEnergy 3 establishes connectivity to attacker-owned command and control (C2) servers, the subsequent installation of additional malware plugins designed to harvest credentials, and the harvesting of credentials using these new plugins.[47]
- Internal reconnaissance and lateral movement on both the targeted network and the industrial control system network to identify targets and the subsequent development of a malicious firmware update for identified serial-to-Ethernet converters.[48]
- Ensuring planned disruptions to incident response efforts by placing the KillDisk malware on a network share and setting a policy on a domain controller to retrieve the malware and execute it upon system reboot. The attackers also scheduled an Uninterruptible Power Supply (UPS) disruption to cause a coordinated outage of UPS for telephone communication servers and data center servers.[49]
- Using legitimate access and credentials to open breakers and disrupt power distribution, creating a blackout across three distribution areas.[50]
- Delivering malicious firmware updates to the serial-to-Ethernet converters used to enable network access to the breakers, bricking the converters, and effectively cutting off the communication paths needed for remote recovery. A DoS attack on telephone call centers was also initiated in coordination with the above-mentioned scheduled UPS outage, in a further attempt to hinder response efforts.[51]
- Finally, critical system data were destroyed through the schedule execution of KillDisk, which among things destroyed system log data on impacted systems.[52]

The incident illustrates that targeted campaigns consist of many discreet steps. Sometimes, referred to as a "Kill Chain", these steps are all necessary for a successful attack. This is why it is important to understand both the lifecycle of an attack and also the lifecycle of defensive efforts against such attacks (see Chapter 10, OT Attack and Defense Lifecycles").

Industroyer

Industroyer (also referred to as Crash Override) caused a blackout in parts of Kyiv, Ukraine, in 2016, nearly 1 year after the blackouts caused by BlackEnergy. Like BlackEnergy, Industroyer was a sophisticated campaign that included preparatory stages, establish C2, the development of targeted payloads, and an attempt to destroy key systems through the use of wiper malware. Unlike BlackEnergy, the Industroyer malware was designed specifically to attack the electrical grid, which it did by leveraging several of the industrial protocols used by the grid.[53] There were four modules used: one for each of the four protocols leveraged. These included: IEC 60870-5-101, IEC 60,870-5-104, IEC 61,850, and OLE for Process Control Data Access (OPC-DA).[54] These protocols are discussed in Chapter 6, "Industrial Network Protocols."

The ability for Industroyer to interact directly with relays over these protocols differentiates it from BlackEnergy, which relied largely on the harvesting of legitimate credentials and misuse of other legitimate network systems. As such, Industroyer has the ability to be used more easily as a part of less coordinated campaigns. While C2 remains a necessary function for Industroyer, the malware consisted of two command and control functions, for persistence in case the primary C2 was discovered and disabled.[55]

TRISIS/TRITON

TRISIS (also known as TRITON) was first discovered in 2017 by the analysts at Dragos, Inc.[56] It is special in that it is the first known malware to target a SIS. TRISIS specifically targeted Schneider Electric's Triconex SIS, although it is feasible that other SISs could be targeted using similar methods. As such, TRISIS represents a new era of cyber-physical threats, where the safety controls put in place to prevent physical failures could potentially be manipulated by a cyberattack.

As malware, TRISIS does not seem that special. Unlike some of the other attacks discussed here, it is fairly simple in design. It consists of a compiled python script, using a python compiler (py2exe) to allow it run on target machines in an industrial environment, including SIS targets.[57] It consists of four components: two embedded binaries that are designed to prepare the target for two external binaries. The external binaries contain replacement logic that is intended to alter the function of the SIS.[58]

Because the specific instrumentation and control logic of an SIS is extremely specific to the system it is protecting, TRISIS must be tailored to a specific target. In the initial analysis, Dragos points out that the script takes its target as a command-line argument that is passed to it on execution, implying that the script is specific to the specific target SIS device (in this case the Triconex 3008 processor module).[59] Other variants of TRISIS have been discovered in the wild, and the initial variant has also been discovered far outside of the initial targets, indicating that the scope of SIS attacks may be larger than first realized.

TRISIS requires access to a system that is able to communicate to the SIS, which requires an attacker to first gain access to such a system. Without prior knowledge of the environment, an attacker must first gain access and determine the specific SIS equipment to target, develop specific logic for TRISIS to implement, test that logic, and then deliver that specific logic for TRISIS to install and modify.[60] On its own, this makes TRISIS less concerning. However, if utilized with other attack frameworks as part of a larger targeted campaign that provides command and control to the attacker, TRISIS could be easily customized and weaponized, allowing an attacker to cause significantly greater impact. This is why SIS systems (and any critical high-value targets) should be segmented within specific and heavily monitored security zones, to limit access via the network and/or other vectors (TRISIS has also been discovered on USB drives being carried into process control environments).[61]

Industroyer2

Industroyer2 is a newer variant of Industroyer involved in another attack on against the power grid in Ukraine in April 2022.[62] Industroyer2 is in some ways simplified from the original Industroyer, yet in some ways more sophisticated. Unlike Industroyer, Industroyer2 consisted of a single Windows executable, rather than a modular framework capable of importing malware plugins. Industroyer only included one industrial protocol, IEC 60870-5-104, and there were enough code similarities that analysts at ESET concluded that Industroyer2 was built using the source code from the original Industroyer's IEC 60870-5-104 payload.

While Industroyer2 is not modular, it is highly configurable. A detailed configuration capability was hardcoded into the body of the Industroyer executable, allowing it to utilize the 104 protocol in an extremely flexible manner.[63] The lack of external configuration files means that each attack using Industroyer2 needs to be individually crafted and compiled; it also means that the entire attack is self-contained without the need to establish network communications to function.[64] According to Mandiant, "[Industroyer2] shows a nuanced understanding of the victim environment … The actor's successful implementation of IEC-104 to interact with the targeted devices indicates a robust understanding of the protocol and knowledge of the victim environment. For example, in the samples we analyzed the actor manipulated a selected list of Information Object Addresses (IOAs), which are used to interact with power line switches or circuit breakers in a remote terminal unit (RTU) or relay configuration."[65]

Together, BlackEnergy, Industroyer, and Industroyer2 show the progression of targeted energy attacks over several years.

Incontroller/pipedream

Incontroller, also known as Pipedream, represents a powerful and highly flexible cyber-physical attack platform. It includes three modular components for industrial control manipulation: TAGRUN, an OPC-UA tool for reconnaissance; CODECALL, a Modbus tool with the ability to interact with various PLCs using device-specific modules, and OMSHELL, a tool targeting Omron PLCs.[66]

Each of these components is modular and extensible and highly capable. TAGRUN is able to brute force credentials to access OPC-UA servers and then scan for OPC-UA servers over the network, read the structure of those servers, and finally both read and write specific tag data.[67] CODECALL is able to identify specific Schneider Electric

Modicon PLCs, as well as other Modbus-enabled devices on a network. It can connect to those devices using Modbus or Codesys (Schneider's proprietary protocol used by the Modicon PLCs). Using Modbus, CODECALL can request device IDs, read and write specific device registers, and manipulate command macros. When using Codesys, CODECALL can brute force PLC credentials, manipulate the filesystem (including downloading, uploading, and deleting files), manipulate network configurations, send custom raw packets over the network, and crash the device.[68] OMSHELL is able to scan for compatible Omron devices, and manipulate them in a variety of ways including activating telnet, establishing a backdoor, capturing network traffic, wiping, and resetting the device, and connecting and controlling any attached servo drives.[69]

Incontroller is interesting for several reasons. First, the presence of an OPC-UA toolkit allows it to gather information regarding the industrial environment via OPC-UA servers that could be deployed in higher levels of the Purdue model (levels 3.5 or 4. See Chapter 5, "Industrial Network Design and Architecture" and Chapter 9, "Establishing Zones and Conduits"). There are also additional toolkits used by Incontroller to attack and exploit Windows systems, likely targeting HMI systems and engineering workstations,[70] as well as providing backdoor and reconnaissance capabilities for command and control.[71] This makes Incontroller the first cyber-physical attack framework since Stuxnet to provide end-to-end capability in both the initial cyberattack phases (required to reach the industrial control environment), and subsequent cyber-physical attack phases (manipulating the industrial control environment once it can be accessed).

Incontroller is extremely flexible: consisting of multiple toolkits, each of which is modular and configurable; these tools can be used all together in a coordinated end-to-end attack or individually. At the time of this writing, Incontroller has not been associated with any known cyber-physical attack campaign or incident.[72] However, it "represents an exceptionally rare and dangerous cyber attack capability" that is "comparable to Triton ... Industroyer ... and Stuxnet."[73]

Attack trends

While the techniques and tactics of industrial cyber threats will typically follow a logical progression (see Chapter 10, "OT Defense Lifecycle & Defensive Methods"). Trends can be discovered through the analysis of known cyberincidents. However, the exact nature of attack is difficult to predict; adversaries will continue to adapt to what is most effective. This could include, but is not limited to, a shift in the initial infection vectors, the quality of the malware being deployed, its behavior, and how it spreads through networks and organizations.

Early industrial threats evolved by trending "up the stack," with exploits moving away from network-layer and protocol-layer vulnerabilities and more toward application-specific exploits, even more recent trends show signs that these applications are shifting away from the exploitation of Microsoft platform products (i.e., operating system exploitation) toward the almost ubiquitously deployed client-side applications like web browsers (Internet Explorer, Firefox, Safari, Chrome), Adobe Acrobat Reader, and Adobe Flash Player.

Newer industrial threats have continued this trend through the direct manipulation of industrial device firmware, industrial protocols, and the development of modular cyber-physical attack frameworks.

Web-based applications became popular both for initial infections and for C2. The use of social networks, such as Twitter, Facebook, Google groups, and other cloud services, also became popular because they are widely used, highly accessible, and difficult to monitor. Even more interesting is that many users access these services on mobile and portable devices that typically contain no additional security software. Many companies continue to embrace social networking for marketing and sales purposes, often to the extent that these services are allowed open access through corporate firewalls. This is further compounded by privacy concerns relating to what corporate IT is actually allowed to monitor within the social media sessions. Issues around privacy are outside the scope of this book, but it is worth noting that regulations vary widely from country to country, and that the expansion of corporate networks across borders could introduce latent security vulnerabilities that should be accounted for.

In a properly segmented industrial network, web applications and social media should be of no concern. Connections to the Internet from within an industrial network should be prohibited or at the very least strictly monitored and controlled. Unfortunately, many industrial networks remain connected to the Internet, posing an increased risk.

One quality of cybersecurity that remains consistent is that it is always changing. As awareness increased and industrial networks became more protected and difficult to infiltrate, threat actors looked for new vectors. The malware itself also evolved. There is continued evidence among incident responders and forensics teams of the existence of deterministic malware and the emergence of mutating bots. Stuxnet is a good example again, as it was one of the earliest examples of malware that contained robust logic and that could operate differently depending upon its environment. Stuxnet spreads, attempts to inject PLC code, communicates via C2, lies dormant, or awakens depending upon changes to its environment. Since Stuxnet, malware has continued to evolve.

Evolving vectors

As industrial control systems evolved and IT and OT began to converge (see Chapter 2, "Industrial Cyber Security History and Trends: The Convergence of OT and IT"), attack vectors have also evolved. In the early days of largely analogue systems and physical "air gap" security, the human element was the primary vector. Insider threats from a rogue employee were a primary concern, with the social engineering of friendly insiders being a close second.

As industrial systems became digitized, and as serial communications shifted to routable networks, the primary vector quickly shifted to the network. Most of the attack examples listed above utilize the network either as an initial attack vector or for C2 capabilities that are used to exfiltrate data and drop new malware.

As industrial network security has improved (something that the authors believe is at least partly because of the original publication of this book in 2011), attackers have been forced to adapt. Dragonfly's watering hole attacks using infected vendor files was likely an attempt to bypass the network perimeter of the ICS: instead of breaching the ICS directly, the attackers tricked an authorized user into downloading their malware, presumably onto an engineering or service laptop that would then be used by authorized users to install what they believed to be legitimate software into ICS systems. With the evolution of BlackEnergy to Industroyer to Industroyer2, we also saw a shift: reliance on network-based attacks was reduced; redundant C2 channels were introduced on the assumption that one would be detected and blocked; and ultimately payloads were developed with no network dependence at all.

A similar reaction to improved industrial network security is the use of removable media, most commonly USB thumb drives, and other portable media. Information has to flow into and out of an industrial network in order for the ICS to operate. Systems need to be patched, information needs to be managed, processes need to be refined, work orders need to be distributed, etc. All of this requires information flow. If information flow via the network is controlled too strictly, alternate methods of moving information will be necessary. USB thumb drives, and the ubiquity of USB interfaces, makes USB removable media a popular method of data transfer in industrial systems.

Unfortunately, attackers realize this also and so USB-based attacks have been developed. Initially, a USB drive might simply be a way to transfer malware in lieu of a network. However, newer and more dangerous USB attacks have been discovered. In a report by researchers at Ben-Gurion University, 29 different USB-based attacks were identified, falling into four main categories: Programable Microcontrollers; Maliciously Reprogrammed Peripherals; Not-Reprogrammed Peripherals; and Electricical.[74] Together they include everything from methods of hiding malware on USB storage volumes to utilizing the USB communications standards themselves as an attack vector. Perhaps, the most well-known USB attack is the "USB Rubber Ducky," a Programable Microcontroller, which is used for keyboard injection or "Human-Interface Device" (HID) attacks. See "Rogue Access Devices," "Key-logging," and "Keystroke Injections" above for more information on these common USB attacks methods.

How serious are USB-based threats? In 2019, Honeywell began publishing an annual USB Industrial Cybersecurity Report, which focuses solely on malware detected on USB removable media while entering or exiting an industrial control environment. The findings in 2022 (the most recent report published at the time of this writing) indicate that USB storage devices are being intentionally used to penetrate highly isolated networks such as control systems. This is inferred from the high percentage of malwares that: are designed to leverage USB devices for propagation; attempt to establish remote access and command-and-control capability; and are capable of causing disruption to an industrial control system (52%, 51%, and 81%, respectively).[75]

Supply chain vulnerabilities

Sometimes, vulnerabilities are found in commonly used operating systems, applications, or software packages that are widely used in industrial (and other) environments. When this happens, otherwise "hard" systems can suddenly become permeated with vulnerabilities. In industrial environments, this is especially problematic due to the difficulties in patching OT systems. A few examples of large-scale vulnerabilities that permeated the supply chain include vulnerabilities in the Portable Document Format (pdf), OpenSLL, and the java logging package Log4j.

Adobe Portable Document Format

Adobe Portable Document Format (PDF) is widely used as a means of reliably transferring documents with consistent formatting and a degree of data protection. Due to the popularity of the file format, PDF is a prime target for attackers looking to develop new trojan malwares. This allows the attack surface to expand significantly as there are far greater desktops to attack than servers. In some cases, the exploits utilize features within PDFs to call and execute code to perform malicious actions, rather than exploiting a specific bug or vulnerability. This occurs by either calling a malicious website or by injecting the code directly within the PDF file. For example:

- An email from a trusted source contains a compelling message; a properly targeted spear-phishing message. There is a PDF document attached to the email.
- This PDF uses a feature, specified in the PDF format, known as a "Launch action." Security researcher Didier Stevens successfully demonstrated that Launch actions can be exploited and can be used to run an executable embedded within the PDF file itself.[76]
- The malicious PDF also contains an embedded file named Discount_at_Pizza_Barn_Today_Only.pdf, which has been compressed inside the PDF file. This attachment is actually an executable file, and if the PDF is opened and the attachment is allowed to run, it will execute.
- The PDF uses the JavaScript function exportDataObject to save a copy of the attachment to the user's local computer.
- When this PDF is opened in Adobe Reader (JavaScript must be enabled), the exportDataObject function causes a dialogue box to be displayed asking the user to "Specify a file to extract to." The default file is the name of the attachment, Discount_at_Pizza_Barn_Today_Only.pdf. The exploit requires that the users' naïveté and/or their confusion regarding a message (which can be customized by the malware author[77]), they do not normally see to cause them to save the file.
- Once the exportDataOject function has completed, the Launch action is run. The Launch action is used to execute the Windows command interpreter (cmd.exe), which searches for the previously saved executable attachment Discount_at_Pizza_Barn_Today_Only.pdf and attempts to execute it.

- A dialog box will warn users that the command will run only if the user clicks "Open."

This simple and effective hack is readily available in open-source toolkits like Kali Linux and the Social Engineering Toolkit (SET) and has been used to spread known malware, including ZeusBot.[78] Although this attack vector requires user interaction, PDF files are extremely common, and when combined with a quality spear-phishing attempt, this attack can be very effective. Quality is typically measured by how trust is established with the recipient and their likelihood of opening the attachment.

Another researcher chose to infect the benign PDF with another Launch hack that redirected a user to a website, but noted that it could have just as easily been an exploit pack and/or embedded Trojan binary.

There are numerous other Adobe Reader-based vulnerabilities that employ alternate methods to compromise a victim's local computer. Adobes, and other popular client application developers, continue to struggle in keeping up with vulnerability disclosures and the creation of exploit code due to the widespread use and dependence on these applications.

Macros

Macros are interpreted programs built into many applications, most notably in Microsoft Office applications. Macros are powerful tools used for automation and can extend the functionality of a word processor or spreadsheet considerably. However, they also represent a vector for attack: the macro virus. By creating documents or document templates with malicious macros, the otherwise-legitimate document becomes a vessel for malware.[79]

Macro viruses became so prolific that Microsoft took steps to prevent the accidentle use of macros as early as Windows 7, and in newer releases of Microsoft's applications macros are disabled by default. Now, macros are only effective if the attacker can convince users to turn on macros so that their malicious macro can run.[80]

Macros are of interest to industrial operators because the systems and applications used are often older. In addition, documents that originate from inside an industrial organization (perhaps a work order or a procedural update) tend to be trusted.

Tip

Consider disabling macros and implementing cybersecurity controls that prevent macros from executing (e.g., Application Whitelisting, or AWS) and/or that prevent access to files that contain macros (e.g., policy-based file access controls). If using document macros is necessary as part of daily operations, more advanced security controls may be needed.

Secure sockets layers

"A missing bounds check in the handling of the TLS heartbeat extension can be used to reveal up to 64k of memory to a connected client or server."[81] This, the first sentence in a 2014 security advisory from OpenSLL that in total was just over 100 words long had profound impact on the Internet and on the tech community. CVE-20014-0160, dubbed "Heartbleed," represents a critical flaw in OpenSSL versions 1.0.1 and 1.0.2 that can be used to reveal private keys used and decrypt sensitive data transferred within SSL sessions.[82]

Beyond its CVSS score of 7.5, Heartbleed is important because of the popularity of OpenSSL, which at the time put an estimated 66% of active sites based on the market share of impacted web servers, and including 17% of SSL web servers using certificates issued by trusted authorities.[83] In other words, Heartbleed put the secure connections used every day by web browsers around the world.

The vulnerability and the exploitation thereof are both relatively simple. The attacker simply provides a heartbeat to the SSL server, with a modified payload variable. By setting the payload_length to a value larger than the payload provided, the attacker tricks the server into copying additional bytes of memory into the heartbeat response. While this limits the data that can be obtained to what happens to be in memory at the time, this could include private keys, unencrypted messages, or other sensitive data.[84]

Log4j

In April 2022, a vulnerability in the popular java logging framework "Log4j" was disclosed with a CVSS score of 10 (the maximum severity score possible).[85] Similar to Heartbleed, the extensive use of the Log4j package in many software applications put millions of software applications at risk of exploitation. Of particular note is that:

1. The vulnerability had existing for many years prior to discovery and disclosure
2. The vulnerability is exploitable, with evidence of wide-spread active exploitation.

The seriousness of Log4Shell has been widely acknowledged because of the simplicity of the exploitation and large attack surface it represents. The maximum CVSS score of 10 indicates that the vulnerability can be exploited easily, over the network, without authentication. Successful exploitation allows arbitrary code execution and can allow attackers to completely take control of target systems.[86]

Log4j is used in a broad range of servers, network appliances, IoT devices, and even many cybersecurity appliances. As expressed by the Washington Post in, "The fact that log4j is such a ubiquitous piece of software is what makes this such a big deal. Imagine if a common type of lock used by millions of people to keep their doors shut was suddenly discovered to be ineffective. Switching a single lock for a new one is easy, but finding all the millions of buildings that have that defective lock would take time and an immense amount of work."[87] It also makes the vulnerability a tempting target for attackers. According to Check Point software, 60 different variations of the original exploit were discovered in a single 24-h period immediately following the disclosure.[88]

While initial attacks seemed primarily focused on bitcoin mining, there was evidence of state-backed hackers using the vulnerability to try to break into government and business targets as well.[89]

Ransomware and industrial control systems

Ransomware refers to malware that is intended to disable a system by encrypting some or all data on that system. Ransomware enables extortion of its victims via a promise of full system recovery once a payment has been made. If the attackers chose to honor that promise, fallowing the attacker can re-enable infected systems by providing decryption keys. Ransomware, like most cyberattacks, has evolved into more complex attack campaigns. They typically include attempts at lateral propagation, so that entire networks can be disabled in a coordinated fashion and may even seek to discover and disable online backup and recovery systems to make recovery more difficult. More recent campaigns often include various degrees of data theft, adding the threat of leaking sensitive information in an effort to increase the attacker's leverage over the victim.

Ransomware deserves some special consideration because it can be highly disruptive even if industrial systems are not infected. In 2021, Colonial Pipeline, an American oil pipeline that delivers fuel to much of the Southeastern United States, suffered a ransomware cyberattack that impacted computerized equipment managing the pipeline. While this was an IT incident and not an industrial control attack, the decision was made by Colonial Pipeline to shut down pipelines as a precaution to prevent ransomware from infecting industrial systems, which could have caused a much longer outage.[90] The ransomware gang DarkSide, who was responsible for the attack, claimed no intention of disrupting critical infrastructure. The motive seemed limited to greed, with a reported $4.4 million ransom being paid.[91]

Ransomware is obviously a threat against industrial operations even if the industrial networks were not at risk, simply because companies who operate industrial systems may be perceived as viable targets for financial extortion. The impact to the business systems and networks that supported the pipeline was sufficient to justify a manual shutdown of the operation. Some ransomware, such as Snake (also referred to as EKANS), specifically targets industrial systems. It leverages insecure configurations of RDP as a primary attack vector, likely with an understanding that RDP is widely used in many operational environments. The attack targets many specific processes, including ICS-and SCADA-specific processes, as well as remote management and Windows backups—presumably in order to disrupt process operations and inhibit recovery efforts.[92] Ransomware that could infect actual process control assets, such as PLCs, RTUs, et. al., are also possible. Although no known examples of such an attack existed at the time of this writing, the malicious encryption of PLCs using a ransomware worm called

"LogicLocker" was demonstrated in 2017 by researchers at the Georgia Institute of Technology.[93]

Industrial application layer protocols

Adobe Reader exploits are highly relevant because many computing products—including ICS products—distribute manuals and other reference materials using PDF files and preinstall these on the ICS hosts. What is often the case as well is that the ICS software developers preinstall the Adobe Reader application, which oftentimes remains unpatched through traditional methods because it is not included with other vendor software update and hotfix notices. There are more directly relevant attacks that can occur at the application layer—industrial application attacks. Attack frameworks increasingly leverage these industrial protocols to achieve malicious outcomes. (See "Weaponized Industrial Cyber Threats," above).

"Industrial applications" are the applications and protocols that communicate to, from, and between supervisory, control, and process system components. These applications serve specific purposes within the ICS, and by their nature are "vulnerable" because they are designed around control: either *direct* control of processes or devices (e.g., a PLC, RTU, or IED), or *indirect* control, via supervisory systems like a DCS or SCADA that are used by human operators to supervise and influence processes or devices.

Unlike typical application layer threats, such as in the case of Adobe Reader, industrial application layer threats do not always require that a specific vulnerability be exploited. This is because these applications are designed for the purpose of influencing industrial control environments. They do not need to be infected with malware in order to gain the control necessary to cause harm, since they can simply be used as they are designed but with malicious intent. By issuing legitimate commands, between authorized systems and in full compliance with protocol specifications, an ICS can be told to perform a function that is outside of the owner's intended purpose and parameters. This method can be thought of as the *exploitation of functionality* and when considered in the context of ICS security represents a problem that is not typically addressed through traditional IT security controls.

Digital Bond published one example of an industrial application layer attack in 2012 under the project name "Basecamp." The research documented how the EtherNet/IP protocol could be manipulated to control a Rockwell Automation ControlLogix PLC. It should be noted that it was not a ControlLogix vulnerability that was exploited, but the underlying protocol, and as such, this exploit is widely applicable due to the prevalence of the EtherNet/IP protocol in ICS supplied by various vendors. A number of attack methods were disclosed, all sharing the common exploitation of EtherNet/IP:[94]

- **Forcing a system stop.** This attack effectively shuts off the CIP service and renders the device dead by sending a CIP command to the device. This puts the device into a "major recoverable fault" state.[95]

- **Crashing the CPU.** This attack crashes the CPU due to a malformed CIP request, which cannot be effectively handled by the CIP stack. The result is also a "major recoverable fault" state.[96]
- **Dumping device boot code.** This is a CIP function that allows an EtherNet/IP device's boot code to be remotely dumped.[97]
- **Reset device.** This is a simple misuse of the CIP system reset function. The attack resets the target device.[98]
- **Crash device.** This attack crashes the target device due to a vulnerability in the device's CIP stack.[99]
- **Flash update.** CIP, like many industrial protocols, supports writing data to remove devices, including register and relay values, but also files. This attack misuses this capability to write new firmware to the target device.[100]

EtherNet/IP is not the only protocol that can be exploited in this way. In 2013, Adam Crain of Automatak and independent researcher Chris Sistrunk reported a vulnerability with certain implementations of the DNP3 protocol stack, which was found to impact DNP3 master and outstation (slave) devices from a large number of known vendors. The weakness was an input validation vulnerability received from a DNP outstation station that could put the master station into an infinite loop condition.[101] This was not a specific device vulnerability, but a larger vulnerability concerning the implementation of a protocol stack, and because many vendors utilized a common library, it impacted a large number of products from multiple vendors. Of particular concern is that this vulnerability can be exploited via TCP/IP (by someone who has gained logical network access) or serially (by someone who has gained physical access to a DNP3 outstation).

Both of these examples represent weaknesses in protocols that were designed decades ago and are now being faced with new security challenges that were unforeseen at the time of their development. Since these also involve community-led open-source or licensed protocols that are not managed by a single vendor, their deployment can be very wide spread making it difficult to deploy patches and hotfixes that can be implemented in a timely manner. While vulnerabilities of this type are cause for concern, they can typically be mitigated through proper network and system design and through the implementation of appropriate cybersecurity controls (which, hopefully, is why you are reading this book). To put this another way, it is going to be a lot easier and less costly to deploy appropriate security controls to mitigate the risk from these open protocols versus attempting to retrofit and/or replace the affected ICS equipment.

An easy way to look at this is though the ICS devices themselves may be "insecure by design," the overall ICS can be sufficiently secured from cyber threats using a "secure by redesign" approach, rather than a "secure by replacement" one. After all, a "secure" device today could likely have vulnerabilities disclosed in the future that makes it "insecure" at that time. This is why industrial security is always focused on the holistic "system-level" security rather than that of individual ICS components.

Antisocial networks: A new playground for malware

While social networks do not seem to have a lot to do with industrial networks (there should never be open connectivity to the Internet from an industrial zone, and certainly not to social networking sites), it is surprisingly relevant. Social networking sites are increasingly popular, and they can represent a serious risk against industrial networks. How can something as benign as Facebook or Twitter be a threat to an industrial network? Social networking sites are designed to make it easy to find and communicate with people, and people are subject to social engineering exploitation just as networks are subject to protocol and application exploitation.

They are at the most basic level a source of gathering personal information and end user's trust that can be exploited either directly or indirectly. At a more sophisticated level, social networks can be used actively by malware as a C2 channel. Fake accounts posing as "trusted" coworkers or business colleagues can lead to even more information sharing or provide a means to trick the user into clicking on a link that will take them to a malicious website that will infect the user's computer with malware. That malware could mine additional information, or it could be walked into a "secure" facility to impact an industrial network directly. Even if a company has strict policies on the use of laptops accessing such websites are these same companies as strict with the laptops used by their vendors and service subcontractors when connected to these same industrial networks? These same vendor/subcontractor computers are commonly connected directly to secure industrial networks. This is why it is equally important to consider the "insider" threats, and not focus entirely on external "outsider" originated attacks.

No direct evidence exists that links the rise in web-based malware and social networking adoption; however, the correlation is strong enough that any good security plan should accommodate social networking, especially in industrial networks. According to Cisco, "Companies in the Pharmaceutical and Chemical vertical were the most at risk for web-based malware encounters, experiencing a heightened risk rating of 543% in 2Q10, up from 400% in 1Q10. Other higher-risk verticals in 2Q10 included energy, oil, and gas (446%), education (157%), government (148%), and transportation and shipping (146%)."[102]

Apart from being a direct infection vector, social networking sites can be used by more sophisticated attackers to formulate targeted spear-phishing campaigns, such as the "pizza delivery" exercise. Users may postpersonal information about where they work, what their shift is, who their boss is, and other details that can be used to engineer a social exploitation through no direct fault of the social network operators (most have adequate privacy controls in place). Spear phishing is already a proven tactic, yet it is easier and even more effective when combined with the additional trust associated with social networking communities.

Tip

Security awareness training is an important part of building a strong security plan, but it can also be used to assess current defenses. Conduct this simple experiment to both increase awareness of spear phishing and gauge the effectiveness of existing network security and monitoring capabilities:

1. Create a website using a free hosting service that displays a security awareness banner.
2. For this exercise, create a Google Mail account using the name (modified if necessary) of a group manager, HR director, or the CEO of your company (again, disclosing this activity to that individual in advance and obtaining necessary permissions). Assume the role of an attacker, with no inside knowledge of the company; look for executives who are quoted in press releases, or listed on other public documents. Alternately, use the Social Engineering Toolkit (SET), a tool designed to "perform advanced attacks against the human element," to launch a more thorough social engineering penetration test.
3. Again, play the part of the attacker and use either SET or outside means, such as Jigsaw.com or other business intelligence websites, to build a list of email addresses within the company.
4. Send an email to the group from the fake "executive" account, informing recipients to please read the attached article in preparation for an upcoming meeting.
5. Perform the same experiment on a different group, using an email address originating from a peer (again, obtain necessary permissions). This time, attempt to locate a pizza restaurant local to your corporate offices, using Google map searches or similar means, and send an email with a link to an online coupon for buy-one-get-one-free pizza.

Track your results to see how many people clicked through to the offered URL. Did anyone validate the "from" in the email, reply to it, or question it in any way? Did anyone outside of the target group click through, indicating a forwarded email?

Finally, with the security monitoring tools that are currently in place, is it possible to effectively track the activity? Is it possible to determine who clicked through (without looking at web logs)? Is it possible to detect abnormal patterns or behaviors that could be used to generate signatures and detect similar phishing in the future?

The best defense against a social network attack continues to be security and situational awareness. Security awareness helps prevent a socially engineered attack from succeeding by establishing best-practice behaviors among personnel. Situational awareness helps to detect if and when a successful breach has occurred, where it originated, and where it may have spread to—in order to minimize the damage or impact from the attack and mitigate or remediate any gaps uncovered in security awareness and training.

Social networks can be used as a C2 channel between deployed malware and a remote server. One case of Twitter being used to deliver commands to a bot is the @upd4t3 channel, first detected in 2009 that uses standard 140-character tweets to link to base64-encoded URLs that deliver infosteeler bots.[103]

Caution

Always inform appropriate personnel of any security awareness exercise to avoid unintended consequences and/or legal liability, and NEVER perform experiments of this kind using real malware. Even if performed as an exercise, the collection of actual personal or corporate information could violate your employment policy or even state, local, or federal privacy laws.

This use of social networking as a malicious vector is difficult to detect, as it is not feasible to scour these sites individually for such activity, and there is no known way to detect what the C2 commands may look like or where they might be found. Application session analysis on social networking traffic could detect the base64 encoding once a session was initiated in the case of @upd4t3. The easiest way to block this type of activity, of course, is to block access to social networking sites completely from inside industrial networks. The wide adoption of these sites within the enterprise (for legitimate sales, marketing, and even business intelligence purposes) however makes it highly likely that any threat originating from or directly exploiting social networks can and will compromise the business enterprise. Special security considerations must be employed for this reason when evaluating the risk an organization faces from social networking.

Polymorphic and adaptive malware

Polymorphic or metamorphic malware is malware with mutation logic capable of adapting its own code, typically in response to external variables. Adaptive malware uses conditional logic to direct activity based on its surroundings until it finds itself in the perfect conditions in which it will best accomplish its goal (spread, stay hidden, deploy a weapon, etc.).

Polymorphic malware often uses mutations to avoid detection: for example, to eliminate malicious code in response to detection attempts. Most modern advanced malware frameworks have both mutation and adaptation capabilities.

Once again, Stuxnet provides a solid example. The goal of Stuxnet was to find a particular ICS by spreading widely through local networks and "sneaker" networks. It then only took secondary infection measures when the target environment (Siemens SIMATIC WinCC/PCS7) was found. It then checked for particular PLC models and versions (Siemens models S7-315-2 and S7-417). Once these models were discovered, it looked for a specific make and model of VFDs (Fararo Paya model KFC750V3 and Vacon NX) before it injected process code into the PLC. If unsuitable targets were infected, it would lay dormant waiting for other hosts to infect.

Malware mutations are also already in use. Stuxnet at a basic level will update itself in the wild (even without a C2 connection), through peer-to-peer checks with other hosts also infected, and if a newer version of Stuxnet bumps into an older version, it updates the older version allowing the infection pool to evolve and upgrade in the wild.[104]

Further mutation behavior involves self-destruction of certain code blocks with self-updates of others, effectively morphing the malware and making it more targeted as well as more difficult to detect. Mutation logic may include checking for the presence of other well-known malware and adjusting its own profile to utilize similar ports and services knowing that this new profile will go undetected. In other words, malware is getting smarter and at the same time, harder to detect.

Dealing with an infection

Ironically, upon detecting an infection, you may not want to immediately clean the system of infected malware. This is because there may be subsequent levels of infection that exist yet are dormant and may be activated as a result. There could also be valuable information, such as the infection path used and other compromised hosts as in the case of Stuxnet. A thorough investigation should instead be performed, with the same sophistication as the malware itself.

The first step should be to logically isolate the infected host so that it can no longer cause any harm. Harm to not only other logical assets that may be on the shared network but also the physical assets that the ICS host may be controlling. Allow the malware to communicate over established C2 channels but isolate the host from the rest of the network and remove all access between that host and any sensitive or protected information. A well-established network segmentation philosophy based on common security criteria needs to be deployed in order to effectively isolate infected hosts. This topic is covered further in Chapter 5, "Industrial Network Design and Architecture" and Chapter 9, "Establishing Zones and Conduits." Collect as much forensic detail as possible in the form of system logs, captured network traffic, supplementing where possible with memory analysis data. Important information can be gathered that may result in the successful removal of the infection by effectively sandboxing the infected system.

When you suspect that you are dealing with an infection, approach the situation with diligence and perform a thorough investigation:

- Remember to consider the safe and reliable operation of the manufacturing process as the primary objective. Extra care must be given to ICS components in their operating mode for this reason, and is why it is important to have a documented and rehearsed incident response plan in place.
- Always monitor everything, collecting baseline data, configurations, and firmware for comparison.
- Analyze available logs to help identify scope, infected hosts, propagation vectors, and so on. Logs should be retrieved from as many components on the network as possible, including those that have not been compromised.
- Sandbox and investigate infected systems.
- Be careful to not unnecessarily power-down infected hosts, and valuable information may be resident in volatile memory.

- Analyze memory to find memory-resident rootkits and other threats that may be residing in user memory.
- Clone disk images when possible to preserve as much of the original state as possible for off-line analysis.
- Reverse engineer-detected malware to determine full scope and to identify additional attack vectors and possible propagation.
- Retain all information for disclosure to authorities.

Note

Information collected from an infected and sandboxed host may prove valuable to legal authorities, and depending upon the nature of your industrial network, you may be required to report this information to a governing body.

A "bare metal reload" may be necessary where a device is completely erased and reduced to a bare, inoperable state depending on the severity of the infection. The host's hardware must then be reimaged completely. Clean versions of operating systems, applications, and asset firmware should be kept in a safe, clean environment for this reason. This can be accomplished using secure virtual backup environments, or via secure storage on trusted removable media that can then be stored in a locked cabinet, preferably in a separate physical location from the asset archived. It is important to ensure that the images used for system restoration are free and clean of any malware or malicious code that may have triggered the initial incident when using a backup and recovery system.

Tip

The ability to perform forensics on a compromised system can be an advanced task. To help in this, the National Institute of Standards and Technology has established the Computer Forensics Tool Testing (CFTT) project and offers a "Computer Forensics Tool Catalog." Information can be found at: http://www.cftt.nist.gov.

Tip

If you think you have an infection, you should know that there are security firms that are experienced in investigating and cleaning advanced malware infections. Many such firms further specialize in industrial control networks. Before allowing anyone access to your ICS assets, it is encouraged to request and validate actual system experience—preferably on an ICS similar to yours. These firms can help you deal with infection as well as provide an expert interface between your organization and any governing authorities that may be involved.

Summary

Cyberthreats are increasing at an alarming rate, making the technologies that everyone now takes for granted the easy criminal path into theft, espionage, and sabotage. Industrial control systems account for less than 1% of the total vulnerabilities listed by the OSVDB, yet the trends associated with ICS cyber-attacks should be alarming. The rate of cyberincidents directly impacting industrial systems has been steadily increasing over the past 30 years according to the Repository of Industrial Security Incidents (RISI).[105] RISI's analysis also reveals that, although malware infections still account for a large number of cyberevents (28% in 2013), it has been steadily decreasing over the past 5 years indicating that ICS users are becoming more aware of the methods to provide malware from affecting ICS architectures. These data also confirm that the vectors involved in ICS cyberevents are shifting to more sophisticated mechanisms that are able to avert detection by traditional defenses, pivot through segmented networks, and exploit weaknesses in the underlying design of the ICS architecture.

Anyone who believes that they can prevent 100% of the possible cyberevents within a particular system is misinformed and likely to be disappointed. A well-rounded cyber-security program is based on a thorough understanding of the threats that face industrial architectures, and blends security defenses that not only focus on event prevention but also postbreach detection and forensic capabilities to contain an event and minimize as best as possible the negative consequences to the manufacturing or industrial process that the ICS is designed to control.

Endnotes

1. K. Stouffer, J. Falco, K. Scarfone, National Institute of Standards and Technology, Special Publication 800-82 (Final Public Draft), Guide to Industrial Control Systems (ICS) Security, Computer Security Division, Information Technology Laboratory, National Institute of Standards and Technology Gaithersburg, MD and Intelligent Systems Division, Manufacturing Engineering Laboratory, National Institute of Standards and Technology Gaithersburg, MD, September 2008.
2. M.J. McDonald, G.N. Conrad, T.C. Service, R.H. Cassidy, SANDIA Report SAND2008-5954, Cyber Effects Analysis Using VCSE Promoting Control System Reliability, Sandia National Laboratories Albuquerque, New Mexico and Livermore, California, September 2008.
3. A. Giani, S. Sastry, K.H. Johansson, H. Sandberg, The VIKING Project: An Initiative on Resilient Control of Power Networks, Department of Electrical Engineering and Computer Sciences, University of California at Berkeley, and School of Electrical Engineering, Royal Institute of Technology (KTH), Berkeley, CA, 2009.
4. MITRE ATT&CK. Document from the Internet, cited March 15, 2023. https://attack.mitre.org/resources/getting-started/
5. Andy Greenberg. How Hackers Can Use 'Evil Bubbles' to Destroy Industrial Pumps. Wired. July 29, 2017.
6. Honeywell Forge. Honeywell Cybersecurity Report: USB Hardware Attack Platforms. October 2020. Honeywell, Inc.
7. Ibid.
8. mg.O.MG Keylogger Cable. 7 August 2020. https://mg.lol/blog/keylogger-cable/
9. Dillon Beresford. Exploiting Siemens SIMATIC S7 PLCs. Prepared for Black Hat USA+2011. Las Vegas, NV. 2011.

10. Ralph Langner. Forensics on a complex cyber attack – lessons learned from Stuxnet. Presentation at the 2011 Applied Control Solutions (ACS) Conference. September 20, 2011. Washington, DC.
11. Dillon Beresford. Exploiting Siemens SIMATIC S7 PLCs. Prepared for Black Hat USA+2011. Las Vegas, NV. 2011.
12. Dillon Beresford. Exploiting Siemens SIMATIC S7 PLCs. Prepared for Black Hat USA+2011. Las Vegas, NV. 2011.
13. SearchSecurity. Definition: Blended Threat. Document from the Internet. Cited Sep 4, 2012. Available from: http://searchsecurity.techtarget.com/definition/blended-threat
14. G. McDonald, L.O. Murchu, S. Doherty, E. Chien, Symantec. Stuxnet 0.5: The Missing Link, Version 1.0, February 26, 2013.
15. Ibid.
16. Ind.ustrial Control Systems Cyber Emergency Response Team (ICS-CERT), ICSA-10-238-01—STUXNET MALWARE MITIGATION, Department of Homeland Security, US-CERT, Washington, DC, August 26, 2010.
17. E. Chien, Symantec. Stuxnet: a breakthrough., November 2010 (cited: November 16, 2010).
18. Open-Source Vulnerability Database (OSVDB). ID 66,441: Siemens SIMATIC WinCC SQL Database Default Password. (cited: December 20, 2013).
19. WinCC Database Problem. (cited: December 20, 2013).
20. N. Falliere, L.O Murchu, E. Chien, Symantec. W32.Stuxnet Dossier, Version 1.1, October 2010.
21. E. Byres, A. Ginter, J. Langill. "How Stuxnet Spreads - A Study of Infection Paths in Best Practice Systems," Version 1.0, February 22, 2011.
22. ICS-CERT. Joint Security Awareness Report (JSAR-12-241-01B) Shamoon/DistTrack Malware - Update B. Document from the Internet. April 30, 2013. Cited December 22, 2013. Available at: https://ics-cert.us-cert.gov/jsar/JSAR-12-241-01B-0
23. Ibid.
24. Kelly Jackson Higgins. 30,000 Machines Infected In Targeted Attack On Saudi Aramco. Dark Reading. August 2012. Document from the Internet. Cited December 22, 2013. Available at: http://www.darkreading.com/attacks-breaches/30000-machines-infected-in-targeted-atta/240006313
25. Kaspersky Labs. Virus News: Kaspersky Lab Experts Provide In-Depth Analysis of Flame's C&C Infrastructure. Document from the Internet. June 4, 2012. Cited Sep 18, 2012. Available from: http://www.kaspersky.com/about/news/virus/2012/Kaspersky_Lab_Experts_Provide_In_Depth_Analysis_of_Flames_Infrastructure
26. Kaspersky Labs. Virus News: Kaspersky Lab Experts Provide In-Depth Analysis of Flame's C&C Infrastructure. Document from the Internet. June 4, 2012. Cited Sep 18, 2012. Available from: http://www.kaspersky.com/about/news/virus/2012/Kaspersky_Lab_Experts_Provide_In_Depth_Analysis_of_Flames_Infrastructure
27. "Dragonfly: Cyberespionage Attacks Against Energy Suppliers," Symantec Security Response v.1.21, July 7, 2014 (v1.0 first published June 30, 2014).
28. Ibid.
29. Ibid.
30. Joel T. Langill. Defending Against the Dragonfly Cyber Security Attacks. Belden. Oct. 22, 2014.
31. "Dragonfly: Cyberespionage Attacks Against Energy Suppliers," Symantec Security Response v.1.21, July 7, 2014 (v1.0 first published June 30, 2014).
32. Ibid.
33. Ibid.
34. "IR-ALERT-H-16-043-01AP Cyber-Attack Against Ukrainian Critical infrastructure," US Department of Homeland Security Industrial Control System Computer Emergency Response Team, March 7, 2016, accessed July 12, 2016, https://info.publicin-telligence.net/NCCIC-UkrainianPowerAttack.pdf
35. Ake Styczynski, Nate Beach—Westmoreland. When The Lights Went Out: A Comprehensive Review of the 2015 Attacks on Ukrainian Critical Infrastructure. Booz Allen Hamilton Inc. 2019.

36. "IR-ALERT-H-16-043-01AP Cyber-Attack Against Ukrainian Critical infrastructure," US Department of Homeland Security Industrial Control System Computer Emergency Response Team, March 7, 2016, accessed July 12, 2016, https://info.publicin-telligence.net/NCCIC-UkrainianPowerAttack.pdf
37. Ibid.
38. Ibid.
39. Ibid.
40. Ibid.
41. Ibid.
42. Ibid.
43. Ibid
44. Ake Styczynski, Nate Beach—Westmoreland. When The Lights Went Out: A Comprehensive Review of the 2015 Attacks on Ukrainian Critical Infrastructure. Booz Allen Hamilton Inc. 2019.
45. Ibid.
46. Ibid.
47. Ibid.
48. Ibid.
49. Ibid.
50. Ibid.
51. Ibid.
52. Ibid.
53. Cherepanov, Anton. "Industroyer: Biggest threat to industrial control systems since Stuxnet". www.welivesecurity.com. ESET. June 17, 2017.
54. Ibid.
55. Ibid.
56. Dragos, Inc. TRISIS Malware Analysis of Safety System Targeted Malware. December 2017.
57. Ibid.
58. Ibid.
59. Ibid.
60. Ibid.
61. Honeywell, Inc. Global Analysis Research and Defense Report: Industrial Cybersecurity USB Threat Report 2020. July 2020.
62. Daniel Kapellmann Zafra, Raymond Leong, Chris Sistrunk, Ken Proska, Corey Hildebrandt, Keith Lunden, Nathan Brubaker, "INDUSTROYER.V2: Old Malware Learns New Tricks". April 25, 2022. Updated December 02, 2022. Mandiant. https://www.mandiant.com/resources/blog/industroyer-v2-old-malware-new-tricks
63. Welivesecurity, by ESET. "Industroyer2: Industroyer reloaded This ICS-capable malware targets a Ukrainian energy company". April 12, 2022. ESET Research. https://www.welivesecurity.com/2022/04/12/industroyer2-industroyer-reloaded/
64. Ibid.
65. Daniel Kapellmann Zafra, Raymond Leong, Chris Sistrunk, Ken Proska, Corey Hildebrandt, Keith Lunden, Nathan Brubaker, "INDUSTROYER.V2: Old Malware Learns New Tricks". April 25, 2022. Updated December 02, 2022. Mandiant. https://www.mandiant.com/resources/blog/industroyer-v2-old-malware-new-tricks
66. Nathan Brubaker, Keith Lunden, Ken Proska, Muhammad Umair, Daniel Kapellmann Zafra, Corey Hildebrandt, Rob Caldwell. "INCONTROLLER: New State-Sponsored Cyber Attack Tools Target Multiple Industrial Control Systems". Mandiant. April 13, 2022. Cited March 2022. https://www.mandiant.com/resources/blog/incontroller-state-sponsored-ics-tool
67. Ibid.
68. Ibid.

69. Ibid.
70. Ibid.
71. Ibid.
72. Rob Lee. (February 2023). PIPEDREAM – Most Flexible & Capable ICS Malware To Date [Conference presentation]. S4X23 2023 Conference. Miami, FL, United States. https://www.youtube.com/watch?v=H82sbIwFxt4
73. Nathan Brubaker, Keith Lunden, Ken Proska, Muhammad Umair, Daniel Kapellmann Zafra, Corey Hildebrandt, Rob Caldwell. "INCONTROLLER: New State-Sponsored Cyber Attack Tools Target Multiple Industrial Control Systems". Mandiant. April 13, 2022. Cited March 2022. https://www.mandiant.com/resources/blog/incontroller-state-sponsored-ics-tool
74. Nir Nissim, Ran Yahalom, Yuval Elovici, "USB-based attacks" Computers & Security, Volume 70, 2017. https://doi.org/10.1016/j.cose.2017.08.002
75. Honeywell, Inc. Global Analysis Research and Defense Report: Industrial Cybersecurity USB Threat Report 2022. August 2022.
76. D. Stevens, Escape from PDF., March 2010 (cited: November 4, 2010).
77. J. Conway, Sudosecure.net. Worm-Able PDF Clarification., April 4, 2010 (cited: November 4, 2010).
78. 86 Security Labs, PDF "Launch" Feature Used to Install Zeus., April 14, 2010 (cited: November 4, 2010).
79. "Frequently Asked Questions: Word Macro Viruses". Microsoft. Archived from the original on 2011-06-04. Retrieved 2006-06-18.
80. Microsoft. "Macro Malware". https://learn.microsoft.com/en-us/microsoft-365/security/intelligence/macro-malware?view=o365-worldwide
81. OpenSSL Security Advisory [07 Apr 2014] https://www.openssl.org/news/secadv/20140407.txt
82. Goodin, Dan (8 April 2014). "Critical crypto bug in OpenSSL opens two-thirds of the Web to eavesdropping". Ars Technica. Archived from the original on 5 July 2017. Retrieved 14 June 2017.
83. Mutton, Paul (8 April 2014). "Half a million widely trusted websites vulnerable to Heartbleed bug". Netcraft. Archived from the original on 19 November 2014. Retrieved 24 November 2014.
84. Robert Erbes, IOActive Insights. "Bleeding Hearts". APRIL 10, 2014. https://ioactive.com/bleeding-hearts/
85. "Apache Log4j Security Vulnerabilities". Log4j. Apache Software Foundation. Retrieved 12 December 2021.
86. CISA. National Vulnerability Database, CVE2021-44,228 vulnerability detail. https://nvd.nist.gov/vuln/detail/CVE-2021-44228
87. Hunter, Tatum; de Vynck, Gerrit (20 December 2021). "The 'most serious' security breach ever is unfolding right now. Here's what you need to know". The Washington Post.
88. "Protect Yourself Against The Apache Log4j Vulnerability" CheckCheck Point Software Technologies. https://blog.checkpoint.com/2021/12/11/protecting-against-cve-2021-44228-apache-log4j2-versions-2-14-1/. Last updated: 20.12.2021 01:30 a.m. PST. Initially published: 10.12.2021.
89. Hunter, Tatum; de Vynck, Gerrit. "The 'most serious' security breach ever is unfolding right now. Here's what you need to know". The Washington Post. 20 December 2021.
90. Sergiu Gatlan, Colonial Pipeline reports data breach after May ransomware attack. Bleeping Computer. August 16, 2021.
91. Robertson, Jordan; Turton, William (May 8, 2021). "Colonial Hackers Stole Data Thursday Ahead of Shutdown". Bloomberg News. Archived from the original on May 9, 2021. Retrieved May 9, 2021.
92. Alexander Ivanyuk Snake/EKANS Ransomware Attacks Industrial Control Systems: Acronis Stops It. Acronis. February 10, 2020.
93. Formby, D., Durbha, S., & Beyah, R. (n.d.). Out of Control: Ransomware for Industrial Control Systems. Retrieved from http://www.cap.gatech.edu/plcransomware.pdf
94. Ruben Santamarta. Attacking ControlLogix. Digital Bond Project Base Camp. 2012.
95. Ibid.
96. Ibid.
97. Ibid.

98. Ibid.
99. Ibid.
100. Ibid.
101. Advisory (ICSA-13-291-01). DNP3 Implementation Vulnerability. ICS-CERT. Original release date: November 21, 2013.
102. Cisco Systems, 2Q10 Global Threat Report, 2010.
103. Nazario, Arbor networks. Twitter-based Botnet Command Channel., August 13, 2009 (cited: November 4, 2010).
104. J. Pollet, Red Tiger, Understanding the advanced persistent threat, in: Proc. 2010 SANS European SCADA and Process Control Security Summit, Stockholm, Sweden, October 2010.
105. Report "2013 Report on Cyber Security Incidents and Threats Affecting Industrial Control Systems," Repository of Industrial Security Incidents (RISI), Published June 15, 2013.

8 Risk and Vulnerability Assessments

Information in this chapter

- Cyber Security and Risk Management
- Methodologies for Assessing Risk within Industrial Control Systems
- System Characterization
- Threat Identification
- Vulnerability Identification
- Risk Classification and Ranking
- Cyber-Physical Threat Modeling
- Cyber Security HAZOP
- Risk Reduction and Mitigation

The concept of cybersecurity goes hand-in-hand with how an organization views and manages risk. Risk is often correlated to the vulnerabilities that may or may not exist with the organization's business enterprise, including risk to and from business systems, IT infrastructure, automation and control systems, and physical business assets that may be directly under the control of one of the aforementioned systems.

The overall process of implementing cybersecurity controls is meant to reduce business risk. However, if one does not understand their exposure to and tolerance of risk, then the overall effectiveness of these controls may be somewhat less than expected. The deployment of cybersecurity in terms of security policies, administrative procedures, business processes, and technological solutions is meant to target specifically identified areas of risk and reduce the impact to an organization should a cyber event occur targeting one of the business assets. If an organization fails to identify areas of risk, how can it properly select, implement, and measure security controls that are meant to reduce these risks?

This topic could fill an entire book. It is not practical to attempt to cover all aspects of risk and vulnerability management in a single chapter. Instead, this chapter will focus on the highlights associated with implementing a risk and vulnerability assessment process specifically designed for industrial systems. Detailed resources and references are provided throughout this chapter. A supplemental ebook with additional tools is available at SCADAhacker.com/ebooks.

Industrial Network Security. https://doi.org/10.1016/B978-0-443-13737-2.00002-6

Cybersecurity and risk management

Why risk management is the foundation of cyber security?

The concept of "functional safety" within most industrial facilities is a cornerstone in the overall operation of the facility, as well as an important key performance indicator (KPI) used in evaluating a company. The deployment of functional safety is well defined by leading international standards including IEC 61508/61511 and ANSI/ISA 84.00.01, which are based around the process of identifying risk in terms of process hazard analysis (PHA), hazards and operability analysis (HAZOP), etc. and then using methods to specifically reduce these risks through the deployment of mechanical and instrumented systems. The concept of "operational security" closely aligns with functional safety in terms of risk identification, risk reduction through the deployment of security controls, and risk management through continuous and periodic monitoring of the industrial security systems. These ideas are documented in several standards on operational security (see Chapter 14, "Standards and Regulations").

The easiest way to understand the importance of risk and how it relates to not only the selection of cybersecurity controls and methods, but also its overall effectiveness is to answer one simple question. Given a FIXED amount of MONEY, and a FIXED period of TIME to secure an ICS, what would you do?

There are many cybersecurity controls "catalogs" that will list hundreds of various procedure and technological solutions that can be implemented (see Chapter 14, "Standards and Regulations"). The first step here must be to understand and establish an acceptable level of risk or what is called "risk tolerance." It is possible to manage this "unmitigated" risk in one of four ways:

1. Mitigation (you manage)
2. Transferal (others manage)
3. Avoidance (no one manages)
4. Acceptance (stakeholder's manage)

Risk mitigation is the process of reducing these catalogs of controls down to an effective list that is designed to help reduce specific risks to an organization. It should be obvious at this point, and by the fact that you are reading this book, that the risks facing organizations is constantly changing, and that with this dynamic landscape comes the possibility that risks may appear tomorrow that did not exist today. This is why cyber risk management is considered a continuous process of identification, assessment, and response and not something that can be addressed once and left unvisited for long periods of time.

To look at how risk directly impacts industrial environments that depend on ICS to maintain a safe, efficient, and profitable environment, let us begin with a high-level identification of risk. What is the greatest threat facing your company's industrial systems?

1. People's Liberation Army Unit 61398
2. On-Site Control Systems Engineer
3. Anonymous "Hacktivists" Group
4. Vendor Site Support Specialist
5. Package Equipment Supplier

These risks cover a broad range of threats that include both internal and external sources, which may use targeted or nontargeted methods, with both intentional and unintentional motives. Nearly 80% of the incidents impacting ICS are "unintentional," yet only 35% of these events were originated from an "outsider."[1] Many organizations are resistant to objectively consider the actual threats to their industrial systems and risk they represent. Another report confirms that in the analysis of 47,000 incidents (not necessarily incidents against ICS), 69% of these events originated from internal threats acting carelessly rather than maliciously.[2] Embedded devices and network appliances were targeted in 34% of the incidents impacting ICS, while Windows-based ICS and enterprise hosts were targeted 66% of the time.[3]

When the top security controls deployed include antivirus software, firewalls, anti-spyware software, VPNs, and patch management,[4] it is clear that these controls do not necessarily align with your most likely threats. It is also obvious at this point that the security controls that are necessary to protect against each of these threats may be quite different. It seems logical that with fixed budgets and schedules that risk should be prioritized and controls selected based on this ranking.

What is risk?

There are numerous definitions of risk, depending on the entity used to define it, yet they all tend to contain several common elements. The definition that seems most aligned with the concepts of risk applied to industrial security is from the International Organization for Standardization (ISO) who defines risk as "the potential that a given threat will exploit vulnerabilities of an asset … and thereby cause harm to the organization." From this definition, it is illustrated that risk is a function of:

- The *likelihood* of a given Threat Event
- Exercising a particular "*potential*" Vulnerability of an asset
- With resulting Consequences that impact operation of the asset

There are two modifiers highlighted ('likelihood' and 'potential') that will be addressed shortly. A fundamental concept of risk management is that you can reduce or mitigate risk by addressing any one or all of these three elements. Many believe that the easiest method of reducing risk is through the identification and elimination of vulnerabilities that may potentially be exploited. The best example of this is through the deployment of a patch management program to regularly update asset software to remove identified security flaws and program anomalies that could impact performance. It is also possible that one could reduce risk by "containing" an event and limiting the extent of resulting damage. This method of risk reduction is often overlooked and can in fact be less expensive and more effective when compared with other more obvious controls. An example of limiting damage following an initial breach is network segmentation and the creation of security zones and conduits (see Chapter 9, "Establishing Zones and Conduits") that is designed to limit the ability of a threat to propagate within the industrial network(s). Another example of limiting consequences following an initial attack is through more granular communication egress control—such as configuring "outbound" rules on host-based firewalls to minimize the extent to which a compromised host can function after a breach.

The **threat event** actually consists of components that all can significantly impact risk, including:

- Threat source or actor to carry out the event
- Threat vector to initiate the event
- Threat target which the event attacks

As before, addressing one or more of these elements can reduce risk. Vectors, such as communication paths or unprotected USB ports, can have security controls deployed that further restrict the entry points used to initiate an attack. The term "reducing the attack surface" refers to the method by which targets that could be compromised are protected or eliminated all together. An example of this might be to disable unused communication services within an ICS controller that depend upon weak or vulnerable industrial protocols.

The terms threat source and threat actor are often used interchangeably and essentially refer to the human aspect of the attack. There are three characteristics of any threat source that must exist in order for a cyberattack to occur. These include:

- Capability to carry out the attack
- Intent to cause harm
- Opportunity to initiate the event

There is a large number of tools, both open-sourced and commercial, that provide the ability to attack ICS assets with little or no capability or specific system knowledge. What is often missing here is the intent of the source to actually cause damage or harm. Like the attack tools available, resources like Shodan and information-exchange communities like expert exchange provide sufficient opportunity for would-be attackers to identify and attack potential ICS targets. It is very difficult for an organization to reduce risk by focusing on outside sources because much of this is not in their direct control. However, if the attack originates from an inside source, or if an outside attacker gains a foothold, from which additional attacks could be leveraged from the inside, the threat becomes more manageable.

So how does the on-site control system engineer (i.e., insider) pose a threat to ICS? It is obvious that the insider in this case has extensive *capability* and sufficient *opportunity* to initiate the attack. The "malicious" insider possesses ample *intent* to cause harm. What intent does the "unintentional" insider possess when performing an accidental action that causes harm to the ICS? The actual intent in this case is very low. However, due to other surrounding factors that are very high (in-depth system knowledge, elevated access privileges, direct access to ICS assets, use of unauthorized tools, intentional bypassing of security policies, etc.), the resulting net risk is very high. This is the primary reason that an insider such as the on-site control systems engineer or ICS vendor site support specialist are likely targets in the early phases of a blended attack, since someone masquerading as an insider can be very difficult to detect and mitigate.

Vulnerabilities, both disclosed and latent or undisclosed, pose a real and obvious risk to industrial networks. A total of 832 vulnerabilities have been disclosed affecting ICS through July 2014, with more than 10% of the total discovered in the preceding 6 months.[5] More than 80% of all ICS vulnerabilities have been discovered since Stuxnet was reported in 2010.[6] It has become clear that security research and vulnerability identification of ICS components has taken on an important role. Traditional information security conferences like Black Hat and DEFCON now include ICS presentation content, dedicated tracks, and associated training workshops.

Information security focuses on assets that commonly comprise IT business systems, the data contained on these systems, and information as it is generated, transmitted, and stored. The **consequences** that result from a successful cyberattack can be large. The actual cost of the recent data breach at retailer target in 2013 was still unknown at the time of publishing,[7] but some are estimating the cost to target alone could exceed US$1 billion.[8] Target expects to spend US$100 million to upgrade their point-of-sale payment terminals following the breach.[9]

Consider now that operational security must manage risk to not only the direct ICS assets, but also those assets that are under the control of the ICS including the physical plant or mill, mechanical equipment, employees working in the facility, the surrounding community, and the environment. Consequences that result from a cyberattack on an ICS are less likely to have direct impact to the system itself but rather cause the plant under control to operate improperly which may impact product quality or production rates, possibly even tripping or shutting the plant down. Mechanical damage may occur leading to costly repair or replacement and extended plant downtime. Hazardous materials could be released directly impacting the surrounding community often resulting in fines. Events could directly result in loss of human life.

Figure 8.1 illustrates the relationships between the concepts and terms previously mentioned and how each interdepends on others as part of the overall risk process.

Standards and best practices for risk management

There are a variety of nationally and globally recognized standards and best practices that focus on the concept of risk management. Most of these documents, however, form a foundation for "information security risk" rather than "operational security risk." In other words, these documents do not form the basis of a risk management framework that may be used to identify and disclose important risk factors necessary to support federal regulations (e.g., those risks typically reported in a company's Annual Report, Form 10-K, or similar) but rather only those risks facing IT systems. It should be clear from the previous section that operational security risk extends beyond the physical and logical ICS assets to the physical plant that is under control of the ICS components.

Some of the organizations that maintain recognized documents include the European Union Agency for Network and Information Security (ENISA), International Organization for Standardization (ISO), the U.S. National Institute of Standards and Technology

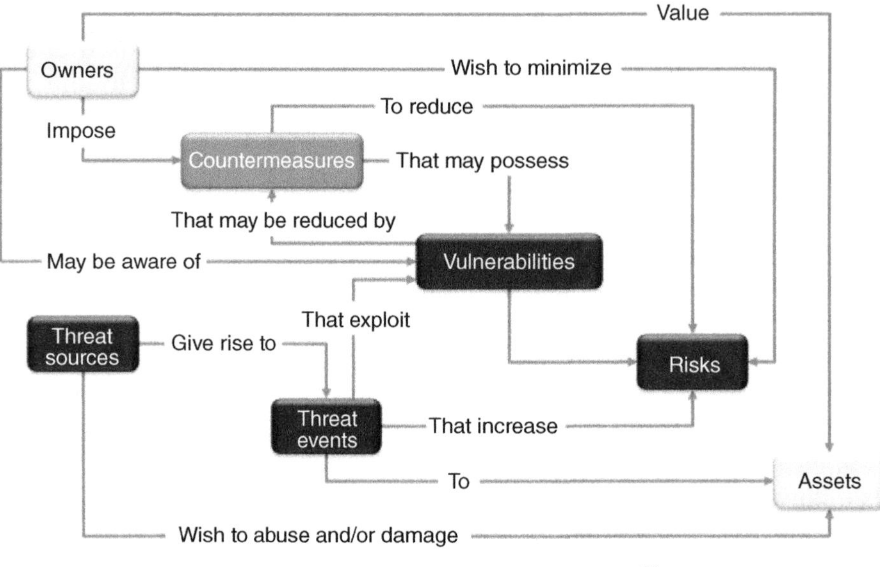

FIGURE 8.1 Understanding risk relationships.[10]

(NIST), and many others. See Chapter 14, "Standards and Regulations," for more information on industry best practices for conducting ICS assessments.

Table 8.1 lists a few of the current standards and best practices pertaining to risk management frameworks and assessment techniques.

Table 8.1 Risk Methodology Standards and Best Practices

Organization	Publication number	Description
BSI	100-3	Risk analysis based on IT-Grundschutz
CERT	OCTAVE	Operationally critical threat, asset, and vulnerability evaluation
ENISA		Principles and inventories for risk management/risk assessment methods and tools
ISO/IEC	27005	Information security risk management
ISO/IEC	31000	Risk management
ISO/IEC	31010	Risk assessment techniques
NIST	800-161	Supply chain risk management practices for federal information systems and organizations
NIST	800-30	Guide for conducting risk assessments
NIST	800-37	Guide for applying the risk management framework to federal information systems
NIST	800-39	Managing information security risk: organization, mission, and information system view

Many of these documents contain similar requirements using slightly different vocabularies or minor sequence alterations. It becomes clear that many of these documents offer the same basic guidance addressing key requirements including:

- Asset identification
- Threat identification
- Vulnerability identification
- Existing security controls identification
- Consequence identification
- Consequence analysis
- Risk ranking
- Security controls recommendations

Few documents have been drafted and approved for direct applicability within manufacturing environments and upon the industrial systems commonly used. It is necessary for this reason to alter these methodologies in order to tailor the objectives and deliverables to more closely align with these industrial systems and the operational security risk reduction goals desired. Figure 8.2 represents one hybrid methodology that has been developed to illustrate the steps necessary to perform an effective ICS cyber risk assessment. Each of these components will be discussed in the remainder of this chapter.

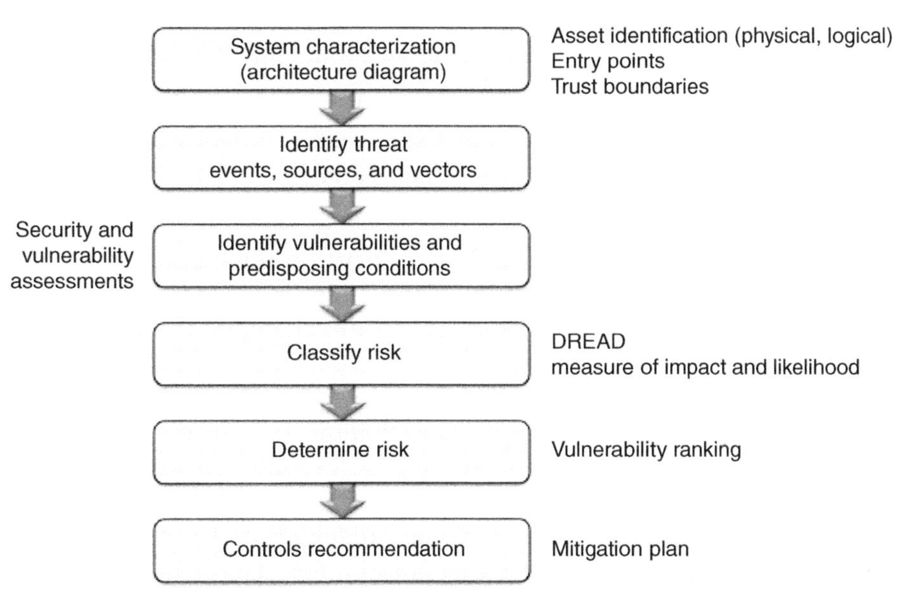

FIGURE 8.2 Methodology for assessing risk to industrial control systems.

Methodologies for assessing risk within industrial control systems

The methodology illustrated in Figure 8.2 defines the process that will be used to identify threats and vulnerabilities that could compromise the operation of not only the ICS, but also the equipment directly and indirectly under its control. This methodology blends elements of a traditional risk assessment with security testing. The risk assessment elements will define the overall "strategy" used to select security controls based on presumed risk, while the security test will define the "operations" of the system to verify the completeness that security and associated controls exist within the system under consideration. It can sometimes be confusing the difference between assessing risk and assessing security. This should become clearer shortly.

Security tests

The benefit one receives from any security test is commonly thought to be proportional to the number of vulnerabilities that the test identifies. These vulnerabilities may either be due to the (lack of) security capabilities of the system under consideration or the thoroughness of the assessment. The objective should be to establish a methodology that is based on criteria that help drive consistency from assessment to assessment and that allows common vulnerabilities that may exist across multiple systems to be uncovered.

Vulnerabilities are discovered, disclosed, and patched daily, along with new exploits and mitigation techniques targeting these weaknesses. Any assessment, audit or test that is conducted therefore only represents a snapshot in time. This is the motivation behind a "repetitive" process that is triggered by external events that may include:

- Changes to the system like a component upgrade or system migration
- Changes to the threat landscape such as the release of a new attack frameworks or capabilities, such as Industroyer, Incontroller, et. al.
- Elapsed periods of time

The purpose of these security tests will be to focus on the identification of not only system vulnerabilities but also the security controls that may (or may not) be deployed and whether or not they are still effective against the changing threat landscape. This area of focus is shown in Figure 8.3.

The goals of the security test will be to assess the current level of security the system under consideration provides in a particular installation. This means that it is important to look at not only the system-specific details (e.g., ICS vendor, network vendor, software and hardware revisions, etc.) but also site-specific factors (e.g., geographical location; compliance with corporate policies, procedures, guidelines and standards; service level agreements (SLA); project-specific documentation; etc.). This will then facilitate the identification of vulnerabilities within the system under consideration. These vulnerabilities may not necessarily be technical flaws, but could be procedural or engineering

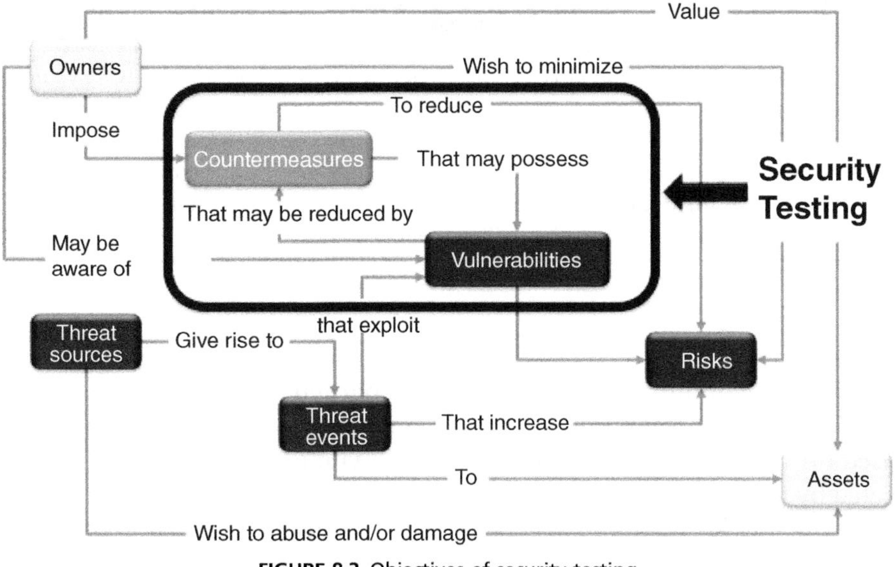

FIGURE 8.3 Objectives of security testing.

errors. Once these vulnerabilities are identified they will then be ranked in terms of severity and actions will be developed to remediate or mitigate these weaknesses.

Vulnerabilities can be found either by evaluating the system in the form of an assessment or by attempting to attack the system in a manner consistent with what a hacker or external threat may do to compromise the system commonly referred to as a "penetration test" or "ethical hacking" exercise. Penetration tests provide an accurate representation of how the system appears to a potential attacker and what actions might be required for the attack to be successful. They are also valuable in demonstrating whether or not a component or system can in fact be compromised through the discovery of exploitable vulnerabilities. The results or return on investment from the penetration test are likely to be heavily dependent on the skills and capability of the tester. These types of tests do not typically identify a high percentage of the actual vulnerabilities that exist within the system and could negatively impact the system.

Shift the attention now from the external threats to the internal ones. It has been mentioned earlier that the insider threat typically has the potential for much greater impact to the ICS and also possesses significant knowledge of the particular system. The purpose of these security tests is not to exploit a system but to determine the relative level of security a system possesses and identify ways to improve the overall level of security that remains. This is the primary reason that the details to follow will be based on assessing the vulnerabilities a system possesses from the point of view of the insider, who like an outsider also represents a credible threat actor.

Caution

Penetration testing or "ethical hacking" is rarely performed on operational ICS systems and networks due to the risks to ICS operation. It was mentioned that most of these types of tests aim to identify exploitable vulnerabilities. The primary goals of safety and reliability mean that no test shall have any risk of impact to the operation of a component or the system under test. To perform adequate penetration tests in a safe manner, a dedicated non-production test environment should be utilized.

Security audits

Security audits are commonly performed to test a particular system against a specific set of policies, procedures, standards, or regulations. These criteria are commonly developed based on knowledge of "known" threats and vulnerabilities. They are also further complicated by the fact that once a new, emerging, or sophisticated threat is discovered, it can take time for the documents to be adjusted from any deficiencies that the threat may have exploited. Audits do not typically uncover unexpected or latent vulnerabilities for this reason.

Audits can be conducted using either active collection techniques that require direct access to the system(s) under consideration or passive techniques that commonly employ questionnaires and checklists. For this reason, audits usually do not require as many resources to conduct as a more thorough security assessment or test.

Security and vulnerability assessments

Security and vulnerability assessments provide ICS users and businesses with a well-balance cost versus value security evaluation mechanism. There are both "theoretical" and "physical" methodologies that can be used—both discussed shortly. The premise of this type of assessment is to look at the entire solution for the system under consideration. This means that for each ICS system and subsystem all servers, workstations, and controllers are included. Third-party equipment such as field instruments, analytical systems, PLCs, RTUs, IEDs, custom application servers, etc. are included. Semi-trusted or demilitarized zones are considered in the assessment, as well as all communication to trusted and untrusted zones.

The active and passive network infrastructure is included covering switches, routers, firewalls, wiring closets, patch panels, and fiber-optic routing. Remote access is included (if applicable) covering not only access from users external to the facility (e.g., remote engineering access, remote vendor support) but also communications that originate outside the local control zone(s) but still may remain within the plant perimeter (e.g., engineering access via administration buildings, patch management systems, security monitoring appliances, etc.).

User identification, authentication, authorization, and accounting functionality is also included to help uncover potential weaknesses in identity and authorization management (IAM) systems like Microsoft Active Directory and RADIUS.

It is not practical to perform complete vulnerability assessments against 100% of the hosts within an ICS architecture. Vulnerability assessments therefore tend to focus on a subset of critical nodes. The results typically yield accurate results because many policies that are deployed within industrial networks apply to all hosts. If you assess one host and find that it is not patched in a timely manner, it is likely that all hosts within that architecture will possess similar vulnerabilities. Another consideration is that there is a large amount of duplication and redundancy within industrial networks, so that assessing a small subset of hosts can actually reflect a large percentage of the composite architecture.

Caution

Vulnerability assessments are performed at the component level and therefore are designed to identify if known vulnerabilities exist in the target of evaluation. It may be a safe alternative to bypass any tests against online ICS devices (particularly embedded devices like controllers) when a simple review of the vulnerability tool can reveal if it is capable of discovering any vulnerabilities.

Establishing a testing and assessment methodology

The challenge in establishing a repeatable methodology for testing and assessing an ICS lies in the lack of any consistent industry guidance. The two primary frameworks discussed that are commonly deployed in IT environments—penetration tests and vulnerability scans—each have positive aspects that should be applied to a credible ICS process. However, there are significant gaps that remain that must be addressed before attempted an on-line assessment of an operational ICS. The following recommendations are provided to assist in improving these processes to suit the particular needs of an organization.

Tailoring a methodology for industrial networks

It is now time to tailor what we have learned into a specific methodology that can be used to suit the particular system under evaluation; whether it be a DCS used in a petroleum refinery or petrochemical plant, or an SCADA system used in a wastewater treatment facility. The overall focus of a security test targeting an industrial network needs to cover a broad range of technologies and components. It shall evaluate the security of all ICS perimeters, including not only local area networks, but wireless networks, remote networks connected via remote access methods, modems, and potential "sneaker nets" that typically do not appear on network architecture diagrams.

The information obtained will be used to evaluate the overall network architecture and understand the basic organization of security zones and conduits, how firewalls have been deployed on the conduits between various zones—including the existence of one or more "functional" semi-trusted or demilitarized zones. Communication channels (conduits) between ICS field networks, field controllers, and supervisory equipment will be analyzed. The objective will be to look for weaknesses that could allow unauthorized access to the industrial networks.

It is important to include "social" aspects in the evaluation. A great deal can be learned from how the various personnel that interact with the industrial systems use the components to perform their assigns responsibilities. This should include key functional roles, including operational personnel who are ultimately responsible for interacting with the ICS to control the facility, engineering personnel who administer and configure the ICS, and maintenance personnel (including possible vendor support staff) who service and support the ICS.

The idea of understanding thoroughly the system under evaluation cannot be stated enough. The exact definition of a penetration test varies, though it is widely accepted that the goal of any pen test is to "breach security and penetrate the system." In other words, to successfully exploit a vulnerability or weakness. Failed attempts within an operational industrial network can cause instability, performance issues, or a system crash. These may lead to not merely a DoS condition but a potentially serious loss-of-view or loss-of-control situation within the ICS that may result in serious impact to the manufacturing process. The general rule is that pen tests should never be performed on an active, on-line ICS component, but rather limited to off-line, lab, or development systems.

Caution

It is always important to remember the priorities of an industrial system when performing any on-line activities on an ICS or industrial network:

- Human health and safety
- Availability of all components on the system
- Integrity (and timeliness) of data communication

Security assessments and tests should never impact any of these priorities!

Theoretical versus physical tests

There may be industrial systems that need a timely assessment; however, the risk to operational integrity is too great to allow even the slightest risk that the tests will impact manufacturing operations. These situations may require a "theoretical" assessment to be performed, which can provide some level of security assurance regarding the system under evaluation without physically contacting any ICS component. This type of assessment is based on a standardized method of completing questionnaires based on a given security baseline in a sort of "interview" format. Accurate results can only be expected when the assessment is conducted as a group exercise and consists of a knowledgeable, cross-functional team representing engineering, operations, maintenance, procurement, HSE, etc.

Theoretical assessments can also be used as an initial mechanism to raise awareness within organizations that are beginning an internal cybersecurity program. The results of these assessments can be very valuable in understand major gaps and implementing subsequent, more in-depth analysis.

The U.S. Dept. of Homeland Security (DHS) Industrial Control System Cyber Emergency Response Team (ICS-CERT) has developed the Cyber Security Evaluation Tool (CSET) as a tool for conducting off-line assessments. The CSET provides a step-by-step process of assessing an ICS based on security practices that are compared against a set of recognized industry standards. The answers provided generate output in the form of a prioritized list of recommendations with actionable items to improve the security of the system under evaluation based on the standards baseline. Figure 8.4 illustrates a sample output generated by CSET.

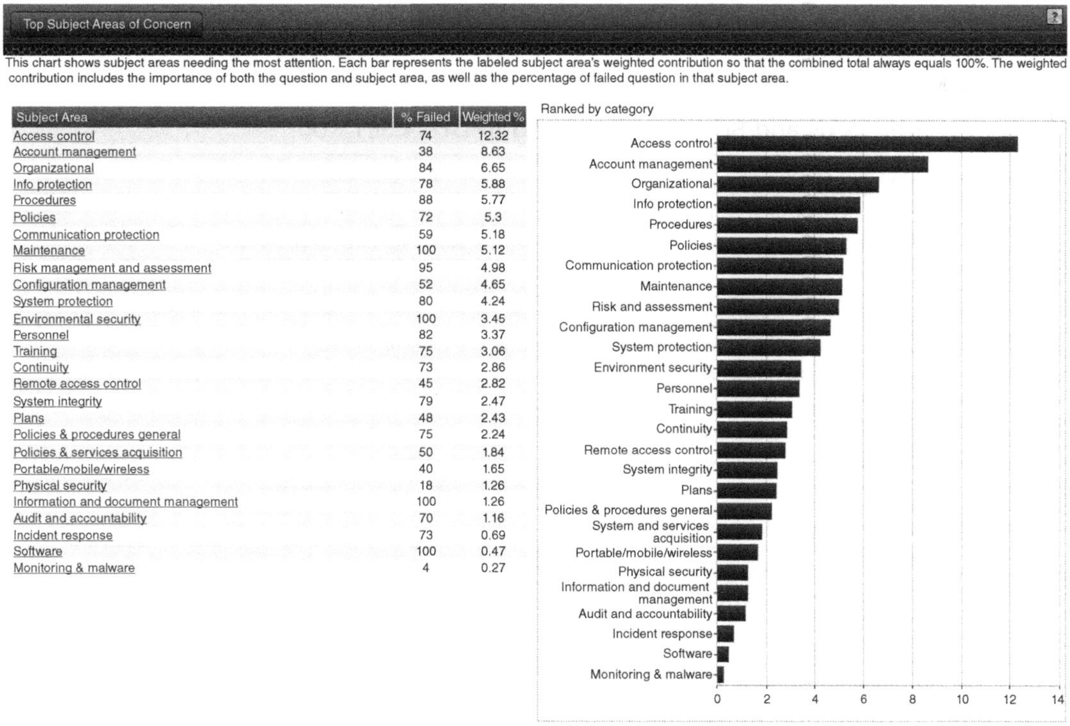

Top Subject Areas of Concern

This chart shows subject areas needing the most attention. Each bar represents the labeled subject area's weighted contribution so that the combined total always equals 100%. The weighted contribution includes the importance of both the question and subject area, as well as the percentage of failed question in that subject area.

Subject Area	% Failed	Weighted %
Access control	74	12.32
Account management	38	8.63
Organizational	84	6.65
Info protection	78	5.88
Procedures	88	5.77
Policies	72	5.3
Communication protection	59	5.18
Maintenance	100	5.12
Risk management and assessment	95	4.98
Configuration management	52	4.65
System protection	80	4.24
Environmental security	100	3.45
Personnel	82	3.37
Training	75	3.06
Continuity	73	2.86
Remote access control	45	2.82
System integrity	79	2.47
Plans	48	2.43
Policies & procedures general	75	2.24
Policies & services acquisition	50	1.84
Portable/mobile/wireless	40	1.65
Physical security	18	1.26
Information and document management	100	1.26
Audit and accountability	70	1.16
Incident response	73	0.69
Software	100	0.47
Monitoring & malware	4	0.27

Ranked by category

FIGURE 8.4 Sample CSET output.

The value of the CSET tool to many organizations is that it provides a high-level of consistency when performing evaluations, since the same questions are asked given the same set of standards requirements. A future release of the CSET tool will also support the ability for the user to input their own question set to assess systems against in-house or custom security practices that may not align exactly with the standards and best practices included with the tool (see Table 8.2 for a list of included standards).

On-line versus off-line physical tests

Physical tests that utilize actual hardware and software that comprise the components included with the system under evaluation can either be performed on an actual running industrial network that is in operation or in an environment that is not connected to a physical process and performing real-time control operations. There are advantages and disadvantages of each technique, each of which must be evaluated by an organization against the established test goals prior to commencing the activities.

The most significant benefit of an on-line test is that it represents a completely functional and operational ICS architecture that includes all the systems, networks, and data integration. Off-line environments typically reflect a small subset of the overall architecture and can omit key components that are a valuable piece of an assessment including complete network topology and connections with third-party systems and applications.

Table 8.2 Standards and Best Practices Used in DHS CSET Tool

General control system standards
NIST SP800-82—Guide to industrial control systems security
NIST SP800-53—Recommended security controls for federal information systems—Appendix I
Sector-specific standards
CFATS—Risk-based performance standards guidance 8 (cyber)
INGAA Control systems cyber security guidelines for the natural gas pipeline industry
NEI 0809 cyber security plan for nuclear power reactors
NERC—CIP reliability standard CIP-002-009
NISTIR 7628 guidelines for smart grid cyber security
NRC—Regulatory guide 5.71—Cyber security programs for nuclear facilities
DHS – TSA—Pipeline security guidelines
Information technology specific standards
NIST SP800-53—Recommended security controls for federal information systems—Appendix I
Requirements mode only standards
DHS - Catalog of control system security—Recommendations for standards developers
Council on cyber security - consensus audit guidelines (20 critical controls)
Dept. of defense—Instruction 8500.2—Information assurance implementation
ISO/IEC 15408—Common criteria for information technology security evaluation

The reason that off-line tests are discussed is that there will be circumstances where it is not possible to perform on-line tests against critical, high-risk ICS components. In these situations, off-line tests can be performed yielding reasonable results from a "component-level" point-of-view. The accuracy of the test results can be greatly improved if an on-line backup image of the critical component is obtained and then loaded on an off-line platform. This will not only allow additional, more rigorous tests such as possible component testing to target the off-line host but will also evaluate the reliability of the backup-restore utilities (also a vital security control). Table 8.3 provides some additional advantages and disadvantages of on-line and off-line test methods.

Another important characteristic of a security test is understanding the difference between observing the systems with minimal knowledge of the actual system configuration (topology, applications, authentication credentials, etc.) or looking into the system and collecting as much information as possible that may reveal less obvious or latent weaknesses. The primary goal of an ICS security test should be to secure the system as best as possible, rather than only securing those vulnerabilities that may be visible to a potential attacker. A system is considered more resilient to future attacks when a test is conducted in the latter manner. This is the reason the preferred practice for ICS security assessments is to follow a "white box" approach. Table 8.4 provides some of the key differences between these types of tests. The benefits of white box testing over black box will be discussed in more detail in the section on "vulnerability identification."

System characterization

Once the premise of the security test that will be conducted has been defined as "physical" and "on-line," the first activity performed is to characterize or identify all physical and logical assets that comprise the system under evaluation. Asset inventory and documentation is difficult and as a result can often contain gaps. This is why

Table 8.3 On-line Versus Off-line Testing Considerations

On-line tests	Off-line tests
Represents realistic network configurations	Can contain realistic configuration of ICS components
Contains volatile ICS components	Can include virtualization technologies
Include complete architecture, including third-party components	Difficult to include all third-party components
Could be used to test susceptibility of network vulnerabilities to attack	Lacks realistic network architecture
Can test less critical third-party components for vulnerabilities	Best at testing ICS components and their vulnerabilities
	Can be used to test ability to exploit vulnerabilities (ethical hacking)

Table 8.4 White Box Versus Black Box Testing Considerations

White box	Black box
Intent of assessment is to identify security vulnerabilities that could lead to an exploit; not ability to exploit	Realistically represents system in way attacker sees system
Requires asset owner to disclose significant information for successful test	Protects asset owner intellectual property
Provides most comprehensive look at vulnerabilities and risk	Does not provide complete exposure to risk
Often includes false positives	

documentation that is obtained prior to the commencement of the test should only act as a starting point and should always be validated for accuracy. A security test is designed to secure a target system by identifying security weaknesses within the architecture. It is very difficult to assess assets that are not identified or known beforehand!

System characterization and asset identification are best performed using a zone concept. This approach provides the ability to take an architecture and create a zone perimeter which will be called a "trust boundary" at this time. Once this trust boundary is established, it is then important to delineate all of the external entry points that require penetration of the perimeter. The concepts of zones/conduits and trusted/untrusted relationships is discussed in Chapter 9, "Establishing Zones and Conduits". Figure 8.5 represents the reference architecture of a single zone that will be used to discuss the concepts of trust and entry points.

The reference architecture contains three physical assets within the zone (SCADA HMI, Engineering Workstation and Controller) and one asset on the conduit (Firewall). Entry points from trusted users can also be used as attack vectors from untrusted users or potential attackers. This is why this important first step is to understand all of the

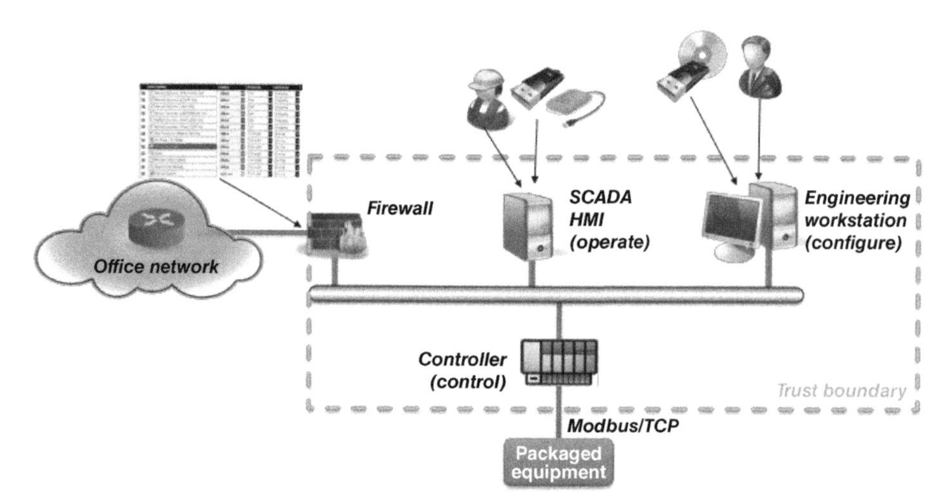

FIGURE 8.5 Trust boundary and entry points.

Table 8.5 System Characterization—Identifying Entry Points

Entry point name	Entry point description	Data flows associated with entry point	Assets associated with entry point
Firewall	Internal firewall between office and control networks	AD authentication (LDAP)	Engineering workstation
		AD authentication (LDAP)	Operator workstation
		File sharing (SMB)	Engineering workstation
		File sharing (SMB)	Operator workstation
		Historical data (OPC)	Operator workstation
Modbus port on controller	Modbus port on embedded controller to packaged equipment	Modbus/TCP	Controller
Keyboard	Keyboard on EWS	Keyboard input	Engineering workstation
Keyboard	Keyboard on OWS	Keyboard input	Operator workstation
CD/DVD drive	CD/DVD drive on EWS	Software, data files	Engineering workstation
CD/DVD drive	CD/DVD drive on OWS	Software, data files	Operator workstation
USB port	USB port on EWS	Software, data files, backup	Engineering workstation
USB port	USB port on OWS	Software, data files, backup	Operator workstation
Wireless	WLAN/Bluetooth on EWS	Software, data files	Engineering workstation
Wireless	WLAN/Bluetooth on OWS	Software, data files	Operator workstation

mechanisms that are possible to introduce "content" into the assets, as well as those assets that are currently deployed and utilized, to understand the initial attack surface of the architecture. A practical example of an unused or hidden entry point to an asset may be the built-in wireless capabilities (802.11, Bluetooth, etc.) of the SCADA HMI and the Engineering Workstation that the platform possesses that may not be currently in use. Table 8.5 summarizes these entry points, as well as the data or "content" that is typically introduced via these mechanisms.

A different way of looking at security is to consider the relationship between the asset and the controls that are deployed to protect the asset. In the vast majority of cases, security controls are specified and implemented to protect specific "logical" assets rather than "physical" ones. As an example, consider the installation of antivirus software (AVS) on a host computer. The primary security objective of the AVS is to prevent the unauthorized execution of malicious code on the platform. The reason is that malicious code is often designed to target the information contained on a computer such as credentials, local files, etc. (the "logical" assets of the host), rather than the computer itself (the "physical" asset in this case).

There are always exceptions to any general rule. The Shamoon attack of 2012 was able to render local hard disks inoperable by corrupting the master boot record (MBR) that left the computer inoperable.[11] The fact remains that most security controls are protecting logical assets within an architecture, and for this reason, it is important to have an understanding of the logical assets that are contained within a particular physical asset. Table 8.6 provides some examples of common logical assets within industrial networks.

Table 8.6 System Characterization—Identifying Logical Assets

Physical asset	Logical asset	Threat event (threat to logical asset)
Firewall	Firmware	Modify firmware to change behavior of firewall
Management port	Modify firmware, modify configuration, elevation of privilege	
Identification and authentication services	Elevation of privilege	
Log files	Modify logs to remove audit trail	
Communication interfaces	Denial-of-service	
Configuration	Modify configuration to change the behavior or the firewall	
Network	Switch ports	DoS, laptop connection injects malware, elevation of privilege
Switch Configuration	Modify switch configuration to change behavior of switch	
Controller	Static control logic Configuration	Modify configuration to change the behavior of controller
Control logic algorithm library	Modify control algorithms to change the behavior of the control algorithms	
Dynamic control data	Modify dynamic data to change the results of Control Algorithms	
I/O database	Modify I/O data to change the results of control algorithms	
Controller firmware	Modify the controller firmware to change the behavior of the controller	
Modbus interface	DoS, send elicit instructions	
Ethernet interface	DoS, inject code (malware), send elicit instructions	
Engineering workstation	Windows OS	DoS, elevation of privilege
Stored files	Copy sensitive information, modify or delete files	
Engineering and configuration apps	Modify stored Configurations, send commands to controller, modify online Configuration	
DLL's	Man-in-the-Middle attack	
Ethernet interface	DoS, inject code (malware), gain remote access	
Keyboard	DoS, elevation of privilege, modify anything	
CD/DVD drive	Inject code (malware), copy sensitive information	
USB interface	Inject code (malware), copy sensitive information	
Modem	DoS, inject code (malware), gain remote access	
Operator workstation	Windows OS	DoS, elevation of privilege
Stored files	Copy sensitive information, modify or delete files	
HMI application	Send commands to controller	
DLL's	Man-in-the-Middle attack	
Ethernet interface	DoS, inject code (malware), gain remote access	
Keyboard	DoS, elevation of privilege, modify anything	

Table 8.6 System Characterization—Identifying Logical Assets—cont'd

Physical asset	Logical asset	Threat event (threat to logical asset)
CD/DVD drive	Inject code (malware), copy sensitive information	
USB interface	Inject code (malware), copy sensitive information	
Modem	DoS, inject code (malware), gain remote access	

Data collection

Documentation is validated, and the system assets are characterized or identified via a variety of data collection methods. As an assessor becomes more familiar with the system(s) under evaluation, it will become easier to rapidly identify the critical physical and logical assets that will form the basis of a hardware and software inventory. On-line sources are a vital part of this activity, as this will identify all devices connected to the industrial network (and DMZs if included in the test). This will not only validate and update existing documentation but can uncover hidden and undocumented devices and appliances that could represent significant risk to the industrial architecture. On-line data collection will provide the ability to accurately identify all open communication ports and running applications/services on a particular device. This information will later be used to evaluate potential attack vectors within the system.

There are a variety of scanning tools that exist, both open-sourced and commercial, to assist with this activity. Scanning tools can however have catastrophic effects on some ICS components and should never be used without extensive, off-line testing, nor without approval from the business owners. The most dangerous tools tend to be "active," which are highly automated and typically inject data onto the network. These active scanners are typically unfriendly to ICS components and are recommended for use only in off-line environments or during manufacturing outages until thoroughly tested. Passive test tools can be used which are less risky and pose minimal threat of impact to the ICS. These tools will be discussed later in "Scanning of Industrial Networks."

Tip

There have been numerous documented incidents where active, automated tools have been used on industrial networks and have resulting in ICS shutdowns. The business impact of such a shutdown may not only damage the credibility of the individual(s) performing the test but also that of the security program itself and seriously undermine the program's business value.

There is an extensive amount of information that can be obtained from a pool of offline resources. This includes technical documentation for the various components comprising the ICS such as vendor manuals, project specific drawings, specifications, build books, and maintenance records. System configuration data can provide extensive information regarding hardware configuration, software applications, versions, firmware, etc. This configuration data are readily available for most ICS platforms, network appliances, third-party appliances, and corporate interfaces and should be requested in advance of the physical test start. Prior assessments, whether internal or external, can provide a valuable source of information that may not be appropriate for standard system documentation but is vital to improving the outcome of any security testing.

Scanning of industrial networks

Device scanners

There are different types of "scanners" that can be used depending on the purpose of the scan. The most basic types are designed to identify devices and may offer additional capabilities that include the identification of specific applications and communication services available on these hosts. The Network Mapper or nmap is one of the most popular device scanners used and is available for most common operating systems. It has evolved in the open-sourced community and includes capabilities of host discovery, host service detection, operating system detection, evasion and spoofing capabilities, and the ability to execute customized code via the Nmap Scripting Engine (NSE).

Basic device identification tools like ping are built-in to most commercial operating systems; however, there are limited capabilities of executing this command across a large number of possible hosts. The ping command utilizes the Internet Control Message Protocol (ICMP) to generate requests to target devices. The results of this application can be inaccurate, as many hosts now block ICMP messages via host-based applications. Security appliances rarely forward ICMP messages, making this application ineffective when used against a typical ICS zone-based architecture.

Nearly all devices depend on the Address Resolution Protocol (ARP) to translate layer-2 hardware addresses (MAC) to layer-3 IP addresses. This type of traffic is common and continuous in all networks, as it is the primary mechanism used for devices to establish and maintain communication within the same LAN subnet. There are tools based on ARP, including arping and arp-scan, that can be effectively used to identify hosts on a network and in some cases, can even identify hosts across security perimeters protected by firewalls.

The nmap tool does all of its data collection via network-based, external packet injection, and analysis. This means that it sends a large amount of network toward a host and analyzes the responses. This tool may be a realistic representation of how an attacker views the hosts on the network but is in fact a very poor tool when used as a method of identifying system assets. The concept of a white box test suggests that tools should be used to characterize "actual" features of a system and not just what is "identifiable." The Network Statistics or netstat tool is another command-line feature that is available on most operating systems. The parameters may change slightly between operating systems, but the useful of this command comes from its ability to display a number of host-based network features including "active" and "listening" network connections, application and associated service/communication port mapping, and routing tables. This tool can become a valuable asset when trying to identify the applications and services that are running on a particular host (as required by many regulations and standards including NERC CIP). It has the ability to identify active sessions with remote hosts and the services used by these hosts—vital information in establishing a network data flow mapping. This is a command-line tool and therefore does not inject packets on the network that could compromise time-sensitive network communication between ICS components making this a "friendly" and "passive" tool.

Vulnerability scanners

Vulnerability scanners form the next major type of commonly used network security scanners. There are a variety of both open-sourced (e.g., OpenVAS) and commercial (e.g., Tenable Nessus, Qualys Guard, Rapid7 Nexpose, Core Impact, SAINT scanner) products available. These applications are designed to identify vulnerabilities that may exist within a target by comparing these hosts against a database of known vulnerabilities. The ability to detect vulnerabilities can vary widely from product to product, as the vulnerability databases are managed by the application and not a common repository.

It was mentioned earlier that the number of vulnerabilities disclosed targeting ICS and industrial network components is growing. It is essential that the tool chosen for vulnerability assessment within the industrial networks is capable of identifying vulnerabilities for the targeted hosts. It would make little sense to deploy a tool that was not able to recognize ICS components when conducting a vulnerability scan on an industrial network.

Vulnerability scanners often include features that allow them to perform device scanning that occurs in advance of service and application identification that comprises the actual vulnerability analysis. These tools are often capable of accepting input from other dedicating device scanners in order to improve the efficiency of the vulnerability scans. More detailed information on vulnerability scanners will be provided later in "Vulnerability Identification."

Traffic scanners

Traffic scanners form another class of scanning tool that is commonly used in security testing activities. These tools are designed to collect raw network packets and provide them for subsequent analysis that may include host identification, data flows, and firewall rule set creation. The basic form of traffic scanner is the tcpdump (formerly ettercap) for Linux and windump for Windows. These command-line tools are designed primarily for the purpose of capturing and saving network traffic.

Wireshark is an application that is commonly used for analysis of network traffic in the form of pcap files. Though Wireshark can be used for raw packet collection, it is not recommended to use this application for this purpose due to both security and memory performance issues. Wireshark provides the ability to filter traffic based on various criteria, create conversation lists for a numerous of network protocols, and extract payloads that may exist within data streams.

Wireshark utilizes protocol "dissectors" so that the protocols used in the various OSI layers can be dissected and presented before passing to the next layer, allowing specific protocol details at each layer to be visualized in the Wireshark GUI. A sample of some of the built-in Wireshark dissectors for industrial protocols is shown in Table 8.7.

Microsoft has developed the Microsoft Message Analyzer, which is the successor to Microsoft Network Monitor. This application provides many of the capture and visualization features of Wireshark. As the name implies, this tool is more than a network traffic analyzer but rather a multifunction tool that allows event logs and text logs to be imported and analyzed, along with trace files that can be collected locally or imported from other tools like Wireshark and tcpdump. The Microsoft Network Monitor does not support the dissection of industrial protocols like Wireshark but has features that make it a valuable tool in any security testers application toolkit.

Live host identification

Several examples are provided below to illustrate how some of the various tools can be used to perform live host identification on an industrial network. All of these examples are run on a Linux host using the root account. These commands should always be

Table 8.7 Wireshark Industrial Protocol Dissectors

Protocol description
Building automation control networks
Bristol standard asynchronous protocol
Common industrial protocol
Component network over IP
Controller area network
ELCOM communication protocol
EtherCAT
Ethernet for control automation technology
Ethernet POWERLINK
Ethernet/IP
FOUNDATION fieldbus
GOOSE
HART over IP
IEC 60870-5-104
IEEE C37.118 Synchrophasor Protocol
Kingfisher RTU
Modbus
OMRON FINS
OPC Unified architecture
PROFINET
SERCOS
TwinCAT
ZigBee

practiced and tested in an off-line environment prior to executing on an operation system. Many of these tools contain numerous options where simple typographic errors can have drastic impact on the execution of the tool.

"Quiet"/"friendly" scanning techniques

The first example demonstrates how **arping** is used to send a single ARP request (-c 1) to one target (192.168.1.1) via a specific network interface (-i eth0):

```
# arping –i eth0 –c 1 192.168.1.1
```

The next example shows how the **arp-scan** command can be used to scan the entire subnet (-l) that corresponds to the configuration of a particular network interface (-I eth0) [notice that this command uses a capital "I" were the previous used a lowercase "i"], sending requests every 1000 ms (-i 1000) and providing verbose output (-v):

```
# arp-scan –I eth0 –v –l –i 1000
```

The **arp-scan** command can also specifically designate a network to scan (192.168.1.0/24) using CIDR notation and does not necessary have to be configured on the local network interface (-I eth0). This makes this tool very useful to scan general network ranges without actually receiving an address on the target network.

```
# arp-scan –I eth0 –v –i 1000 192.168.1.0/24
```

The next example uses the **tcpdump** command to initiate a packet capture that does not attempt to resolve addresses to hostname (-n) using a specific network interface (-i eth0) that writes the output to a file (-w out.pcap) and only includes traffic with a specific IP destination address (dst 192.168.1.1) and communication port (and port 502):

```
# tcpdump –n –i eth0 –w out.pcap dst 192.168.1.1 and port 502
```

Can you identify what is wrong with the previously example? What traffic does it actually capture? Since the command only captures traffic with a specific destination address, it will never see the return responses that would consist of packets that now have the same IP address as the source (src). A modified example that captures both sides of the communication includes a new filter (dst x or src x) and looks like this:

```
# tcpdump –n –i eth0 –w out.pcap dst 192.168.1.1 or src
192.168.1.1
```

Potentially "noisy"/"dangerous" scanning techniques

There may be times when the use of more active tools is required for a security test. This may include off-line tests, tests that occur during production outages, or after testing and understanding the predicted response of the target device. The first example uses the **nmap** command to perform a ping sweep (-sn) on a single subnet (192.168.1.0/24):

```
# nmap –sn 192.168.1.0/24
```

Additional options can be added to the **nmap** command to probe a target using an SYN scan (-sS) omitting name resolution (-n) and setting the timing of the scan (-T3) that provides service version identification (-sV) and operating system identification (-O) using a range of TCP ports (1-10240) against a subnet range of targets (192.168.1.0/24) and saving the output to a file in XML format (-oX out.xml):

```
# nmap –sS –n –T3 –sV –O –p 1-10240 –oX out.xml 192.168.1.0/24
```

A very powerful command-line tool to create and send specific packets on to the network is the **hping3** command. This is a Linux tool that can be very useful in testing firewalls and the performance of the rule sets against various criteria. This tool is classified as noisy since it does inject traffic on to the network, so it should be checked for compatibility with the target hosts before deploying in an operational network.

The first example sends a single packet that only contains the TCP header flag SYN set (-S) to a single target (192.168.1.1) using the port for Modbus/TCP (-p 502):

```
# hping3 –S –p 502 192.168.1.1
```

This next example performs a function similar to the **nmap** –sS option described above, by scanning a range of ports (–scan 1-10000) on a single target (192.168.1.1). The second example redirects the output into the "grep" application and only displays lines that contain the string "S..A" signifying that the response contained a packet with the TCP header SYN + ACK flags set:

```
# hping3 —-scan 1-10000 192.168.1.1
# hping3 –scan 1-10000 192.168.1.1 | grep S..A
```

Port mirroring and span ports

Most networks today are built using switches that provide a single collision domain between the host and the switch that it is connected. The switch is then responsible for maintaining a local hardware address (MAC) table and forwarding traffic as needed to the access ports that contain the desired MAC destination address. This means that the only types of traffic that can be monitored from a computer's network interface is the traffic specifically destined for the computer and local network broadcast and multicast traffic. It is necessary to enable a feature called "port mirroring" or "span ports" on the adjacent switch that will forward all network traffic within the switch to not only the desired target's access port, but also the mirrored port. This modification requires privileged access to the switch, and will forward a significant amount of traffic to the mirrored port. Attention should be paid to disabling this feature when it is no longer needed.

The example below illustrates the steps to create a span port on a Cisco Catalyst 2960 switch that mirrors traffic from Fast Ethernet ports 1-23 to port 24:

```
C2960# configure terminal
C2960(config)# monitor session 1 source interface range fe 0/1 –
23
C2960(config)# monitor session 1 destination interface fe 0/24
```

This technique allows a security tester to connect to each switch and collect a representation of the network traffic that exists locally within or transfers via uplinks through the switch. This is often a beneficial step in a security test that can provide a snapshot of actual network traffic that is collected passively and can be used for additional analysis and reporting.

Most industrial networks will consist of a number of network switches that may be configured in a redundant manner. This will require that samples be collected from all switches and then consolidated to create a single snapshot of the complete industrial network. The mergecap utility installed with Wireshark provides the capability to take multiple libpcap-formatted files and merge them into a single file for subsequent analysis:

```
# mergecap –w outfile.pcap infile1.pcap infile2.pcap …
infilen.pcap
```

There are always going to be some level of risk when performing scans of industrial networks that actively inject new traffic and target network-based hosts. Table 8.8 has been provided as a final reminder that actions typically performed on IT networks (where the primary targets are Windows-based hosts) are different from those provided on OT or industrial networks. Many of the results from common IT actions can be obtained using alternative techniques.

> **Caution**
>
> Always remember than any tool used in an online ICS environment should be thoroughly tested for potential impact prior to use in a production environment. The procedures for any on-line test should also include an action plan that should address the steps to be taken in the event of an unexpected consequence occurring during the test.

Command line tools

Up to this point, the majority of the tools discussed were run from an "assessment console" or other computer that traditionally is loaded with a hardened version of Linux and is connected to the industrial network. Many of the tools mentioned were friendly or minimally invasive to most ICS components; however, they all still injected new traffic on to the network. This may not be allowed in some environments because there is even the slightest chance that these actions could negatively impact the availability and performance of the ICS. There are alternatives that will allow the same, if not more data to be collected, yet via local interaction with the keyboard and monitor rather than remotely over the network. These tools are installed on most systems, allowing a robust assessment to be conducted with existing equipment, and can significantly improve the ability to thoroughly analyze Windows hosts. These tools also support the ability to write

Table 8.8 Minimizing the Risk of Network Scans to ICS

Target	Typical IT action	Suggested ICS action
Hosts, nodes, networks	Ping sweep	• Visually example router configuration files • Print local route and arp table • Perform physical verification • Conduct passive network listening • Use of IDS on network • Specify a subset of targets to programmatically scan
Services	Port scan	• Do local port verification (netstat) • Scan a duplicate, development, or test system on a nonproduction network
Vulnerabilities within a service	Vulnerability scan	• Perform local banner grabbing with version lookup in CVE • Scan a duplicate, development, or test system on a nonproduction network

the output to editable files that can then be merged and combined with other data for easy analysis and reporting.

There are a variety of options available, most depending on the version of operating system installed on the target. For the purposes of this section, these tools will be focused on a Windows-based ICS host platform and the tools discussed will be those available as early as Windows XP Professional and Windows Server 2003.

Tip

Every tester needs to have a solid library of reference texts that can be called upon to assist in performing ICS security tests. The Windows Command-Line Administrator's Pocket Consultant[12] provides one of the most comprehensive reference guides to Windows command-line utilities that are often forgotten in the world of the Windows GUI!

ipconfig is a common Windows command-line tool that not only displays all current network configuration values but can also be used to refresh Dynamic Host Configuration Protocol (DHCP) and Domain Name System (DNS) settings. Information provided by **ipconfig** includes:

- Hardware (MAC) Address
- IP Address (IPv4 and IPv6)
- Subnet Mask
- Default Gateway

- DHCP Server
- DNS Server
- NetBIOS over TCP/IP Enabled/Disabled

The example below uses the/all option to provide a complete report of network settings. The output is also redirected (>) to a text file (host.ipconfig.text) for collection and use.

```
C:>ipconfig / all > host.ipconfig.text
```

The Network Statistics (**netstat**) command is the authoritative method to determine what applications are running on a computer and how they map to associated communication ports and service names. It displays sessions that are both local and remote to the host, as well as active connections. Information provided by **netstat** includes:

- Active TCP connections
- Ports on which the computer listening
- Ethernet statistics
- IP routing table
- IPv4 and IPv6 Statistics

There are several parameters that can be supplied with the command. The following example requests all active connections (-a) and the associated TCP/UDP ports in numerical form (-n) on which the computer is listening, along with the executable associated with the connection (-b). The output has again been redirected (>) to a text file (host.netstat.text). This command requires elevated privileges when user account control (UAC) is enabled on Windows. The second example adds an additional parameter limiting the information to the TCP protocol (-p TCP). The third and fourth examples show the output can be piped (|) into a second utility (**findstr**) that can parse the output similar to the Linux "grep" command and only provide those connections that are active ("ESTABLISHED") or waiting ("LISTENING"). As with most commands, additional details can be found by adding/? after the command with no parameters.

```
C:>netstat –anb > host.netstat.text
C:>netstat –anbp TCP > host.netstat.text
C:>netstat –anb | findstr "ESTABLISHED"
C:>netstat –anb | findstr "LISTENING"
```

The Network Statistics commands may not return the name of an executable associated with a running service when running on some platforms but rather the process identification (PID) for the service. This requires the **tasklist** command to be executed providing a list of all running applications and services with their associated PID.

```
C:>tasklist > host.tasklist.text
```

It is valuable during a security test to collect detailed configuration information about each host included in the activity. The System Information (**systeminfo**) command provides valuable information that supports the hardware and software inventory activities (see below), as well as:

- Operating system configuration
- Security information
- Product identification numbers
- Hardware properties (RAM, disk space, network interface cards)

The example below shows the **systeminfo** command with the output redirected (>) to a text file (host.systeminfo.text) for retention.

```
C:>systeminfo > host.systeminfo.text
```

The Window Management Instrumentation Command-line (**wmic**) utility provides a powerful set of systems management features that can be executed independently, interactively, or as part of a batch file. Access to the Windows Management Instrumentation (WMI) system allows comprehensive system information to be extracted and stored in a variety of formats that support retention and analysis (CSV, HTML, XML, text, etc.). The example below uses **wmic** to query the system and provide a listing of all installed software (product get) output in HTML format (/format:htable) and saved as a file (/output:"host.products.HTML").

```
C:>wmic /output:"host.products.html" product get /format:htable
```

Some other examples of how **wmic** can be used include: local group management (group), network connections (netuse), quick fix engineering (qfe), service application management (service), local shared resource management (share), and local user account management (useraccount).

```
C:>wmic /output:"host.group.html" group list full /format:htable
C:>wmic /output:"host.netuse.html" netuse list full
/format:htable
C:>wmic /output:"host.qfe.html" qfe list full /format:htable
C:>wmic /output:"host.service.html" service list full
/format:htable
C:>wmic /output:"host.share.html" share list full /format:htable
C:>wmic /output:"host.useraccount.html" useraccount list full
/format:htable
```

A summary of the wmic command-line tool has been shown below.

```
C:>wmic /?
[global switches] <command>

The following global switches are available:
/NAMESPACE           Path for the namespace the alias operate
against.
/ROLE                Path for the role containing the alias
definitions.
/NODE                Servers the alias will operate against.
/IMPLEVEL            Client impersonation level.
/AUTHLEVEL           Client authentication level.
/LOCALE              Language id the client should use.
/PRIVILEGES          Enable or disable all privileges.
/TRACE               Outputs debugging information to stderr.
/RECORD              Logs all input commands and output.
/INTERACTIVE         Sets or resets the interactive mode.
/FAILFAST            Sets or resets the FailFast mode.
/USER                User to be used during the session.
/PASSWORD            Password to be used for session login.
/OUTPUT              Specifies the mode for output redirection.
/APPEND              Specifies the mode for output redirection.
/AGGREGATE           Sets or resets aggregate mode.
/AUTHORITY           Specifies the <authority type> for the
connection.

/?[:<BRIEF|FULL>]    Usage information.
For more information on a specific global switch, type: switch-
name /?
The following alias/es are available in the current role:

ALIAS                 - Access to the aliases available on the
local system
BASEBOARD             - Base board (also known as a motherboard)
management
BIOS                  - Basic input/output services (BIOS)
management
BOOTCONFIG            - Boot configuration management
CDROM                 - CD-ROM management
COMPUTERSYSTEM        - Computer system management
CPU                   - CPU management
CSPRODUCT             - Computer system product information from
SMBIOS
DATAFILE              - DataFile Management
DCOMAPP               - DCOM Application management
DESKTOP               - User's Desktop management
DESKTOPMONITOR        - Desktop Monitor management
DEVICEMEMORYADDRESS   - Device memory addresses management
DISKDRIVE             - Physical disk drive management
DISKQUOTA             - Disk space usage for NTFS volumes
DMACHANNEL            - Direct memory access (DMA) channel
management
```

```
ENVIRONMENT          - System environment settings management
FSDIR                - Filesystem directory entry management
GROUP                - Group account management
IDECONTROLLER        - IDE Controller management
IRQ                  - Interrupt request line (IRQ) management
JOB                  - Provides access to the jobs scheduled using
schedule service
LOADORDER            - Mgmt of system services that define
execution dependencies
LOGICALDISK          - Local storage device management
LOGON                - LOGON Sessions
MEMCACHE             - Cache memory management
MEMLOGICAL           - System memory management (config layout &
avail of mem)
MEMPHYSICAL          - Computer system's physical memory
management
NETCLIENT            - Network Client management
NETLOGIN             - Network login information (of a particular
user) management
NETPROTOCOL          - Protocols (and their network
characteristics) management
NETUSE               - Active network connection management
NIC                  - Network Interface Controller (NIC)
management
NICCONFIG            - Network adapter management
NTDOMAIN             - NT Domain management
NTEVENT              - Entries in the NT Event Log
NTEVENTLOG           - NT eventlog file management
ONBOARDDEVICE        - Mgmt of common adapter devices built into
the motherboard
OS                   - Installed Operating System/s management
PAGEFILE             - Virtual memory file swapping management
PAGEFILESET          - Page file settings management
PARTITION            - Management of partitioned areas of a
physical disk
PORT                 - I/O port management
PORTCONNECTOR        - Physical connection ports management
PRINTER              - Printer device management
PRINTERCONFIG        - Printer device configuration management
PRINTJOB             - Print job management
PROCESS              - Process management
PRODUCT              - Installation package task management
QFE                  - Quick Fix Engineering
QUOTASETTING         - Setting information for disk quotas on a
volume
RECOVEROS            - Info that will be gathered from mem when
the os fails
```

```
RECOVEROS             - Info that will be gathered from mem when
the os fails
REGISTRY              - Computer system registry management
SCSICONTROLLER        - SCSI Controller management
SERVER                - Server information management
SERVICE               - Service application management
SHARE                 - Shared resource management
SOFTWAREELEMENT       - Mgmt of elements of a software product
installed on a system
SOFTWAREFEATURE       - Management of software product subsets of
SoftwareElement
SOUNDDEV              - Sound Device management
STARTUP               - Mgmt of commands that run automatically
when users log on
SYSACCOUNT            - System account management
SYSDRIVER             - Management of the system driver for a base
service
SYSTEMENCLOSURE       - Physical system enclosure management
SYSTEMSLOT            - Mgmt of physical connection points (ports,
slots, periph)
TAPEDRIVE             - Tape drive management
TEMPERATURE           - Data management of a temperature sensor
TIMEZONE              - Time zone data management
UPS                   - Uninterruptible power supply (UPS)
management
USERACCOUNT           - User account management
VOLTAGE               - Voltage sensor (electronic voltmeter) data
management
VOLUMEQUOTASETTING    - Associates disk quota setting with specific
disk volume
WMISET                - WMI service operational parameters
management
```

For more information on a specific alias, type:

```
alias /?
```

```
CLASS     - Escapes to full WMI schema.
PATH      - Escapes to full WMI object paths.
CONTEXT   - Displays the state of all the global switches.
QUIT/EXIT - Exits the program.
```

For more information on CLASS/PATH/CONTEXT, type:

```
(CLASS | PATH | CONTEXT) /?
```

Hardware and software inventory

The command-line tools that were just discussed form the basis of the toolset that can be used to create a hardware and software inventory. These inventories are a vital first step in any security program that helps ensure accurate documentation of the industrial network and its connected equipment, as well as a quick reference that can be used when security vulnerabilities are published or software updates are available. The development of these inventories may be one of the most valuable deliverables from a physical security test. The steps to developing these inventories are outlined as follows:

1. Use **arp-scan** to identify all network-connected hosts. This command must be run on each layer three broadcast domain or subnet. This can also be accomplished in a passive manner by obtaining a consolidated network capture file obtained using **tcpdump** and importing this into Wireshark. Wireshark contains several *Statistics* features, including the ability to display *Endpoints*. This list represents all devices that are actively communicating on the network. This method does not identify nodes that were not communicating on the network when the capture files were collected.
2. Confirm that the identified hosts are authorized for the industrial network. If not, physically inspect the node and determine appropriate actions. Update the system architecture drawings with any newly discovered information.
3. Collect host platform information for each network-connected device. This should include base hardware and operating system information, network configuration details, BIOS revisions, firmware details, etc. This can be obtained using the **systeminfo** command, or via a third-party Simple Network Management Protocol (SNMP) application. For non-Windows-based devices, this typically requires specific, manual activities depending on the device. Some may offer web services that display information via a standard web browser (many PLC vendors offer these web pages as standard features), while others may require the engineering or maintenance tools for the device to be used to collect this information.
4. Collect application information for each network-connected device. This should include application vendor, name, revision, installed patches, and anything else that characterizes what and how the application has been installed on the target. This can be obtained using the **wmic** command with the **product get** option.
5. Consolidate this information into a spreadsheet or portable database, depending on size. The data provided is sensitive in nature, and as such these documents should be appropriately classified and controlled per local policy.

Data flow analysis

It is not uncommon that asset owners, and sometimes, ICS vendors do not completely understand the underlying communications and associated data flow that exists between host that comprise an ICS. Many are unclear of the value of such an exercise and therefore do not put a priority on its creation. It is important as systems are migrated from previous "flat" architectures to those that are segmented into various security zones that the communication channels that exist between these zones are documented. If they are not understood, it can become very difficult to manage the security conduits that are used to connect these zones. This is likely the reason that misconfiguration of firewalls occurs—failure to understand the data flow through the firewall, and failure of the suppliers to provide sufficient documentation on data flow requirements.

The steps required to create a data flow diagram are rather simple and will allow any asset owner, system integrator, or supplier to create this for any system. There are two pieces of data that are required. The first is a snapshot of the network traffic for the system operating under normal conditions. This can be collected as described previously using **tcpdump**. Multiple network capture files may be required, which can be merged into a single file for analysis using the **mergecap** utility.

Wireshark is used to then open the consolidated capture file and to perform simple analysis of the network via the *Statistics* features using *Conversations*. The output shown in Figure 8.6 reflects the host-to-host sessions that were active when the network captures were collected. The TCP tab would then reveal the TCP ports used for these sessions. The output shows that there were 91 active host-to-host sessions that utilized 1113 different TCP port pairs and 101 UDP port pairs. Additional filtering could be used to eliminate the multicast traffic on 224.0.0.0/8 and 225.0.0.0/8 to reduce the pairings even further.

The Network Statistics (**netstat**) command is also used to develop a mapping of the local host services and what network devices are using these services. The added value of this method is that it will provide some indication of the applications and service names

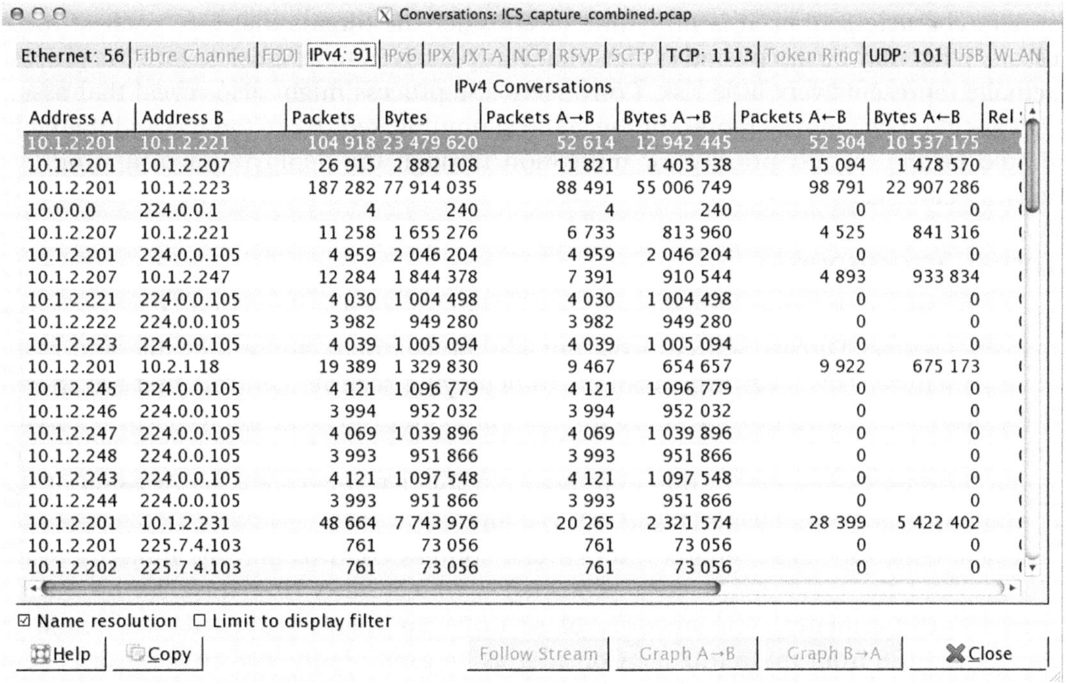

Conversations: ICS_capture_combined.pcap

Ethernet: 56 | Fibre Channel | FDDI | IPv4: 91 | IPv6 | IPX | JXTA | NCP | RSVP | SCTP | TCP: 1113 | Token Ring | UDP: 101 | USB | WLAN

IPv4 Conversations

Address A	Address B	Packets	Bytes	Packets A→B	Bytes A→B	Packets A←B	Bytes A←B
10.1.2.201	10.1.2.221	104 918	23 479 620	52 614	12 942 445	52 304	10 537 175
10.1.2.201	10.1.2.207	26 915	2 882 108	11 821	1 403 538	15 094	1 478 570
10.1.2.201	10.1.2.223	187 282	77 914 035	88 491	55 006 749	98 791	22 907 286
10.0.0.0	224.0.0.1	4	240	4	240	0	0
10.1.2.207	10.1.2.221	11 258	1 655 276	6 733	813 960	4 525	841 316
10.1.2.201	224.0.0.105	4 959	2 046 204	4 959	2 046 204	0	0
10.1.2.207	10.1.2.247	12 284	1 844 378	7 391	910 544	4 893	933 834
10.1.2.221	224.0.0.105	4 030	1 004 498	4 030	1 004 498	0	0
10.1.2.222	224.0.0.105	3 982	949 280	3 982	949 280	0	0
10.1.2.223	224.0.0.105	4 039	1 005 094	4 039	1 005 094	0	0
10.1.2.201	10.2.1.18	19 389	1 329 830	9 467	654 657	9 922	675 173
10.1.2.245	224.0.0.105	4 121	1 096 779	4 121	1 096 779	0	0
10.1.2.246	224.0.0.105	3 994	952 032	3 994	952 032	0	0
10.1.2.247	224.0.0.105	4 069	1 039 896	4 069	1 039 896	0	0
10.1.2.248	224.0.0.105	3 993	951 866	3 993	951 866	0	0
10.1.2.243	224.0.0.105	4 121	1 097 548	4 121	1 097 548	0	0
10.1.2.244	224.0.0.105	3 993	951 866	3 993	951 866	0	0
10.1.2.201	10.1.2.231	48 664	7 743 976	20 265	2 321 574	28 399	5 422 402
10.1.2.201	225.7.4.103	761	73 056	761	73 056	0	0
10.1.2.202	225.7.4.103	761	73 056	761	73 056	0	0

☑ Name resolution ☐ Limit to display filter

Help | Copy | Follow Stream | Graph A→B | Graph B→A | Close

FIGURE 8.6 Performing data flow analysis with Wireshark.

associated with the communication channels that are identified between hosts. The Wireshark method only reveals the TCP and UDP port numbers. A common method is a hybrid of both techniques that provides for the quick creation of an overall diagram using Wireshark, with additional details regarding the communications established using **netstat**.

Threat identification

The methodology described in Figure 8.2 continues with the identification of threats covering threat events, threat sources/actors, and threat vectors. This step is likely the most difficult step in the entire process and, for that reason, is commonly omitted. This is because it can be very difficult to describe all aspects of the unmitigated risk that is present for a particular industrial environment. It was described earlier that cybersecurity controls are applied to logical assets rather than physical assets. The identification of physical and logical assets occurred during the system characterization phase. These assets must now be mapped to specific threats that can later be assessed as to whether appropriate controls are in place to secure these assets from the identified threats.

Threat mapping can be performed in one of several fashions, including organization by physical asset, by threat source (outsider, insider), or by intent (intentional, unintentional). The easiest method for most learning the process is to first create an organization by physical asset, which is then expanded to logical assets after completing system characterization. It is now time to consider the threats that face each of these assets. What may be discovered is that what was perceived to be a risk before the process actually represents very little risk. Conversely, the process might also reveal that assets that are often overlooked represent the greatest unmitigated risk to the ICS and therefore should be the highest priority for mitigation through the deployment of appropriate security controls.

Threat actors/sources

Many develop industrial cyber security programs under the (unqualified) assumption that the greatest Threats Sources exist outside the company and are hostile and malicious in nature. This leads organizations to deploying security controls that are specifically designed to help prevent these threats from compromising the ICS. These threats are real and do face some risk to industrial networks; however, they typically do not represent the greatest risk to the architecture. Table 8.9 provides a list of some of the common threat actors facing IT and OT systems.

Documented incident reports from several sources confirm that the majority of incidents, and the greatest risk to a protected architecture is from insiders or trusted partners. Unfortunately, the majority of security controls deployed do very little to protect the ICS from these threats. Consider as an example the on-site control system engineer who configures and administers the ICS. His job is very demanding causing

Table 8.9 Common Threat Actors/Sources[13]

Adversarial
Outside individual
Inside individual
Trusted insider
Privileged insider
Ad Hoc group
Established group
Competitor
Supplier
Partner
Customer
Nation state
Accidental
User
Privileged user
Administrator
Structural
Information technology equipment
Environmental controls
Software
Environmental
Natural disaster (e.g., Fire, flood, tsunami)
Man-made disaster (e.g., Bombing, overrun)
Unusual natural event (e.g., Solar EMP)
Infrastructure failure (e.g., Telecommunications, electrical power)

him to find ways to improve efficiency and productivity by installing a suite of untested and unqualified applications on his engineering workstation. He also knows that the corporate antivirus software and host-based firewall often interfere with his applications, and since he is the administrator, he disables these features from his workstation. The original malicious payload may have originated from an external source, but it is now insider (the engineering) who is now going to initiate the event. What controls are left to protect the entire system from a cyber event caused by the insertion of his infected USB flash drive into his engineering workstation that has elevated privileges and global industrial network access? This is how Stuxnet infected its target! This is the reason why an objective risk process is necessary.

This is not a simple exercise, and for that reason, it may be beneficial to begin the threat identification activities by focusing on four different threat sources: intentional (malicious) outsider, intentional (malicious) insider, unintentional outsider, and unintentional (accidental) insider.

Threat vectors

The threat vector identifies the method by which the threat source will impact the target. This directly corresponds to the Entry Points in the context of the methodology established in this section. The reason for introducing the concept of an Entry Point as a means of identifying Threat Vectors is that it provides a mechanism of looking beyond traditional IT access mechanisms (e.g., USB flash drives, networks, etc.) and introduces more of the human factor including the use of policies and procedures. Entry Points are also intentionally identified before diving into the threat identification phase to allow individuals to consider less obvious mechanisms (e.g., an unused wireless LAN adapter).

The establishment of the trust boundary provides a vital role of scoping and limiting the potential entry points or vectors entering a zone. Consider as an example an industrial network that is connected to the business network via a firewall. The entry point into the ICS in this case is the network connection through the firewall. The business network on the other hand will have its own set of entry points and threat vectors that could potentially allow unauthorized access from untrusted zones (i.e., the Internet) to the trusted business zones. This is not in scope when evaluating the entry point into the ICS zones. What this has effectively done is consider unauthorized external traffic on the business network the same as authorized local traffic, since the security controls used on the conduit into the ICS (the firewall in this case) must handle all traffic accordingly. This approach provides necessary resilience when unauthorized external actors have masqueraded as potentially trusted insiders.

Table 8.10 provides basic guidance on the selection of possible ICS entry points and threat vectors.

Table 8.10 Common Threat Vectors

Direct
Local area network—wired
Local area network—Wireless
Personal area network (NFC, Bluetooth)
USB port
SATA/eSATA port
Keyboard/mouse
Monitor/projector
Serial port
Webcam
Electrical supply
Disconnect switch
Indirect
Application software (via media)
Configuration terminal (via serial port)
Modem (via serial port, internal card)
Human (via keyboard, webcam)

Threat events

The threat event represents the details of the attack that would be carried out by a particular threat source. When the source is an adversarial one, the threat event is typically described in terms of the tactics, techniques, and procedures (TTP) used in the attack. Multiple actors could possibly use a single event, and likewise, a single actor could use multiple events. This is why the first attempt at developing an ICS risk assessment worksheet can quickly become a very complex task; however, there is a high likelihood of reusability on subsequent assessment exercises. The list that is initially developed could contain numerous events that are later determined to be unrelated or not relevant to the particular system under evaluation. It is best, however, to not eliminate any information during the early steps of the exercise.

The "Guide to Conducting Risk Assessments" published by the U.S. National Institute of Standards and Technologies provides a comprehensive appendix of threat events that can be used in conducting an ICS assessment. Some of the relevant events from this list have been provided in Table 8.11.

Identification of threats during security assessments

It is likely that during a security assessment that threats will be discovered and will need to be added to the spreadsheet for tracking and measuring risk throughout the exercise. These threats are typically found when analyzing the data that are collected early in the process that could reveal any of the following:

- Infected media discovered from antivirus logs
- Infected desktop or laptop workstations discovered from Windows Event logs
- Static data corrupted discovered from local disk evaluation
- Data copied to untrusted location discovered from network resource usage
- Accounts not deactivated discovered from local/domain account review
- Stolen credentials discovered when used to access unauthorized hosts
- Overload communications network discovered when reviewing network statistics

The tasks associated with threat identification will not only improve one's overall awareness of the system, its operation, and the environment which it operator in but also will provide useful information that can later be combined with identified weaknesses to prioritize the action plan and mitigating controls that will be selected to secure the industrial systems.

Vulnerability identification

The activity of vulnerability identification is the next step in the process and is the basis for performing a detailed evaluation of the complete ICS as defined by the security test rules of engagement. This activity will combine automated tools such as vulnerability

Table 8.11 Common Threat Events[14]

Adversarial threat events
Perform network reconnaissance/scanning
Perform organizational reconnaissance and surveillance
Craft spear phishing attacks
Create counterfeit/spoof website
Craft counterfeit certifications
Inject malicious components into the supply chain
Deliver malware to organizational systems
Insert subverted individuals into organizations
Exploit physical access to organization facilities
Exploit poorly configured or unauthorized systems exposed to the Internet
Exploit split-tunneling
Exploit multitenancy in a cloud environment
Exploit known vulnerabilities
Exploit recently discovered vulnerabilities
Exploit vulnerabilities using 0-day attacks
Violate isolation in multitenant environment
Compromise software of critical systems
Conduct attacks using unauthorized ports, protocols and services
Conduct attacks levering traffic/data movement allowed across perimeter
Conduct denial-of-service attack
Conduct physical attack on organization facilities
Conduct physical attack on infrastructure supporting organizational facilities
Conduct session hijacking
Conduct network traffic modification (man-in-the-middle) attack
Conduct social engineering campaign to obtain information
Conduct supply chain attacks
Obtain sensitive information via exfiltration
Cause degradation of services
Cause integrity loss by polluting or corrupting critical data
Obtain unauthorized access
Coordinate a multi-state (hopping) attack
Coordinate cyberattacks using external (outside), internal (insider) and supply chain vectors
Non-adversarial threat events
Spill sensitive information
Mishandling of critical information by authorized users
Incorrect privilege settings
Communications contention
Fire (Arson)
Resource contention
Introduction of vulnerabilities into software products
Disk error

scanning applications, with manual analysis of data collected throughout the exercise. A vulnerability is not just the presence of unpatched software designed to correct published vulnerabilities but also the use of unnecessary services and applications that cannot be determined by simply scanning for the presence (or absence) of software. Vulnerabilities may exist in the form of improper authentication, poor credential management, improper access control, and inconsistent document. A rigorous vulnerability assessment looks at all of these and more.

The assessment phase depends a great deal on automated vulnerability scanning software. It also involves the review of relevant application, host, and network configuration files. The implementation of any existing security controls is reviewed and documented for effectiveness, and the overall physical aspects of the ICS are inspected. The idea behind such a thorough process is to attempt to review and discover many of the more common ICS vulnerabilities. Some of the more common ICS vulnerabilities are shown in Table 8.12.

The potential vulnerabilities as shown in Table 8.12 are meant to serve as a form of reminder when performing the actual assessment. The objective is to identify backdoors or "holes" that may exist in the industrial network perimeter. Devices with little or no security features and those that are susceptible to attack need to be identified so that they can be placed in special security zones and secured separately. Networks are reviewed to uncover possible opportunities for communications hijacking and man-in-the-middle (MitM) attacks. Every network-connected ICS component is assessed to discover improper or nonexistent patching of both software and firmware that could potentially compromise the network. Suppliers can also be included in the assessment to ensure that insecure coding techniques and software development lifecycles do not introduce unnecessary risk.

Vulnerability scanning

Vulnerability scanning is the process of methodically reviewing the configuration of a set of hosts by attempting to discover previously identified vulnerabilities that may be present. Automated tools are available, with some of these described earlier under "vulnerability scanners." It is also possible to perform this exercise manually if the use of an automated tool against a critical host is not allowed due to the potential for any negative impact to the performance and availability of the host.

Manual vulnerability scanning consists of collecting information using some of the command-line tools described earlier and individually comparing the revision information of the operating system, applications, and services against databases of known vulnerabilities. Two of the popular databases of vulnerabilities are the National Vulnerability Database[15] (NVD) hosted by NIST and the Open-Source Vulnerability Database[16] (OSVDB). There are more than 100,000 vulnerabilities tracked between these two databases, with most vulnerabilities also tracked against a "common enumeration" system known as common vulnerabilities and exposures (CVE).

Table 8.12 Common ICS Vulnerabilities

Category	Potential vulnerabilities
Networks	Poor physical security
	Configuration errors
	Poor Configuration management
	Inadequate port security
	Use of vulnerable ICS protocols
	Unnecessary firewall rules
	Lack of Intrusion detection capabilities
Configuration	Poor account management
	Poor password policies
	Lack of patch management
	Ineffective anti-virus/Application whitelisting
Platforms	Lack of system hardening
	Insecure embedded applications
	Untested third-party applications
	Lack of patch management
	Zero-days
ICS applications	Poor code quality
	Lack of authentication
	Use of vulnerable ICS protocols
	Uncontrolled file sharing
	Zero-days
	Untested application integration
	Unnecessary active directory replication
Embedded devices	Configuration errors
	Poor configuration management
	Lack of device hardening
	Use of vulnerable ICS protocols
	Zero-days
	Insufficient access control
Policy	Inadequate security awareness
	Social engineering susceptibility
	Inadequate physical security
	Insufficient access control

An example of a simple manual vulnerability assessment is detailed below:

1. The **wmic** command is used with the product get option to list all of the installed applications running on a Windows 2003 Server host.
2. The SCADA application software is shown as "IGSS32 9.0" with the vendor name "7-Technologies" and a version of 9.0.0.0.
3. Using OSVDB, "igss" is entered in the Quick Search field, and several results are returned. Selecting the most recent item, a link is provided to an advisory

published by ICS-CERT that confirms that the installed version of software has a published vulnerability.

4. The advisory contains information on how to download and install a software patch from the software provided.

It is apparent that this process can be very time-consuming, and that a great deal of cross-referencing must be performed. The use of automated tools simplifies this process by systematically assessing the target and quickly comparing the information extracted against a local database of documented vulnerabilities. Vulnerability scanning applications depend on external data to maintain a current local database, so the application should be updated before conducting any assessments. It is also recommended to always include the update sequence number or data used when generating a vulnerability report with the security test.

As mentioned earlier, there are several commercial vulnerability scanners available. The important feature to consider when using a particular product—commercial or open-sourced—is the ability to assess the applications that are installed on the target system. Even if there are no application-specific vulnerabilities in the database (as would be the case with many embedded ICS devices), the scanner may still be able to provide useful information regarding active services and potential weaknesses associated with those services.

What is important when using a vulnerability scanning application is to obtain as accurate of results as possible. The way that this is most often performed is via an "authenticated scan." This performs an effective "white box" assessment of the target by authenticating remotely on the device and then performing a variety of internal audits, including registry reviews and network statistics. These results provide an accurate reflection of the true security posture of the target and not just what is visible to a potential attacker. An authenticated scan is also more "friendly" on the target and does not typically inject as much hostile traffic into the network interfaces against various listening services. Figure 8.7 shows an example of the Nessus vulnerability scanner from Tenable Network Security where a "black box" unauthenticated scan yielded only four high severity vulnerabilities, while a scan against the same target using authentication yielded 181 high severity vulnerabilities.

The most common method of vulnerability scanning utilizes active mechanisms that place some packets on the network. The "aggressiveness" of the scan can be control in many applications, but as with any active technique, close attention must be paid to the potential impact of the scanner on the target.

Passive vulnerability scanners are available that collect the information needed for analysis via network packet capture rather than packet injection. Unlike active scanners that represent a "snapshot" view of the vulnerabilities on the target, passive methods provide a continuous view of the network. They are able to enumerate the network and detect when new devices are added. This type of scanner is well suited for industrial

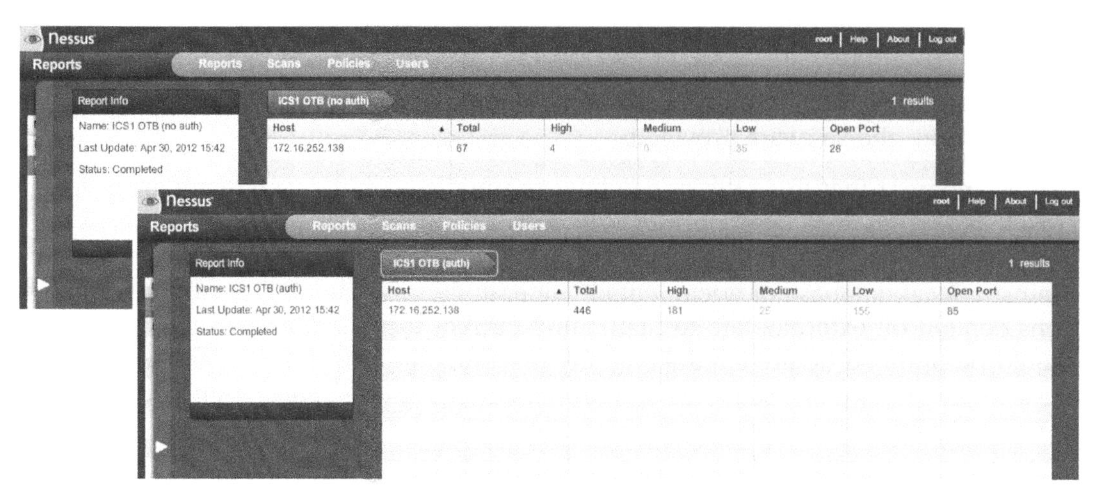

FIGURE 8.7 Authenticated versus unauthenticated vulnerability scan results.

networks because of the static nature of the network topology and the regular traffic patterns and volumes that exist.

Host-based vulnerability scanners are also available; however, they would not likely be accepted within the ICS zones on industrial networks due to the fact that they must be installed on the target. These scanners do facilitate compliance auditing of configurations and content inspection, so they do fit a need. A good example of a host-based scanner would be the Microsoft Baseline Security Analyzer (MBSA).

It should be obvious at this point that vulnerability scanners are only capable of assessing a target against vulnerabilities that are known. In other words, it offers no guidance of any "zero-day" or those vulnerabilities that exist that have been discovered, but the presence has not been communicated. This is why a strong defense-in-depth security program must depend on the ability to prevent, detect, respond, and correct against not only the threats that are known today but also those threats that may appear tomorrow.

Caution

A vulnerability scanner should never be used on an on-line ICS and industrial network without prior testing and approval from those directly responsible for the operation of the ICS.

Tip

Just because a system has no vulnerabilities does not mean that it has been configured in a secure manner.

Configuration auditing

Vulnerability scanners are designed to assess a particular target against a set of known software vulnerabilities. Once the device has updated its firmware, installed the security updates for the operating system, and/or confirmed that the application software does not have any known weaknesses, the target is now considered safe … right? Wrong! The absence of software vulnerabilities does not mean that the software has actually been installed, configured, and even hardened in a manner that helps to reduce the possibility of a breach.

This is known as configuration "compliance auditing" and compares the current configuration of a host against a set of acceptable settings. These settings may be determined by an organization's security policy, a regulatory standard, or a set of industry-recognized benchmarks. Organizations that provide configuration benchmarks include NIST,[17] Center for Internet Security,[18] National Security Agency,[19] and Tenable Network Security[20]. The repository of compliance and audit files provided by Tenable is an aggregate of many available from other parts (such as CIS, NSA, CERT, etc.) as well as custom developed files that are designed to provide a measure of compliances against published recommendations from BSI (Germany), CERT (Carnegie Melon University), and others.

The Nessus vulnerability scanner provides the ability to import predesigned or customized files that can be applied against target systems. These audits can be performed on the configuration of operating systems, applications, antivirus software, databases, network infrastructure, and content stored on file systems. Figure 8.8 shows the output from a typical compliance audit. Tools are available that support the creation of audit files from existing policy inf files.

The U.S. Dept. of Energy funded a project and partnered with DigitalBond to develop a set of security configuration guidelines for ICS.[21] The project developed Nessus audit configuration files for more than 20 different ICS components (see Table 8.13). These audit files provide a method by which asset owners, system integrators, and suppliers can verify that the systems have been configured in an optimal, preagreed manner against a consistent set of metrics. These audit files are no longer maintained but were still available free-of-charge on the DigitalBond website archives at the time of publication.[22]

Vulnerability prioritization

Not all vulnerabilities that are discovered during a security test are necessarily exploitable. The development of exploits can prove to be valuable in determining if the vulnerabilities represent a real threat; however, the cost should be weighted against the benefits when considering this activity. What proves more effective is an objective method of rating the severity of vulnerabilities as they are discovered within a particular architecture. A vulnerability that exists on an Internet-facing corporate web server does

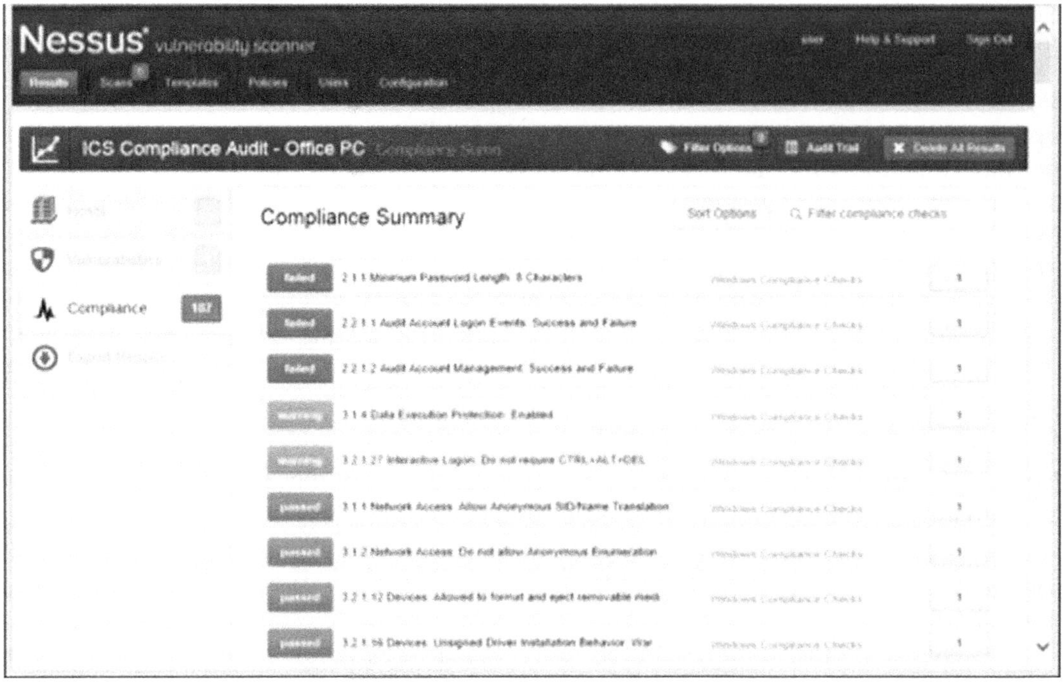

FIGURE 8.8 Compliant auditing report example.

Table 8.13 Bandolier Project ICS Details

Vendor	Platform
ABB	800xA PPA
Alstom grid	e-terraplatform
CSI (Control Systems International)	UCOS
Emerson	Ovation
Matrikon	Security Gateway Tunneller
OSIsoft	PI Enterprise Server
Siemens	Spectrum Power TG
SISCO	AX-S4 ICCP Server
SNC-Lavalin ECS	GENe SCADA
Telvent	OASyS DNA

not represent the same amount of risk as that vulnerability existing on a web server on a protected security zone that is nested deep within the organization. The outcome of this rating exercise can then be used to prioritize the corrective action plan following any site security test, allowing more severe (aka those representing a higher net risk to the organization) vulnerabilities to be mitigated before less severe ones are considered.

Common vulnerability scoring system

The Common Vulnerability Scoring System (CVSS) is a free, open, globally accepted industry standard that is used for determining the severity of system vulnerabilities. The CVSS is not owned by any organization but rather is under the custodial care of the Forum of Incident Response and Security Teams (FIRST). Each vulnerability is provided with one to three different metrics that produce a score on a scale of 0–10 that reflect the severity as the vulnerability is applied in different situations (see Figure 8.9). Each score consists of a "vector" that represents the value used for each component in calculating the total number. This scoring system allows vulnerabilities to be prioritized based on the actual risk they pose to a particular organization.

The Base metric and score are the only mandatory component of the CVSS and is used to present the characteristics of a vulnerability that are constant with time and across different user environments. This score is commonly provided by the party responsible for disclosing the vulnerability and is included with many advisories and security alerts. The Base score in the context of risk management, can be thought of as a measure of "gross unmitigated" risk.

The temporal metric and score provide refinement of the severity of the vulnerability by including the characteristics of a vulnerability that change over time but not across different user environments. An example of how this number can change over time is that a vulnerability is initially disclosed, and there are no public exploits available (Exploitability may be "unproven" or "proof of concept"). When a tool like the Metasploit Framework from Rapid7 makes an automated exploit module available, the temporal score would increase to reflect this change (Exploitability may now be "functional" or "high"). The same can apply to the availability of a patch or update to correct the vulnerability. The patch may not be immediately available but is published at some time

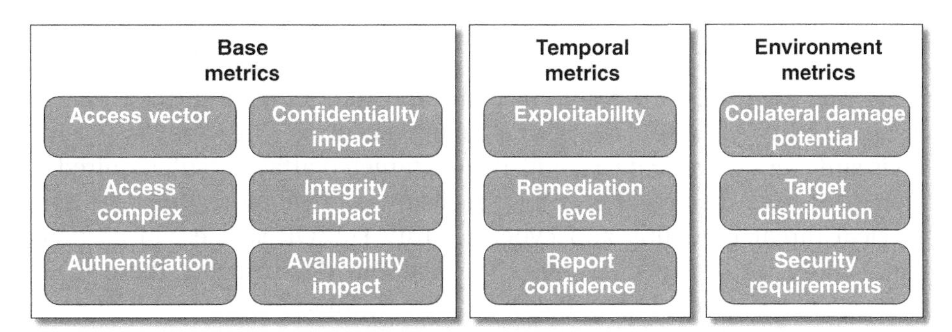

FIGURE 8.9 Common vulnerability scoring system (version 2).

in the future. The temporal score does not consider any unique characteristics of a particular user or installation.

The environmental metric and score reflects the characteristics of the vulnerability that are relevant and unique to a particular user's environment. This when calculated offers the best indication of "net unmitigated" risk to those systems and environments that possess the vulnerability.

There are quantitative formulas[23] that can be used to calculate the individual scores based on each vector. The NVD website of NIST provides an on-line calculator[24] that can be used.

Process vulnerabilities

Again, when considering cyber-physical outcomes, the vulnerability of the control system itself must be considered. The term "vulnerability" as thought of by most security professionals does not apply here: a control system is designed for automation using "command and control by design"; if a controller tells a pump to operate at a specific parameter, the system will oblige, without any consideration of potential software or firmware vulnerabilities of the controller to the pump. Process vulnerability is typically thought of in terms of safety and process design: safety instrumentation could identify a hazardous operational parameter and make adjustments to the system (e.g., there is too much pressure so input flow rates could be reduced); a physical safety control could be implemented to prevent a system from entering a hazardous condition (e.g., a mechanical limiter to prevent the input flow from exceeding unsafe parameters; or a compensating safety control could release excess pressure).

Industrial cybersecurity expert Ralph Langner introduced the concept of "cyber fragility" and "cyber robustness" within an automation process. Cyber fragility refers to the "deficient ability of an automated process to withstand variations of normal conditions for data processing, data storage, and data transmission during the process's lifetime."[25] Cyber robustness refers to the opposite: the "ability to withstand changes in procedure or circumstance, and the ability to cope with variations in the operating environment with minimal or no damage, alteration, or loss of continuality."[26] Langner posits that in a properly designed process automation system, there should be sufficient robustness that a targeted cyber-physical attack such as Stuxnet would be unsuccessful because the process itself would able to withstand intentionally introduced variations and contingencies.[27] While this is a sound theory, ensuring this robustness is in fact in place and is sufficient to mitigate a sophisticated targeted attack is still required. The risk assessment processes discussed can help measure and quantify this. The process of cyber-physical threat modeling and cyber security HAZOP studies, in particular, can help identify areas of unexpected cyber fragility and determine if there is sufficient robustness in place (see "Cyber-Physical Threat Modeling" and "Cyber Security HAZOP," below).

Risk classification and ranking

The process of Risk Classification and Ranking provides a means for evaluating the threats and vulnerabilities identified so far and creating an objective method to compare these against one another. This activity supports the creation of the budget and scheduled required to implement the security program of the industrial systems. Classification and ranking is important in making an "effective" security program that addresses the goals of both business operations and operational security.

Consequences and impact

The data collection aspect of the security test is complete, and it is now time to prioritize the results through classification and ranking. The process to this point as shown in Figure 8.2 has resulted in a set of physical and logical assets that have been matched against one or more threats as defined by the actor (person or persons who would initiate the attack), the vector (entry point used to introduce the malicious content of the attack), and the event (methods used to perform the attack). The assets have also been assessed to determine if there are any vulnerabilities or flaws that could possibly be exploited by an attacker. Remembering the earlier definition of risk, the last piece of information needed is a determination of the consequences or impact to operations that would occur should the cyber event occur. The term "operations" has been used here instead of "industrial systems" because remember the primary purpose of an ICS is to control a manufacturing facility and not merely to process information.

Once the risk assessment team shifts their focus from "impact to the system" to "impact to the plant or mill," the severity of the unmitigated risk can become significant. Table 8.14 provides some examples of the consequences that could occur should any ICS

Table 8.14 Common ICS Consequences

Common ICS Consequences
Impact to quality
Customer reputation
Loss of production
Loss of intellectual property
Economic (micro) impact
Mechanical stress or failure
Environmental release
Catastrophic equipment failure
Localized loss of life
Generalized panic
Economic (macro) impact
Widespread loss of life

component fail to perform their intended function. These consequences can have local (plant), regional (surrounding community), or global (national, multinational) impact.

Many would challenge that a single cyber event could have global consequences. It was reported in a U.S. Dept. of Homeland Security's National Risk Profile that old and deteriorating infrastructure in the U.S. could pose significant risks to the nation and its economy.[28] Now consider natural gas pipelines as part of this deteriorating infrastructure and how there have been more than 2800 "significant" gas pipeline accidents in the U.S. since 1990.[29] The ICS monitors and controls the parameters associated with the mechanical integrity of these pipelines. What is the attractiveness of this target? Adversaries would have little chance of victory if the battle was fought on a traditional military battleground, but in cyberspace, the odds shift dramatically and the ICS is a critical target in any cyberwar launched against infrastructure.

How to estimate consequences and likelihood?

The challenge that many face in risk classification is how to apply a measure of the "likelihood" that a cyberattack will occur. Traditional IT information risk processes consider likelihood as a measure of time—will this event happen in 1 month, 1 year, or longer. If straight quantitative methods of calculating risk were used, a very serious threat with multiple vulnerabilities could quickly be subdued by applying a low likelihood number and assuming that the event does not occur until some point in the future. Can you see the flaw here? If the same event that can occur today can also occur next year, does it not mean that the cost associated with the consequences would be greater? Absolutely—factors such as inflation, cost of capital, population growth, and many others will cause the cost of the event to grow, yet this is not fed back into the initial calculation model. Using the previous pipeline as an example, if you do nothing to maintain the pipeline, then it is going to fail at some point in the future. The consequences are likely to be greater than today as well because the pipeline was originally built in a rural area, but in 20 years, it is now part of a residential area.

These situations illustrate the need to use some other form of estimating the likelihood of a cyber event, and the consequences should the event occur. The DREAD model, named from the first letter of each of the five rating categories, was developed by Microsoft as part of their Software Development Lifecycle (SDL) to provide a method to classify security bugs. This model (shown in Table 8.15) provides an indirect means of calculating consequences and likelihood by looking at these factors in a different way. For example, rather than asking if the threat of a vulnerability being exploited is likely to occur in the next 6 months, why not consider how easy it is to obtain the knowledge (exploit code) necessary to exploit the vulnerability. If the information is readily available via the Internet or open-source tools, the likelihood that this vulnerability will be exploited is much greater than if no proof-of-concept code has ever been developed. Similarly, the vulnerability is far more likely to be exploited if the necessary skill level of the attacker is low (e.g., a script kiddie could perform the attack).

Table 8.15 DREAD Model[30]

	Rating	High	Medium	Low	Indirectly measures
D	Damage potential	Attacker can subvert the security; get full trust authorization; run as administrator; upload content	Leaking sensitive information	Leaking trivial information	Consequences
R	Reproducibility	Attack can be reproduced every time; does not require a timing window; no authentication required	Attack can be reproduced, but only with a timing window and a particular situation; authorization required	Attack is very difficult to reproduce, even with knowledge of the security vulnerability; requires administrative rights	Likelihood
E	Exploitability	Novice programmer could make the attack in a short time; simple toolset	Skilled programmer could make the attack, then repeat the steps; exploit and/or tools publicly available	Attack requires and extremely skilled person and in-depth knowledge very time to exploit; custom exploit/tools	Likelihood
A	Affected users	All users; default configuration; key assets	Some users; nondefault configuration	Very small percentage of users; obscure feature; affects anonymous users	Consequences
D	Discoverability	Published information explains the attack; vulnerability is found in the most commonly used feature; very noticeable	Vulnerability is in a seldom-used part of the product; only a few users should come across it; would take some thinking to see malicious use	Bug is obscure; unlikely that users will work out damage potential; requires source code; administrative access	Likelihood

The DREAD model provides a "qualitative" method of assigning a value to each of the five classifications that can be useful for group assessment exercises where it can be difficult to get consensus on an exact figure (dollar amount, number of months, etc.). A number value can be assigned to each ranking allowing the DREAD model to be implemented as a spreadsheet that is used along with the asset, threat and vulnerability data that has been previously obtained. The Six Sigma Quality Function Deployment (QFD) is an appropriate methodology to introduce at this point, as this can be applied directly to the DREAD model transforming the qualitative parameters (high, medium, low) into quantitative values that can be analyzed statistically.

Risk ranking

The application of QFD to the DREAD model will allow the data to be consolidated and used alongside the asset, threat, and vulnerability data. Figure 8.10 illustrates part of an

Intent	Threat Source	Physical Asset	Logical Asset	Entry Point	Threat Event (Threat to Asset)	Vulnerability (General)	D	R	E	A	D	Risk Score
Intentional	Outsider	Firewall	-	-	Make a physical change (reboot, pwr)	Physical security breach	10	10	10	10	10	10
Unintentional	INSIDER	Firewall	-	-	Make a physical change (reboot, pwr)	Human error	10	10	10	10	10	10
Intentional	Outsider	Firewall	Firmware	Management Port	Modify stored data (mem, hist, file)	Logical network security breach	5	5	5	10	1	5.2
Unintentional	INSIDER	Firewall	Firmware	Management Port	Modify stored data (hist, prog, firm)	Human error	5	10	10	5	10	8
		Firewall	Management Port	Physical Access								0
Intentional	Outsider	Firewall	Ident / Auth Services (Credentials)	Logical Network Access	Steal information	Logical network security breach	10	5	5	10	10	8
Unintentional	INSIDER	Firewall	Ident / Auth Services (Credentials)	Logical Network Access	Disclose information	Spyware, file sharing	10	1	10	5	10	7.2
Intentional	Outsider	Firewall	Log Files	Logical Network Access	Modify stored data (mem, hist, file)	Logical network security breach	5	5	5	10	10	7
Unintentional	INSIDER	Firewall	Log Files	Logical Network Access	Modify stored data (hist, prog, firm)	Human error	1	1	1	1	1	1
Intentional	Outsider	Firewall	Communication Interfaces	Logical Network Access	Cause a network disturbance	Logical network security breach	5	10	10	10	10	9
Unintentional	INSIDER	Firewall	Communication Interfaces	Logical Network Access	Cause a network disturbance	Infected laptop, network utils (scan), net loop	1	10	5	5	10	6.2
Intentional	Outsider	Firewall	Configuration - ACL / Rules	Management Port	Modify a program/configuration	Logical network security breach, file sharing	10	5	5	10	1	6.2
Unintentional	INSIDER	Firewall	Configuration - ACL / Rules	Management Port	Make a program/config error	Human error	5	1	1	5	5	3.4
Intentional	Outsider	Firewall	Runtime Data - Routing Info	Management Port	Steal information	Logical network security breach, file sharing	1	1	1	5	1	1.8
Intentional	Outsider	Firewall	Runtime Data - IP / MAC Adrs	Management Port	Steal information	Logical network security breach, file sharing	1	1	1	1	1	1
Intentional	Outsider	Network Switch(es)	-	-	Make a physical change (reboot, pwr)	Physical security breach	10	10	10	10	10	10
Unintentional	INSIDER	Network Switch(es)	-	-	Make a physical change (reboot, pwr)	Human error	10	10	10	10	10	10
Intentional	Outsider	Network Switch(es)	Firmware	Management Port	Modify stored data (mem, hist, file)	Logical network security breach	1	5	5	1	1	2.6
Unintentional	INSIDER	Network Switch(es)	Firmware	Management Port	Modify stored data (hist, prog, firm)	Human error	1	10	10	10	10	8.2
		Network Switch(es)	Management Port	Physical Access								0
Intentional	Outsider	Network Switch(es)	Log Files	Logical Network Access	Modify stored data (mem, hist, file)	Logical network security breach	1	5	10	1	10	5.4
Unintentional	INSIDER	Network Switch(es)	Log Files	Logical Network Access	Modify stored data (hist, prog, firm)	Human error	1	1	1	1	1	1
Intentional	Outsider	Network Switch(es)	Communication Interfaces	Logical Network Access	Cause a network disturbance	Logical network security breach	5	10	10	10	10	9
Unintentional	INSIDER	Network Switch(es)	Communication Interfaces	Logical Network Access	Cause a network disturbance	Infected laptop, network utils (scan), net loop, switch port (qos)	5	10	5	10	10	8
Intentional	Outsider	Network Switch(es)	Configuration	Management Port	Modify a program/configuration	Logical network security breach	1	5	5	5	10	5.2
Unintentional	INSIDER	Network Switch(es)	Configuration	Management Port	Make a program/config error	Human error	5	10	10	10	10	9
Intentional	Outsider	Network Switch(es)	Runtime Data - IP / MAC Adrs	Management Port	Steal information	Logical network security breach, file sharing	1	5	5	1	5	3.4
Intentional	Outsider	Controller	-	-	Make a physical change (reboot, pwr)	Physical security breach	10	10	10	10	10	10
Unintentional	INSIDER	Controller	-	-	Make a physical change (reboot, pwr)	Human error	10	10	10	10	10	10
Intentional	Outsider	Controller	Static Control Logic Configuration	Engineering Apps	Modify a program/configuration	Logical network security breach	5	1	1	5	1	2.6
Unintentional	INSIDER	Controller	Static Control Logic Configuration	Engineering Apps	Make a program/config error	Human error	10	10	10	10	10	10
Intentional	Outsider	Controller	Control Logic Algorithm Library	Engineering Apps	Modify stored data (mem, hist, file)	Logical network security breach	5	5	1	5	1	3.4
Unintentional	INSIDER	Controller	Control Logic Algorithm Library	Engineering Apps	Modify stored data (hist, prog, firm)	Human error	10	10	10	10	10	10
Intentional	Outsider	Controller	Dynamic Control Data	Engineering Apps	Modify stored data (mem, hist, file)	Logical network security breach	5	5	1	5	1	3.4
Unintentional	INSIDER	Controller	Dynamic Control Data	Engineering Apps	Modify stored data (hist, prog, firm)	Human error	10	10	10	10	10	10
Intentional	Outsider	Controller	I/O Database	Engineering Apps	Modify a program/configuration	Logical network security breach	5	1	1	5	1	2.6
Unintentional	INSIDER	Controller	I/O Database	Engineering Apps	Make a program/config error	Human error	10	10	10	10	10	10
Intentional	Outsider	Controller	Firmware	Engineering Apps	Modify stored data (mem, hist, file)	Logical network security breach	5	5	1	5	1	3.4
Unintentional	INSIDER	Controller	Firmware	Engineering Apps	Modify stored data (hist, prog, firm)	Human error	10	10	10	10	10	10

FIGURE 8.10 Risk and vulnerability assessment worksheet.

example spreadsheet for the complete process used against the reference architecture shown in Figure 8.5. The mapping was accomplished using values of 10 = high, 5 = medium and 1 = low. This was done in order to provide adequate numerical separation between a high or "significant" item and a low or medium event. With this numbering scheme, two medium ratings would equal one high. Other possibilities include using a 1,3,7 system so that three medium ratings would exceed one high, and so on. The numbers 1,2,3 have not been used because this would place inappropriate weighting on low or "insignificant" items compared to high ones.

The use of a spreadsheet tool such as Microsoft Excel will allow the values calculated from the DREAD model to be compared across all of the listed items, forming the basis of a ranked list of items and priority of events to address those security weaknesses discovered during the security test. The spreadsheet example in Figure 8.10 utilizes the *Conditional Formatting* features of Excel and applies a shading scale over the range of 0–10. This provides easy visual recognition.

Cyber-physical threat modeling

Much of what has been discussed so far concerns measuring risk to the computers and networks associated with industrial control systems. This is because cybersecurity risk is a discipline born from information security needs and not operational ones. These methods are valid and applicable to measuring risk in industrial environments because much of the computing and networking infrastructure in OT systems is similar to that used in IT systems. However, OT and IT networks are not the same. Industrial control and automation systems use common technology in some areas but in much different ways. There are also numerous assets that do not share common technology and must be considered independently. Risk assessments and risk management, therefore, can be considerably more valuable if viewed within the context of cyber-physical relationships.

To reiterate, cyber-physical threats refer to cyber activities (e.g., manipulation of digital systems) that can result in a specific and undesired physical outcome (e.g., sabotage of a process). These attacks typically consist of two phases: one to gain access to industrial process control systems, and one that manipulates those systems to achieve a specific physical outcome. However, the cybersecurity controls that monitor most computer and network assets within the industrial network lack visibility to the operations of the process control system itself. That is, while a cybersecurity monitoring tool might be able to alert on suspicious login activity, the presence of known malware, etc., those tools are not aware of a variance in production quality or yields, or the failure of a sensor or actuator within the automation system. Likewise, alarm management tools used by operators might be able to alert on aberrations in the process, they lack visibility to cybersecurity events. Because of this disconnect, correlating activity across both phases of a cyber-physical attack can be cumbersome, complex, and likely to produce low-confidence results with an abundance of false positives (this is discussed more in Chapter 13, "Security Monitoring of Industrial Control Systems").

For many, this correlation is one performed by humans: the process engineer who has also learned cybersecurity out of necessity and has access to the tools and domain expertise to understand both sides of the cyber-physical threat. If possible, process alarm tools and cybersecurity monitoring tools could be integrated, and these types of correlations could be done programmatically; however, at the time, this writing there are no commercial tools available to do this. It may be worth the effort to perform this type of custom integration work in-house within your organization, if the resources are available, because of the insight it can give you to how digital (cyber) and process (physical) systems interact and influence each other.

It is especially useful when considering the relatively new discipline of cyber-physical threat modeling: identifying how a cyberattack might infiltrate an industrial environment and manipulate process control to cause a specific desired outcome. In other words, mapping the entire cyber-physical attack from start to finish, across all attack phases.

How does one model a cyber-physical threat?

Building off of the DREAD model, we can apply similar methods to cyber-physical threats. What is the physical damage potential of a specific, unwanted physical condition? Simply interrupting a process might trigger an operator action and cause a moment of irritation. Closing a valve in a pipeline to build pressure sounds destructive, but there are likely safety controls (and process alarms) that will automatically mitigate any damage. Finding a specific overlooked condition that might cause a specific unwanted physical outcome, however, could have potentially devastating impact. Cyber-physical threat modeling is an exercise in identifying these outlier physical conditions and tracing them back to any potential cyber causes.

One way to identify potential cyber-physical threats is to involve a Safety Manager and introduce cybersecurity threats into safety assessments and HAZOP studies. A typical cybersecurity threat model attempts to identify and understand all possible threats and potential mitigations with the goal of protecting something of value. HAZOP studies perform a similar function for the industrial environment. A typical HAZOP study will meticulously evaluate the design and operation of a process control system to identify potential hazards or deviations from normal operations that could potentially lead to accidents, incidents, or other unwanted outcomes. While the approaches are similar, HAZOP studies are based on hard facts and physics, while cybersecurity threat models are based on suppositions and unknowns. A cyber-physical threat model must conjoin the two and encompass both: to determine what digital systems could be reached, breached, and manipulated, and what process inputs, logic, and/or outputs could then be manipulated to prevent a previously unconsidered physical condition or hazard: essentially a "cybersecurity threat model but with physics."

Leveraging the DREAD model: In a cyber-physical attack, the industrial attack phase will typically utilize the inherent capabilities of the industrial control system and the process control and instrumentation that is in place. The reproducibility, exploitability, and discoverability of this attack phase are therefore a given: if the attack gets to this phase, the control system itself freely enables the remainder of the attack. There may not be a clear "affected user," as it may not be immediately obvious who the user is in a process control system. There may not be a user or an application that is being affected but rather a physical system and a physical outcome. However, there is a need to measure the extent or scope of that condition (e.g., the safety perimeter of a reactor or the population that could be affected by a blackout), and so the model remains valid. Once high-severity threats have been identified on the physical side of the cyber-physical attack, that attack can then be walked back to the initial attack phase. What controllers, PLCs, HMIs, RTUs, or other devices are capable of producing the physical threats identified using DREAD? How could those devices be access and manipulated digitally? The answers to these questions map out the cyber-portion of the cyber-physical attack. Together, they make up a cyber-physical threat model.

In Figure 8.11, the "unwanted condition" is an increase in liquid temperature in a storage tank (1), which is carried out using a water-hammer attack. This attack manipulates the valve (2) and the pump (3) to create a condition where the liquid in the pipe comes to a sudden stop. When this happens, the energy needs to go somewhere, and heat is generated as a result.[31] In extreme cases, the "stop energy" is sufficient to turn water into steam on the surface of the valve, which creates a pressure dynamic that can in turn create a vacuum condition in the pipe. This can be sufficient to cause

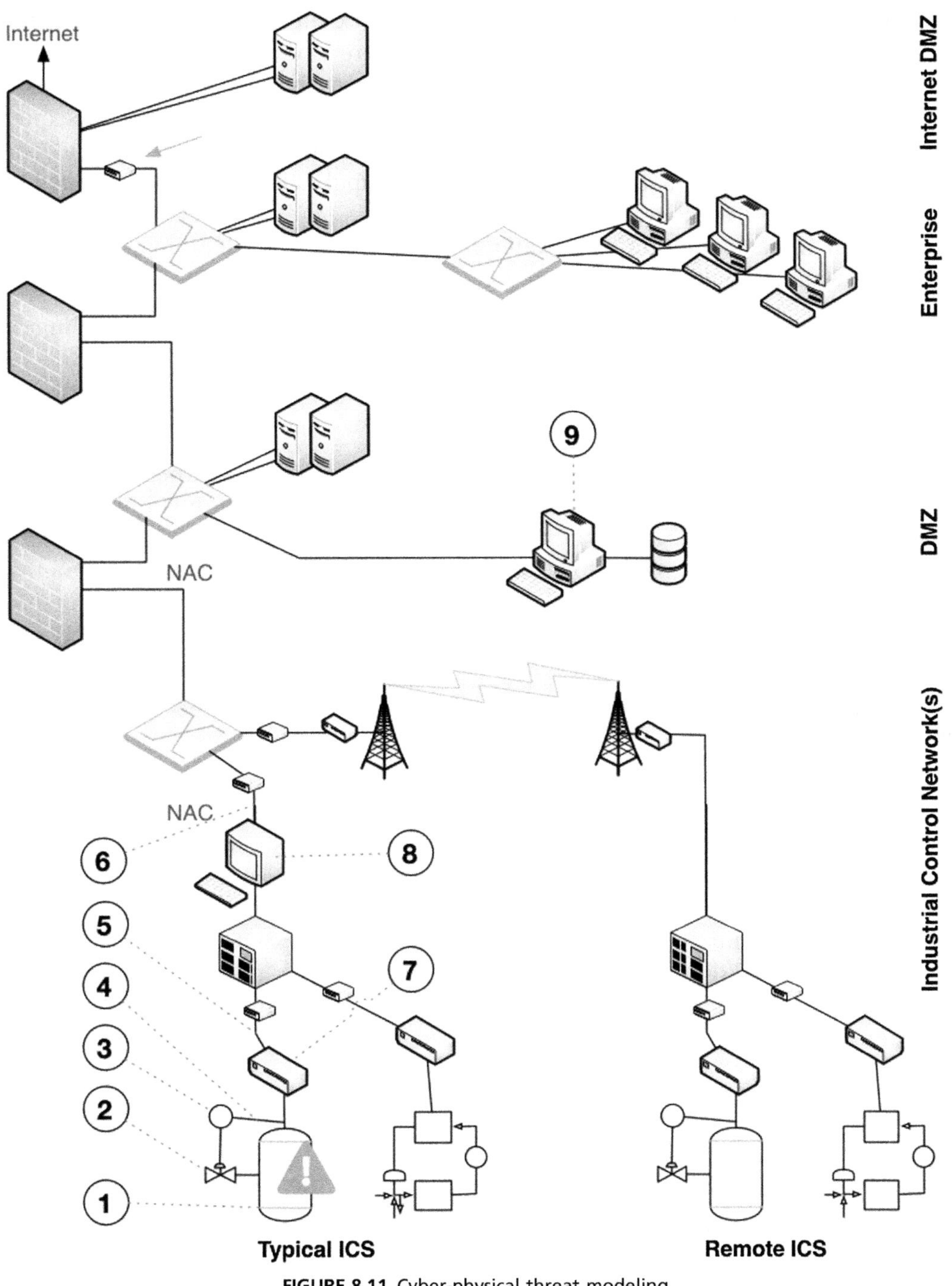

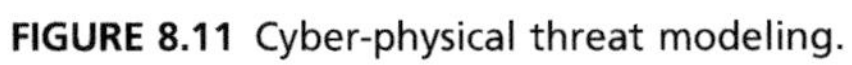
FIGURE 8.11 Cyber-physical threat modeling.

implosions of the infrastructure.[32] This type of attack highlights the need for cyber-physical threat modeling. Most physical infrastructure is designed to withstand stress in expected conditions: for example, an excess of pressure in a pipe could cause the infrastructure to fail or rupture. However, a lack of pressure (vacuum) is not an expected condition. As such, the physical tolerances can be insufficient to prevent a failure. Similarly, the unexpected nature of such a condition makes it likely to be excluded from typical safety and HAZOP evaluations.

Once the threat condition (the water hammer attack) is identified, we can start evaluating what it would take to cause this condition:

- Manipulation of the valve (2) and pump (3) could be done through packet injection on various network and/or fieldbus segments (4, 5, 6)
- The controller could be manipulated (7)
- The controller could be manipulated via an HMI or operator workstation (8)

These are all industrial-phase attacks; an attacker must first reach these targets before any such attack could occur. For simplicity, let's assume that there is only one established network communication path from the DMZ into the industrial network, that allows an engineering station (9) to write to control assets within the industrial network. Also for the sake of simplicity, assume that engineering workstation (9) is prevented from communicating to other systems in the DMZ or in the business network. This would limit the initial attack phase to threats against this one workstation. Therefore, to prevent a successful cyber-physical attack leveraging a water hammer attack against tank (1), there is a much narrower scope to consider:

- Hardening system (9) and implementing strong antimalware to prevent malicious code from executing a cyber-physical attack.
- Monitoring of communications between system (9) and the industrial control network. This could include monitoring of industrial control traffic, to alert operators whenever system (9) attempts to alter parameters or logic of controller (7).
- Monitoring of industrial network segments (4, 5, 6) for specific control codes that could result in closing valve (2) while increasing pressure created by pump (3).
- Alteration of the process design to prevent the physical outcome of the attack. For example, using valves that are unable to close quickly enough to allow a water hammer attack.

Note that this example is highly simplified: in a real scenario, there would likely be multiple targets in the DMZ or business network that could be sued to successfully pivot an attack into the industrial control network. Indeed, the practice of cyber-physical threat modeling is only feasible because in a well-designed industrial network, there should be a limited number of inbound vectors. Initial attack phases are required to breach a target's more exposed business networks and information systems, but there should be only a few "pivot points" to enter the industrial environment. These vectors consist primarily of:

- A limited number of allowed network connections, to enable communications between industrial applications and business information systems. These communications should be made via proxy interfaces in the layer 3.5 DMZ (between the ICS and business environments) and should be controlled and monitored.
- Physical access via mobile computing devices, laptops, removable, or other transient technologies that could be carried into the environment. This could also include insider threats, an infiltrated supply chain, etc.

Without these well-known pivot points and beachhead systems, the cyber-physical threat models would be extremely complex and difficult to define.

Using simulations versus labs for threat modeling

To identify a potential cyber-physical threat, there must be a degree of creativity. Unlike traditional HAZOP study, the conditions leveraged by an attacker might be highly unlikely and may never occur naturally. To facilitate this thinking, experimentation is helpful. How does one experiment on a process control system without causing a real physical catastrophe? Two options include lab environments and dynamic digital twins. "A process digital twin allows operators to apply traditional cybersecurity threat modeling to physical systems, uncovering possible unwanted scenarios that might occur as the result of a cyber-physical attack. By imagining yourself as the attacker and using the digital twin to develop new potential attack scenarios, your digital "evil" twin provides the understanding that is necessary to develop effective defenses. The end-to-end attack scenario – including all assets, data flows, physical process parameters, and process logic."[33]

Using a twinned control environment, what-if-scenarios can be used to identify unwanted conditions in much the same way that they would be used to improve process efficiency. In one example, manipulation of one portion of a process with a low SIL was able to heat the infrastructure components of a chemical mixing system in such a way that higher SIL portions of the process would heat to unsafe levels, producing toxic vapors.[34] Obviously, this would be difficult to do in a real environment without introducing real hazards.

Labs can also be useful in cyber-physical threat modeling but less so for identifying the end conditions (unless you have an extremely robust lab that closely mirrors production systems). They are extremely useful for "walking back" the end condition to identify pivot points and initial cyberattack vectors, however. Together, a well-equipped lab and dynamic simulation software will greatly facilitate cyber-physical threat modeling.

Cybersecurity HAZOP

The next step in the journey of industrial cybersecurity risk assessment is the Cybersecurity HAZOP study or csHAZOP. Developed by long-time industrial cybersecurity advocate and expert Sinclair Koelemij, csHAZOP builds upon the concept of cyber-physical threat modeling, to provide a detailed assessment of cyber-physical risk. Like cyber-physical threat modeling, this type of risk assessment extends the more traditional cybersecurity to include the physical domain of the production process. Unlike a cyber-physical threat model, which is focused solely on identifying threats, csHAZOP also consider other factors of risk, including the likelihood and impact of a given cyber-physical threat, to produce a standard risk measurement.[35]

The cyber-physical risk measured by csHAZOP "connects the dots" between the cyber security posture of the process automation functions and the safety and security of the complete process control installation. In Figure 8.12, we can easily see three distinct stages of a cyber-physical risk assessment: the first, illustrated at the top of Figure 8.12, involve the identification and definition of "incident scenarios"; the last, illustrated at the bottom of 8–12, involves the reduction of the risk associated with that threat scenario; while the middle stage, illustrated within the dotted line at the center of Figure 8.12, involves the actual quantification of the risk. In this stage, any of the above mentioned risk methodologies is suitable (see "Methodologies for Assessing Risk Within Industrial Control Systems," above).

In this way, it defines and quantifies the relationship between a cyberattack and an industrial control incident (e.g., death, dismemberment, environmental damage, societal harm, *et. al.*). This is an important distinction as there may be legal or regulatory criteria that apply to these losses, and therefore, they must be measured quantitatively in accordance with those regulations.[36] This is because despite the losses being the same, a traditional HAZOP quantifies risk based on *probabilistic* determinations, and a cyber-physical risk assessment quantifies risk based on *intentional* rather than probabilistic causes. Therefore, csHAZOP must be quantified independently.[37]

Both cyber-physical threat models and csHAZOP studies are complex in nature, but the benefits can be significant:

- By identifying potential hazard conditions, rather than more traditional cyber security risk assessments which tend to limit considerations to digital systems. With the added context of potential hazard conditions, cyber risk assessments can be quantified, which in turn facilitates incorporation into other organizational risk management efforts.[38]
- By identifying system dependencies, these assessments will facilitate other parts of an industrial cyber security plan, such as identifying communication flows for use in the development of a strong zone and conduit map.
- Mapping cyber-physical threats against system dependencies will help select the most appropriate cyber security controls and countermeasures and provide valuable insight on where and how to deploy them.
- By identifying potentially new hazard conditions that might otherwise have been overlooked during traditional HAZOP, operators have the opportunity to reevaluate existing safeguards and potentially improve those safeguards.

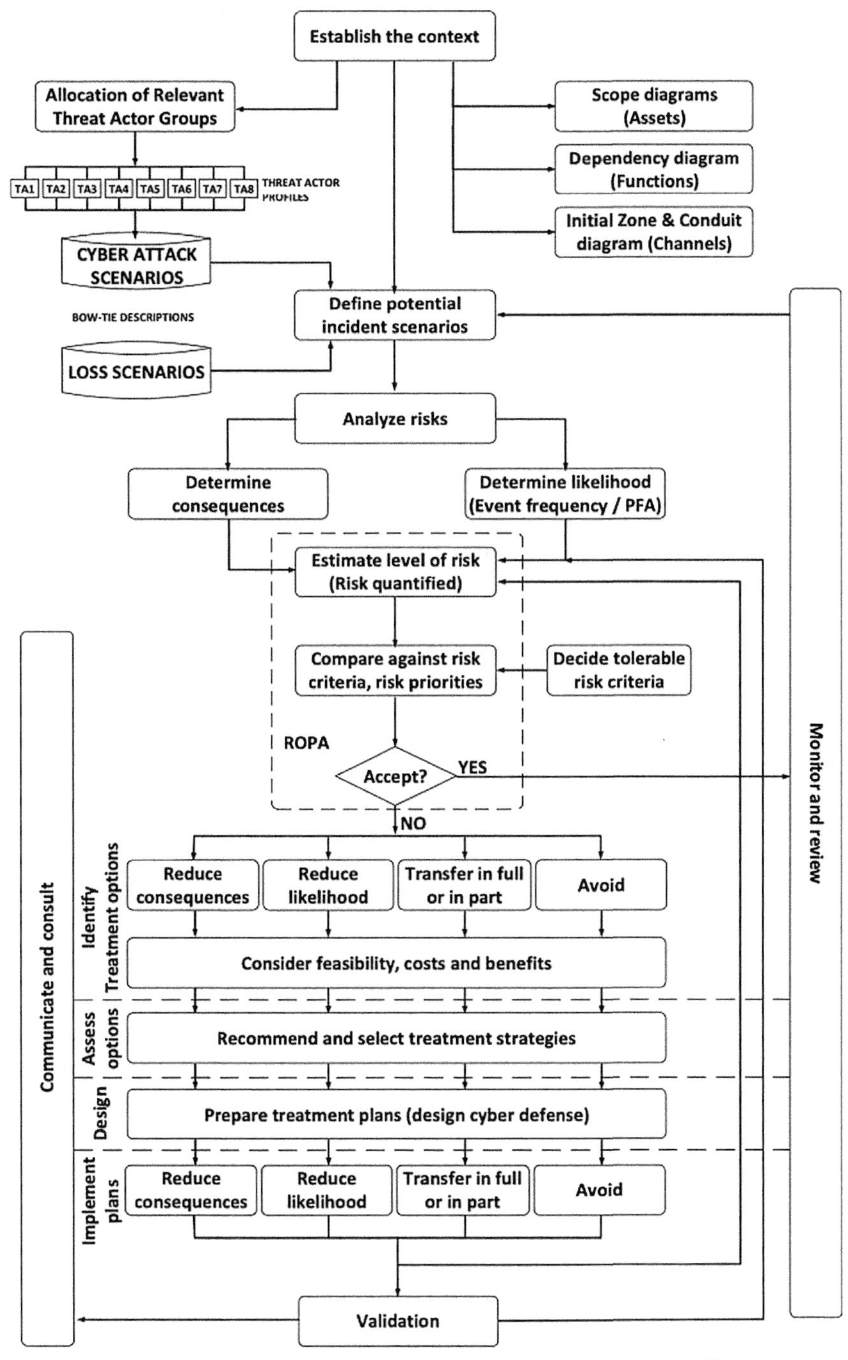

FIGURE 8.12 A cyber-physical risk method based on csHAZOP.[41]

The fundamentals of performing a csHAZOP are:

- Understand the threat, who is going to attack, what are the capabilities, what is the intent, and what are the opportunities for the threat actor.[39] Having performed a degree of cyber-physical threat modeling will facilitate this stage by helping to determine what hazard conditions could be initiated by a cyberattack.
- Understand how the system is exposed to the threat actor, what functions, assets, channels are exposed to an attack, and how do the different functions depend on each other for data and security.[40]

So, apart from analyzing and documenting the technical system for which we are going to execute the risk analysis, we need to understand the overall threat profile of the plant and also understanding which threat actors showed an intent to attack the cyber-physical installation.

Unlike the simpler cyber-physical threat modeling process, which focuses solely on the threat and how it could occur, csHAZOP requires an expanded threat profiling method. It requires a qualitative process to score threats based on historical rational, motivation rational, capability rational, resource rational, focus, and geopolitical tensions. The method of obtaining the score is less relevant than the consistency of the scoring, as this score indicates how likely it is that the threat actor (group) targets system. A similar method for determining the probability that the threat will be successful. Together they can be used to quantify risk using the formula:

```
Risk = (Probability of an Attack) x (Probability of failure from
an attack ) x (The consequence severity of the attack)
```

The "probability of failure from an attack" is comparable with the "probability of failure on demand" or "PFD" parameter used in process safety: it measures the probability of whether protection barriers are able to adequately perform their task. However, there are important differences between cyber security barriers (security measures) and process safety barriers (safeguards). Activation of a process safety barrier will create a safe process state if it executes as designed, whereas the activation of a security barrier will typically only reduce the probability of success for the attack if used on its own. This is why defense in depth is an important consideration for all cybersecurity programs: multiple cybersecurity measures used together will further reduce the probability of success. It is important to understand for the purposes of risk assessments, however, that even the best layered defenses are unable to guarantee total security.

Risk reduction and mitigation

The methodology discussed in this chapter provides a consistent, repeatable means to assess the security implemented around an ICS and the industrial networks. The process has yielded a prioritized list of items in terms of net "unmitigated" risk to the ICS and the plant under its control. Some risks may have been mitigated to an acceptable level following the security and vulnerability assessment. The final activity for those remaining risk items is to apply a range of cybersecurity controls or countermeasures to the assets within the ICS in order to reduce or mitigate these risks. The selection and implementation of security controls is discussed elsewhere in this book, and is available through numerous standards and controls catalogs.

Improving the cyber resilience of an ICS should be one of the many benefits obtained through the implementation of an industrial security program. This resiliency is accomplished when security controls are selected that span the Security Life Cycle shown in Figure 8.13. The life cycle is used to illustrate the continuous process of addressing cybersecurity that not only begins with threat deterrence and prevention, but equally balances threat detection and correction necessary to identify cyber events in a timely manner and to respond accordingly in order to minimize the consequences of the event and return the manufacturing facility to normal operation in a safe and timely manner.

Organizations often devote large portions of their security budget on mechanisms to prevent an attack from occurring.[42] External parties often notify these same organizations that lack a balanced investment in controls to detect an event of a breach long after the attack.[43] Security should be considered as a long-term "strategic" investment rather than a short-term or one-time "tactical" expense. Those that invest and build manufacturing facilities understand the long-term life cycle of the capital investment, so it makes sense that the industrial security used to protect these same facilities is treated in a similar manner and receives continuous attention (and budget) like other operational expenses (maintenance, improvements, training, etc.).

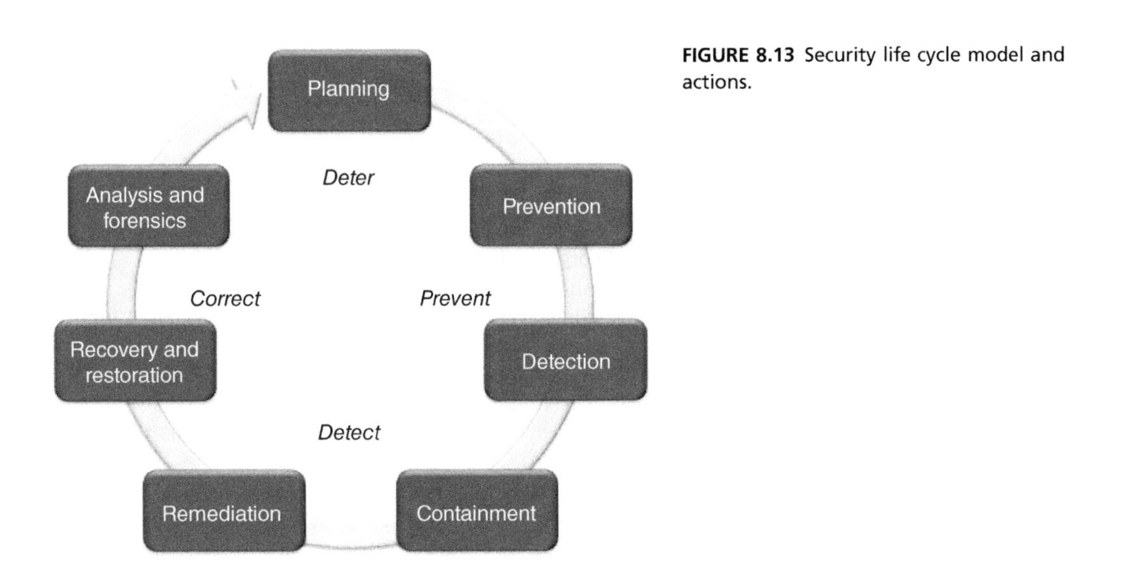

FIGURE 8.13 Security life cycle model and actions.

The security life cycle can be used as an effective tool when mapping security controls to each phase and will help identify potential short- and long-term weaknesses in the security strategy that could affect the overall resilience of the security program.

Summary

The implementation of an industrial cybersecurity program is an investment of both time and money. The primary objective is always to secure not just the industrial networks and those systems that utilize it but to also aide in securing the plant or mill that depends on these systems to remain operational in a safe, efficient, and environmentally responsible manner. Risk management is a daily part of those that lead and manage these facilities, and the management of cyber risk is a vital component. Threats to these industrial sites can originate inside and outside the organization where vulnerabilities can be exposed and exploited both maliciously and accidently. The consequences from these events can span simple "inconveniences" all the way to catastrophic mechanical failures that may result in plant shutdowns, fires, hazardous releases, and loss of life.

The evolution of industrial automation and control systems and industrial networks over the past 40 years has left many organizations with systems that possess numerous security weaknesses that cannot be replaced overnight. The process of upgrading and migrating the plethora of integrated industrial systems comprising an ICS can take years. A balanced approach to industrial security must be followed that provides the balanced and objective evaluation of risk in terms of threats, vulnerabilities, and consequences in order to align an organization's short- and long-term goals while workings toward a more safe and secure industrial automation environment.

Endnotes

1. Repository of Industrial Security Incidents, "2013 Report on Cyber Security Incidents and Trends Affecting Industrial Control Systems", June 2013.
2. "2013 Data Breach Investigations Report", Verizon, April 2013.
3. Repository of Industrial Security Incidents, "2013 Report on Cyber Security Incidents and Trends Affecting Industrial Control Systems", June 2013.
4. "15th Annual Computer Crime and Security Survey", Computer Security Institute, 2010/2011.
5. Open-Source Vulnerability Database, http://osvdb.org, cited July 1, 2014.
6. "ICS Vulnerability Trend Data", http://www.SCADAhacker.com/resources.html, cited July 1, 2014.
7. "Data breach costs still unknown: Target CEO", CNBC, http://www.cnbc.com/id/101694256, cited July 28, 2014.
8. "Analyst sees Target data breach costs topping $1 billion", TwinCities Pioneer Press, http://www.twincities.com/ci_25029900/analyst-sees-target-data-breach-costs-topping-1, cited July 28, 2014.
9. "The Target Breach, By the Numbers", Krebs on Security, http://krebsonsecurity.com/2014/05/the-target-breach-by-the-numbers/, cited July 28, 2014.
10. "Common Criteria for Information Technology Security Evaluation – Part 1: Introduction and general model", Version 3.1 - Revision 4, September 2012.

11. "The Shamoon Attacks", Symantec Security Response, http://www.symantec.com/connect/blogs/shamoon-attacks, cited July 29, 2014.
12. Wm. Stanek, "Windows Command-Line Administrator's Pocket Consultant", Microsoft Press, second Edition, 2008.
13. "Guide to Conducting Risk Assessments", Special Publication 800-30, National Institute for Standards and Technology, September 2012.
15. "National Vulnerability Database Version 2.2", National Institute of Standards and Technology/U.S. Dept. of Homeland Security National Cyber Security Division, http://nvd.nist.gov, cited July 30, 2014.
16. "Open-Source Vulnerability Database", http://osvdb.org, cited July 30, 2014.
17. "National Checklist Program Repository", National Institute of Standards and Technology, http://web.nvd.nist.gov/view/ncp/repository, cited July 30, 2014.
18. "CIS Security Benchmarks", Center for Internet Security, https://benchmarks.cisecurity.org, ited July 30, 2014.
19. "Security Configuration Guides", National Security Agency/Central Security Service, http://www.nsa.gov/ia/mitigation_guidance/security_configuration_guides/index.shtml, cited July 30, 2014.
20. "Nessus Compliance and Audit Download Center", Tenable Network Security, https://support.tenable.com/support-center/index.php, cited July 30, 2014.
21. "Bandolier", DigitalBond, http://www.digitalbond.com/tools/bandolier, cited July 30, 2014.
22. "Nessus Compliance Checks Reference", Revision 53, Tenable Network Security, July 2014.
23. "A Complete Guide to the Common Vulnerability Scoring System Version 2.0", Forum of Incident Response and Security Teams, http://www.first.org/cvss/cvss-guide.html, cited July 30, 2014.
24. "Common Vulnerability Scoring System Version two Calculator", National Institute of Standards and Technology – National Vulnerability Database, http://nvd.nist.gov/cvss.cfm?calculator&version=2, cited July 30, 2014.
25. Ralph Langner. Robust Control System Networks: How to Achieve Reliable Control After Stuxnet. Momentum Press, LLC. New York, NY. 2012.
26. Ibid.
27. Ibid.
28. "DHS Says Aging Infrastructure Poses Significant Risk to U.S.", Public Intelligence, http://publicintelligence.net/dhs-national-risk-profile-aging-infrastructure/, cited July 31, 2014.
29. "Aging Natural Gas Pipelines Are Ticking Time Bombs, Say Watchdogs", FoxNews, http://www.foxnews.com/us/2011/02/28/aging-natural-gas-pipelines-ticking-time-bomb-say-experts/, cited July 31. 2014.
30. "Threat Modeling", Microsoft Developer Network, http://msdn.microsoft.com/en-us/library/ff648644.aspx, cited July 31, 2014.
31. Jason Larsen. Physical Damage 101: Bread and Butter Attacks. Black Hat conference, Las Vegas, NV, 2015.
32. Ibid.
33. Knapp, Eric D. Digital Evil Twins for Cyber-Physical Threat Modeling. Honeywell, Inc. September 2023.
34. Ibid.
35. Sinclair Koelemij. *Cyber Physical Risk Assessment.* S4x24 Conference Whitepaper. Honeywell, Inc. November 2022.
36. Ibid.
37. Ibid.
38. Ibid.
39. Ibid.
40. Ibid.
41. Sinclair Koelemij. *Cyber Physical Risk Assessment.* S4x24 Conference Whitepaper. Honeywell, Inc. November 2022.
42. "15th Annual Computer Crime and Security Survey", Computer Security Institute, December 2010.
43. "Risk Intelligence Governance in the Age of Cyber Threats", Deloitte Consulting, January 2012.

9 Establishing Zones and Conduits

Information in this chapter

- Security Zones and Conduits Explained
- Identifying and Classifying Security Zones and Conduits
- Recommended Security Zone Separation
- Establishing Security Zones and Conduits

The concepts of defense in depth, as discussed up to this point, have focused on the separation of devices, communication ports, applications, services, and other assets into groups called "security zones." These zones are then interconnected via "security conduits" that much like the conduit used to house and contain wire and cable and are used to protect one or more communication paths or channels. The logic is simple—by isolating assets into groups and controlling all communications flow within and between groups, the attack surface of any given group is greatly minimized.

This concept was originally defined in the Purdue[1] Reference Model for Computer Integrated Manufacturing (CIM), which defines the hierarchical organization of CIM systems. The concept was later incorporated into ISA-99 as the "Zone and Conduit Model," which was later incorporated into the IEC-62443 standard.[2]

Security zones, or simply zones from this point onward, can be defined from either a "physical" perspective or a "logical" one. Physical zones are defined according to the grouping of assets based on their physical location. Logical zones are more like virtual ones in that the assets are grouped based on a particular functionality or characteristic.

Security conduits are actually a special type of zone that groups "communications" into a logical arrangement of information flows within and between various zones. Conduits can also be arranged according to physical (network cabling) and/or logical (communication channels) constraints.

The zone and conduit model has been embraced for a reason. When properly implemented, zones and conduits limit digital communications in such a way that each zone will be inherently more secure. In other words, it is more resilient to negative consequences in the event of a threat exploiting a particular vulnerability within the zone. It therefore provides a very strong and stable foundation upon which to build and maintain a cybersecurity policy, and by its nature supports other well-known security principles, including the Principle of Least Privilege (where users can only access systems to which they are authorized), and the Principle of Least Route (where a network node is only given the connectivity necessary to perform its function).

Unfortunately, zones are often defined only in very broad terms, separating the industrial network into as few as two or three zones (for example: a control system zone, a

Industrial Network Security. https://doi.org/10.1016/B978-0-443-13737-2.00013-0

business zone, and a demilitarized zone between the other two). Likewise, conduits are often defined too broadly as "all communications paths within a single zone" or "all communications paths between two zones." As zones and conduits become more granular, there will be a corresponding improvement in security (Figure 9.1). It is therefore important to carefully identify zones in the early stages of the cybersecurity lifecycle.

In some cases, such as in nuclear facilities, a five-tier system is mandated, based upon specific regulations (in this case, the Nuclear Regulatory Commission guidelines defined in RG 5.71).[3] These guidelines should be treated as a minimum benchmark for zone separation. In most cases, the zones can—and should—be defined much more precisely.

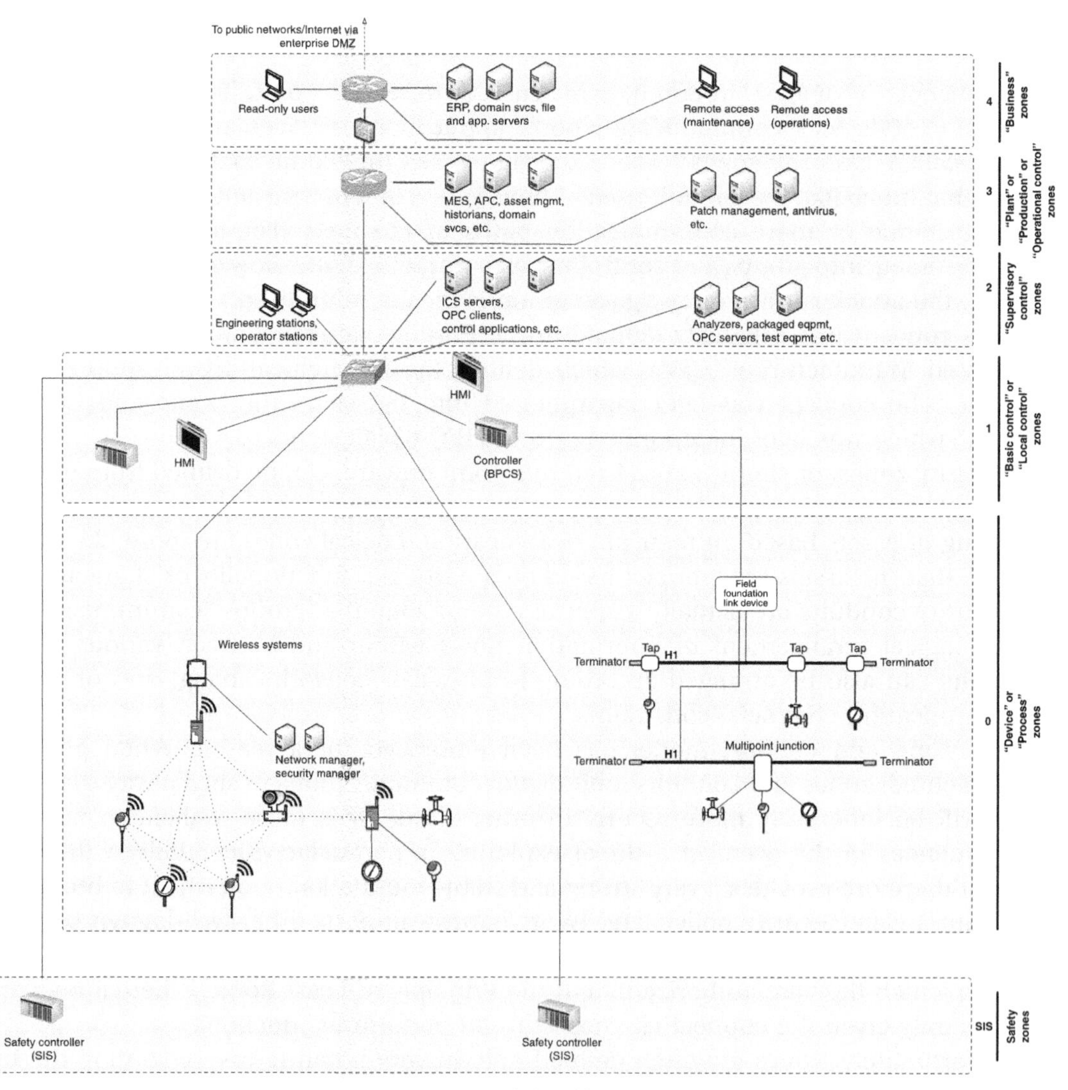

FIGURE 9.1 Security zones defined by integration levels.

Once defined, zones and conduits will help to pinpoint areas where network and host security and access controls may be required. This is because, by limiting communications to defined conduits, each conduit represents a potential network attack vector. If implemented poorly, zones and conduits will result in a well-organized architecture; if implemented properly, they will result in a highly secure architecture. This is not to say that a zone or a conduit is defined by its security controls, but rather that zones and conduits can facilitate the proper selection, placement, and configuration of security controls. Network security controls—such as firewalls, Network IDS and IPS devices (**NIDS** and **NIPS**), router access control lists (ACLs), application monitors, and/or similar security products—will be highly effective when implemented against a well-organized architecture with clear policies that are defined around zones and conduits. As with perimeter defenses, internal defenses should be configured in concert with the authorized parameters of established and documented zones and conduits.

Another way to look at the design and implementation of zones and conduits is how it can be used to provide a more resilient security architecture. Consider a grouping of assets that cannot be protected individually with antimalware defenses like antivirus and application whitelisting. These assets can be logically grouped into a zone, and the antimalware defenses are implemented on the conduit(s) into this zone. This is one effective way asset owners are able to continue operation of legacy and even unsupported systems (e.g., Windows XP) through the creation of zones of related assets, and then applying strong security controls on the conduits entering these zones.

This chapter will cover the identification and classification of zones and conduits. Network and host defenses that can be deployed to directly support the zones and conduits are discussed in Chapter 11 "Implementing Security and Access Controls." It is also important to define the expected behavior within and between zones and to monitor all activities within and between each zone—both for the obvious alerts that might be generated by perimeter and host security products and for behavioral anomalies. Baselining activity is covered in Chapter 12, "Exception, Anomaly, and Threat Detection," while monitoring is covered in Chapter 13, "Security Monitoring of Industrial Control Systems."

Security zones and conduits explained

The concepts behind zones and conduits can be confusing and are often misunderstood by those that believe it is simply a new term for the Purdue Reference Model originally released in the late 1980s and adopted as the ISA Standards and Practice SP95 (also known as IEC-62264). One should realize that the motivation behind the Purdue Model and SP95 was the integration of enterprise and automation applications and the associated exchange of information. These concepts are quite different than those behind the grouping and classification of assets based on particular security criteria.

Each industrial architecture is unique, not because of the selection of equipment but how each system is deployed in a particular environment (end products manufactured, geographical location, staffing, etc.) and how each system is integrated with other ancillary systems to form a complete, integrated, industrial control architecture. A good analogy to security zones is to consider how many industrial facilities maintain separation of basic control and safety-related assets. This separation occurs, not just because of existing laws and regulations, but because of the underlying layers of protection that each of these systems provides, and how the relative protection of each system is unique. This "safety level" can be applied to each system so that appropriate measures can be in place to ensure that each system performs as intended without unintentional consequences or interactions between systems to impact their basic functionality.

In terms of security, a similar concept can be applied. Assets at a particular site are grouped based on their relative security requirements or "security level." These zones are then created as either "external" ones, or when multiple layers of protection are required, they can be "nested" inside one another. This allows security controls to be deployed to zones (and the assets they contain) based on the unique security requirements of each. This will be further expanded later when discussing how zones and conduits are classified based on their assets.

Information needs to flow into, out of, and within a given zone. Even in standalone or "air-gapped" systems, software updates and programming devices are typically used to maintain the system. These all represent entry points into the zones, called conduits.[4]

Identifying and classifying security zones and conduits

One of the greatest challenges in establishing proper security zones and conduits is the creation of a set of base requirements or "goals" that are used to determine if a particular asset should be placed in a given zone. There is no single answer to the method on which this is based—after all, rarely are two ICS installations identical, and therefore, their relative security levels are also never the same.

These requirements or goals typically can be broken down into two broad categories. The first is based on communications and how each asset interacts with other assets outside a particular zone. To explain this in another way, consider a company employee (a process engineer) who uses his/her office computer in the administration building and his/her engineering workstation in the control room. This user is an asset, but which "zone" is he/she a member of? Or is this user in fact a "conduit" between zones? These assets are also typically connected to an industrial network that provides the ability for the electronic exchange of information. This communication can further be designated as "local" or within the same zone and "remote" or outside the zone.

Physical access to assets was explained earlier, and is another means of classifying the assets within a particular security zone. Consider a control room that houses plant operators, technicians, and control system engineers. Though these individuals are all within the physically secure control room, they do not necessarily possess the same level of "trust" with respect to each other. This leads to the creation of embedded zones where a higher security level zone (used by the engineer) is embedded in a lower-level zone (used by the operators) reflecting the relative trust and security of the users.

Assets may exist outside of a particular security zone. This does not mean that these assets are at a necessarily higher or lower-level, but rather a level that is "different" from other assets in the given zone. One of the best examples of this type of zoning exists when you have a particular grouping of assets that utilize a vulnerable or insecure network-based protocol (e.g., Telnet). These protocols are necessary to perform specific functions within a zone that is not meant to contain "hostile" or "untrusted" assets. A manufacturing facility may have multiple areas or work cells that deploy similar equipment and associates zones. In order to properly secure this zone, the conduit(s) into this zone restricts communications prohibiting the use of these less-secure protocols.

Recommended security zone separation

As mentioned, zones may be defined broadly ("control" vs. "business" zones) or narrowly, creating zones for highly granular functional groups of assets. The zone and conduit model can be applied at almost any level—the exact implementation will depend upon the network architecture, operational requirements, identified risks, and associated risk tolerance, along with many other factors. The following are some recommendations on how to define discrete zones.

Note: When defining highly granular zones, it should be assumed that there will be an overlap that prevents adequate zone and conduit enforcement. For example, a zone created by physical control subsystems is likely to overlap with zones defined logically by specific protocols, and it may be architecturally difficult to separate the two. This is usually okay, and is why most standards and guidance documents reference a broader definition of zones. The process of examining the various ways in which assets can be logically grouped, and how communication can be controlled, is still important and highly beneficial. This will help to identify previously unrecognized areas of risk, and where more granular zones can be defined and controlled. It will also help to improve the overall security posture of the end-to-end network.

When assessing the network and identifying potential zones, include all assets (physical devices), systems (logical devices like software and applications), users, protocols, and other items. Attempt to separate two items, such as a protocol from an asset. If the two can be separated without impacting either item's primary function, they belong to two functional groups, and are therefore excellent candidates for their own zones. For example, if some SCADA systems use the DNP3 protocol, create a list of all devices currently communicating over DNP3. Assess each to see if DNP3 is necessary to its function or not (it may support multiple protocols and may be actively using a different protocol to perform its functions). If not, remove it from the functional group, and if possible disable the unused protocol on the SCADA server as well. The result will be a list of all assets legitimately using that protocol (see "Protocols").

Similarly, consider which assets are connected to each other on the network, both physically and logically. Each represents a functional group based on network connectivity (see "Network Connectivity") and data flow. Again, assess each item in question individually, and if it does not need to belong, remove it from the group.

A functional group can be based on almost anything. Common functional groups to consider when defining zones in industrial networks include safety, basic process control, supervisory controls, peer-to-peer control processes, control data storage, trading communications, remote access, ability to patch, redundancy, malware protection, and authentication capability. Other groups, such as user groups and industrial protocol groups, can be considered.

Network connectivity

Functional groups based on network segmentation are easy to understand because networks by nature connect devices together. How the different devices are connected on the network clearly qualify those items that belong to an interconnected group and those that are excluded by an enforceable network connection or conduit. Networks should be considered both physically (what devices are connected to other devices via network cables or wireless connections) and logically (what devices share the same routable network space, subnet or access control list).

Physical network boundaries are easy to determine using a network map. Ideally (although not realistically), all control system networks should have a hard physical boundary in the form of a unidirectional flow that prevents traffic from entering a more secure zone from a less secure one. Realistically, there will be interconnection points consisting of a single link, preferably through a firewall and/or other defensive devices.

> **Caution**
>
> Wireless networks are easy to overlook as physical network connections. Without network-level authentication on the wireless LAN, any two devices with wireless antennae, regardless of whether they have logical connection to the "active" wireless network in question, should be considered "physically" connected. The separation provided by basic authenticated wireless access is a logical separation.

Logical network boundaries are defined by the use of devices operating on OSI Layer 3 (routers, advanced switches, firewalls) to separate a physical network into multiple address spaces. These devices provide a logical demarcation between each network. This forces all communications from one logical network to another to go through the layer 3 device, where ACLs, rule sets, and other protective measures can be implemented.

Note that virtual LANs (VLANs) are a type of logical boundary, but one that is enforced at layer 2 rather than Layer 3. VLANs use a standardized tag in the Ethernet packet header to determine how they are handled by a layer 3 device. Traffic destined for the same VLAN is switched, while traffic destined for a different VLAN is routed. VLANs, however, are not recommended for security, as it is possible to modify the packet header to hop VLANs, bypassing the router.[5]

Control loops

A control loop consists of the devices responsible for a particular automated process (see Chapter 4, "Introduction to Industrial Control Systems and Operations"). Applying this list of devices to a functional group is relatively simple. In most instances, a control loop will consist of a sensor (such as a switch or transducer), a controller (like a PLC), and an actuator (such as a relay or control valve), as illustrated in Figure 9.2.

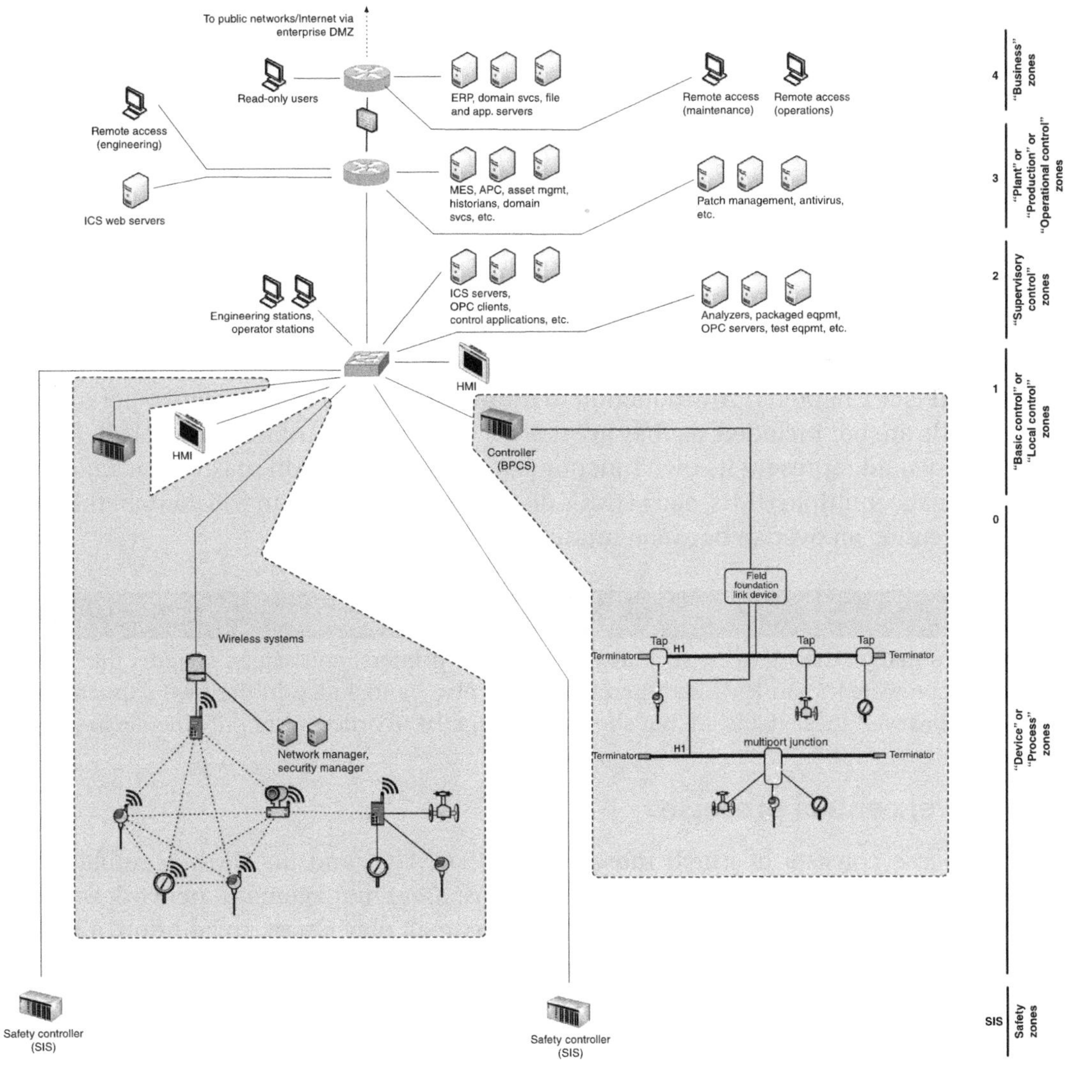

FIGURE 9.2 Zones defined by process.

Where defining a functional group based on network connectivity is a broad example that might result in a handful of functional groups, building a functional group based on a control loop is a very precise example. The functional groups created will be numerous, and each will contain a relatively small number of devices (a specific PLC or remote terminal unit [RTU] and a collection of relays and intelligent electronic devices [IEDs]). One of the most practical examples of how this is used in industrial architectures today is in the use of digital field networks (e.g., FOUNDATION Fieldbus) and how particular control loops are placed on dedicated network segments based on classification of risk and functionality.

Supervisory controls

Each control loop is also connected to some sort of supervisory control—typically a communications server and one or more workstations—that are responsible for the configuration (engineering workstation [EWS]) and monitoring and management (operator workstation HMI) of the automated process. Because the HMI is responsible for the PLC, these two devices belong to a common functional group. However, because the HMI is not directly responsible for those IEDs connected to the PLC, the IEDs and PLC are not necessarily in a common functional group as the HMI (they belong to a common functional group based on some other common criteria, such as protocol use). Figure 9.3 shows an example of two such zones within the broader "basic control" zone.

All PLCs controlled by the HMI are included, as are any "master" HMI, communication servers, or control management systems that might have responsibility or control over the initial HMI (see Chapter 4, "Introduction to Industrial Control Systems and Operations"). Other HMIs are not included, as they are not the responsibility of the initial HMI. Rather, each HMI would represent its own functional group. If a common master controller is in use to manage multiple HMIs, each HMI's distinct functional group will contain the same master, creating an overlap between multiple functional groups.

Note

There are many other devices, such as motor drives, printers, and safety, systems that may also be connected to an HMI and therefore might also be included in the HMI's functional group. However, these items are not shown in Figure 9.3 in order to simplify the illustration.

Plant-level control processes

Every process consists of much more than a PLC, I/O, and an HMI. Manufacturing systems, industry-specific applications, historians, asset management, network services, engineering and operations workstations, and so on all play a part. In addition, a master controller, master terminal unit (MTU), or SCADA Server may be used to manage multiple HMIs, each responsible for a specific part of a larger control process (see Chapter 4, "Introduction to Industrial Control Systems and Operations"). This same

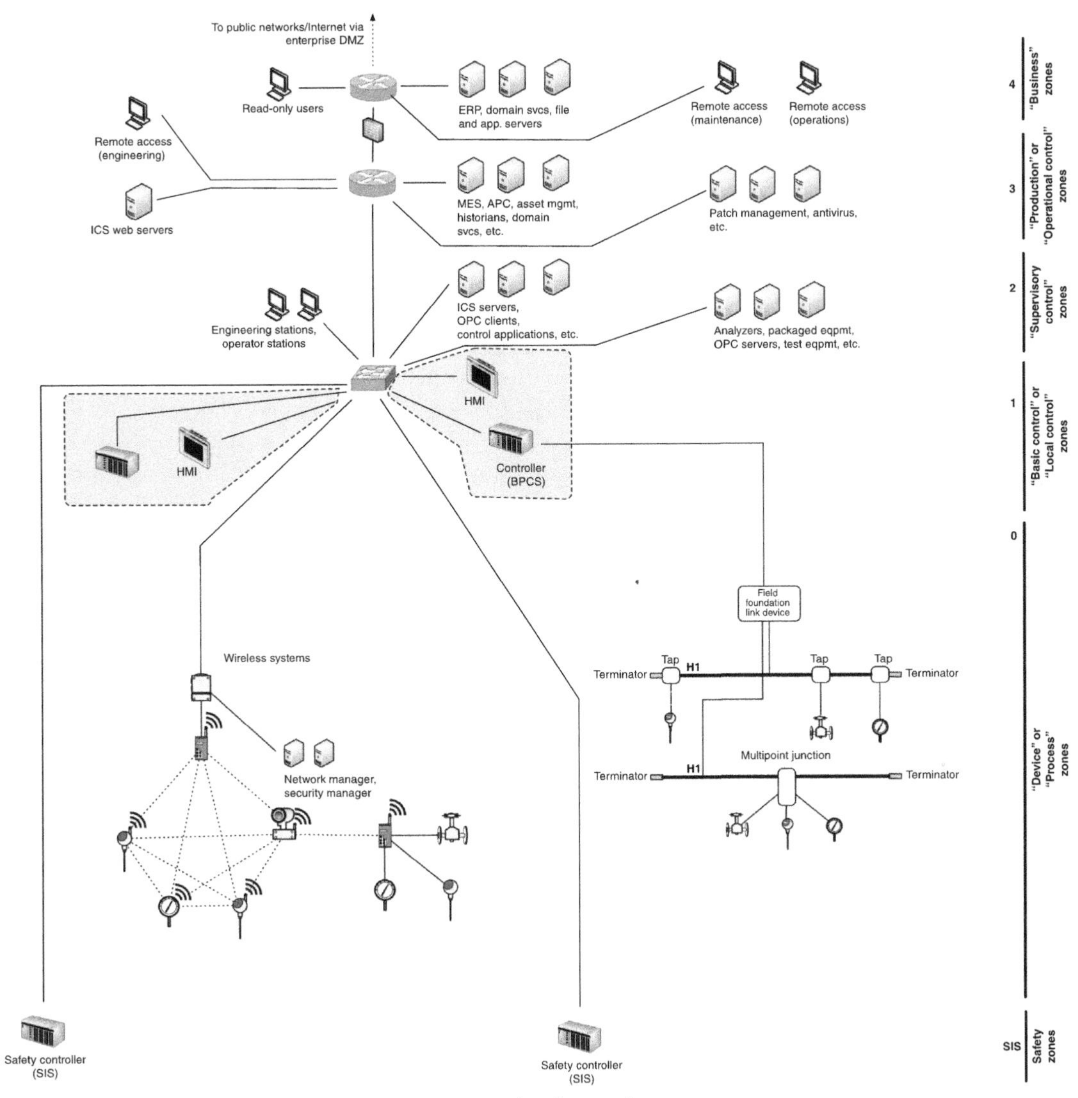

FIGURE 9.3 Example of supervisory zones.

master device now represents the root of yet another functional group—this time containing all relevant HMIs. Figure 9.4 shows how basic control zones might extend to include other relevant systems that span "integration levels."

This example also introduces the concept of process communication and historization. If a device or system interfaces with an ICCP server, for example, in order to communicate bulk electrical load to another electrical entity, the ICCP server should also be included in the same functional group. Similarly, if the process information from the device or system is fed into a data historian, that system should likewise be included.

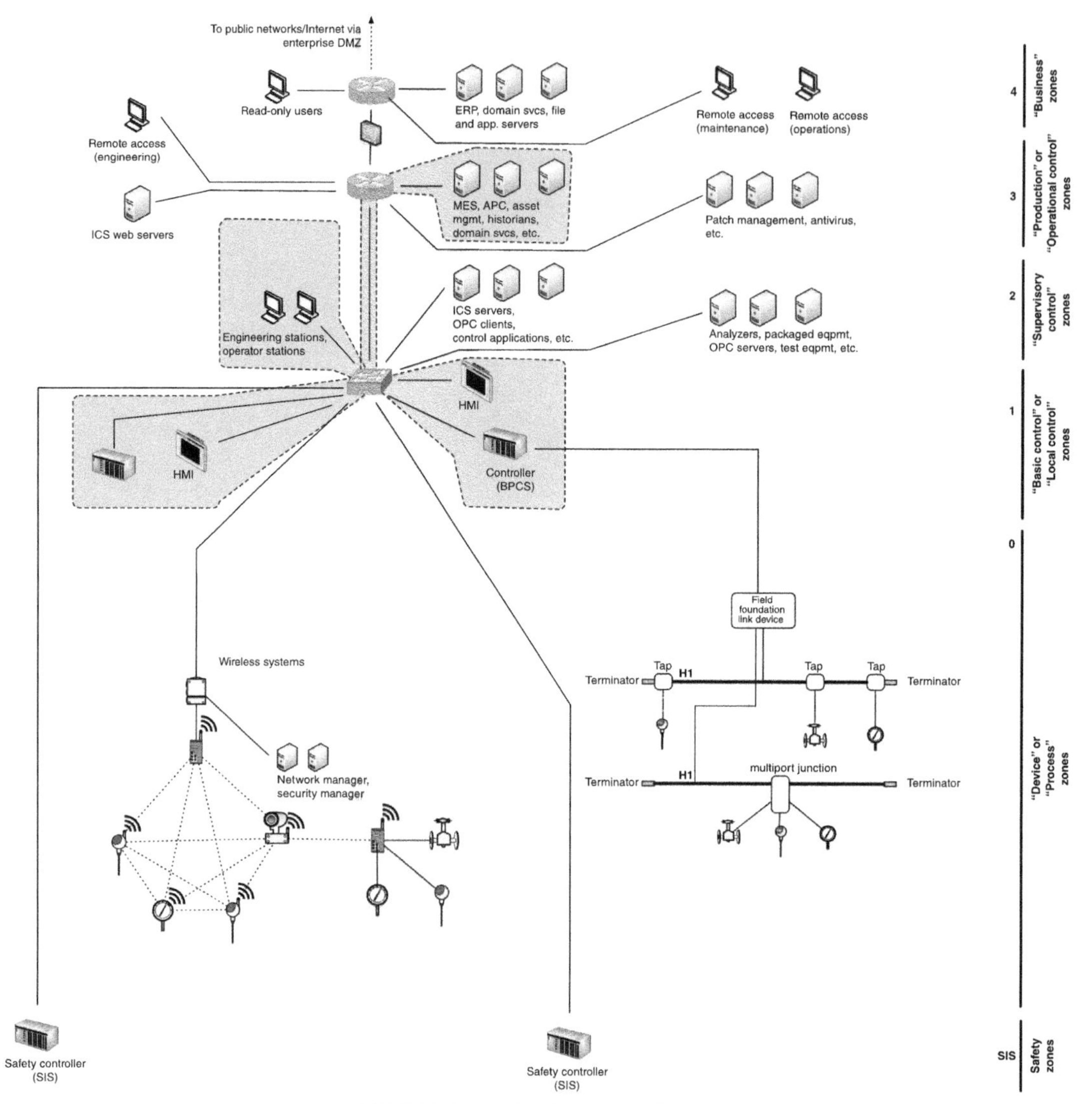

FIGURE 9.4 Example of plant level zones.

Control data storage

Many industrial automation and control system devices generate data, reflecting current operational modes, status of the process, alarms, and other vital manufacturing information. This information is typically collected and "historized" by a data historian (see Chapter 4, "Introduction to Industrial Control Systems and Operations"). The data historian system may collect data from throughout the control system network, supervisory network, and in some cases the business network, as illustrated in Figure 9.5.

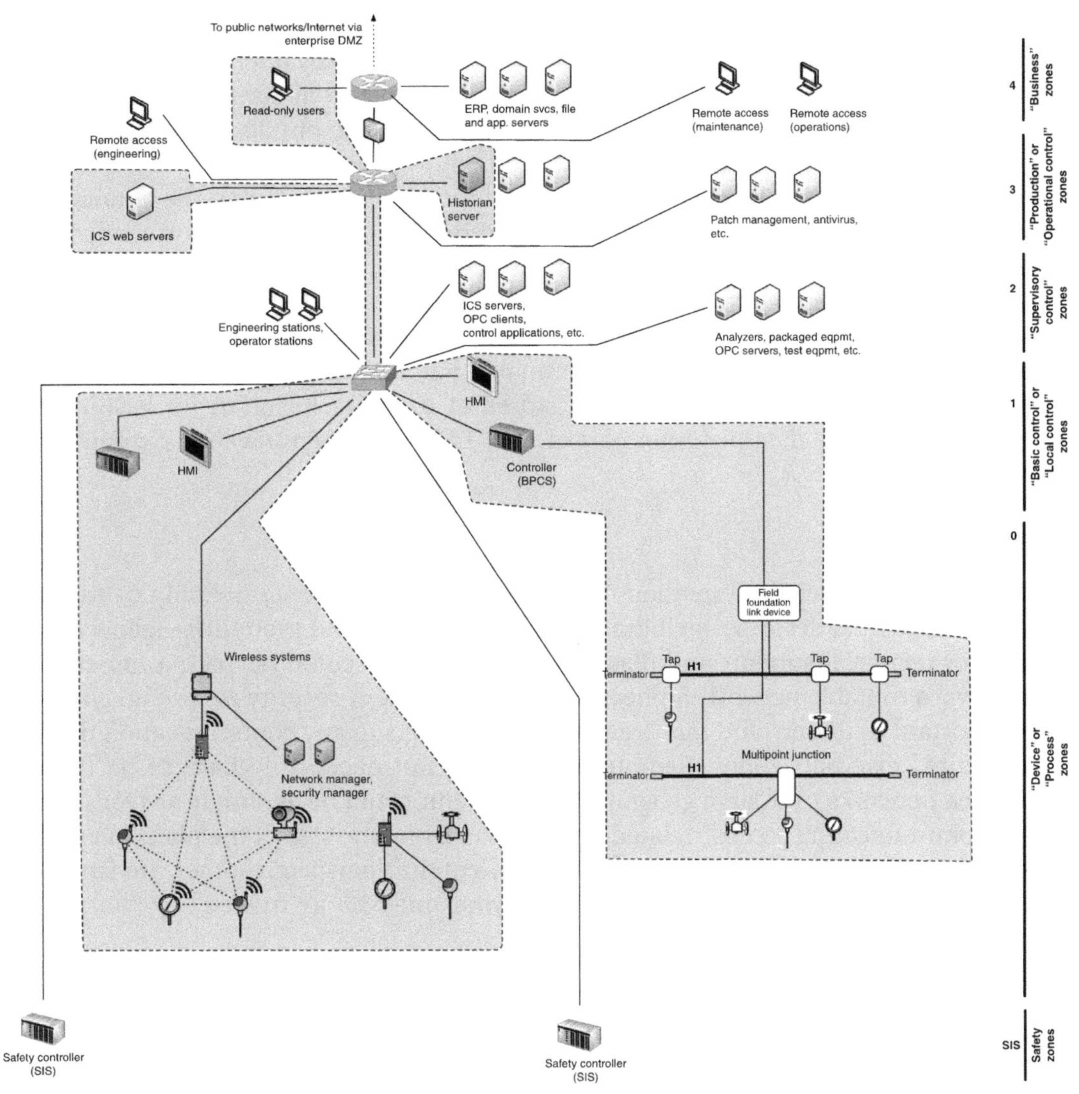

FIGURE 9.5 A zone containing all devices feeding into and utilizing data from a historian.

Not shown here are other devices, such as network attached storage (NAS) devices, storage area networks (SAN), and other devices that may be present to support the data storage requirements of a historian, especially in larger industrial operations.

Trading communications

The need to communicate between control centers (common within the electric transmission and pipeline sectors) is sufficient enough to justify a specialized industrial protocol, developed specifically for that task. The Inter-Control Center Communication

Protocol, or ICCP (see Chapter 6, "Industrial Network Protocols") connections require explicitly defined connections between clients and servers. Any operation utilizing ICCP to communicate with a field facility and/or a peer company will have one or more ICCP servers and one or more ICCP clients (these can be a single physical server or multiple distributed servers).

One thing to remember when assessing this functional group is that the remote client devices are all explicitly defined, even if owned by another company and hosted at its facility. These remote clients should be included within the functional group, as they have a direct relationship to any local ICCP servers that may be in use.

Because ICCP connections are typically used for trading, access to operational information is necessary. This could be a manual or automated informative process, which most likely involves the historized data stores of the data historian (or a subsystem thereof), making the data historian part of the "Trading Communications" zone in this example.

Remote access

ICCP is but one specialized method of remotely accessing a system. Many control systems and industrial devices—including HMIs, PLCs, RTUs, and even IEDs—allow remote access for technical support and diagnostics. This access could be via dial-up connection, or via a routable network connection. In the context of security zones and conduits, it is important to understand that "remote access" refers to any communication through conduits to "external" zones. Remote access does not necessarily have to be through wide-area networks over large geographical areas but could be as simple as two security zones communicating control-related information from one side of the plant to another. When looking at the problem from a zone-and-conduit perspective, they are similar in terms of two "trusted" zones connected via what may be a "trusted" or "untrusted" conduit.

Remote access to control system devices, if it is provided, should be controlled via specialized virtual private networks (VPNs) or remote access servers (RAS), and should only allow explicitly defined, point-to-point connections from known entities, over secure, and encrypted channels. These remote access "conduits" should be further secured with enhanced access control methods including end-point policy enforcement, application layer firewalls, and point-to-point authorization. These explicitly defined users, the devices that they access, and any VPN or RAS systems that are used constitute a remote access functional group, as illustrated in Figure 9.6.

By functionally isolating remote connections, additional security can be imposed. This is extremely important in order to avoid an open and inviting vector to an attacker.

Users and roles

Either a user or another system ultimately accesses every system. Until now, functional groups have been built around the latter—explicitly defining which devices should

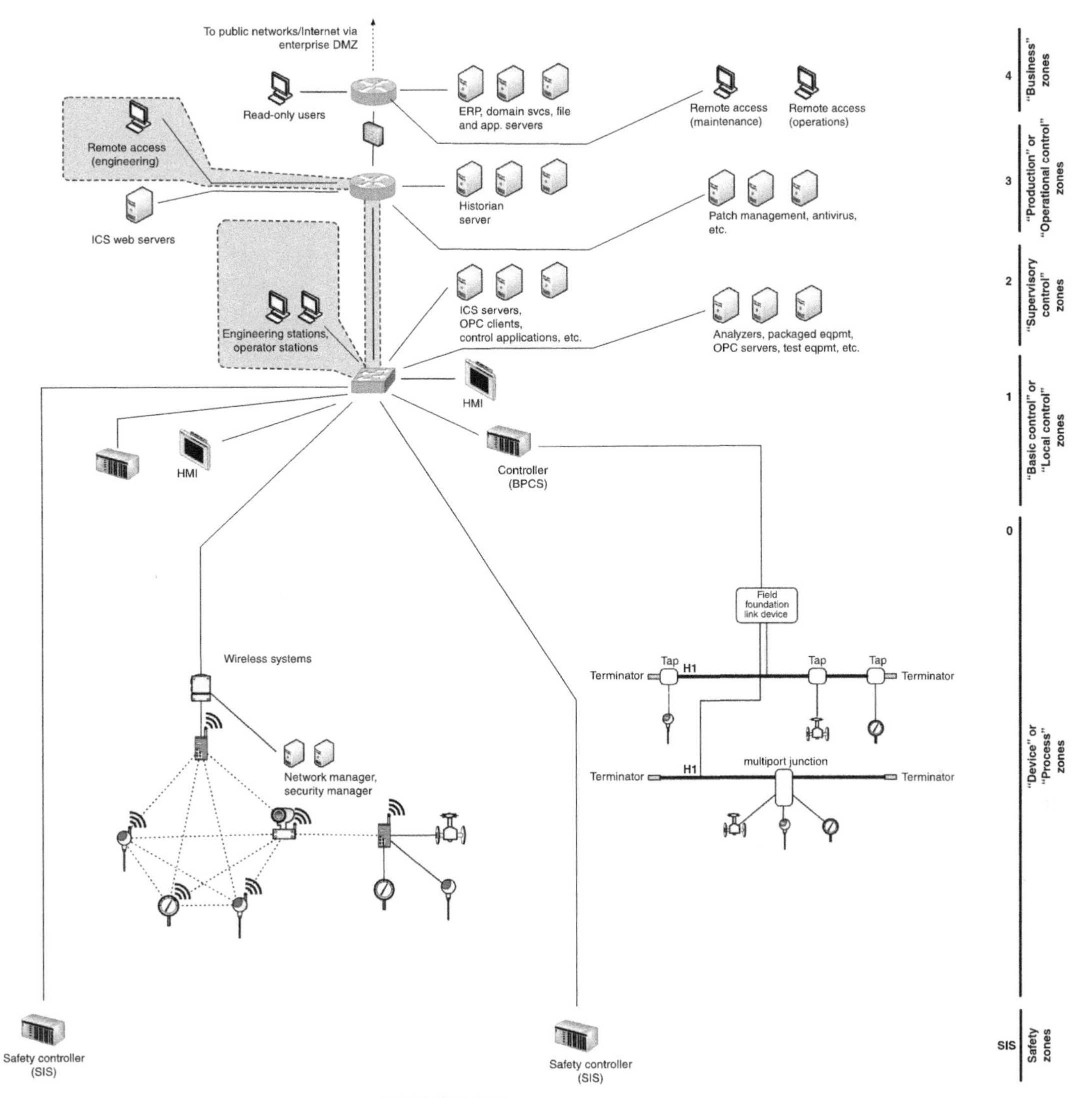

FIGURE 9.6 Remote access zones.

legitimately be communicating with other devices. For human interaction, such as an operator accessing an HMI to control a process, it is just as important to define which users should legitimately be communicating with which devices. This requires a degree of Identity and Access Management (IAM), which defines users, their devices, and their roles. The most well-known example of an IAM solution is Microsoft's Active Directory services, although many other commercial IAM systems exist. Figure 9.7 illustrates the concept of a functional group containing a user and those devices that the user is allowed to interface.

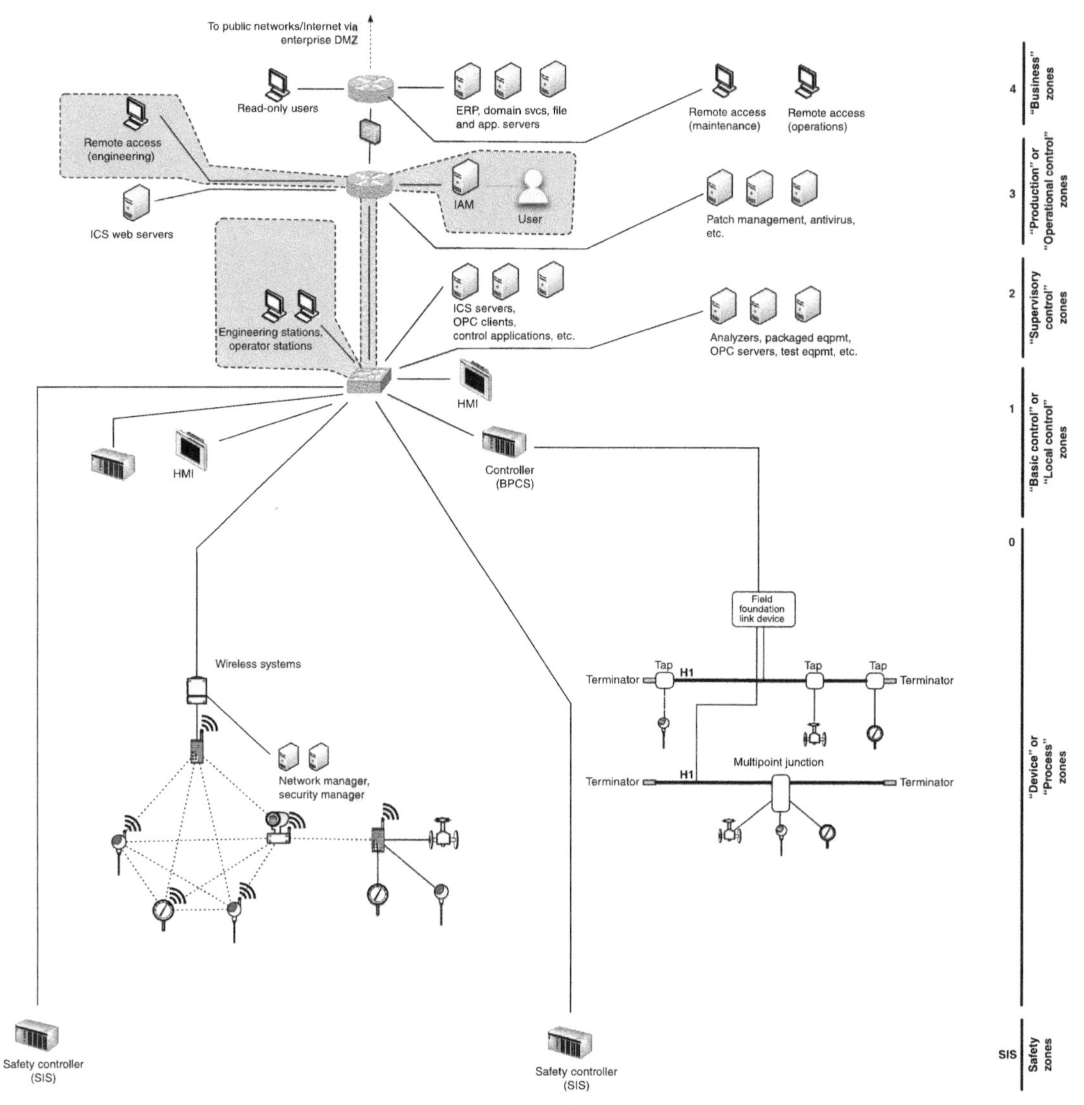

FIGURE 9.7 A zone example based on a user.

Mapping roles and responsibilities to devices can be tedious but is very important, as the resulting functional group can be used to monitor for unauthorized access to a system by an otherwise legitimate user. This is one of the primary reasons many ICS architectures are moving toward a role-based access control (RBAC) infrastructure. RBAC provides a mechanism to configure specific access privileges to specific roles and then assign individual users to these roles. Typically, the responsibilities associated with a given role do not change over time; however, the roles assigned to a particular user can change. An employee with control system access to a certain HMI, upon termination of his or her employment, might decide to tamper with other systems. By placing a user in a functional group with only those devices he or she should be using, this type of activity could be easily detected and possibly prevented (remember, defining functional groups is only the first step to define zones, and once actual zones are defined, they still need to be properly implemented and secured. See "Implementing Network Security and Access Control" and "Implementing Host Security and Access Control" in Chapter 11).

Protocols

Protocols that a device uses in industrial networks can be explicitly defined in order to create functional groups based on protocols. Only devices that are known to use DNP3, for example, should ever use DNP3, and if any other device uses DNP3, it is a notable exception that should be detected quickly and prevented outright if possible. The areas where a specific industrial protocol are commonly used has already been discussed in Chapter 6, "Industrial Network Protocols." The specific devices using specific industrial protocols should now be identified and recorded, in order to build one more important functional group, as shown in Figure 9.8.

Criticality

Zone-based security is about isolating common influencing factors into functional groups so that they can be kept separate and secure from other non-influencing factors. In terms of functional safety in the plant, this concept has been communicated in terms of the "Safety Integrity Level" (SIL). This SIL allows the safety capability of the component to be quantified in order to ensure that similar devices can be deployed in a system and provide sufficient assurance of functionality when demanded. A similar concept known as "security level (SL)" has been developed by ISA as part of the ISA-62443 security standards to provide a measure for addressing the relative security of a particular security zone or conduit.

When applied as part of the security lifecycle, a "target security level" is determined during initial system design. This initial level is then used to select components that have a particular "capability security level," so that components and systems can be selected that help ensure all assets within a particular zone meet the same SL. Once the system is commissioned, a final "achieved security level" can be determined through physical assessment to ensure that the system has been properly installed and commissioned, and that the system meets the desired security level once it is in operation.[6]

The ISA-62443 standard provides a basis for achieving a particular Security Level through the deployment of security controls defined as foundation requirements (FR) and associated system requirements (SR).[7] Each SR contains a baseline requirement and zero or more requirement enhancements (RE) necessary to strengthen the security assurance. These baseline requirement and REs are then mapped to one of four desired SLs.

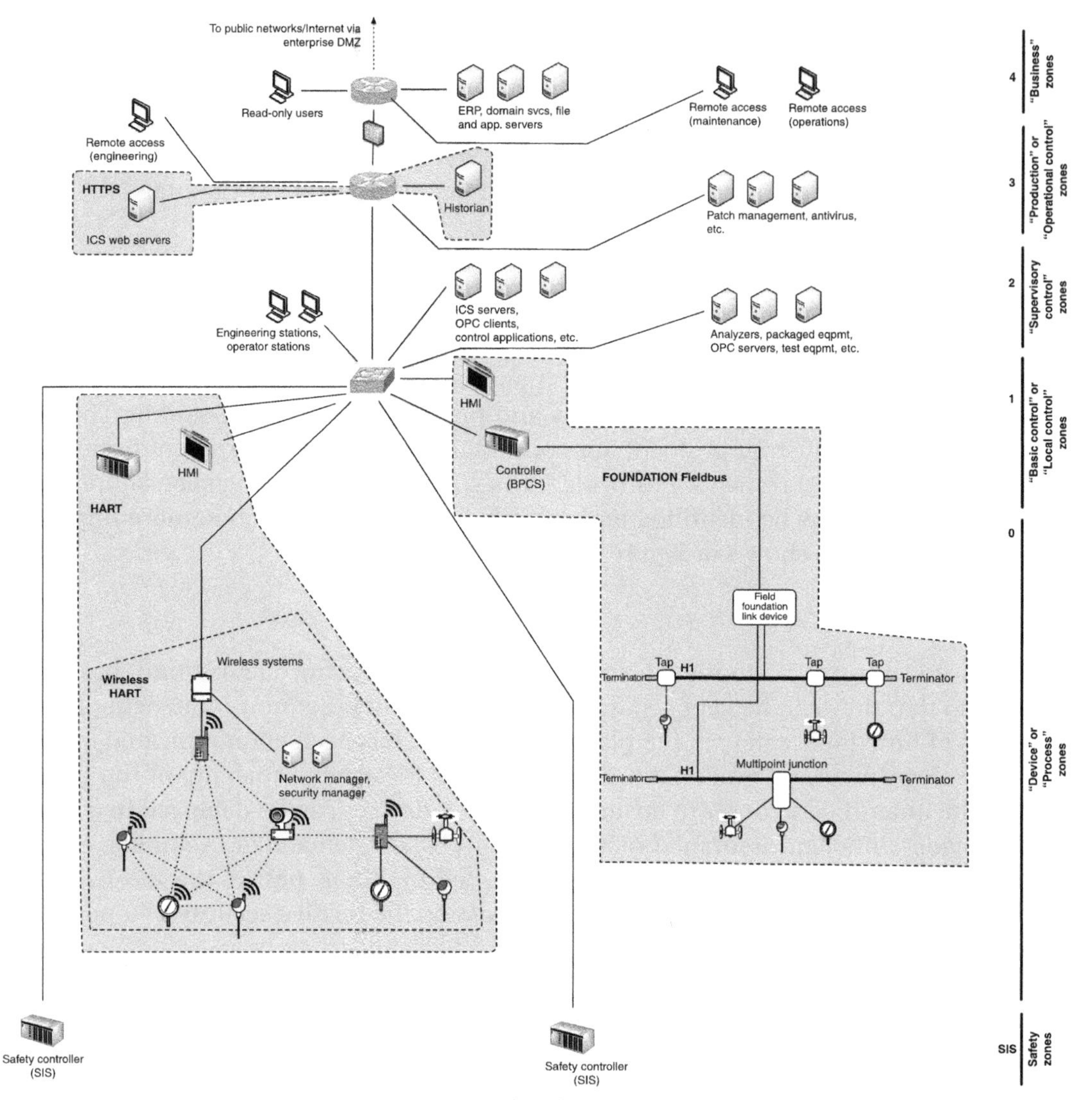

FIGURE 9.8 Zones based on protocol use.

The Nuclear Regulatory Commission (NRC) dictates within CFR 73.54 that the criticality of assets be determined so that they can be separated into five logical security zones.[8] The NRC security zones are a good example of zone-based security, as the NRC regulatory Guide 5.71 provides clear guidance of how stronger security measures should be used as the criticality of the zone increases.

Critical assets, as defined by the North American Electric Reliability Corporation (NERC), are those that can impact the operation of the bulk electric system.[9] They might include control centers, transmission substations, generation systems, disaster recovery systems, black start generators, load shedding systems and facilities, special protection systems, and so on.[10] They can be identified using a simple methodology (see Chapter 2, "About Industrial Networks"). Determining the criticality of a zone is a similarly straightforward process and uses a similar methodology.

Critical assets are extrapolated to the critical function group(s) to which they belong, which may or may not contain other critical and/or noncritical assets. A good rule of thumb is that any zone that contains a critical asset is a critical zone. If noncritical assets are also present in the zone, they must either rise to meet the minimum security requirements of the critical zone or be moved into a separate zone.

Tip

While grading the importance of an asset for compliance can be construed as a means to measure accountability (and fines), it also allows us to improve threat detection and measure the severity of an event should one occur. By taking the time and making the effort to identify critical assets and zones, it is also possible to greatly improve the threat detection capability, by configuring security monitoring tools to weigh the perceived severity of suspicious activities, ranking them in order of consequence and priority. This is discussed in more detail in Chapter 13, "Security Monitoring of Industrial Control Systems."

However, simply defining functional groups around criticality to identify zones will result in very few zones (a total of five, using the NRC guidelines). In contrast, the more zones that are defined the stronger the security of the industrial network as whole, and so a broader methodology—which identifies many more distinct zones and subzones—is recommended. Therefore, functionally defined zones should be assessed within the context of their criticality and vice-versa. In this way, the most critical systems will be protected by an additional layer of separation—for example, the protections between critical and noncritical zones and then additional protection between systems within each zone.

Granular zoning provides the following benefits:

- It will help to minimize the scope of an incident, should one occur, by further separating systems according to the Principle of Least Route. If an asset is compromised, it will only be able to impact a limited number of systems as the ability to communicate to other zones via defined conduits is restricted.
- It will help to secure critical devices from the insider threat, such as a disgruntled employee who already has legitimate physical and logical access to the parent zone since only limited communication channels are permitted between zones.
- It will help to prevent lateral attacks from one critical system to the next—if all critical systems are grouped together solely because they are all "critical," a successful breach of one critical system puts the entire critical infrastructure at risk.

Tip

Carefully document and characterize each zone and all of the devices, services, protocols, and users within it. This is a vital security measure since these lists will come in handy when implementing perimeter defenses (see Chapter 11, "Implementing Security and Access Controls") and also when monitoring zone behavior (see Chapter 13, "Security Monitoring of Industrial Control Systems").

Establishing security zones and conduits

It was mentioned earlier that conduits are a special type of security zone, so when it comes to understanding how zones and conduits are created, it makes sense to discuss these together. Conduits are essentially a type of zone that only contains communication mechanisms as its assets. When the word "zone" is used in the context of this section, it shall be assumed to include "conduits" unless stated otherwise.

It was explained earlier that physical and logical assets are grouped into zones. In terms of conduits, these assets are communication assets, such as active and passive network infrastructure (cables, switches, routers, firewalls, etc.) as well as the communication channels that are transmitted over these cables (industrial protocols, remote procedure calls, file sharing, etc.). It was also discussed that early in the security lifecycle, these zones are assigned a relative security level that is used to create the foundation for the security requirements and associated characteristics that will be applied to all assets contained within the zone. These characteristics include:

- Security policies
- Asset inventory
- Access requirements and controls
- Threats and vulnerabilities
- Consequences in the event of a breach or failure
- Technologies authorized and not authorized
- Change management process
- Connected zones (conduits only).

As each of the characteristics of a zone are defined, the allocation of assets within the zone become obvious, including the possible creation of nested subzones for particular assets that may be align with other assets within the particular zone. It will then become possible to establish a comprehensive asset inventory that lists physical components, such as computers, network appliances, communication links, and spare parts, as well as logical components like operating systems, applications, patches, databases, configuration files, and design documentation just to name a few.

The assets now contained within a zone are then evaluated for threats and vulnerabilities in order to determine the resulting risk to the zone should these assets cease to perform their intended function. This information will become vital in identifying possible security countermeasures that could be used to reduce the risk resulting from a threat exploiting a vulnerability and then selecting the appropriate controls necessary to both meet the security level for the zone while considering the cost versus risk trade-off. These concepts were discussed in more detail in Chapter 8, "Risk and Vulnerability Assessments."

Zones are established considering the technologies that are both allowed and disallowed within the zone. Each type of technology possesses inherent vulnerabilities (both known and unknown) and with these vulnerabilities a certain amount of risk. These technologies must be aligned with security zones in order to prevent one technology from compromising the entire zone. One example many industrial users now face is the concept of "bring you own device" or BYOD within the critical control zones. It is clear that these devices bring with them a certain amount of risk, but by creating dedicated security zones for such devices, it becomes possible to enforce a particular security policy through other controls that may be deployed on the communication channels of the conduit from this zone to other more critical zones.

It is probably clear up to this point how one would take a particular computing asset or embedded device and place it in a particular security zone. What may not be so clear is how to create conduits and assign "communication" assets to these special zones. The easiest place to start is to consider that in most industrial architectures, the physical network is the conduit. Before saying to yourself, "that was easy," it is important to note that the industrial network only acts as the conduit for "external" communication channels between other assets and zones; it does not represent the channels used to communication between applications and processes that exist within a single asset. These "internal" conduits will become important as the concept of system and host hardening is considered later in this book.

The idea that threats and vulnerabilities exist for computing assets is equally important to communication assets. It is well known that many industrial protocols in use today contain vulnerabilities that, if not properly addressed through appropriate security controls, could introduce considerable risk to not only the device(s) using these protocols, but other devices that may exist within the same zone. It is also important to evaluate the vulnerabilities that may exist within the active network infrastructure, including switches, routers, and firewalls since the loss of any of these components can introduce significant risk to not only the network (conduit), but all zones connected via this conduit. This is why a thorough risk and vulnerability assessment must also be performed for security conduits in order to ensure that appropriate countermeasures have been deployed on the conduit to ensure that the conduit meets the desired security level (See Chapter 8, "Risk and Vulnerability Assessments").

Using microsegmentation to establish zones and conduits

Microsegmentation is extremely useful when establishing zones and conduits, by providing a great deal of flexibility in terms of how networks are segmented and by enabling specific security policies and controls to be implement on each micro-segment (See Chapter 5, "Industrial Network Design and Architecture: Microsegmentation").

Assuming that your network infrastructure supports microsegmentation, setting up zones and conduits becomes a much simpler process:[11]

1. Determine which devices share similar security requirements and criticality.
2. Use microsegmentation to logically group these devices together into micro-segments. Focus on the most critical groups first (those that contain assets vital to reliable process operations.
3. Apply specific and appropriate security policies to each micro-segment, in accordance with that group's security requirements.
4. Remember that unique security policy can and should be applied to each micro-segment.

Creating a zone and conduit map

The documentation of security conduits—and the communication channels contained within them—is a vital piece of information necessary to accurately deploy security controls throughout the architecture. This document will be used to not only configure upper-level appliances like routers and firewalls that manage access between zones, but also next-generation technologies like application monitoring, intrusion prevention systems, and event monitoring and correlation technologies. This will be discussed further in Chapter 11, "Implementing Security and Access Controls."

One of the leading root causes of compromises to secure industrial networks is from misconfiguration of appliances placed on conduits that connect less-trusted "external" zones to more-trusted "internal" zones. These configuration errors commonly result from attempting to configure the communication access control without sufficient documentation of the content of each of the desired communication channels crossing the conduit. This will be discussed further during "System Characterization" in Chapter 8, "Risk and Vulnerability Assessments."

Summary

Zones and conduits are abstract concepts designed to group similar devices and control communications between groups, in order to improve security and to minimize the impact of a cyberincident by making it more difficult for malware to propagate unrestricted laterally and hinder an attacker from pivoting between systems. Zones can be used to identify broad groups or highly focused subsystems, supporting the specific operation, business, and technology requirements of a given system. As can be seen in Figure 9.9, which shows how different zones built around different requirements can overlap, this can unfortunately lead to confusion if zones and conduits are not defined carefully and consistently. Once the difficult work is done, the benefits are tangible. The overall infrastructure will become more secure by segmenting systems into zones and controlling communication between zones using controllable communication conduits.

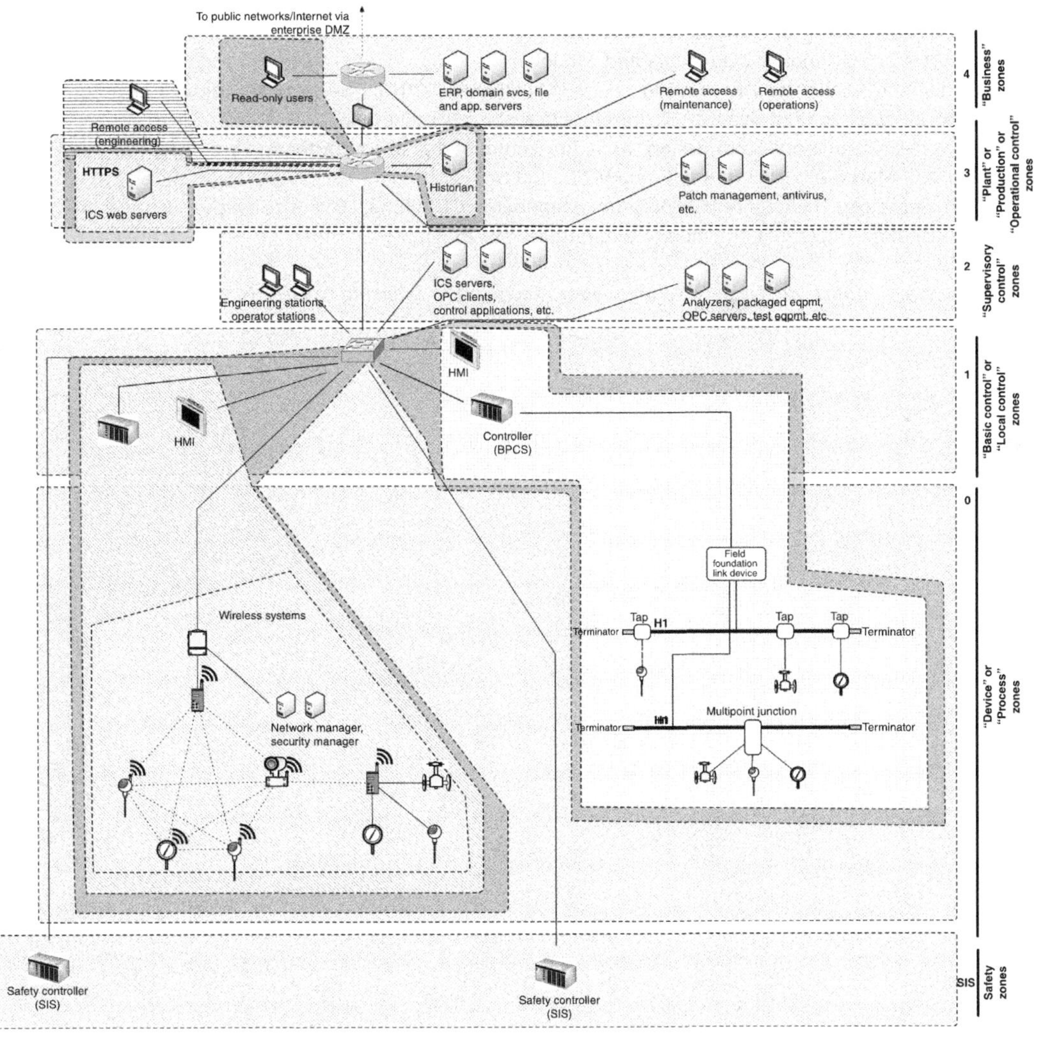

FIGURE 9.9 Overlapping zones based on different criteria.

Endnotes

1. Theodore J. Williams. A Reference Model For Computer Integrated Manufacturing (CIM): A Description from the Viewpoint of Industrial Automation. Purdue Research Foundation. North Carolina. 1989.
2. International Society of Automation (ISA), ISA-99.00.01-2007, "Security for industrial automation and control systems: Terminology, Concepts and Models," October 2007.
3. U.S. Nuclear Regulatory Commission, Regulatory Guide 5.71 (New Regulatory Guide), Cyber Security Programs for Nuclear Facilities, January 2010.
4. International Society of Automation (ISA), ISA-99.00.01-2007, "Security for industrial automation and control systems: Terminology, Concepts and Models."
5. D. Taylor, Intrusion detection FAQ: are there vulnerabilities in VLAN implementations? VLAN Security Test Report, The SANS Institute., July 12, 2000 (cited: January 19, 2011).
6. International Society of Automation (ISA), ISA-99.00.01-2007, "Security for industrial automation and control systems: Terminology, Concepts and Models".
7. International Society of Automation (ISA), ISA-62443-3-3-2013, "Security for industrial automation and control systems: System Security Requirements and Security Levels".
8. U.S. Nuclear Regulatory Commission, 73.54 Protection of digital computer and communication systems and networks., March 27, 2009 (cited: January 19, 2011).
9. North American Reliability Corporation, Standard CIP-002-3. Cyber Security—Critical Cyber Asset Identification., December 16, 2009 (cited: January 19, 2011).
10. Ibid.
11. Honeywell. *Industrial Network Security: Best Practices For Securing Critical Infrastructure Networks*. 2023.

10 OT Attack and Defense Lifecycles

Information in this chapter

- Attack Lifecycles and Kill Chains
- Defense Lifecycles
- The Importance of Understanding Lifecycles

Malware has been around since the dawn of personal computing: the first virus being the Creeper virus in 1971, a proof-of-concept virus that spread through ARPANET-connected computers[1], and the first personal computer being attributed to the Kenbak-1 computer[2] also in 1971. However, malware has advanced significantly since then, and cyberattacks very rarely depend on a single piece of malware. Instead they have evolved into campaigns that consist of multiple steps. Instead of singular actions that result in a single outcome, attack campaigns coordinate multiple actions to produce a more sophisticated outcome. The sophistication of both attack and defense techniques has grown to a degree where each involves multiple steps that occur, intertwined in a dance of attack and defense. The stronger the defense posture of an organization, the more sophisticated the moves of the attacker must become. Likewise, when faced with a more capable attacker, the defender must be able to anticipate the next move in order to prevent it or be well positioned to react quickly if it cannot be prevented. This is only possible with an understanding of how both attack and defense efforts progress.

Attack lifecycles and kill chains

The many steps of an attack that "enable access and provide sufficient information to devise an effect"[3] are often referred to as a campaign. SANS analysts Michael J. Asante and Robert M. Lee,[4] have framed the attack campaign within the "Cyber Kill Chain" model, which is a military model originally created by Eric M. Hutchins, Michael J. Cloppert, and Rohan M. Amin of Lockheed Martin.[5] The result is a two-phase attack lifecycle: to create a cyber-physical outcome, the attacker must first breach the industrial network; and then begin the second phase of executing a cyber-physical attack on the industrial control system itself.[6] Each of these phases is a dance of its own, with a necessary progression from the initial planning to the final execution.

Phase 1 obtaining access to industrial networks[7]

1. Planning (reconnaissance)
2. Preparation (Weaponization, targeting)

Industrial Network Security. https://doi.org/10.1016/B978-0-443-13737-2.00004-X

3. Intrusion (Delivery, exploitation, installation and modification)
4. Enablement (Command and Control, or C2)
5. Execution (Action)

Phase 2 manipulation of industrial networks[8]

1. Attack Development & Tuning (Develop)
2. Validation (Test)
3. Industrial Control System Attack (Deliver, Install/Modify, and ICS Attack Execution)

It can be helpful to visualize each phase of an attack as a linear progression, as illustrated in Figures 10.2 and 10.3. Each step within each phase takes time, and each step must be completed to some degree of efficacy before the next stage can begin.

Understanding this progression can be extremely helpful as defenders need to respond differently as an attack proceeds. Knowing how far the attack has already progressed, and what the next necessary steps are likely to be, helps the defender to make more informed decisions (see "Defensive Lifecycles" below). If, in doing so, a defender can prevent an adversary from completing a necessary step within the attack lifecycle, it may be possible to prevent an incident entirely. If a defender is able to force an attacker to rush through important steps, the attack may be less effective. For example, by interrupting or hindering C2 communications during "enablement," the defender could prevent the attacker from obtaining data needed to fully develop and validate their attack before executing it (Figure 10.1).

In practice, "understanding where an attacker is within the progression of the attack lifecycle" is not always obvious. As a defender, we do not always see what an attacker is

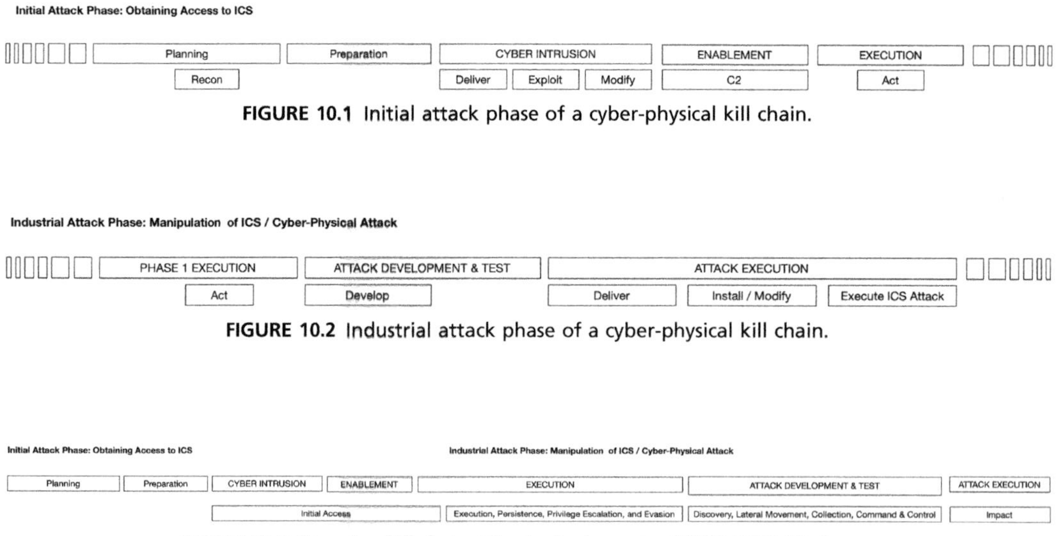

FIGURE 10.1 Initial attack phase of a cyber-physical kill chain.

FIGURE 10.2 Industrial attack phase of a cyber-physical kill chain.

FIGURE 10.3 Mapping kill chain ICS attack phases to MITRE ATT&CK stages.

intending to do; we only see evidence of their actions (and only if there are detection and monitoring tools in place to allow us to see). To make this easier, there is another important model that should be considered: the MITRE ATT&CK frameworks. The ATT&CK frameworks map attack behaviors to specific stages of an attack lifecycle. The frameworks are described by MITRE as a "curated knowledge base for cyber adversary behavior" that are specifically designed to illustrate the various stages of the attack lifecycle,[9] and there are three frameworks available at the time of this writing—one for enterprise, mobile, and ICS attacks. Each focuses on the specific tactics and techniques used against these target categories.

Unlike the Kill Chain model, the ATT&CK frameworks go into significantly more detail, providing the specific techniques used by an attacker. While this makes ATT&CK more complex, it also makes it extremely useful in day-to-day security management and response efforts. This is because ATT&CK tactics and techniques can be mapped directly to specific security events and indicators of compromise. In fact, many cybersecurity monitoring tools (i.e., SIEM) will map events to the ATT&CK framework automatically, letting cybersecurity professionals immediately gain the context of the attack progression.

The two ATT&CK frameworks that are relevant here (ATT&CK for enterprise and ATT&CK for ICS) can be loosely mapped to Purdue levels 2 through 4 and 0 through 2 respectively and can also be loosely mapped loosely to the initial- and industrial-attack phases referenced in the Kill Chain model.[10] In other words, the ATT&CK for enterprise model consists of early intrusion into business systems and initial pivots into operations management, while ATT&CK for ICS is focused on techniques and tactics used once the adversary has access to supervisory control, control, and process I/O.

The terms "tactics" and "techniques" refer to the tactical goals of an adversary and the specific actions and/or the results of specific actions taken by the adversary to achieve that goal, respectively. Tactics—what the attackers are trying to do—remain fairly consistent over time, while techniques—the way the attackers try to achieve that—can evolve and adapt very rapidly.[11]

The tactics listed in Table 10.1 begin with "initial access" and then progress to "execution," "persistence," "privilege escalation," and "evasion." Attacks then progress to "discovery," "lateral movement," "collection," and "command and control," before finishing with "inhibit response function," "impair process control," and "impact." They overlap somewhat with the two phases of the kill chain, with "initial access" mapping to the "planning," "preparation," and "intrusion" steps within the initial attack phase, and the remaining tactics mapping to the industrial phase of the cyber-physical kill chain, as illustrated in Figure 10.3.

Each tactic can be achieved by a number of techniques, as illustrated in Figure 10.4. Many techniques also include sub-techniques.

Table 10.1 MITRE ATT&CK for ICS Tactics

ID	Name	Description
TA0108	Initial Access	The adversary is trying to get into your ICS environment.
TA0104	Execution	The adversary is trying to run code or manipulate system functions, parameters, and data in an unauthorized way.
TA0110	Persistence	The adversary is trying to maintain their foothold in your ICS environment.
TA0111	Privilege Escalation	The adversary is trying to gain higher-level permissions.
TA0103	Evasion	The adversary is trying to avoid security defenses.
TA0102	Discovery	The adversary is locating information to assess and identify their targets in your environment.
TA0109	Lateral Movement	The adversary is trying to move through your ICS environment.
TA0100	Collection	The adversary is trying to gather data of interest and domain knowledge on your ICS environment to inform their goal.
TA0101	Command and Control	The adversary is trying to communicate with and control compromised systems, controllers, and platforms with access to your ICS environment.
TA0107	Inhibit Response Function	The adversary is trying to prevent your safety, protection, quality assurance, and operator intervention functions from responding to a failure, hazard, or unsafe state.
TA0106	Impair Process Control	The adversary is trying to manipulate, disable, or damage physical control processes.
TA0105	Impact	The adversary is trying to manipulate, interrupt, or destroy your ICS systems, data, and their surrounding environment.

Initial Access 12 techniques	Execution 9 techniques	Persistence 6 techniques	Privilege Escalation 2 techniques	Evasion 6 techniques	Discovery 5 techniques	Lateral Movement 7 techniques	Collection 11 techniques	Command and Control 3 techniques	Inhibit Response Function 14 techniques	Impair Process Control 5 techniques
Drive-by Compromise	Change Operating Mode	Hardcoded Credentials	Exploitation for Privilege Escalation	Change Operating Mode	Network Connection Enumeration	Default Credentials	Adversary-in-the-Middle	Commonly Used Port	Activate Firmware Update Mode	Brute Force I/O
Exploit Public-Facing Application	Command-Line Interface	Modify Program	Hooking	Exploitation for Evasion	Network Sniffing	Exploitation of Remote Services	Automated Collection	Connection Proxy	Alarm Suppression	Modify Parameter
Exploitation of Remote Services	Execution through API	Module Firmware		Indicator Removal on Host	Remote System Discovery	Hardcoded Credentials	Data from Information Repositories	Standard Application Layer Protocol	Block Command Message	Module Firmware
External Remote Services	Graphical User Interface	Project File Infection		Masquerading	Remote System Information Discovery	Lateral Tool Transfer	Data from Local System		Block Reporting Message	Spoof Reporting Message
Internet Accessible Device	Hooking	System Firmware		Rootkit	Wireless Sniffing	Program Download	Detect Operating Mode		Block Serial COM	Unauthorized Command Message
Remote Services	Modify Controller Tasking	Valid Accounts		Spoof Reporting Message		Remote Services	I/O Image		Change Credential	
Replication Through Removable Media	Native API					Valid Accounts	Monitor Process State		Data Destruction	
Rogue Master	Scripting						Point & Tag Identification		Denial of Service	
Spearphishing Attachment	User Execution						Program Upload		Device Restart/Shutdown	
Supply Chain Compromise							Screen Capture		Manipulate I/O Image	
Transient Cyber Asset							Wireless Sniffing		Modify Alarm Settings	
Wireless Compromise									Rootkit	
									Service Stop	
									System Firmware	

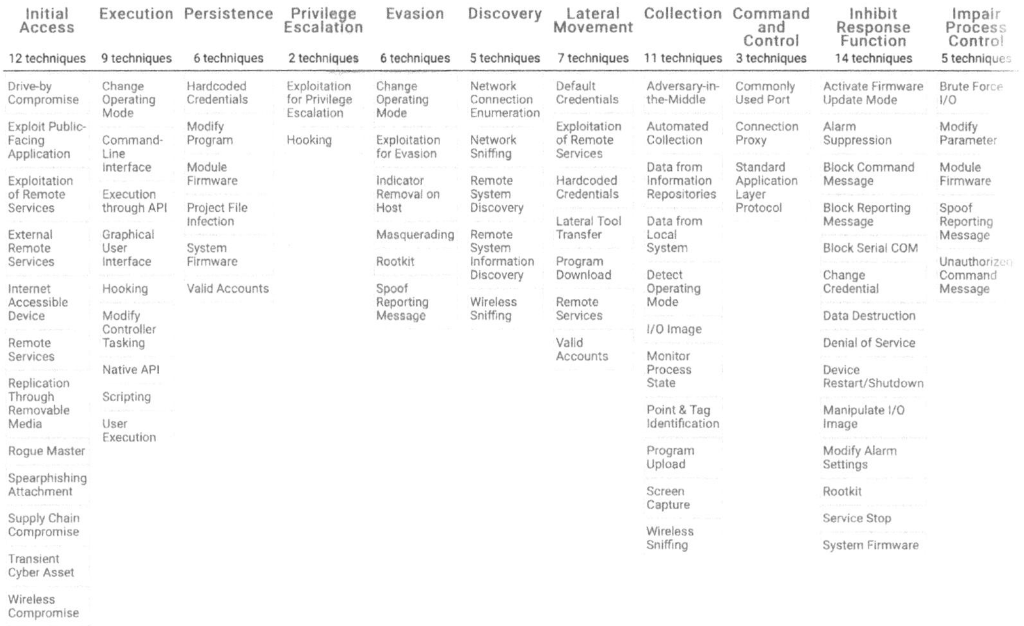

FIGURE 10.4 MITRE ATT&CK for ICS matrix. *© 2023 The MITRE Corporation. This work is reproduced and distributed with the permission of The MITRE Corporation.*

Tip

Many security monitoring tools will map specific security events to MITRE ATT&CK techniques. This can help operationalize security monitoring efforts by providing a clue as to where a specific event or group of events falls within the attack lifecycle. However, it should also be understood that consistent tactics and techniques are used in both ATT&CK for enterprises (i.e., phase one) and ATT&CK for ICS (i.e., phase two). There are also numerous examples of when an adversary might utilize common techniques at multiple stages of an attack. For example, "spoof reporting messages" is a technique used when attempt to evade detection in the early stages of an attack, but it is also used in the later stages of an attack while attempting to "impair process control." Additional context of the event(s) must be considered to determine at what actual point within the overall cyber-physical attack lifecycle a specific technique truly applies.

Obtaining access to industrial networks

As discussed in Chapter 7, the initial attack phases of a cyber-physical attack are primarily concerned with gaining access to the industrial network environment. This typically requires establishing an initial foothold outside of the industrial network (typically the business network of the organization) and then pivoting into the industrial environment.

The initial attack phases starts with planning and preparation. Initial reconnaissance helps to identify targets and plan a successful infiltration of that target. This enables the adversary to begin laying the foundation for the execution of an attack that will penetrate the industrial network. The initial will typically include extensive data exfiltration efforts and the establishment of command and control so that attackers can learn about their target environment and develop a more targeted attack to ultimately penetrate the industrial network, as shown in Figure 10.5.

Planning

The first step of an attack is all about information gathering. Reconnaissance can include any number of activities, including: researching the target company or its employees (e.g., using OSINT) or identifying weaknesses in a target's attack surface (e.g., identifying Internet-facing assets using Shodan). Using publicly available information (including

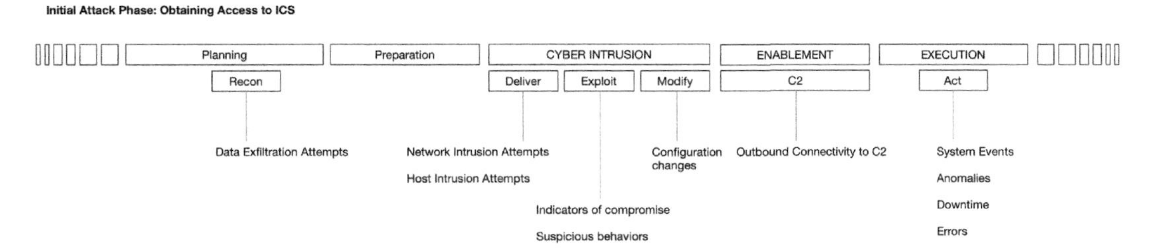

FIGURE 10.5 Attack tactics mapped to initial phase of an ICS attack kill chain.

social media), details of an industrial control system, such as the DCS equipment vendor(s) used, can be determined.[12] Social engineering can be highly effective for information gathering. It is also important to remember that a cyber-adversary targeting industrial networks may be heavily resourced, and reconnaissance efforts can include everything up to and including professional espionage and spycraft. At this early stage, anything that can be used to support an attacker in their attempt to infiltrate their target and execute an attack is useful to the attacker.[13]

Preparation

The next step is preparation. As defined by SANS, this step often includes using malware (e.g., a weaponized Word document containing a Macro virus) for the purpose of enabling the attacker to progress to further steps.[14] This might seem confusing to some because it is easy to think of malware as "the attack" and the execution of that malware the successful "execution" of the attack. In simpler times, and perhaps within the context of less sophisticated cyberthreats, this might be true. However, in the context of an attack campaign, there will likely be many types of malware used along the way. Here, it is used to help identify targets and identify potential weaknesses in those targets that might be exploitable, in order to facilitate the next step: intrusion.[15]

Intrusion

The Intrusion step means gaining access to the defender's environment. This could be directly accessing the defender's network (e.g., through compromised VPN credentials) or a system one the defender's network (e.g., infecting a domain server using malware planted on a USB drive). This step typically consists of some sort of delivery mechanism, enabling an initial infection that will in turn allow the attacker to develop and deliver additional capabilities in the "enablement" step. To do this, malware must be successfully delivered so that it can perform its intended malicious function: typically the installation of additional malware, with additional capabilities; however, not all exploits require malware and the attacker could simply modify or misuse legitimate system functions (e.g., PowerShell).[16]

Enablement

The enablement step is perhaps the most important. To develop and validate the end-goal of the attack, the adversary needs to learn more about the target environment. This requires an active command and control capability, to allow the attacker to interact with the environment.[17] The methods of establish C2 are varied and often creative. Leveraging existing communication paths is easy and effective but also more likely to be detected if the defender is preforming any degree of network monitoring. In the "USB Hardware Threat Report" published by Honeywell, Inc., numerous methods of implanting covert communications were identified, providing an impressive array of wireless and even cellular access directly to the attached host and bypassing the defender's network security controls entirely.[18]

Execution

This is the culmination of the initial phase of a cyber-physical attack. It is important to understand that there may by multiple initial attacks to achieve different goals associated with the overall cyber-physical attack. That is, several instances of the initial phase could be occurring simultaneously or asynchronously to achieve different goals. For example, one attack might attempt to exfiltrate credentials for remote access VPNs necessary to reach an HMI system within a substation, as used in the Black Energy attack, while another might focus on compromising a level 3.5 DMZ firewall to enable deeper penetration to industrial networks (see Chapter 7, "Hacking Industrial Control Systems").

Manipulation of industrial networks

The successful completion of one or more initial attacks will provide the attacker with access to the industrial network and will begin the second phase of a cyber-physical attack. This "industrial phase" shares some of the same tactics at a high level, as shown in Figure 10.6, with an important distinction: the target is now the process control environment, and the goal is to manipulate that environment to create a specific cyber-physical outcome. The industrial phase includes additional stages of reconnaissance, often in the form of data exfiltration using established command and control channels. This is necessary to learn as much as possible about the industrial control system in order to develop and execute a targeted attack against it.[19]

Development and test

Once inside the industrial network, the first thing an attacker needs to do is develop their second- or industrial-phase attack: that is, the "physical" side of the "cyber-physical" attack. To develop an attack that will create a specific physical impact against a unique ICS operation—with specific assets, process logic, physical and environmental parameters, etc.—the minutia of the entire process control system must be taken into consideration. To identify potential hazard conditions that could be created also requires

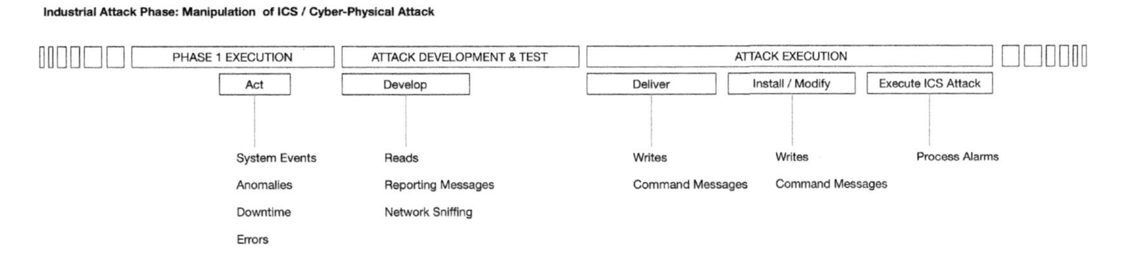

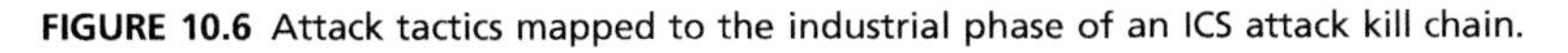

FIGURE 10.6 Attack tactics mapped to the industrial phase of an ICS attack kill chain.

a significant amount of testing and validation. This leads an attacker with few choices: they can develop on their target's system, which is likely to be discovered; or they can exfiltrate sufficient information to develop and test their attack offline in a lab environment.[20] A more audacious attacker might develop their attack live on a softer or weaker targets first in order to develop attack capabilities, with the intention of leveraging their findings on their true target at a later date, although this would also likely be discovered and allow the intended target to prepare for such an attack. With lab testing being the only viable option, attackers face the same challenges that defenders do when performing cyber-physical threat modeling: truly replicating an industrial environment in its entirety is expensive and challenging. Obtaining access to process simulation software to virtualize this process may be an option for well-resourced adversaries, but even then an extensive amount of data exfiltration would be needed. (See Chapter 8, "Risk and Vulnerability Assessments: Cyber-Physical Threat Modeling").

Delivery, installation, and modification

As with the initial phase, once there is an attack, steps need to be taken to deliver and install malware or to modify existing systems.[21] In the industrial attack phase, however, these steps are arguably simpler, as the control system itself is the "existing system" that can be directly modified to achieve the attacker's goals. The delivery and installation of malware on a system within the industrial network may facilitate this and can provide a modular framework from which specific capabilities can be initiated by that attacker, as was the case with Industroyer.[22]

Execution

The attack is realized in the final step of the second phase of the cyber-physical attack. It is here that process control systems are altered in precise and calculated ways to achieve the attacker's specific, intended impact. This step could involve process logic changes, set point changes, manipulation of variables, direct manipulation of actuators, or any combination thereof.[23] The more complex the outcome, the more aspects of the system may need to be manipulated to achieve the attacker's desired outcome. If the target control system is designed with a consideration for cyber resilience, it will be more difficult for an attacker to sufficiently manipulate the system.

Defense lifecycles

Cyberdefenses have a similar lifecycle and consist of multiple tactics and techniques that occur in a natural progression that is important to understand. Perhaps, the most recognized cyber security defensive framework is the NIST Cyber Security Framework or NIST CSF. Introduced in 2015, the NIST CSF is a useful and concise reference. Currently available as v1.1 at the time of this publication, the NIST CSF breaks defensive efforts down into five stages:

- Identify
- Protect
- Detect
- Respond
- Recover

While often described as a list of distinct defensive steps, the CSF is cyclical and constantly repeats itself with a goal of continuous improvement. Like attack kill chains, these defensive efforts can also be made in concurrence, which each stage potentially introducing new awareness that can instigate these concurrent efforts. For example, in reaction to the detection of a new type of threat, response, and recovery efforts will be instigated. If multiple threats are detected together as part of a cyber-attack campaign, multiple response and recovery efforts might be justified as part of an overall response and recovery goal. At the same time, the identification of new type of threat might instigate the implementation of new security controls, which in turn might improve detection efficacy, which could lead to additional detections, ad infinitum.

To simplify this process, we can loosely map defensive efforts to the attack lifecycle to show the interdependencies of attack and defensive efforts. This is illustrated in Figures 10.7 and 10.8.

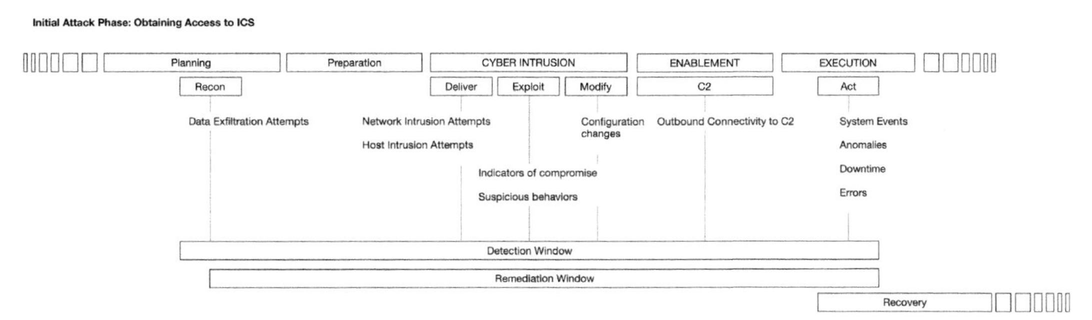

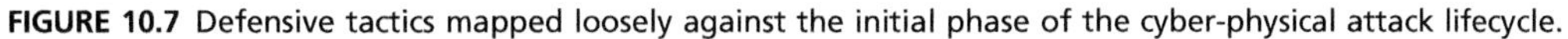

FIGURE 10.7 Defensive tactics mapped loosely against the initial phase of the cyber-physical attack lifecycle.

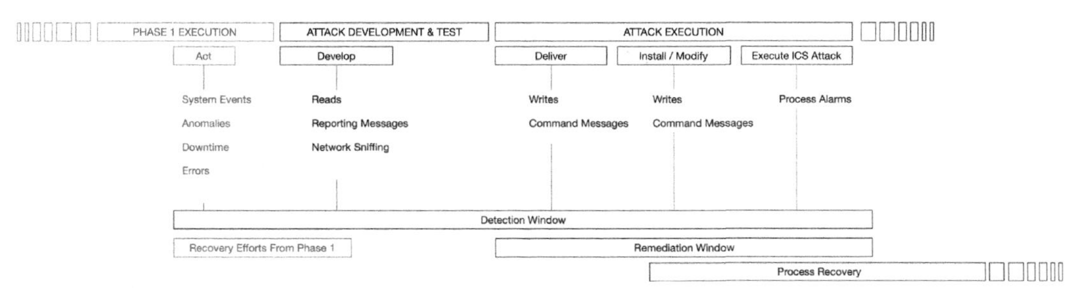

FIGURE 10.8 Defensive tactics mapped loosely against the industrial phase of the cyber-physical attack lifecycle.

Identify

The first function of defense as defined by NIST is "identify," and like most of the functions in NIST CSF is covers a lot: identifying threats, vulnerabilities, risk, assets, users, policies, etc, in order to "develop an organizational understanding to manage cybersecurity risk."[24]

To facilitate the target outcomes of this function (asset management; business environment; governance; risk assessment; and risk management strategy[25]):

- Maintain an inventory of all assets. There are many commercial products that claim to automate this process. These tools can be useful for maintaining accurate inventories and detecting new assets that connect the industrial network; however, due to limitations in the efficacy of these tools, the initial inventory should be built by performing a manual assessment, and regular assessments should be performed periodically even if automated tools are in place.
- Perform periodic vulnerability assessments, prioritizing any systems that could enable access to industrial control assets. Industrial control systems, once accessed by an attacker, do not require specific exploits in order to perform industrial-phase cyber-physical attacks—the attacker can simply use the control system as designed. Cyber-physical threat modeling will help identify the assets to prioritize.
- Threat assessment can be challenging and multifaceted. For purposes of broader policies and identification of risk, a quantifiable understanding of the threat landscape is required. This could include participation in information sharing initiatives, subscription to threat intelligence programs, and similar strategic efforts. However, understanding threats is also required a tactical level, which requires a real-time understanding of the cyberthreats facing your environment. This requires at least some degree of threat detection capability as well. While the CSF breaks out detection as a discreet category, the results of detection efforts need to be considered as part of the overall identification of risk.
- For detailed guidance on risk assessment and risk management in industrial systems, please refer to Chapter 8, "Risk and Vulnerability Assessment."

Protect

Protection requires the implementation of various safeguards to minimize and contain the impact of a potential cybersecurity incident. These safeguards range from: administering policies (training personnel to increase cybersecurity awareness, implementing procedures to improve cybersecurity hygiene and best practices, etc.), to maintenance efforts, which can improve system resilience, to implementing hard cybersecurity controls (identity management, access controls, antimalware technology, network security controls, etc.), which can potentially interfere with interrupt the cybersecurity attack process.[26] This stage can be challenging because it uses the term "protective technology" very broadly, and there is a vast and diverse market of cybersecurity controls that are available to chose from. Many organizations question which protective technologies they should invest in. The only correct answer is "whichever control(s) are needed to progress from the current level of cybersecurity protection to the desired level of cybersecurity protection." This requires awareness of where an organization is on their individual cybersecurity journey and requires ongoing assessment of cybersecurity risk (see Chapter 8, "Risk and Vulnerability Assessment").

To facilitate this stage:

- Harden assets to minimize the attack surface. This means uninstalling unnecessary applications, disabling unused ports and services, and configuring each asset with the end goal of limiting the functionality of that asset to *only* its intended purpose. The Center for Internet Security has published numerous benchmarks to facilitate hardening efforts, available at https://www.cisecurity.org/cis-benchmarks
- Establish zones and conduits to isolate systems based on security levels. The greater the degree of network segmentation that is implemented, the more likely that a cybersecurity event can be effectively contained.
- Implement strong identity management and access control (IMAC). The best segmentation efforts provide little benefit without also having enforceable access controls in place. Consider a "zero trust" model where all access is denied by default: any access to networks, assets, and even specific applications can be controlled and managed on a per-user basis.
- Active cybersecurity controls should be used to enforce policies. This should be considered before implementing "protective technology," to ensure that technology investments are aligned with actual needs.
- Also consider other aspects of the CSF when planning protective technology investment. Does a specific cybersecurity control support detection efforts? Does it provide information about the infrastructure that can contribute to identification efforts? Does that same information facilitate response and recovery efforts?

Detect

Detection is critical to observing active stages of a cyberattack. The earlier that detection occurs in relation to the attack lifecycle, the more time is available for the defender to recover and respond. The target outcomes of this function include anomalies and events, security monitoring, and detection processes,[27] which are all discussed in detail in Chapters 12, 13, and 11, respectively.

When considering detection functions, remember the following:

- Detection of active threat activity requires a combination of people, process, and technology. Technology is required to detect intrusion attempts, malware activity, anomalies, process deviations, etc. Without processes in place to support this technology, and without the people in place to follow through on those policies, there is little value to be realized
- Detection efficacy varies, but is never 100%. Always assume that there is threat activity occurring, but that it is not being detected.

- Detection technology can support other areas of the CSF by providing valuable data points, but only if implemented and operationalized correctly. Always ensure that logging and alerting features are fully enabled, and that the event data that is produced is fully utilized (e.g., collected, analyzed, monitored, and stored appropriately).
- Anomaly detection solutions that look for abnormal industrial protocol behavior are extremely popular at the time of publication. While useful, these solutions will typically produce high rates of false positives. Without adequate resources to review the results, these solutions can be problematic.
- It is easy to think of "detection" technology only within the context of point solutions for detecting malware on a host, or intrusions on a network, etc. However, SIEM, XDR, SOAR, and other solutions are also capable of detecting more complex threats that make up an attack campaign, and some may be able to "connect the dots" to detect campaigns in their entirety. This will facilitate response and recovery efforts.
- The ability to map threat data to the MITRE ATT&CK framework is also extremely useful to support response efforts.

Respond

Response can begin as soon as there is something to respond to. This could theoretically include responses to intelligence gathered during the "identify" stage or responses to weaknesses discovered during the "protect" stage, but it is most commonly in response to activity detected during the "detect" stage. The outcomes defined here include more than what many think of as "incident response" efforts, however: response planning, analysis, and improvements are some of the additional identified outcomes defined by NIST.[28] This supports the framework's overall objective of supporting continuous improvement.

To facilitate this stage:

- Remember that segmentation makes containment easier, enabling compromised systems to be quarantined on the network. Dynamic segmentation (any segmentation that can be reconfigured on a production system without interruption) allows this with minimal (if any) impact on network administration.
- Practice! Response plans that are not put into practice have minimal value. Response exercises should be performed regularly. Introduce unexpected variations in response exercises so that responders are ready to overcome any challenge.
- Understand that even with a well-trained, practiced team, a situation might occur that is beyond their capability. Understand the limits of response personnel and establish an escalation process for these situations. This could include a contract with a dedicated third-party incident response provider, contact procedures for relevant government agencies, etc.
- Document everything: communications, analysis, and improvement efforts can benefit from proper documentation, in particular.

Recover

The recovery process is all about resiliency. Plans for how to recover, training on how to execute those plans, and how to improve associated processes as a result of recovery efforts all fall under the umbrella of the "recover" function.[29]

To facilitate in this stage:

- Remember that recovering servers and workstations is only part of the process. If an industrial attack reached the second phase (the industrial phase) of the kill chain, the process may have been impacted, altered, or even stopped. Process recovery and validation efforts should be factored into the recovery process.
- Validate backups. Recovery without validated backups requires a lot more time and effort. Back up critical systems regularly, validate the backups, and practice the "3-2-1" backup strategy: keep at least three copies of your data; store two copies on different storage media; and store one copy off site.
- Remember that recovery of an industrial automation process consists of far more than bringing servers and workstations back online. Depending on the process, and how and when it was disrupted, there could be larger recovery efforts required, including: replacement of damaged assets, validations of process logic, potential engineering changes, environmental cleanup efforts, etc.

The importance of understanding lifecycles

Attack lifecycles are important to understand, as they directly impact (and are in turn directly impacted by) an organization's defensive efforts, in the "dance" of attack and defense efforts referenced at the beginning of this chapter. As a defender, a primary goal is to minimize the MTTR.

This could be accomplished by:

- Preventing an attack from occurring in the first place, and thereby eliminating the need to recover.
- Prevent an attack from executing successfully
- Detecting an attack in the early stages, in order to mitigate the threat before it can progress. For example, by detecting a threat at the "intrusion" stage, and mitigating the threat before it can become "enabled" (establish C2, exfiltrate data, deploy new payloads, etc.) impact of the "execution" phase can be minimized or avoided altogether.
- Detecting an attack in later stages of the initial phase of a cyber-physical attack can determine the target(s) of the industrial phase and help mitigate the threat to the process.

While the NIST framework is designed for continuous improvement across all functions[30], within the context of an ongoing attack, there are certain activities and outcomes that are reactive in nature: it is not possible to detect threat activity until that activity is performed; it is not possible to respond to an incident that has not happened, and it is not possible to recover without first experiencing a failure. Figures 10.7 and 10.8, illustrate this, showing how the attack lifecycle maps to detection, remediation, and recovery windows. However, the framework does offer guidance on how to perform these functions proactively in support of continuous improvement. Many outcomes of the NIST functions—such as vulnerability identification, hardening, response planning, training exercises, system backups, etc.—can and should be performed during peace time. It is useful to think about these defensive functions within the narrower context of an ongoing attack because it helps visualize the effort required to respond to an attack, which helps understand mean time to recovery (MTTR) and the need to minimize it.

Minimizing MTTR

To minimize overall MTTR, we can focus on three areas where a defensive action is taken in direct respond to an offensive action. In a 2021 study published by the SANS Institute, these areas are identified as compromise to detection (CtD), detection to containment (DtC), and containment to remediation (CtR). Consider a simplified version of the overall attack process:

1. First the attacker must compromise the industrial target. This is the execution of the initial attack phase that gains entry to the industrial network and begins the second phase of the ICS Kill Chain.[31]
2. The attacker then needs to deliver the initial payloads necessary to further develop the targeted industrial attack.[32]
3. Finally, the attacker leverages these payloads to refine their attack, enabling the execution of the targeted ICS attack.[33]

At the same time, defenders must:

1. Detect the initial compromise.[34]
2. Identify the full extent of the compromise and contain it.[35]
3. Remediate the threat once it has been contained.[36]

Remembering that these efforts take time for both attackers and defenders, it can be seen how the minimization of the time it takes between the attacker's first action (compromise) and the defender's first action (detection) could change the overall outcome of the attack. If the CtD time is immediate, the defender can begin containment efforts early, while the attacker is still attempting to develop an effective targeted attack. If the time it takes for the defender to contain the attack once it has been detected, is also immediate, the attacker could be prevented from execution of the final industrial attack. In other words, the faster the defender can react, the greater the chance of breaking the ICS attack kill chain.[37]

Unfortunately, the same study showed that average reaction times were less than optimal: 42% of surveyed operators took more than 2 days to detect an initial compromise, with 23% taking longer than one week, and 14% taking longer than 1 month. Similarly, the time from detection to containment took more than 2 days in 31% of cases, and more than a week in 9% of cases.[38] It can be clearly seen that once a threat is detected, containment can be achieved relatively quickly, while the initial detection can take longer.

SANS further quantified these limitations by identifying the types of industrial assets with the highest perceived risk, and comparing that to the types of data most commonly collected for cybersecurity analysis. The results showed that the highest risk areas had the least amount of data collection, with one exception: Servers running commercial operating systems had the highest instances of data collection at 76%. However, connections to other internal systems, connections to field control networks, and embedded controllers and components had poor collection rates at 54.5%, 38.9%, and 18.7%, respectively.[39]

This is not surprising: there are numerous information security tools available commercially that facilitate data collection from servers running common operating systems; network data collection is slightly more difficult within and between industrial networks due to the prevalence of legacy infrastructure; and embedded devices such as PLCs and IEDs can be extremely difficult to collect data from, and in some cases, these types of assets may not produce security log or event data at all.

Improving detection rates may therefore require the use of specialized monitoring tools. Some examples include:

- Control system alarm management systems that are able to produce data related to process variations or errors that may be relevant to cyber security efforts
- Network monitoring tools deployed inline or on a network span or tap, that are capable of producing network flow data that may be relevant to cybersecurity efforts
- Network monitoring tools deployed inline or on a network span or tap that are capable of identifying process control activity and detect anomalies that may be relevant to cybersecurity efforts
- Host cybersecurity controls that are able to produce highly relevant cybersecurity event data

Unfortunately, the specific cybersecurity controls and detection technologies that are required to close the compromise to detection gap can be difficult to implement in an industrial network. Once implemented, the data generated by these tools can be difficult to collect and analyze in a manner that is useful to industrial operations.

Summary

Attacks against industrial systems require many steps to be successful, and likewise, there are multiple steps that defenders must take when attempting to minimize (or eliminate) the success of an attack. Understanding these individual lifecycles and how they are intertwined is extremely beneficial. Through continuous improvement of cybersecurity functions, the job of the defender will be easier overall. Through the efficient response to specific stages of an attack, threats can be further minimized when they occur, and the overall response and recovery times can be minimized.

Endnotes

1. Val Saengphaibul. *A Brief History of The Evolution of Malware.* Fortigaurd Labs Trheat Research. Fortinet. March 15, 2022.
2. Computer History Museum. *What was the First* PC? 1986.https://www.computerhistory.org/revolution/personal-computers/17/297
3. Michael J. Assante, Robert M. Lee. *The Industrial Control System Cyber Kill Chain.* SANS Institute. 2021.
4. Ibid.
5. Eric M. Hutchins, Michael J. Cloppert and Rohan M. Amin, Ph.D., "Intelligence-Driven Computer Network Defense Informed by Analysis of Adversary Campaigns and Intrusion Kill Chains". https://www.lockheedmartin.com/content/dam/lockheed-martin/rms/documents/cyber/LM-White-Paper-Intel-Driven-Defense.pdf
6. Ibid.
7. Michael J. Assante, Robert M. Lee. *The Industrial Control System Cyber Kill Chain.* SANS Institute. 2021.
8. Ibid.
9. The MITRE Corporation. *MITRE ATT&CK Framework v13.1.* https://attack.mitre.org
10. Otis Alexander, Misha Belisle, Jacob Steele. *MITRE ATT&CK for Industrial Control Systems: Design and Philosophy.* The MITRE Corporation. Project No 01DM105-OT. March 2020.
11. The MITRE Corporation. *MITRE ATT&CK Framework v13.1.* https://attack.mitre.org
12. Michael J. Assante, Robert M. Lee. *The Industrial Control System Cyber Kill Chain.* SANS Institute. 2021.
13. Ibid.
14. Ibid.
15. Ibid.
16. Ibid.
17. Ibid.
18. GARD Threat Research. *2023 USB Industrial Threat Report.* Honeywell, Inc. September 2023.
19. Ibid.
20. Michael J. Assante, Robert M. Lee. *The Industrial Control System Cyber Kill Chain.* SANS Institute. 2021.
21. Ibid.
22. Cherepanov, Anton. "Industroyer: Biggest threat to industrial control systems since Stuxnet". www.welivesecurity.com. ESET. June 17, 2017.
23. Michael J. Assante, Robert M. Lee. *The Industrial Control System Cyber Kill Chain.* SANS Institute. 2021.
24. National Institute of Standards and Tecehnology (NIST). Framework for Improving Critical Infrastructure Cybersecurity. Version 1.1. April 16, 2018.
25. Ibid.
26. Ibid.
27. Ibid.
28. Ibid.
29. Ibid.
30. Ibid.
31. Michael J. Assante, Robert M. Lee. The Industrial Control System Cyber Kill Chain. SANS Institute. 2021.
32. Don C. Weber. Responding to Incidents in Industrial Control Systems: Identifying Threats/Reactions and Developing the IR Process. SANS Institute. May 2020.
33. Ibid.
34. Ibid.
35. Ibid.
36. Ibid.
37. Ibid.
38. Ibid.
39. Ibid.

11

Implementing Security and Access Controls

Information in this chapter

- Network Segmentation
- Implementing Network Security Controls
- Implementing Security Controls on Removable Media
- Implementing Host Security and Access Controls
- From Theory to Practice
- How Much Security is Enough?

Once security zones and the associated conduits connecting these zones have been defined (see Chapter 9, "Establishing Zones and Conduits"), they now need to be properly secured according to the target security level identified. A "zone" is nothing but a logical construct without proper network segmentation and access controls. A "zone" represents a logically and often times physically isolated network of systems that, when proper network segmentation and access controls are in place, will by its nature be more difficult to breach from an outside threat agent, and will better contain incidents in the event a breach does occur.

The process of securing zones can be summarized as follows:

1. Map the logical container of the zone against the network architecture, so that there are minimal network paths or communication channels into and out of each zone. This is effectively creating a zone "perimeter" and from this, "entry/exit points" are identified.
2. Make any necessary changes to the network so that the network architecture aligns with the defined zones. For example, if two zones currently coexist within a flat network, segment the network in order to separate the zones.
3. Document the zone for purposes of policy development and enforcement.
4. Document the zone for purposes of security device configuration and monitoring.
5. Document the zone for the purposes of change management.

In some instances, such as the one illustrated in Figure 11.1, a single zone may consist of multiple, geographically or otherwise separated groups (e.g., by business function). In these cases, the zone is still considered to be a single zone. If there are any network connections between the two (or more) locations, they should be held to the same security requirements (meaning the use of the same set of controls) as the rest of the zone.

Industrial Network Security. https://doi.org/10.1016/B978-0-443-13737-2.00012-9

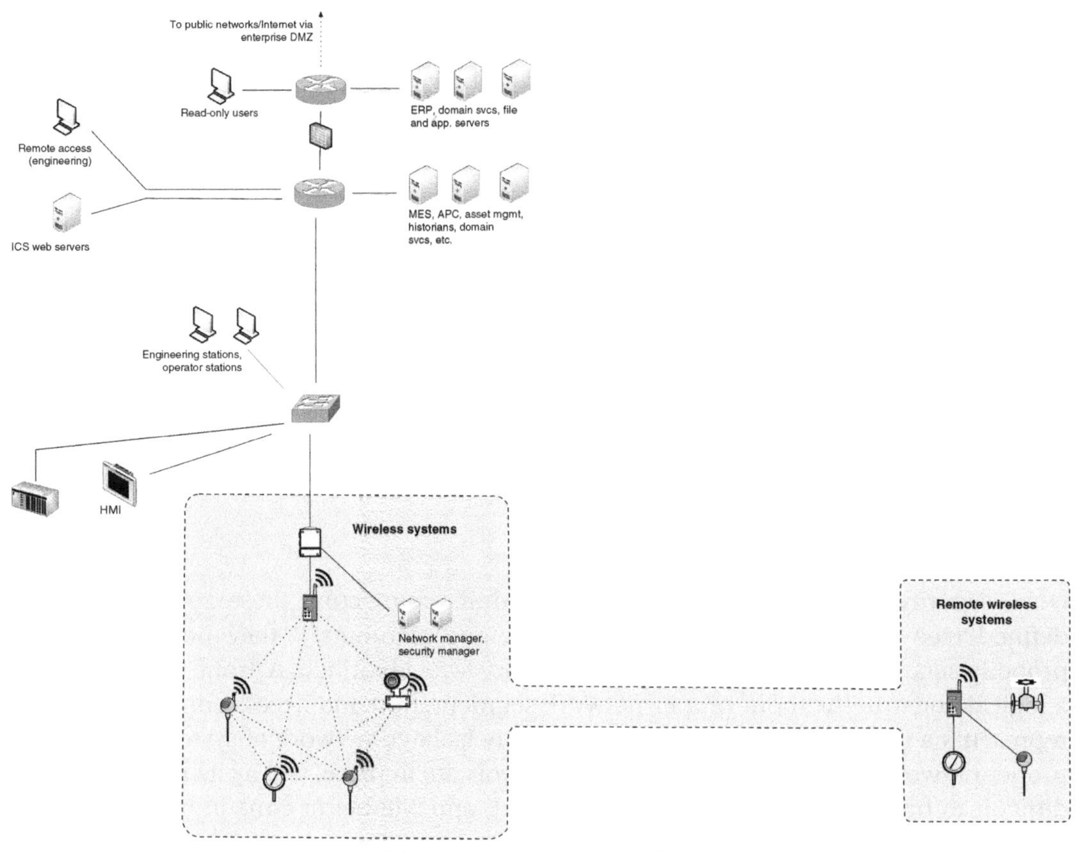

FIGURE 11.1 A geographically split zone.

That is, there should be no communication across those links that do not originate and terminate within the zone, and if outside communication is required (i.e., a communication that either originates or terminates outside of one of the two zones), it must occur through defined and secure access points (note: this is referring to a general point of access, and not a "wireless access point" or WAP). One common method of interconnecting distributed zones is the use of a dedicated virtual private network (VPN) or other encrypted gateways that provide secure point-to-point communications. A dedicated network connection or fiber cable may be used to interconnect extremely critical zones so that physical separation is maintained.

The goal is that each zone be isolated as strictly as possible, with as few conduits as possible between that zone and any other directly adjacent (or surrounding) zone. Figure 11.2 shows how, by providing a single access point in and out of a zone, that point can be secured using a perimeter security device, such as a next-generation firewall. In the event of a single zone that is split (geographically or by another zone), intrazone communication that must traverse another zone can still be allowed—in this case

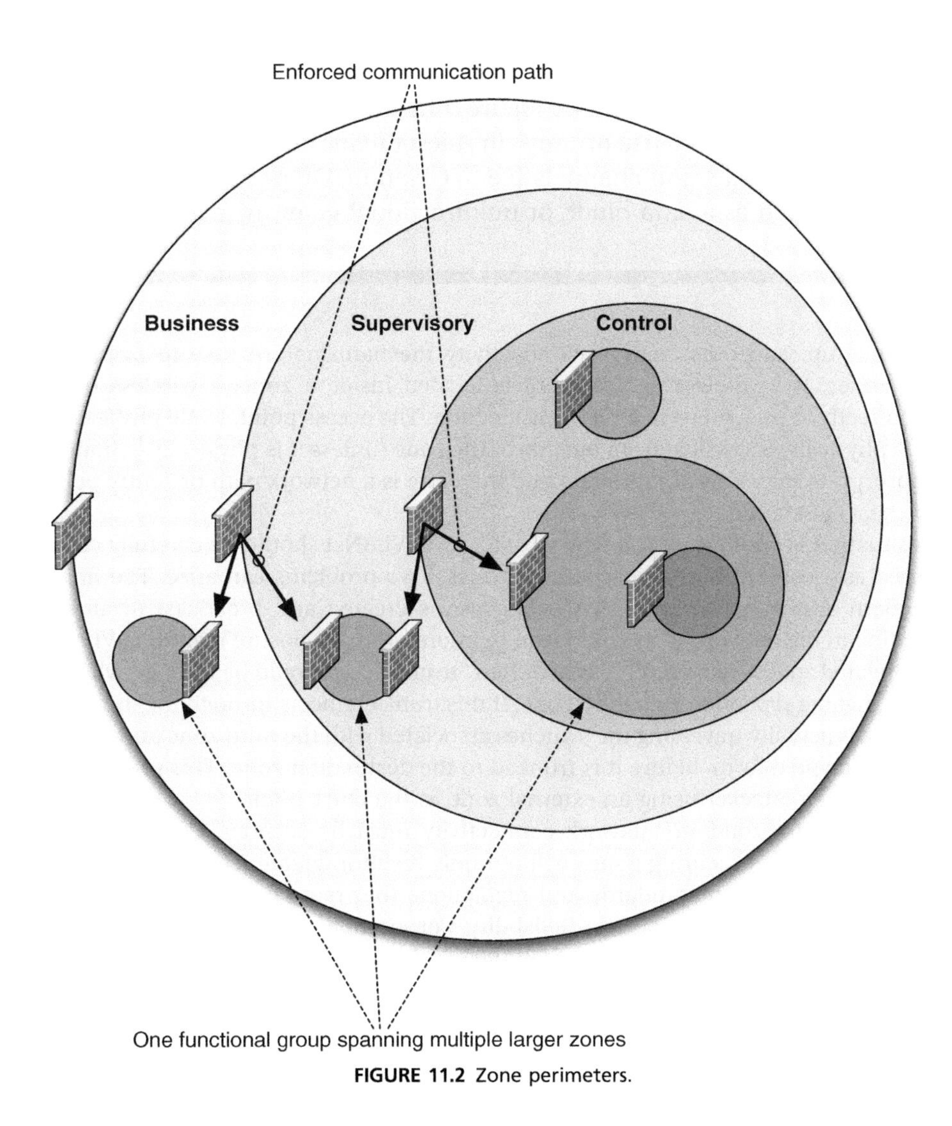

FIGURE 11.2 Zone perimeters.

through the use of VPNs or other encrypted network access control to enforce a point-to-point route between the split zone.

In scenarios where a zone needs to be extended across another zone boundary (i.e., there are two overlapping zones), consider the functional goals of that extension. For example, in many cases, a business user may require access to information originating from within a secure SCADA zone. However, there is no requirement for the business user to communicate back into the SCADA environment. In situations like these, the use of a "semi-trusted" or demilitarized zone (DMZ) is recommended, and the use of strong access controls, such as one-way communications, should be considered to prevent network flows from the less-secure or "untrusted" zone(s) to the more secure "trusted" zone(s). One-way communication can be enforced by provisioning network security controls (e.g., the firewalls shown in Figure 11.2) to disallow inbound traffic. These controls should minimize the use of "any" in ruleset fields and specifically define host IP addresses and communication channels (i.e., TCP and UDP ports). A dedicated network security control, such as a data diode or unidirectional gateway, can also be deployed.

Tip

Wireless, dial-up, and other remote connectivity mechanisms are easy to overlook when securing zones. If a wireless access point is located inside a zone, a wireless user could connect directly to that zone via a Wi-Fi connection. The access point, while physically inside a zone, is physically accessible from outside of the zone (unless it is physically contained with signal absorption materials or jammers) and therefore is a network path or "entry point" that must be heavily secured.

This situation is another reason why virtual LANs (VLANs) should be carefully considered when used as a conduit between separated zones. Two problems can arise. The first is that with modern switch networks, a VLAN database is created and broadcast to all switches participating in the network. This could lead to information disclosure regarding VLAN IDs in use in unrelated zones. Second, VLANs are often "trunked," as would be the case when joining two zones that are separated by a third zone. If this trunk connects through the third zone, the VLAN traffic is actually traversing the switches associated with the third zone and is not in any protected/encrypted form, before it is trunked to the destination zone. This provides an easy entry point for an attacker using an external zone as the entry point.

When securing a zone, *all* network connectivity must be secured. Consideration of all remote entry points in securing zones will not only result in greater security, but it will also facilitate compliance with standards and regulations that require network access controls, such as the North American Electric Reliability Corporation (NERC) Critical Infrastructure Protection (CIP) regulatory requirement CIP-005-3a R1.1, which dictates that "access points to the Electronic Security Perimeter(s) shall include any externally connected communication end point (for example, dial-up modems) terminating at any device within the Electronic Security Perimeter(s)."[1] This requirement has been expanded in CIP-005-5 R1[2] to include additional measures for inbound and output access to the ESP.

Network segmentation

In accordance with the Principle of Least Route (see Chapter 5, "Industrial Network Design and Architecture"), a device that does not physically belong to a zone should not be allowed to directly connect to that zone or to any device within that zone. This is the primary reason networks consist of one or more semi-trusted "DMZs" that act as an intermediate connection between the devices that possess both similar and different functional goals while residing in two different zones (i.e., a business user needing ICS historical data).

In many cases, there will be secondary devices identified that have access to or are connected to a zone, such as a printer or storage device that may provide network connectivity. An example is a network printer that has a Wi-Fi interface, which may be enabled by default. These aberrations are easy to overlook but must be addressed if the zone is to be secured. This is one reason that thorough security risk and vulnerability assessments need be performed (see Chapter 8, "Risk and Vulnerability Assessments").

It may not be possible in other cases to clearly identify the boundaries of a zone in terms of network design. For example, if supervisory, control, and enterprise systems are all interconnected via a flat network (a network that is switched purely at Layer 2, without network routing) or a wireless network, it will not be possible to isolate zones through subnetting. In these cases, some other means of logical network segmentation must be used. For example, VLANs could be used to separate devices that are in different zones by segmenting the network at Layer 2 of the Open Systems Interconnection (OSI) model. Another approach could be to implement a technology known as "variable-length subnet masking" (VLSM), which manipulates the Subnet Mask and Default Gateway parameters of a network interface restricting those devices that can actually communicate at the network layer (OSI Layer 3) without introducing any new Layer 3 devices. Alternately, a next-generation firewall could be used on the conduit between zones to segment the devices at Layer 7 of the OSI model. Each has its strengths, and ideally zone separation, should be enforced at all seven layers; if budgets and operational overhead were of no consideration, this might even be possible. Realize that the use of VLANs and VLSM only provide moderate levels of cybersecurity defense as described in Chapter 5, "Industrial Network Design and Architecture," and is not recommended for networks that require higher levels of security typically accomplished using physical segmentation mechanisms.[3]

The following method is effective for zone separation:

- Identify and document all network connections into or out of each zone (i.e., identify entry/exit points that form conduits).
- For each conduit
  - Start at Layer 1 (the physical layer) and work up to Layer 7 (application layer).
  - For each layer, assess if network segmentation at this layer is feasible for that conduit (see Chapter 5, "Industrial Network Design and Architecture" for details on segmenting networks at different layers).
  - For more critical conduits, aim for greater segmentation—enforce network segmentation through the use of a mixture of Layer 1 data diode or unidirectional gateway, Layers 3–4 switching *and* application segmentation, and next-generation firewalls at Layers 5–7.
  - For each desired layer of segmentation, implement appropriate network security and access controls to enforce that segmentation.
  - Provide sufficient monitoring capabilities with each security control deployed to support event consolidation and reporting mechanisms to assist in potential security breaches.

Zones and security policy development

A distinct milestone is reached once zones and conduits are defined, and the necessary adjustments to the network architecture are made. With defined zones and conduits in place, the organization is armed with the information needed to satisfy several compliance requirements of NERC CIP, Chemical Facility Anti-Terrorism Standards (CFATS), and so on, plus other industry-recognized standards like ISO 27000 and ISA 62443.

Documenting all zones within the context of the organization's security policy provides many benefits, by clearly identifying what systems may be accessed by what other systems, and how. These access requirements will facilitate policy documentation for compliance, security training and review materials, and similar security policy functions required by NERC CIP-003-3,[4] ISA 62443-3-3 FR-5,[5] CFATS Risk-Based Performance Standards Metric 8.2,[6] and Nuclear Regulatory Commission (NRC) 10 CFR 73.54/NRC RG 5.71 section C.3.2.[7]

Documentation of zones also defines how ongoing security and vulnerability assessments should be measured. This is again useful for compliance, including NERC CIP 007-3a R8,[8] ISA 62443-2-1,[9] CFATS Risk-Based Performance Standards Metric 8.5,[10] and NRC CFR 73.54/NRC RG 5.71 section C.13.[11]

Using zones within security device configurations

Documentation can be a function of security as well as compliance. Firewalls, intrusion detection and intrusion prevention systems (IDS/IPS), security information and event management (SIEM) systems, and many other security systems support the use of variables, which are used to map hard security configurations to organizational security policies.

For each zone, the following list should be maintained at a minimum:

- Devices belonging to the zone, by IP address and preferably by MAC address as well.
- Software inventory for devices contained within the zone including basic platform applications (operating system, common support tools, etc.) and specialized applications (ICS applications, configuration tools, device drivers, etc.).
- Users with authority over the zone, by username or other identifier, such as active directory organization unit or group.
- Protocols, ports, and services in use within the zone.
- Technologies that are specifically forbidden from deployment within the zone, such as cloud-based applications that must communicate with disallowed zones, legacy operating systems, insecure wireless technologies, and automated port scanning tools to name a few.

If additional metrics are identifiable, additional lists should be created. Depending on the number of zones that have been defined, this may require several lists—five (device, users, applications, ports/services, technologies) for every established zone. Additional lists could also be maintained; for example, users by shift or users by computer, in addition to users defined solely by zones. However, unless there is a centralized authentication system in use, maintaining these lists may be cumbersome and could increase the likelihood of a misconfiguration being overlooked.

When finished, these variables will appear as follows:

```
$ControlSystem_Zone01_Devices
  192.168.1.0/24
  10.2.2.0/29
  $ControlSystem_Zone01_Users
  jcarson
  jrhewing
  kdfrog
  mlisa
  $ControlSystem_Zone01_Applications
  VendorA SCADA Server - Release 110.1.3
  VendorA SCADA HMI - Release 110.1.3
  VendorA SCADA Engineering Tools - Release 110.1.5
  VendorB Historian - Release 5.1.7
  $ControlSystem_Zone01_PortsServices
  TCP 502 #Modbus TCP
  TCP 20000 #DNP3
  TCP 135, 12000-12100 #RPC/OPC
```

The creation of these variables will assist in the creation of firewall and IDS rules for the enforcement of the zone's perimeter, as discussed under "Implementing Network Security and Access Controls," and will also allow for security monitoring tools to detect policy exceptions and generate alarms, as discussed in Chapter 13, "Security Monitoring of Industrial Control Systems."

Note

In this book, variables are defined using

```
var VariableName [value1, value2, value3, etc.]
```

and referenced using.

```
$VariableName
```

in line with standard Snort IPS/IDS rule syntax. However, depending on the device used, the specific syntax for defining and referencing variables may differ. For example, a variable is defined using Snort as follows:

```
ipvar ControlSystem_Zone01_Devices 192.168.1.0/24
```

Note the use of "ipvar" here, which is used to denote a variable containing IP addresses and lists. "portvar" is used to signify port variables and list, while "var" is used for other variable types.

The same example for an iptables firewall is defined within the iptables configuration file would be written as follows:

```
ControlSystem_Zone01_Devices=192.168.1.0/24
```

To define a useable variable that maps to a range of IP addresses that may further define a zone,

```
ipvar ControlSystem_Zone01_Devices [192.168.1.0/24, 10.2.2.0/29]
```

is used, and then that variable is referenced within a specific rule using.

```
$ControlSystem_Zone01_Devices
```

This is a logical extension of the classic $HOME_NET variable used in many IDS policies, only applied to a specific zone. This allows for exception-based detection of unauthorized behavior within the zone, as seen in the following rule header to detect any traffic with a destination IP of a device within the defined control system zone:

```
alert tcp any any -> $ControlSystem_Zone01_Devices any
```

It is also possible to use "negation" and signify all entities not contained in the variable, as seen in the following rule that will detect any traffic with a destination IP of a device within the defined control system zone and source IP that is "not" in the zone:

```
alert tcp !$ControlSystem_Zone01_Devices any ->
$ControlSystem_Zone01_Devices any
```

With zones defined, and relevant variables defined for each, the zones can now be secured using perimeter and host security devices. More details will be provided on variables later in section "Intrusion Detection and Prevention (IDS/IPS) Configuration Guidelines."

Implementing network security controls

Establishing network security to protect access to a defined zone is actually an enforcement of conduits. The rules used align with the communication channels contained within the conduit. Network security controls protect against unauthorized access to the enclosed systems and also prevent the enclosed systems from accessing external systems from the inside-out. To effectively secure inbound and outbound traffic, two things must occur:

1. All inbound and outbound traffic must be forced through one or more known network connections that are monitored and controlled.
2. One or more security devices must be placed in-line at each of these connections (this could be a security capability built into network communication switches and routers).

For each zone, appropriate security devices should be selected and implemented using the recommendations given next.

Selecting network security devices

At a minimum, some form of network firewall is usually required. Additional security—provided by IDS, IPS, and a variety of specialized and hybrid devices, such as Unified Threat Management (UTM) devices, Network Whitelisting devices, Application Monitors, and Industrial Protocol Filters—may be desired as well, depending upon the specific situation. Typically, the security level or criticality of the zone (see "Criticality") dictates the degree of security that is required. Table 11.1 maps the criticality of a zone to required security measures of NERC CIP and NRC CFR 73.54, as well as recommended enhancements to improve security beyond regulatory requirements.

Table 11.1 recommends that both a firewall and an IPS be used at each security perimeter. This is because firewalls and IPS devices serve different functions. Firewalls enforce what types of traffic are allowed to pass through the perimeter by what is called "shallow packet inspection." Intrusion prevention systems on the other hand perform "deep-packet inspection" (DPI) by closely examining the traffic that is allowed through in order to detect "legitimate" traffic with malicious intent—that is, exploit code, malware, and so on—that is transferred over allowed paths. Using both devices together provides two mutual benefits: first, it allows the IPS to perform inspection of the "content" of all traffic allowed in through the firewall; second, the firewall limits the allowed traffic based on the defined parameters of the security zone, freeing the IPS to focus its resources on just that traffic, and therefore enabling it to enforce a more comprehensive and robust set of IPS rules.

Again, if the network infrastructure supports microsegmentation, this process is greatly simplified. Because micro-segmentation enables more granular control over how devices, applications, and users can communicate, the logical grouping of devices into zones and conduits becomes an issue of device configuration. In addition, because many network devices that support microsegmentation are able to both microsegment traffic *and* implement specific security controls within- and between each microsegment, it also becomes easier to establish and enforce policies.

It is important to understand the distinction between "detection" and "prevention" in the context of intrusion prevention systems. Recall that the most important priorities of

Table 11.1 Perimeter Security Requirements by Criticality

Criticality	Required security	Recommended enhancements
4 (highest)	NRC CFR 73.54: Unidirectional perimeter, NERC CIP 005: Firewall or IDS or IPS	Application layer monitoring, Firewall, IDS and IPS
3	NRC CFR 73.54: Unidirectional perimeter, NERC CIP 005: Firewall or IDS or IPS	Application layer monitoring, Firewall, IDS and IPS
2	NERC CIP 005: Firewall or IDS or IPS	Firewall and IDS and IPS
1	NERC CIP 005: Firewall or IDS or IPS	Firewall and IPS
0 (lowest)	NERC CIP 005: Firewall or IDS or IPS	Firewall and IPS

industrial networks are availability and performance. In other words, the network cannot tolerate accidental dropping of packets between hosts that are located on levels low within the ISA 95 model (i.e., Levels 1–3). This would occur if the security device generates a "false positive" and mistakenly interprets a valid packet as invalid and blocks it from reaching its destination. However, this may not necessarily be the case between industrial and business zones (i.e., Levels 3 and 4). This is the reason IDS is the preferred security appliance within industrial zones (placed "out-of-band" to network traffic) and IPS is used between industrial and business zones, or between semi-trusted DMZs and untrusted business zones (placed "in-line" to all network traffic).

We have also learned that industrial protocols consist of common standards like Modbus and DNP3 but also depend heavily on vendor-specific proprietary protocols that have been optimized for a particular system. It is not common for major IT network security suppliers like Cisco, HP ProCurve, Juniper, Checkpoint, and others to offer solutions for industrial networks. So what options exist to implement advanced DPI analysis with industrial protocols? The answer is a new class of industrial security appliances that are industrial protocol aware and possess the capability to analyze and inspect both open and proprietary protocols. Companies supplying these devices include Tofino/Belden, Secure Crossing, ScadaFence, SilentDefense, and others. At the time this book was written, many other startups were in progress, and readers are encouraged to research the market thoroughly in order to fully understand all of the available options. In addition, OEM-branded solutions or recommended third-party solutions may be available from your control system vendors. Once an appropriate solution is selected and deployed, DPI can then be used to analyze specific industrial protocol functions. Figure 11.3 illustrates the increased security capability of firewalls, IDS/IPS devices, and application session monitoring systems.

In the most critical areas, application-layer session monitoring provides a valuable and necessary level of assurance, as it is able to detect low-level protocol anomalies (such as a base64-encoded application stream inside of an HTTP layer 4 80/tcp session, used by many APTs and botnets) and application policy violations (such as an unauthorized attempt to write a new configuration to a PLC). However, unless monitoring very simple application protocols where the desired contents are distinctly packaged within a single packet or frame, the application session must be reassembled prior to monitoring as illustrated in Figure 11.4.

The most stringent network security device may be the data diode, also referred to as a unidirectional gateway. A data diode is, very simply, a one-way network connection—often a physically restricted connection that uses only one fiber-optic strand from a transmit/receive pair. By only using TX optics on the source side, it is physically impossible for any digital communications to occur in a highly sensitive network area containing control system devices, while supervisory data may be allowed to communicate out of that highly secure zone into the SCADA DMZ or beyond. In certain instances, such as for the storage of highly sensitive documents, the diode may

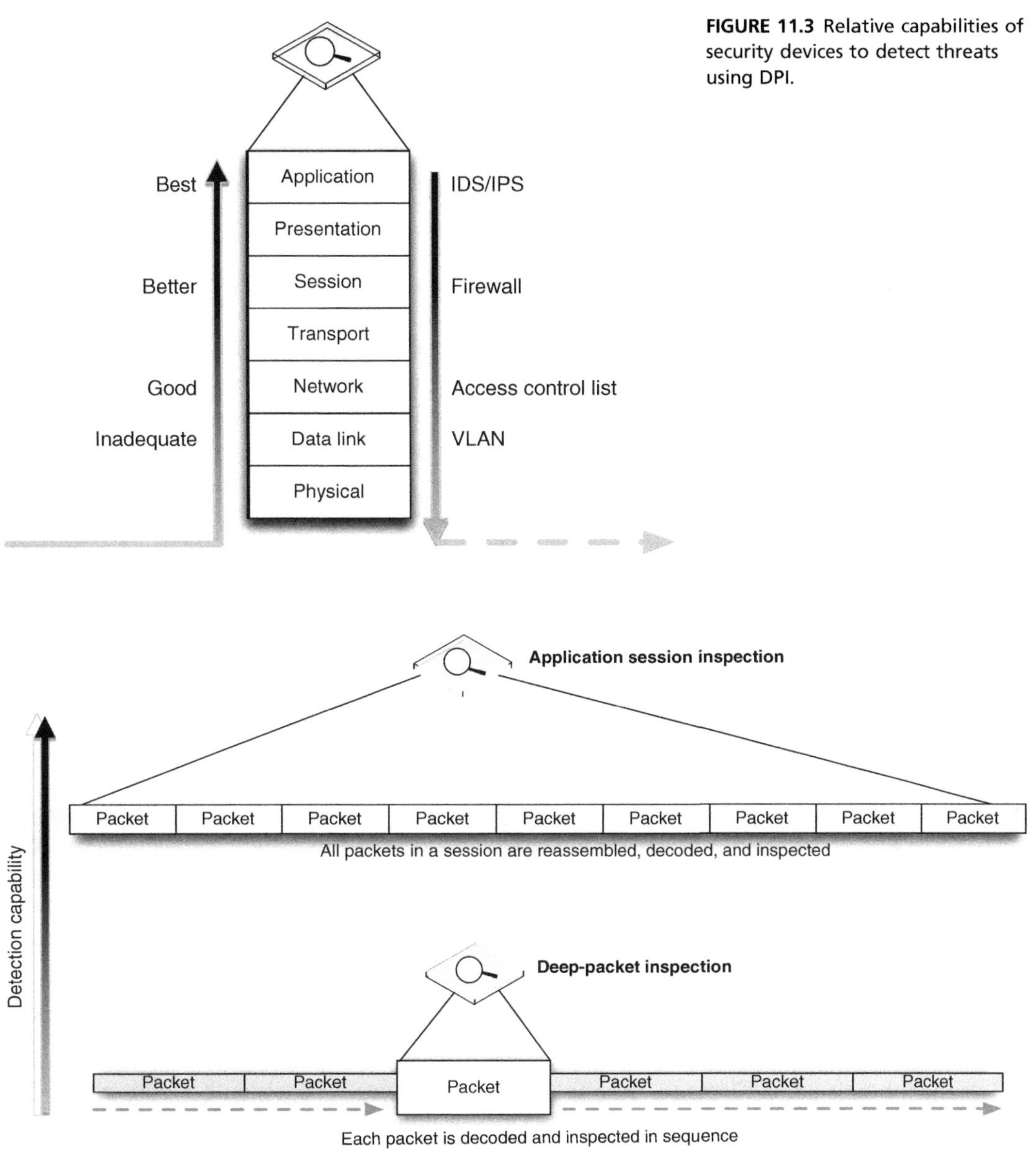

FIGURE 11.3 Relative capabilities of security devices to detect threats using DPI.

FIGURE 11.4 Application session inspection versus deep packet inspection.

be reversed, such that information can be sent into a secure zone that is then physically prevented from communicating that information back outside of the zone. During this "flip" phase, the previous communication flow should be terminated to disable any ability for two-way communication to occur at any point in time through the gateway.

Implementing network security devices

Once appropriate security product(s) have been selected, they must be installed and configured correctly. Luckily, the process of identifying, establishing, and documenting zones will simplify this process. The following guidelines will help to configure firewalls, IDS/IPS devices, and application monitors using the variables defined earlier under "establishing zones."

Firewall configuration guidelines

Firewalls control communication using a defined configuration policy called a "rule set," typically consisting of allow (accept) and deny (drop) statements. Most firewalls enforce a configuration in sequence (either by "lower-to-higher" number or simply from "top-to-bottom"), such that they start with a broadly defined policy, such as deny all, which will drop all inbound traffic by default. Once a packet has satisfied a given rule, no further processing occurs, making rule order very critical. These broad rules are tailored by adding before them subsequent, more focused rules. Therefore, the following firewall policy would only allow a single IP address to communicate outside of the firewall on port 80/tcp (HTTP).

```
Allow 10.0.0.2 to Any Port 80
Deny All
```

Had this rule order been reversed, starting with the "deny all" policy, no traffic would be allowed through the firewall, since all traffic would have been dropped by the first rule.

Note

Firewall rule examples are written generically so that they can be more easily understood. Depending on the firewall used, specific rule syntax may have to be used via a command-line interpreter, while others are configured exclusively via a graphical user interface.

Tip

A variety of tools are available to assist in firewall development consistently across multiple vendors, including the open-source package Firewall Builder. This allows the same GUI and syntax to be used when configuring multiple firewalls.

Note

Firewalls can restrict network access between interfaces using two primary actions: drop or reject. The exact form used in configuring firewalls typically depends on the interface monitored and the potential consequences of the denied traffic. When the "reject" form is used, the firewall actually sends a response back to the originating host informing it that the packet was rejected. This information can be very useful to a potential attacker as it signifies that a particular IP address or service port is actively being blocked and should not be used on untrusted interfaces. The "drop" form, on the other hand, simply discards the matching data and does not send any response back to the originator. This is a more secure mechanism, as the network-based attacker is no longer provided with any information that can be used to further enumerate the network in terms of devices, hosts, and available services.

Tip

Trying to become fluent in numerous firewall vendors' language and configuration tools can be discouraging. For this reason, it is strongly encouraged that generic rule visualization tools like Solarwind's Firewall Browser are used to allow firewall-specific configuration files to be parsed allowing rules and objects to be easily displayed and analyzed.

Determining what rules should be configured is typically easier in an industrial network because the nature of an industrial network is such that there is no need to accommodate the full diversity of applications and services typically found in an enterprise network. This is especially true when configuring a specific firewall against a specific zone-to-zone conduit—the zone will by its nature be limited in scope, resulting in concise firewall policies. In general, the more firewalls deployed on conduits, the simpler the configuration will be on each firewall. This is in contrast to attempting to utilize a single firewall (or firewall pair) and managing all rule sets on a single appliance.

The method of properly configuring a zone firewall is as follows:

1. Begin with bidirectional deny all rules placed at the end of the configuration
2. Configure specific exceptions, using the defined variables

   `$ControlSystem_Zone01_Devices`

   and

   `$ControlSystem_Zone01_PortsServices.`

3. Verify that all allow rules are explicitly defined—in other words, prevent the use of "Any" parameters for IP Address and destination Port/Service entries.

One simple way to configure a firewall is to follow the guidelines of the National Infrastructure Security Coordination Center (NISCC) "Good Practice Guide on Firewall Deployment for SCADA and Process Control Networks," using the defined zone variables as detailed in Table 11.2.[12]

Table 11.2 NISCC Firewall Configuration Guidelines With Zone Variables[36]

NISCC recommendations	Example rule using zone variables	Notes
Start with universal exclusion as a default policy Ports and services between the control system environment and an external network should be enabled and permissions granted on a specific case by case basis		Firewalls should explicitly deny all traffic inbound and outbound as the default policy. Comments used within the firewall configuration file can be used to document special cases, permissions, and other details.
All "permit" rules should be both IP address and TCP/UDP port specific, and stateful if appropriate, and shall restrict traffic to specific IP address or range of addresses	N/A	This guideline can be enforced by using to define rules.
All traffic on the SCADA and DCS network(s) are typically based only on routable IP protocols, either TCP/IP or UDP/IP; thus, any non-IP protocol should be dropped	N/A	By using within all defined rules, only protocols explicitly allowed within that zone will be accepted by the firewall, and all others will be dropped by the overarching deny all rule.
Prevent traffic from transiting directly from the process control/SCADA network to the enterprise network; all traffic should terminate in the DMZ		By configuring a rule on each zone that explicitly denies all traffic to and from any zone that is NOT a neighboring zone will prevent any transitive traffic. All traffic will need to be terminated and re-established using a device local to that zone.
Any protocol allowed between the DCS and the SCADA DMZ is explicitly NOT allowed between SCADA DMZ and enterprise networks (and vice versa)	At the demarcation between the enterprise network and SCADA DMZ: At the demarcation between the DCS and SCADA DMZ:	These rules enforce the concept of "disjointing" protocols, and further prevents transitive communication from occurring across a zone.
Allow outbound packets from the PCN or DMZ only if those packets have a correct source IP address assigned to the PCN or DMZ devices	N/A	Explicitly defined deny all rules combined with explicitly defined known-good IP addresses using $ControlSystem_Zone01_Devices ensures that all outbound packets are from a correct source IP.Firewalls may also be able to detect spoofed IP addresses. In addition, network activity monitoring using a network behavior anomaly detection (NBAD), security information and event management (SIEM), or log management solution may be able to detect instances of a known-good IP address originating from an unexpected device based on MAC address or some other identifying factor (see Chapter 13, "Security Monitoring of Industrial Control Systems")

Table 11.2 NISCC Firewall Configuration Guidelines With Zone Variables[36]—cont'd

NISCC recommendations	Example rule using zone variables	Notes
Control network devices should not be allowed to access the Internet	At the Internet firewall: Deny	Because all devices in all zones have been identified and mapped into variables, these devices can be explicitly denied at the Internet firewall.
Control system networks shall not be directly connected to the Internet, even if protected via a firewall	N/A	Using the zone approach, no control system should be directly connected to the Internet (see "establishing zones").
All firewall management traffic be: **1.** Either via a separate, secured management network (e.g., out of band) or over an encrypted network with two-factor authentication **2.** Restricted by IP address to specific management stations	N/A	This recommendation supports the establishment of a Firewall management zone using the methods described earlier under "identifying and classifying zones." by placing all firewall management interfaces and management stations in a zone, which is isolated from the rest of the network, the traffic can be kept separate and secured.

Intrusion detection and prevention (IDS/IPS) configuration guidelines

IDS and IPS devices inspect network traffic for signs of malicious code or exploits. Intrusion detection refers to passive inspection and is typically placed "out-of-band" of network flow. IDS and IPS examine traffic and compare it against a set of detection signatures, and taking some predefined action when there is a match. The main difference between the two lies in the actions allowed when there is a match. IDS actions can include Alert (generate a custom message and log the packet), Log(log the packet), and Pass (ignore the packet), while IPS actions can also include Drop (drop the packet and log it), Reject (drop the packet and initiate a TCP reset to kill the session), and Drop (drop the packet, but do not log it). In addition, both IDS and IPS rules can use the Activate and Dynamic actions, the former of which activates another rule, and the latter of which remains idle until activated by an Activate rule.[13]

An enabled collection of IDS/IPS detection signatures is referred to as an IDS/IPS policy, and this policy will dictate what types of threats may be detected by the device, as well as the degree and scope of events that will be generated. This collection should align with the list of threats and vulnerabilities that were previously defined for the security zone, as described in "Establishing Security Zones and Conduits" in Chapter 9. While active blocking of malicious traffic is important, the IDS/IPS events that are generated can also be analyzed to provide other important indicators—including attribution, network behavior, payloads, and larger threat incidents (see Chapter 13, "Security

Monitoring of Industrial Control Systems"). Signatures generally follow a format similar to a firewall rule, where there is an identified source and destination address and/or port—with the primary difference being the "action" that is performed in the case of a match. In addition, IDS/IPS signatures may match against specific contents of a packet, looking for patterns within the packet that indicate a known exploit (i.e., a "signature"). Common IDS/IPS signature syntax follows the de facto standards defined by Snort, an open-source IDS project owned by Sourcefire. An example signature is written as follows:

```
[Action] [Protocol] [Source Address] [Source Port] [Direction Indicator] [Destination Address] [Destination Port] [Rule Options]
```

which when written in correct syntax looks like

```
drop tcp 10.2.2.1 any -> 192.168.1.1 80 (flags: <optional tcp header flags>; msg: "<message text>"; content: <this is what the rule is looking for>; reference: <reference to external threat source>;)
```

To highlight the difference between a firewall rule and an IDS/IPS signature, consider the following example:

```
drop tcp 10.2.2.1 any -> 192.168.1.1 80
```

Without any rule options, the previous rule is essentially the same as the firewall rule Deny src-ip 10.2.2.1 dst-port any, which would block all traffic originating from 10.2.2.1 destined for IP address 192.168.1.1 on 80/tcp, effectively prevent that user from accessing web services on the destination (via HTTP on 80/tcp). However, the ability to match packet contents within the rule options enables an IDS/IPS device to control traffic at a much more granular level, such as

```
drop tcp 10.2.2.1 any -> 192.168.1.1 80 (msg: "drop http POST request"; content: "POST";)
```

This rule functions differently, only dropping traffic from the source address in question if the HTTP traffic contains a POST request (used by many web forms or applications attempting to upload a file to a web server over HTTP).

Note

IDS/IPS rule examples are written using Snort syntax, as it is the de facto signature creation language. However, many IDS or IPS devices support proprietary rule syntax, GUI rule editors, or other rule creation methods. Depending on the product used, the example rules in this book may or may not function as intended. All rules should always be tested prior to deployment.

Note

Snort is an open-source IDS/IPS developed by Sourcefire (acquired by Cisco in 2013) that combines signature, protocol, and anomaly-based inspection of network traffic with nearly 400,000 registered users.[14] In 2009, a nonprofit organization called the "Open Information Security Foundation (OISF)" released their first beta version of the Suricata next-generation IDS/IPS engine. This project, funded by the US Department of Homeland Security and a number of private companies, released the first stable version of Suricata in 2010, and continues to develop and evolve this product that offers direct interpretation of standard Snort rules.[15]

As with a firewall configuration, determining the exact IDS/IPS policy to be enforced is the first step in correctly configuring the device. The zone variables defined earlier under "establishing zones" are valuable tools that can be used to write succinct and highly relevant signatures. However, unlike a firewall that ends with a simple deny all rule, an IDS/IPS typically employs a default "allow all" rule, and therefore should be deployed "large"—with many active signatures—and then pruned back to the specific requirements of the zone. A method of properly configuring an IDS/IPS is as follows:

1. Begin with a more robust signature set, with many active rules.
2. If a protocol or service is not allowed in the zone, remove any specific detection signature associated with that protocol or service, and place with a broader rule that will block all traffic from that protocol or service (i.e., drop unauthorized ports and services) in the L3–L4 device (router or firewall) that exists upstream of the IDS/IPS.
3. If a protocol or service is allowed in the zone, keep all detection signatures associated with that protocol or service active.
4. For all active signatures, assess the appropriate action, using Table 11.3.
5. Keep all IDS signatures current and up to date.

Remember that an IDS or IPS can be used in a purely passive mode, to analyze traffic that is allowed, including traffic within a zone (that is, in the conduits between two devices within the same zone, that do not cross a zone perimeter). Passive monitoring will generate alerts and logs that can be useful in many security operations, including forensic investigations, threat detection, and compliance reporting (see Chapter 13, "Security Monitoring of Industrial Control Systems," and Chapter 14, "Standards and Regulations").

IDS/IPS rules should be tailored to the appropriate zone using the variables defined in Chapter 9 "Establishing Zones and Conduits." A typical Snort variable is established using the var command, as follows:

```
var VARIABLE_NAME <alphanumeric value>.
```

A specialized ipvar and portvar variable are used exclusively for IP addresses and ports, respectively.[16] In the zone method described earlier under "Establishing Zones," variables would be defined as

Table 11.3 Determining Appropriate IDS/IPS Actions

Allowed port or service?	Source	Destination	Criticality of service	Severity of event	Recommended action	Note
No	Any	Any	Any	Any	Drop	Any communication not explicitly allowed within the zone should be blocked to disrupt unauthorized sessions and deter an attack.
Yes	Trusted Zone	Trusted Zone	High	Any	Alert	Active blocking of traffic that originates and terminates within a zone could impact operations. For example, a false positive could result in legitimate control system traffic being blocked.
Yes	Trusted Zone	Trusted Zone	Low	Any	Alert or pass	For noncritical services, logging is recommended but not necessary (alert actions will provide valuable event and packet information that could assist in later incident investigations).
Yes	Untrusted Zone	Trusted Zone	High	Low (events from obfuscated detection signatures or informational events)	Alert	Many detection signatures are broad to detect a wider range of potential threat activity. These signatures should alert only to prevent unintentional interruption of control system operations.
Yes	Untrusted Zone	Trusted Zone	High	High (explicit malware or exploit detected by a precisely tuned signature)	Drop, alert	If inbound traffic to a critical system or asset contains known malicious payload, the traffic should be blocked to prevent outside cyber incidents or sabotage.
Yes	Trusted Zone	Semi-trusted Zone (explicitly allowed destination address)	Any	Any	Alert	This traffic is most likely legitimate. However, alerting and logging the event will provide valuable event and packet information that could assist in later incident investigations.
Yes	Trusted Zone	Untrusted Zone (unknown destination address)	Any	Any	Drop	This traffic is most likely illegitimate. Generated alerts should be addressed quickly: If the event is a false positive, necessary traffic could be unintentionally blocked; if the event is a threat, it could indicate that the zone has been breached.

```
ipvar ControlSystem_Zone01_Devices [192.168.1.0/24, 10.2.2.0/29]
var ControlSystem_Zone01_Users [jcarson, jrhewing, kdfrog, mlisa]
portvar ControlSystem_Zone01_PortsServices [502, 135,
12000:12100]
```

These variables can then be used extensively throughout the active detection signatures. For example, a signature designed to detect a known SCADA buffer overflow attack that is available within the Metasploit framework might appear as follows:

```
   alert tcp !$ControlSystem_Zone01_Devices any ->
$ControlSystem_Zone01_Devices 20222 (msg: "SCADA ODBC Overflow
Attempt"; content: <REMOVED - long string in the second
application packet in a TCP session>; reference:cve,2008-2639;
reference:url,http://www.digitalbond.com/index.php/research/ids-
signatures/m1111601/; sid:1111601; rev:2; priority:1;)
```

Note

Many Snort rules reference the $HOME_NET or $MY_NET variable. The use of multiple $ControlSystem_Zone01_Devices variables (one for each defined zone) accomplishes the same purpose, effectively defining a unique $HOME_NET for each zone. The nomenclature of $ControlSystem_Zone01_Devices is deliberately verbose in order to easily identify the variable's contents, so that the examples within this book are easier to understand.

Additional examples include signatures designed to specifically block known infection vectors used by Stuxnet.[17] The first example looks for one of the early delivery mechanisms for the Stuxnet malware that utilized a shortcut image file delivered via a WebDav connection. The second example detects Siemens WinCC connection attempts by logging into the WinCC database via a specific username and password combination, used in early Stuxnet propagation phases:

```
   tcp !$ControlSystem_Zone01_Devices $HTTP_PORTS ->
$ControlSystem_Zone01_Devices any (msg: "Possible Stuxnet
Delivery: Microsoft WebDav PIF File Move Detected";
flow:from_server; content: "MOVE"; offset:0; within:5; content:
".pif"; distance:0; classtype:attempted-user; reference:cve,
2010-2568; reference:osvdb,66387; reference:bugtraq,41732;
reference:secunia,40647; reference:research,20100720-01;
sid:710072205; rev:1;)

tcp any any -> any 1433 (msg: "Possible Stuxnet Infection:
Siemens Possible Rootkit.TmpHider connection attempt";
flow:to_server; content: "Server=|2e
5c|WinCC|3b|uid=WinCCConnect|3b|pwd=2WSXcder";
classtype:suspicious-login; reference:cve,2010-2772;
reference:osvdb,66441; reference:bugtraq,41753; sid:710072201;
rev:2;)
```

Recommended IDS/IPS rules

Basic recommendations for IDS/IPS configuration include active rules to:

1. Prevent any undefined traffic from crossing zone boundaries (where the disruption of the communication will not impact the reliability of a legitimate service).
2. Prevent any defined traffic containing malware or exploitation code from crossing zone boundaries.
3. Detect and log suspicious or abnormal activity within a zone (see "Implementing Host Security and Access Controls" and Chapter 13, "Security Monitoring of Industrial Control Systems").
4. Log normal or legitimate activity within a zone, which may be useful for compliance reporting (see Chapter 14, "Standards and Regulations").
5. Log all traffic originating from remote access clients, which may be useful for compliance reporting and acceptable use confirmation.

> **Caution**
>
> A false positive (a rule that triggers in response to unintended traffic, typically due to imprecisions in the detection signature) can block legitimate traffic, and in a control system legitimate traffic could represent a necessary operational control that may not be frequently used (i.e., plant startup and shutdown activities). Only use IPS and block rules where absolutely necessary and only after extensive testing.

The greater the extent of functional isolation and separation into defined zones, the more concise and effective the IDS/IPS policy will be. Some basic IDS/IPS rules suitable for use in zone perimeters include the following:

- Block any industrial network protocol packets that are the wrong size or length.
- Block any network traffic that is detected inbound to or outbound from any zone where that is not expected or allowed.
- Block any industrial network protocol packets that are detected in any zone where that protocol is not expected or allowed.
- Alert any authentication attempts, in order to log both successful and failed logins.
- Alert any industrial network port scans.
- Alert any industrial network protocol function codes of interest, such as:
- "Write" functions, including codes that write files or that clear, erase, or reset diagnostic counters.
- "System" functions, including codes that stop or restart a device.
- "System" functions that disable alerting or alarming.
- "Read" functions that request sensitive information.

- "Alarm" or "Exception" codes and messages.

Consideration should be given when defining IDS/IPS rules as to whether you want to begin analysis before or after the TCP three-way handshake has taken place—of course this is limited to only those applications and services that depend on TCP as their transport protocol. It is not possible to perform content or deep-packet inspection of data that has not completed the three-way handshake. However, this type of information can be very valuable in determining if a rogue or malicious host is "probing" for potential targets and attempted to enumerate and fingerprint the network under consideration. The example rule given next can be used to identify any traffic that is attempting to communicate with an ICS host via the EtherNet/IP protocol at the onset of the three-way handshake—an initial segment is sent with only the SYN flag set in the TCP header:

```
alert tcp !$ControlSystem_Zone01_Devices any ->
$ControlSystem_Zone01_Devices 44818 (msg: "Attempt to connect to
ICS device from another zone using known service"; flags: S;
<additional options>)
```

While almost any IDS/IPS device may be able to detect and trigger upon industrial network protocols by searching for specific values in a packet, those devices that can perform stateful inspection of application contents including inspection of function codes, commands, and additional payloads will provide more value, and will generally be capable of detecting threats with greater efficacy. Many industrial protocols are not easily parsed by traditional IDS/IPS engines, and often utilize message fragmentation making them very difficult to analyze with consistent results. Therefore, it is recommended that "industrial" products with application inspection capability be used. This class of product will be more capable of analyzing the application layer protocols and how they are used and will be useful for detecting injection attacks, malformed messages, out of sequence behavior and other potentially harmful activity.

Caution

Most IDS/IPS signatures are only able to block known threats, meaning that the IDS/IPS policy must be kept current in order to detect more recently identified attacks (virus, exploits, etc.). Therefore, IDS/IPS products must be included within the overall patch management strategy in order for the devices to remain effective (see "Patch Management" later in this chapter). What makes this difficult for ICS environments is that unless the vulnerability has been publicly disclosed, many IDS/IPS vendors will not have access to the actual payloads that exploit these weaknesses—in other words, it is very difficult for them to develop relevant signatures for ICS components. Products that utilize anomaly-based detection, protocol filtering, and/or "network whitelist" enforcement will be able to provide protection without requiring specific signatures, and therefore, it is only necessary to patch these types of devices if there is a firmware update or similar upgrade to apply.

Anomaly-based intrusion detection

Only signature-based detection has been discussed at this point. Anomaly detection is also supported on many IDS/IPS systems using statistical models to detect when something unusual is happening. This is based on the premise that unexpected behavior could be the result of an attack.

The exact capabilities will vary from product to product, as there is no standard anomaly detection mechanism. Theoretically, anything monitored by the IDS could be used for anomaly detection. Because network flows are highly quantifiable, anomaly detection is often used to identify abnormal behavior in what devices are communicating between each other and how. Referred to as network anomaly detection, these systems are able to detect a sudden increase in outbound traffic, an increase in sessions, an increase in total bytes transmitted, an increase in the number of unique destination IP addresses, or other quantifiable metrics.

Anomaly detection is useful because it does not require an explicitly defined signature in order to detect a threat. This allows anomaly detection systems to identify 0-day attacks or other threats for which no detection signature exists. At the same time, however, anomaly detection tends toward a higher number of false positives, as a benign change in behavior can lead to an alert. Anomaly-based threat detection is typically used passively for this reason by generating alerts rather than actively blocking suspect traffic.

In industrial networks—especially in well-isolated control system zones—network behavior tends to be highly predictable, making anomaly detection more reliable.

Anomaly detection systems may be referred to as "rule-less" detection systems. This is because they do not pattern match against a defined signature, although they do use rules. Unlike a normal IDS rule, anomaly rules are often based on thresholds and/or statistical deviations, such as in the following example:

```
TotalByteCount from $Control_System_Zone01_Devices increases by
>20%
```

An example of a threshold rule would use a hard upper- or lower-limit, most likely derived automatically by the anomaly detection system:

```
TotalDestinationIPs>34
```

As a general guideline, the greater the variation of network traffic being monitored, the greater the chances of anomaly detection rules generating a false-positive result.

Anomaly detection can be used across devices as well, coupled with an information consolidation tool, such as a SIEM system. This system-level anomaly detection is discussed in more detail in Chapter 13, "Security Monitoring of Industrial Control Systems."

Protocol anomaly detection

Another type of anomaly detection looks specifically at the protocol: malformed messages, sequencing errors, and similar variations from a protocol's "known good" behavior. Protocol anomaly detection can be very powerful against unknown or 0-day exploits, which might attempt to manipulate protocol behavior for malicious purposes. However, be very careful when deploying protocol anomaly detection, as many legitimate products from legitimate ICS vendors utilize protocols that have been implemented "out of spec"—either using proprietary protocol extensions or altering the protocol's implementation in a product to overcome some limitation in the "pure" standard. Knowing this, protocol anomaly detection of industrial protocols can be subject to high rates of false positives, unless some effort has been made to "tune" the detection parameters to the nuances of a particular vendor or product.

Application and protocol monitoring in industrial networks

Because many industrial operations are controlled using specialized industrial network protocols that issue commands, read and write data, perform device configuration, and so on using defined function codes, specialized devices can leverage that understanding along with firewall, IDS, and IPS technology to enforce communications based on the specific operations being performed across the network.

In addition to the inspection of industrial protocol contents (e.g., DNP3 function codes), the applications themselves—the software that controls how those protocols are used—can also be inspected. This degree of Application Monitoring, also referred to as Session Inspection, allows the contents of an application (e.g., human–machine interface [HMI], Web Browser) to be inspected even though it might exist across a large number of individual packets. That is, inspection can occur up to and include the contents of a file being transferred to a PLC, a virus definition downloaded from the web browser of an update server, and so on. Application monitors provide a very broad and very deep look into how network traffic is being used and are therefore especially useful in environments where both control systems and enterprise protocols and applications are in use.

Many specialized security devices are available for ICS and other control system environments that use either application or protocol monitoring to this degree. At the time of this writing, these devices include the Tofino Security Appliance and the Secure Crossing Zenwall Access Control Module, as well as other broader-use enterprise application data monitors. The two former devices were designed specifically to identify the operations being performed within industrial protocols and to prevent unauthorized operations. The latter refers to a more general-purpose enterprise security appliance, which is able to support the most common industrial network protocols. Each of these specialized devices has specific strengths and weaknesses, which are summarized in Table 11.4.

Because these devices are highly specialized, configurations can vary widely. In general terms, a firewall capable of industrial protocol inspection may utilize a rule as follows to block any protocol function from writing a configuration or register, or executing a system command (such as a device restart):

```
Deny [$ControlSystem_ProtocolFunctionCodes_Write,
$ControlSystem_ProtocolFunctionCodes_System]
```

Table 11.4 A Comparison of Industrial Security Devices

Security product	Functionality	Strengths	Weaknesses	Rule example
ICS firewall	Traffic policy enforcement	Enables isolation of traffic based on networks, ports and services	Does not block hidden threats or exploits within "allowed" traffic	Allow only TCP port 502 (Modbus TCP)
ICS IDS/IPS	Detects malware and exploits within traffic	Prevents exploitation of vulnerabilities via authorized ports and services	"Block-list" methodology can only detect and block known threats	Block Modbus packets containing known malware code
ICS UTM or hybrid security appliance	Combines firewall, IDS/IPS, VPN, antivirus and other security functions	Combination of security functions facilitates "defense in depth" via a single product	Security functions maintain their component weaknesses (i.e., the whole is equal to but not greater than the sum of its parts)Must be updated in order to remain effective	Allow only TCP port 502 with "read only" function codes
Allow outbound TCP 502 only via encrypted VPN to other SCADA zones				
ICS content firewall or application firewall	Traffic policy enforcement	Enables content-based traffic isolation, based on industrial network protocols	Assesses content of a single packet only (lacks session reassembly or document decode) Difficult to deploy on protocols that utilize packet fragmentation	Allow only "read only" Modbus TCP functions
Deep session inspection (application content monitoring)	Session Reassembly	Functions of an ICS content firewall, plus visibility into full application session and document contents to detect APT threats and insider data theft; provides strong security in hybrid enterprise/industrial areas such as ICS DMZ or other semi-trusted zones such as remote access	Typically limited to TCP/IP inspection, making session inspection less suitable for deployment in pure control system environments	Alert on Modbus TCP traffic on ports other than TCP 502
File/Content decode	Alert on any traffic with base64-encoded content			
File/Content capture				
Network whitelist	Allows only defined "good" traffic	Prevents all malicious traffic by allowing only known, good traffic to pass as defined by a fingerprint of acceptable host and protocol relationships.	Requires proper baselining of correct network behavior	Can make legitimate changes in network operations more difficult

An IDS capable of industrial protocol inspection may utilize a rule as follows, which looks for a specific function code within a DNP3 packet (DNP3 is supported with both TCP and UDP transports):

```
tcp any any -> $ControlSystem_Zone01_Devices 20000 (msg: "DNP
function code 15, unsolicited alarms disabled - TCP";
content:"|15|"; offset:12; rev:1;)
udp any any -> $ControlSystem_Zone01_Devices 20000 (msg: "DNP
function code 15, unsolicited alarms disabled - UDP";
content:"|15|"; offset:12; rev:1;)
```

In contrast, an application monitor performing full session decode may use syntax similar to the following rule to detect windows.LNK files within application traffic, which could indicate a possible Stuxnet delivery attempt.

```
FILTER_ID=189
NORM_ID=830472192
ALERT_ACTION=log-with-metadata
ALERT_LEVEL=13
ALERT_SEVERITY=10
DESCRIPTION=A Microsoft Windows .LNK file was detected
EXPRESSION=(objtype==application/vnd.ms-lnk)
```

Data diodes and unidirectional gateways

Data diodes and unidirectional gateways work by preventing return communications at the physical layer typically over a single fiber-optic connection (i.e., fiber strand). The "transmit" portion generally does not contain "receive" circuitry, and likewise, the "receive" does not possess "transmit" capability. This provides absolute physical layer security at the cost of bidirectional communications. Because the connection in reverse direction does not exist, data diodes are true air gaps, albeit in only one direction.

Because many network applications and protocols require bidirectional communication (such as TCP/IP, which requires a variety of handshakes and acknowledgments to establish, maintain, and complete a session), considerations should be taken when using data diodes in order to ensure that the remaining one-way data path is capable of transferring the required traffic. To accommodate this concern, many data diode vendors implement a software-based solution, where the physical diode exists between two "agents." These agents support a variety of bidirectional applications and their associated communication services, so that the bidirectional requirements can be met fully at each end. The receiving end effectively "spoofs" the behavior of the original transmitter—essentially tricking the application to operate over a one-way link. This allows an additional level of control over the applications and services that can be transmitted over the diode or gateway. An example of enabling DNP3 services over a unidirectional gateway is shown in Figure 11.5. While data diodes are physical layer devices that do not require any specific configuration, the communication servers may need to be correctly configured before these applications work correctly over the diode. Table 11.5 shows the applications and protocols supported using a unidirectional gateway supplied by Waterfall Security.

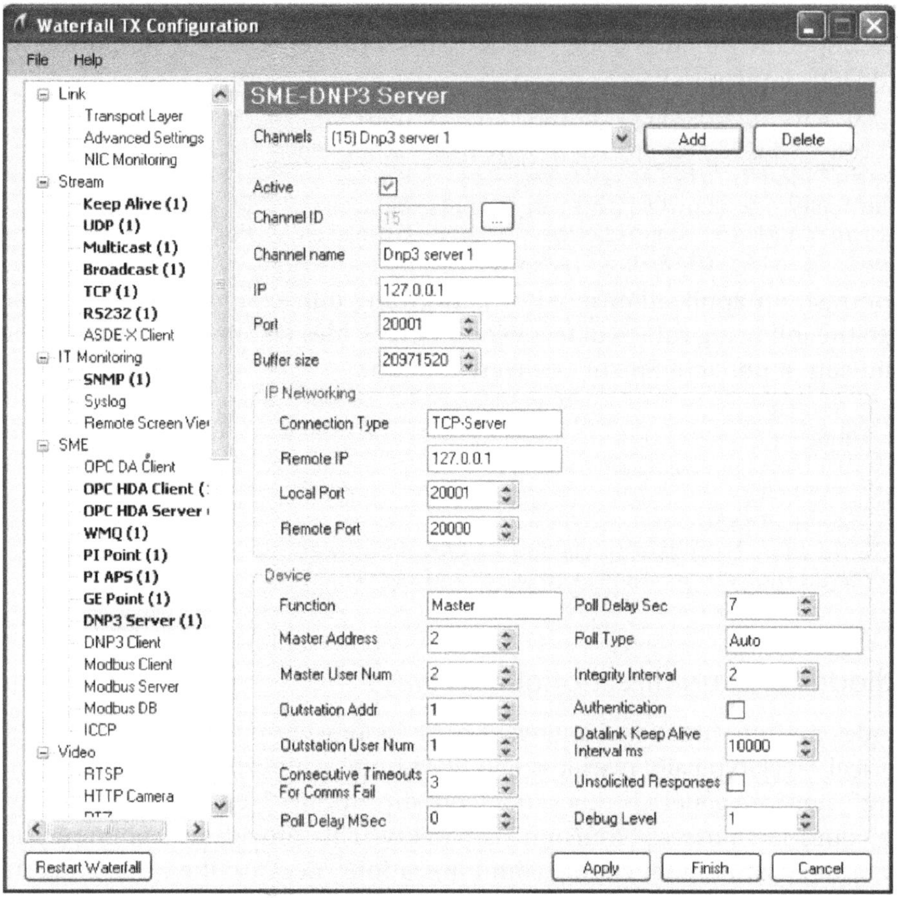

FIGURE 11.5 Enabling DNP3 over a unidirectional gateway.

Table 11.5 Unidirectional Gateway Application/Protocol Support[a]

Application family	Description
Historian	OSIsoft PIGE iHistorianGE OSMWonderware HistorianInstep eDNA
Human–machine interface	GE iFixSiemens SINAUTSiemens WinCC
Control center communications	ICCPIEC 60870-104
Remote access	Remote Screen View
File transfer	FTPFTPSSFTPTFTPRCPCIFS
Monitoring	CA SIMCA UnicenterHP OpenViewSNMPLog TransferSyslog
Video	ISE
Antivirus	OPSWAT MetascanNorton Updater
Middleware	IBM Websphere MQMS Message Queuing
ICS protocols	OPC-UAOPC-DA (Classic)ICCPModbusDNP3Bently-Nevada System 1
Database replication	SQLOracle
General	UDPTCPEmailRemote PrintingMicrosoft BackupTibco EMS

[a]Waterfall Security Solutions, Ltd. www.waterfall-security.com (cited: December 27, 2013).

Implementing host security and access controls

All zones are essentially logical groups of assets. They therefore contain a variety of devices, which may themselves be susceptible to a cyber-attack.

> **Caution**
>
> Not all cyberattacks occur via the network! Devices (network connected or otherwise) may be susceptible to viruses or other threats. This is true not only of devices, such as workstations and servers that use commercial operating systems, but also of specialized "embedded" devices including PLCs, HMIs, and similar devices. Even if the device uses an embedded or real-time operating system, it may be vulnerable to infection. If the device is network connected, it might be at risk from the network; if it is not, does that device possess USB interfaces? Infrared or wireless diagnostics interfaces? Serial communications to a master server or device? A firmware upgrade capability? Some other interface or dependency that could be used as an attack vector? If it does, it is important to harden that device to the best degree possible. Also understand that "the greatest degree possible" might be "not at all" for many embedded devices. However, if a device can be hardened, it should be!
>
> Devices that cannot be hardened or secured through traditional means should be considered for inclusion in dedicated security subzones so that the conduit that connects to this zone can be rigorously controlled and secured using techniques previously described (see Chapter 9, "Establishing Zones and Conduits"). It may not be possible to deploy malware prevention controls directly on a PLC, but they can easily be deployed on the conduit acting as the only entry point into this zone. This approach utilizes compensating security controls in establishing a "zone-based security policy."

Zones consist of specific devices and applications, and conduits consist of a variety of network communication channels between those devices and applications. This means that all zones will contain at least one device with a network interface, and therefore, it is important to secure the device (including OS and applications) and access to that device (including user authentication, network access controls, and vendor maintenance). Host security controls address the questions of who is allowed to use a device, how a device communicates on the network, what files are accessible by that device, what applications may be executed by it, and so on (the monitoring of host activities, such as the communications between hosts within a zone, is also useful for detecting threats). This was discussed in Chapter 9, "Establishing Zones and Conduits" and will be further discussed in Chapter 13, "Security Monitoring of Industrial Control Systems," so it will not be discussed further in this chapter.

This section discusses three distinct areas of host security, including

1. Access control, including user authentication and service availability.
2. Host-based network security, including host firewalls and host intrusion detection systems (HIDS).
3. Antimalware systems, such as antivirus (**AV**) and application allow-listing.

Table 11.6 Common Host Security Methods by Device Type

Device	Methods
HMI or similar device running a modern operating system. Application is not time sensitive.	• Host firewall • HIDS • Anti-virus or application whitelisting • Disable all unused ports and services
HMI or similar device running a modern operating system. Application is time sensitive.	• Host firewall • Disable all unused ports and services • Optional: Application whitelisting (will require testing to ensure imposed latency is acceptable)
PLC, RTU, or similar device running an embedded commercial OS.	• Host Firewall or HIDS if available • External security controls
PLC, RTU, IED or similar device running an embedded operating environment.	• External security controls

Selecting host cybersecurity systems

As a matter of best practices, all host access controls and host network security solutions should be implemented on all networked devices. The problem is that not all network devices are capable of running additional security software, and in some cases, the software may incur latency or unacceptable processor overhead. Table 11.6 shows which devices are typically capable of running the common methods of host security.

Where possible, one option of each type—access control, network security, and antimalware—should be used on each device. Especially where host security options are not possible, an external security control should be implemented.

Tip

ICS vendors are beginning to offer optional security features for their embedded devices, such as PLCs. In 2013, Siemens released a line of enhanced communication processors for their S7-300 and S7-400 line of PLCs that provide integrated firewall and VPN capabilities at the chassis level. Other vendors like Caterpillar/Solar, Honeywell, Invensys, Schneider Electric, and Yokogawa have leveraged OEM solutions to provide advanced security external to the embedded device. Because the available and/or recommended solutions may change over time, always consult your ICS vendor when selecting a security product.

Caution

Major ICS vendors often recommend and/or support the use of particular host security options and may even perform regression testing to validate authorized tools.[18] This is an important consideration, especially when utilizing time-sensitive applications that could be affected by delay. Many control system assets may also use proprietary extensions or modifications of commercial operating systems that may conflict with some host security solutions.[19] Asset vendors should always be consulted prior to the installation of a commercial host security product.

Tip

ICS vendors must be able to guarantee the performance and reliability of their real-time control systems. This is the primary reason many restrict the installation of additional, unqualified, third-party software on certain ICS devices. It is important to realize that this does not mean "one size fits all" and that a policy that applies to specific ICS devices must be followed for all devices contained within the composite ICS architecture. In other words, the restrictions that a vendor may place on their ICS Server may not apply to generic components, such as Microsoft Active Directory Servers. These devices often can be hardened with controls not typically qualified and supported by the ICS vendor but necessary to provide sufficient protection against cyber threats.

Host firewalls

A host firewall works just like a network firewall and acts as an initial filter between the host and any attached network(s). The host firewall will allow or deny both inbound and outbound traffic based on the firewall's specific configuration. Host firewalls are typically session-aware firewalls that allow control over distinct inbound and outbound application sessions. Unlike network-based firewalls that can monitor all traffic entering a network zone via a defined conduit, host-based firewalls can only inspect traffic that is either sent directly to the device or traffic that uses a broadcast address.

As with network firewalls, host firewalls should be configured according to the guidelines presented under "Firewall Configuration Guidelines"—starting with Deny All policies, and only adding Allow rules for the specific ports and services used on that particular asset.

Many organizations believe that hosts should be protected from network-based attacks. In doing so, their attention is paid to only configuring the host-based firewall inbound or "ingress" rules. Recent studies around security controls to protect against advanced targeted attacks (those that are typically the most difficult to prevent) have shown that overall network resilience to cyber events can be improved by also deploying outbound or "egress" rules on these firewalls.[20] This effectively contains or isolates that malware to the compromised host and offers significant defenses against information leakage, C2 communication, and lateral movement and infection. Implementing a simple outbound rule limiting communication to IP addresses within the allowed zones and conduits could have prevented the consequences (C2 communication, payload download, OPC enumeration, etc.) resulting in the installation of trojanized ICS software during the Dragonfly/Havex campaign in 2013–14.

Host IDS

Host IDS (HIDS) work like network IDS, except that they reside on a specific asset and only monitor systems internal to that asset. HIDS devices typically monitor system settings and configuration files, applications, and/or sensitive files.[21] These devices are differentiated from antivirus and other host security options in that they can perform network packet inspection and can therefore be used to directly mimic the behavior of a network IDS by monitoring the host systems network interface(s) to detect or prevent inbound threats. HIDS can be configured using the information presented under "Intrusion Detection and Prevention (IDS/IPS) Configuration Guidelines." Because a HIDS may also be able to inspect local files, the term is sometimes used for other host-based security devices, such as antivirus systems, or propriety host security implementations that provide overlapping security functions.

A HIDS device will generate alerts detailing any violations of the established policy similar to a network IDS. If the system is able to actively block the violation, it may be referred to as a host IPS (**HIPS**).

> **Caution**
>
> Like network-based IDS/IPS systems, host-based products require regular signature updates in order to detect more recently identified threats. These applications should therefore be included in the overall patch management strategy described later in this chapter.

Antivirus

Antivirus systems are designed to inspect files for malware. They use one or more methods of malware detection to analyze files (see "Malware Detection Methods", below). When the antivirus system indicates the presence of a virus, Trojan, or other malware, the suspect file is typically quarantined so that it can be cleaned or deleted, and an event is generated signifying the occurrence. Antivirus systems remain a vital layer of defense; even with other host security controls in place, such as allow-listing or HIDS, there are use cases where active file analysis to detect malware remains necessary. Always remember that no solution is capable of detecting 100% of all possible threats and that a defense-in-depth posture using a variety of controls will provide greater overall protection.

> **Caution**
>
> Some antivirus systems can consume large amounts of systems resources, including CPU and memory utilization. Always consult your control system vendor(s) to ensure that a particular antivirus system is supported and for recommendations on how to best implement and configure supported antivirus systems, in order to prevent potential stability issues.

Caution

Most antivirus systems require regular signature updates, with updates typically being provided every 24 h, or even more frequently when a rapid response is required (e.g., to protect against a particularly dangerous new virus). Antivirus systems should therefore be included in the overall patch management strategy described later in this chapter, with special consideration to allow frequent updates.

Note that the antivirus/antimalware industry evolves quickly. In the past years, pure signature-based detection has been continuously enhanced with new detection technologies including behavioral analysis, memory analysis, and sandboxing. Threat intelligence—leveraging knowledge of previously detected threats—is increasingly used to enhance detection. Machine learning is also increasingly used to detect similarities of new code to previously analyzed malware. The result overall is highly effective malware detection but with some caveats. For example, many features may only be available when connected to an online threat feed or cloud-based service maintained by the antimalware vendor. This obviously introduces challenges when implementing these controls in isolated industrial networks, and compromises will likely need to be made. Is it better to enable outbound connectivity to these services to enhance detection or leave these services disconnected with the understanding that detection efficacy could drop significantly? The answer will depend on the specific circumstances of the industrial network, other compensating controls that might be in place (such as Application whitelisting, below), and the individual risk appetite of the operator (see Chapter 8, "Risk and Vulnerability Assessments").

Application whitelisting/application allow-listing

Application whitelisting, now more commonly referred to "application allow-listing," offers a different approach to host security than traditional HIDS/HIPS, antivirus, and other "blocklist" technologies. A "blocklist" solution compares the monitored object to a list of what is known to be bad. This presents two issues: the first is that the block-list must be continuously updated as new threats are discovered; the second is that there is no way to detect or block certain attacks, such as 0-days, and/or known attacks for which there is no available signatures. The latter is a common problem facing ICS installations and one of the challenges that must be addressed in order to properly secure these vital, fragile systems. In contrast, a "whitelist" solution creates a list of what is known to be good and applies very simple logic—if it is not on the list, block it.

Allow-listing solutions apply this logic to the applications and files on a host. In this way, even if a virus or Trojan successfully penetrates the control system's perimeter defenses and finds its way onto a target system, the host itself will prevent that malware from executing—rendering it inoperable. It can also be used to prevent the installation of authorized files on the file system. This becomes important to providing defenses against exploits that may initially run entirely in memory and are difficult to detect until they place files locally.

Antivirus techniques depend on continuous updates to their signatures or blocklist, which means that the demands on computational components can increase as the number of block-listed entries climbs. This is a major cause for dissatisfaction with AV and why it is not always deployed on ICS devices. Allow-listing is well suited for use in control systems, where an asset should have explicitly defined ports and services. It is also desirable on systems that depend on legacy or unsupported applications and operating systems that can no longer be patched for security vulnerabilities. There is no need to continuously download, test, evaluate, and install signature updates. Rather, the allow-listing only needs to be updated and tested when the applications used on the host system are updated. ICS vendors prefer this approach as well, because the impact to device operation and performance can easily be base-lined after initial software installation, since ICS hosts remain relatively static after commissioning.

Allow-listing can introduce new code into the execution paths of all applications and services on that host because it operates at the lowest levels of an operating environment. This adds latency to all functions of the host, which may cause unacceptable delay for time-sensitive operations, and requires full regression testing.

Caution

Many people think of application allow-listing as a "Silver Bullet," and this is actually an accurate description. Like a silver bullet, which according to legend is effective against werewolves, application whitelisting is effective against malware. However, simply owning a silver bullet will not protect you from werewolves; you will need to use the silver bullet (load it into a gun, fire it at the werewolf, and hit your target) for it to be effective. Similarly, application whitelisting needs to be used appropriately if it is to be effective. That means understand the limitations of the allow-listing solution—does it protect against memory attacks, embedded scripts, macros, and other malware vectors, or does it simply enforce executable processes? It is also important to understand that "not all threats are werewolves"—application whitelisting cannot and will not protect against the misuse of legitimate applications. Example: A disgruntled employee uses an engineering workstation to rewrite the process logic of a controller. Application whitelisting on the engineering workstation would not prevent this, because the software used is authorized—it is simply being misused. Application whitelisting on the controller would also not prevent the activity, because the logic would be written using legitimate application-layer protocols. This is an extremely important consideration in light of the many cyber-physical attacks that have occurred to date which do not exploit the industrial control system using malicious code, but rather use the industrial control system's intended functionality for malicious purposes (see Chapter 7, "Hacking Industrial Control Systems").

Removable media

Removable media has been identified as the second largest cybersecurity threat vector into Industrial Control Systems. Universal Serial Bus (USB) devices are of particular concern due to their ubiquity, low cost, ease of use, and susceptibility to manipulation.[22] Some USB devices, such as thumb drives and other external storage devices, represent a potential carrier for infected files. USB devices can also be maliciously reprogrammed, allowing an attacker to change the behavior of a "good" USB device into something malicious. One examples includes having a USB device present itself to the host computer as a keyboard in order to rapidly type commands (I.e., a "Human Interface Device or HID Attack"), up to and including the creation and execution of scripts, modification of system settings, creation and/or deletion of users, creation and/or deletion of network interfaces, etc. Some USB devices are specifically built to behave maliciously; these devices are widely available, often sold as penetration testing tools. These devices range in capability, from performing simple HID attacks up to and including remote command and control over covert wireless communications.[23]

Probably the most common threat is introduction of malware via USB removable media. USB drives are an effective way to enter industrial networks that are (or should be) isolated from outside attackers with strong network segmentation and perimeter controls: the device is simply carried in. The drive of an unsuspecting employee could be infected or maliciously reprogrammed without their knowledge, allowing an attack to literally "walk" their attack across the air gap into the industrial environment. While this scenario sounds far-fetched, studies of malware found on USB drives being carried into industrial facilities show that 53% of that malware is written specifically to leverage USB drives for infection or propagation, 82% are able to impact industrial networks, inferring a degree of intentional targeting, and 54% attempt to establish remote connectivity once infection has occurred.[24]

Removable media also represents a threat of data exfiltration. A user could (intentionally or by accident) carry sensitive information out of a secure facility on a USB drive.[25] Considering the importance of early reconnaissance to cyber-physical threats faced by industrial operators, data theft is something to be heavily considered (see Chapter 7, "Hacking Industrial Control Systems").

There are several ways to address removable media threats, including policies and procedures, and active security controls.

Policies

- Education and training, so that employees understand the risks of using unauthorized USB devices and train them on proper USB security practices.
- Monitoring logs for evidence of USB activity is an example of a policy that could allow security operations teams to detect the misuse of USB devices and other removable media, including potential data loss.
- AV scan stations with a written policy requiring users to scan USB drives, but without a technical control to enforce this behavior.
- A policy forbidding the use of USB or other removable media entirely, but without a technical control to enforce this behavior.

Controls

- USB device control software that actively enforces which USB devices can be used. Some controls, such as Windows' included domain policies, can be fairly broad, only controlling what types of devices are allowed. Other purpose-built solutions provide more granular control, allowing the user to set policies based on device type, product or vendor IDs, serial number, or other identifiers. Other solutions require a human operator to explicitly allow device connections after the device has been identified by the system, to ensure that the device is what it claims to be—a direct response to malicious USB devices and penetration testing tools.
- USB scanning systems that include an active enforcement element (e.g., a driver or other software installed on computers that will prevent access to USB devices, drives, or even specific files, based on scan results).
- Physical port locks, while heavy-handed, prevent the use of USB ports by physically preventing the insertion of a device. While useful for systems where other solutions are not possible due to compatibility or other reasons, they prevent the use of USB devices entirely and therefore also negate the many benefits of using USB media.

It should be noted that the USB standard is very flexible by design, but it is also based on a system of trust between the USB device (the USB "device controller") and the computer it attaches to (the USB "host controller"). It is therefore trivial for an attacker to spoof USB device identifiers and other behaviors, and for this reason combinations of controls should be considered for systems with high security levels.

External controls

External tools may be required when it is simply not possible to use host-based security tools. For example, certain IDS/IPS, firewalls, and other network security devices that are specialized for control system operations may be used to monitor and protect these assets. Many of these devices support serial as well as Ethernet interfaces and can be deployed directly in front of a specific device or group of devices, including deployment within a specific process or loop.

Other external controls, such as Security Information and Event Management systems, may monitor a control system more holistically, using information available from other assets (such as a master terminal unit or HMI), from other information stores (such as a Data Historian), or from the network itself. This information can be used to detect risk and threat activity across a variety of systems. This will be discussed more in Chapter 13, "Security Monitoring of Industrial Control Systems."

External controls, especially passive monitoring and logging, can also be used to supplement those assets that are already secured via a host firewall, host-based IDS/IPS, antivirus, AWL, and so on.

Patch management

It is by no mistake that the topic of patch management is at the end of this chapter. It should be very clear by now that timely deployment of software updates is vital to maintaining the operation of not only the base ICS components (servers, workstations, devices) but also the security technologies (appliances, devices, applications) that are implemented to help protect them. Risk, in the context of industrial security, can be thought of as a function of threats—including actors, vectors, and targets—and how they exploit system vulnerabilities that result in some form of an undesirable consequence or impact. In simple terms, you can reduce risk by reducing any of these three mentioned components.

Patching as a form of vulnerability management

Patch management, as it has been traditionally defined, addresses the notification, preparation, delivery, installation, and validation of software hotfixes or updates designed to correct uncovered deficiencies. These shortcomings may not only be related to security vulnerabilities but also software reliability and operational issues. Patch management, in the context of risk reduction, is a means of reducing vulnerabilities in an effort to reduce the resulting risk of a particular target. The idea is that if you can remove vulnerabilities from a system, then there is nothing for a threat to exploit and no resulting consequences to your system or plant operation. This sounds simple; since performance and availability are our first priority, and patch management addresses these concerns while at the same time helping to secure the system, it should be deployed on all systems. Right? Not necessarily!

There are many facets to this dilemma, probably all worthy of a book devoted solely to this topic. On the surface it makes perfect sense, but as a long-term strategy, it can be argued that it is a "reactive" approach to security—one of defensive tactics, rather than proactive offensive strategies. After all, you are patching what is "known" to be weaknesses yesterday and today, so even after you deploy the updates, new ones WILL be discovered tomorrow!

Leave no vulnerability unturned

By now it should be clear that ICS architectures consist of a large number of components, including servers and workstations, network appliances, and embedded devices. Each one of these possesses a central processing unit capable of executing code, some form of local storage, and an operating system. In other words, each one of these has the potential to have vulnerabilities that must be patched in order to maintain system performance, availability, and security. This book is entitled "Industrial Network Security" because the network is the foundation upon which the entire ICS is built. This means that if the network infrastructure can be compromised through a single vulnerability in a barrier device like a firewall, then the entire ICS architecture could be at risk. This leads you to realize that network appliances must be included as part of the patch management program, just like familiar Windows OS-based servers and workstations and ICS devices that typically run embedded OSes and proprietary applications. For a patch management program to be effective and provide reasonable risk reduction, it must be able to address the complete array of vulnerabilities that exist within the entire 100% of the architecture.

Vulnerabilities can impact every component within the ICS architecture. There also may be components that cannot be patched, such as those running the Windows XP operating system, which is no longer updated as of April 2014, or others such as those where the vendor has restricted the modifications that can be made to the system once it has been commissioned. So what options are left to reduce the risk of a threat exploiting these systems' vulnerabilities? One effective method is through the deployment of "zone-based security." Figure 11.6 illustrates how a security zone has been created and contains only those devices that cannot be patched or updated while in operation. The only entry points into this security zone are through network connections.

A security conduit is established, and the security controls are implemented on the conduit rather than on the individual assets. As mentioned earlier, industrial firewalls have been deployed to limit network traffic to only that which is allowed including only allowed "functions," such as the revocation of all engineering and update functions. Intrusion prevention has also been installed in the Conduit to analyze all traffic for authorized use and potential ingress of malware or other attempts to exploit target vulnerabilities.

Maintaining system availability

An ICS is typically designed to meet very high levels of availability (typically minimum 99.99% or less than 15 min of downtime per year), which means any downtime resulting

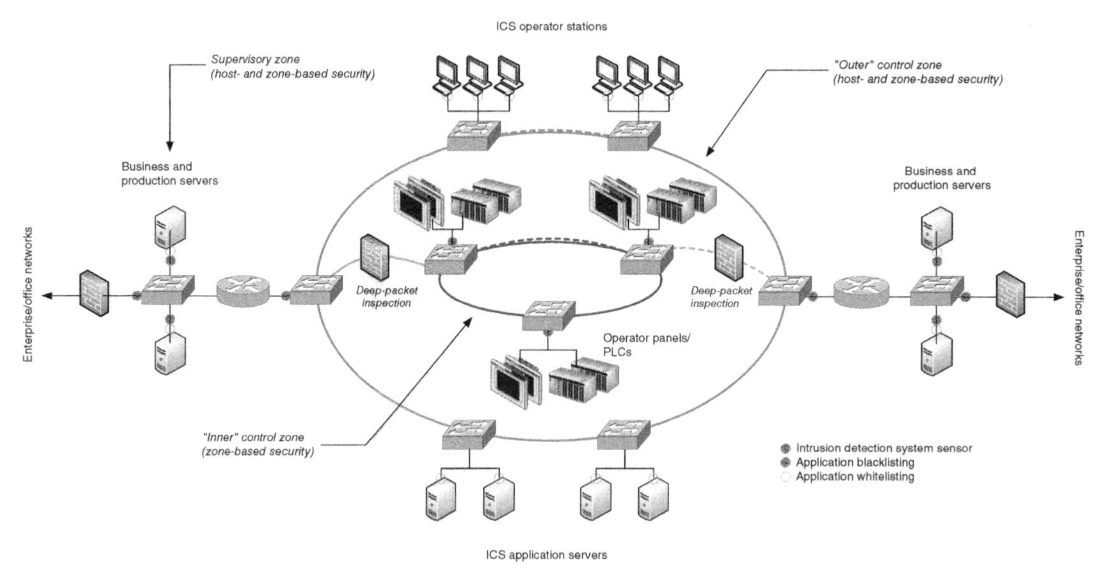

FIGURE 11.6 Zone-based vulnerability management.

from a monthly "reboot" required to activate an OS hotfix is considered unacceptable. Redundancy is common at the lowest levels of an ICS architecture, including devices, network interfaces, network infrastructure, and servers. Why then is it so difficult to perform a reboot on a system that is provided with redundant components? Production facilities do not like to invoke redundancy when it is not absolutely necessary because during the period of time a device is taken out of service, the overall system is left in a nonredundant configuration. Plant management now has to consider the risk of a manufacturing outage due to a known threat (a system operating without redundancy) versus an unknown threat (a cyber event originating from an unpatched system). What do you do if during the routine reboot, the system does not recover? What if you install an AV update and it crashes your server?[26]

Comprehensive predeployment testing

This is the reason that prior to deploying any patch, it is vital to thoroughly test and validate that the updates will not negatively impact the component being patched. The first step involves confirmation from the device vendor or manufacturer that a particular patch is acceptable to install, and equally important, that the patch is tested on an offline system that represents a site's particular configuration. Some vendors of ICS subsystems have deployed assets that are prohibited from having any security software installed or patches applied for fear that they may impact overall system operation. This may sound irrational, but given the fact that many ICS components have been in operation long before cyber security was a concern and will remain in operation for many more years to come without undergoing any major system upgrades, this is a problem that must be acknowledged and addressed.

Luckily the implementation of virtualization technologies makes predeployment validation easy for modeling and testing Windows-based assets, but what about network appliances and embedded devices? These generally cannot be deployed in virtual environments and can represent much greater net risk in terms of consequences resulting from a cyber event. After all, the embedded device is typically the final device that physically connects to the process under control. This leaves organizations with two options, both equally bad: either (1) do not deploy the patches, or (2) do not test the patches before deployment. The problem quickly escalates when you move away from the IT-centric Windows environment to an OT one consisting of a greater percentage of nonstandard embedded devices that do not run standard IT applications and OSes. This is the conundrum that organizations face every day with respect to patch management programs and whether or not they are truly a good method of risk management.

Industrial control systems tend to be heterogeneous in nature, comprising components from multiple vendors all integrated through commercial standards of networking (i.e., EtherNet and IP) and data communications (i.e., OPC, SQL, and OLEDB). This means that to minimize any negative impact to operations and system availability, end-users should test ALL patches and updates before deployment.

Automating the process

Integrated control systems—whether they are SCADA or DCS—are complex and have evolved dramatically since their inception in the 1980s resulting in little consistency from vendor-to-vendor on how their particular application or system is updated. Some vendors may provide complete package updates that require reinstallation of entire applications and suites, while others provide file-level updates and appropriate scripts. Any patch management solution must be able to handle this diversity. It should also be able to handle the management (and hopefully deployment) of patches in the form of firmware updates to the non-Windows components like network appliances and embedded devices (BPCS, SIS, PLC, RTU, IED, etc.). This process must be automated in order to provide a reasonable level of assurance. Automated, not in terms of a "lights out" approach to pushing and installing patches "in the dark," but rather a process of grouping assets based on criticality, duplicity, and redundancy, and allowing updates to be deployed initially on low-risk assets, then, proceeding to medium-risk assets that may not be redundant, but may be duplicated throughout the architecture (such as the HMI). Finally, critical servers are patched, one at a time, after these critical assets have been tested for compatibility in an off-line environment. The patch management solution should also maintain documentation of what updates have deployed to each asset and when. This documentation should align with that established and maintained within each zone as discussed in Chapter 9, "Establishing Zones and Conduits" in terms of both assets and change management procedures.

Finally, do not forget to perform comprehensive backups of the assets prior to performing any patching or updating, as it may be necessary to revert or abort the update if anomalies are detected or incompatibilities arise—up to and including a system not booting. It may also be necessary to abort updates if unplanned external events, like process disturbances, occur that require greater demands in terms of performance and availability of the ICS. When performing firmware updates of embedded devices and appliances, it is important to have equipment on hand, as failed firmware updates can often "brick" the device making it inoperable.

Malware detection methods

There are numerous methods available to detect malware, and many antivirus or anti-malware solutions will support multiple methods together. Understanding how these methods work can help determine which malware detection solutions are most suitable for industrial environments.

Signature-based detection

This method involves using predefined signatures or patterns of known malware to identify and detect malicious files or code. It is most effective for detecting well-known malware. Probably, the simplest method of malware detection, signature-based detection compares files to a database of known malicious "signatures" or patterns. If a file's signature matches the signature of known malware, it is flagged as malicious.[27]

Heuristic analysis

Heuristic analysis involves identifying potentially malicious behavior by analyzing behavioral patterns that might indicate malware activity. Heuristic analysis involves creating rules to identify potential malicious behaviors such as unauthorized modifications to critical system areas, attempts to conceal processes or files, unusual network communication, or abnormal interactions with system components. By detecting the presence of these behaviors, heuristic analysis aims to identify potential threats based on their actions rather than their specific code patterns. Heuristic analysis is effective at detecting new "0 day" malware and evolving malware variants. However, it can also generate false positives. Many legitimate applications make the same system calls, access files, and utilize the network. While heuristic analysis has been around for many years, it has recently benefited from advancements in machine learning models; with the availability of larger, curated malware sample sets, accuracy is improved.[28]

Behavioral analysis

Unlike heuristics, which detects patterns of behavior in code, this method focuses on monitoring the behavior of software or code during execution to identify any actions that are characteristic of malware. That is, instead of determining that code is capable of a behavior, behavioral detection systems let the code execute and watch to see if that behavior happens. Behavioral detection observes the actions and interactions of software or processes as they run. This makes it extremely effective at detecting behavior, although it is also a "last line of defense," as the potential malware is executing, and therefore any failure to detect tit will result in an infection. For this reason, many behavioral analysis systems trigger on abnormal behavior (anything that veers too far from a baseline measurement). This could include unauthorized attempts to modify system files, suspicious network communication, unauthorized access to sensitive data, or abnormal memory or CPU utilization. Like heuristics, behavioral analysis is effective at detecting 0-day threats and can produce high rates of false positives.[29]

Sandboxing

Sandboxing combines behavioral analysis with a safe execution environment. Instead of letting malware execute on the host system, it will run potentially malicious software in an isolated environment, that is, sandbox. Within the sandbox, the malware's behavior can be observed without affecting the actual system. Sandboxes are designed to mimic a real system, allowing the malware to function as it is designed so that it can be fully observed, but keeping it separate so as to prevent potential harm to the host system.

Sandboxing shares the same 0-day efficacy as behavioral analysis; however, because the sandbox prevents harm to the host, malware can be allowed to fully execute rather than relying on broad rules and anomalous behavior. This reduces false positives. However, sophisticated malware will attempt to detect if it is running in a sandbox and alter its behavior to evade detection. To counter this, advanced sandboxes use techniques like deception and dynamic analysis to appear more like real environments.[30]

Memory analysis

Malware memory analysis refers to the process of examining a computer's system memory (RAM) for signs of malicious activity or the presence of malware. Some modern malware variants will try to avoid detection by residing directly in memory rather than on disk, protecting the malware from detection by tools that rely on analyzing files. Memory analysis involves investigating running processes, code injections, hidden artifacts, and other memory-resident elements that could indicate the malware of presence.[31]

YARA

YARA is a pattern matching tool and a rule-based language that is commonly used for identifying and classifying files and processes based on specific patterns or characteristics. It allows users to define custom rules using a syntax that describes the features of files and processes they want to detect. These rules can then be used to scan files, memory, or other data sources to identify instances of malware or other suspicious behavior in a variety of environments and use cases.[32]

YARA rules are composed of a series of strings, conditions, and logical operators that describe the characteristics of the data you want to match. Each YARA rule consists of:[33]

- Rule name: A user-defined name for the rule.
- Metadata: Optional metadata about the rule, such as its author, description, and references.
- Condition: A set of conditions that specify the pattern or characteristics to be matched.
- Strings: Definitions of strings that are to be matched within the data being analyzed.

An example yara rule is provided for reference:

```
rule Sample_Rule {
    meta:
        author = "Your Name"
        description = "Detects a sample malware code pattern"

    strings:
        $malware_string = “This is malware”

    condition:
        $malware_string
}
```

In this example, the rule “Sample_Rule” looks for the string “This is malware” within the data being scanned. Unfortunately, most malware will not self-declare itself as such with the text string “This is malware.” Yara is popular because it is extremely flexible in what it can detect. For example, complex patterns and logic can be used to great effect, detecting very specific malware variants or entire families of malware based on known indicators and behaviors. The following rule from ReversingLabs' open-source yara rules (available at https://github.com/reversinglabs/reversinglabs-yara-rules/) shows a more complex example:

```
rule ByteCode_MSIL_Ransomware_Thanos : tc_detection malicious
{
    meta:

        author               = "ReversingLabs"

        source               = "ReversingLabs"
        status               = "RELEASED"
        sharing              = "TLP:WHITE"
        category             = "MALWARE"
        malware              = "THANOS"
        description          = "Yara rule that detects Thanos
ransomware."

        tc_detection_type   = "Ransomware"
        tc_detection_name   = "Thanos"
        tc_detection_factor = 5

    strings:

        $find_files_p1 = {
            6F ?? ?? ?? ?? 72 ?? ?? ?? ?? 6F ?? ?? ?? ?? 16 FE 01
2B ?? 16 00 13 ?? 11 ?? 2D ?? DD
            ?? ?? ?? ?? 08 6F ?? ?? ?? ?? 28 ?? ?? ?? ?? 2C ?? 08
6F ?? ?? ?? ?? 6C 7E ?? ?? ?? ??
            28 ?? ?? ?? ?? 23 ?? ?? ?? ?? ?? ?? ?? ?? 5A 23 ?? ??
?? ?? ?? ?? ?? ?? 5A 35 ?? 7E ??
            ?? ?? ?? 72 ?? ?? ?? ?? 28 ?? ?? ?? ?? 16 FE 01 2B ??
17 00 13 ?? 11 ?? 2D ?? 00 06 08
            6F ?? ?? ?? ?? 6F ?? ?? ?? ?? 00 00 2B ?? 08 6F ?? ??
?? ?? 28 ?? ?? ?? ?? 2C ?? 7E ??
            ?? ?? ?? 72 ?? ?? ?? ?? 28 ?? ?? ?? ?? 16 FE 01 2B ??
17 00 13 ?? 11 ?? 2D ?? 00 06 08
            6F ?? ?? ?? ?? 6F ?? ?? ?? ?? 00 00 00 DE ?? 26 00 00
DE ?? 26 00 00 DE ?? 00 00 00 11
            ?? 6F ?? ?? ?? ?? 13 ?? 11 ?? 3A ?? ?? ?? ?? DE ?? 11
?? 14 FE 01 13 ?? 11 ?? 2D ?? 11
            ?? 6F ?? ?? ?? ?? 00 DC 00 00 07 72 ?? ?? ?? ?? 6F ??
?? ?? ?? 6F ?? ?? ?? ?? 13 ?? 38
            ?? ?? ?? ?? 11 ?? 6F ?? ?? ?? ?? 0D 00 07 6F ?? ?? ??
?? 6F ?? ?? ?? ?? 72 ?? ?? ?? ??
            6F ?? ?? ?? ?? 2D ?? 07 6F ?? ?? ?? ?? 6F ?? ?? ?? ??
72 ?? ?? ?? ?? 6F ?? ?? ?? ?? 2D
            ?? 07 6F ?? ?? ?? ?? 6F ?? ?? ?? ?? 72 ?? ?? ?? ?? 6F
?? ?? ?? ?? 2D ?? 07 6F ?? ?? ??
            ?? 6F ?? ?? ?? ?? 72 ?? ?? ?? ?? 6F ?? ?? ?? ?? 2D ??
07 6F ?? ?? ?? ?? 6F
        }
```

```
        $find_files_p2 = {
                72 ?? ?? ?? ?? 6F ?? ?? ?? ?? 2D ?? 07 6F ?? ?? ?? ??
6F ?? ?? ?? ?? 72 ?? ?? ?? ?? 6F
                ?? ?? ?? ?? 16 FE 01 2B ?? 16 00 13 ?? 11 ?? 2D ?? 38
?? ?? ?? ?? 00 00 09 72 ?? ?? ??
                ?? 17 6F ?? ?? ?? ?? 6F ?? ?? ?? ?? 13 ?? 38 ?? ?? ??
?? 11 ?? 6F ?? ?? ?? ?? 0C 00 00
                08 6F ?? ?? ?? ?? 6F ?? ?? ?? ?? 72 ?? ?? ?? ?? 6F ??
?? ?? ?? 3A ?? ?? ?? ?? 08 6F ??
                ?? ?? ?? 6F ?? ?? ?? ?? 72 ?? ?? ?? ?? 6F ?? ?? ?? ??
3A ?? ?? ?? ?? 08 6F ?? ?? ?? ??
                6F ?? ?? ?? ?? 72 ?? ?? ?? ?? 6F ?? ?? ?? ?? 3A ?? ??
?? ?? 08 6F ?? ?? ?? ?? 6F ?? ??
                ?? ?? 72 ?? ?? ?? ?? 6F ?? ?? ?? ?? 3A
        }

        $find_files_p3 = {
                6F ?? ?? ?? ?? 6F ?? ?? ?? ?? 72 ?? ?? ?? ?? 6F ?? ??
?? ?? 3A ?? ?? ?? ?? 08 6F ?? ??
                ?? ?? 6F ?? ?? ?? ?? 72 ?? ?? ?? ?? 6F ?? ?? ?? ?? 3A
?? ?? ?? ?? 08 6F ?? ?? ?? ?? 6F
                ?? ?? ?? ?? 72 ?? ?? ?? ?? 6F ?? ?? ?? ?? 2D ?? 08 6F
?? ?? ?? ?? 72 ?? ?? ?? ?? 00 6F
                ?? ?? ?? ?? 2D ?? 08 6F ?? ?? ?? ?? 72 ?? ?? ?? ?? 6F
?? ?? ?? ?? 2D ?? 08 6F ?? ?? ??
                ?? 72 ?? ?? ?? ?? 6F ?? ?? ?? ?? 2D ?? 08 6F ?? ?? ??
?? 72 ?? ?? ?? ?? 6F ?? ?? ?? ??
                2D ?? 08 6F ?? ?? ?? ?? 72 ?? ?? ?? ?? 6F ?? ?? ?? ??
2D ?? 08 6F ?? ?? ?? ?? 72 ?? ??
                ?? ?? 6F ?? ?? ?? ?? 16 FE 01 2B ?? 16 00 13 ?? 11 ??
2D ?? DD ?? ?? ?? ?? 08 6F ?? ??
                ?? ?? 28 ?? ?? ?? ?? 2C ?? 08 6F ?? ?? ?? ?? 6C 7E ??
?? ?? ?? 28 ?? ?? ?? ?? 23 ?? ??
                ?? ?? ?? ?? ?? ?? 5A 23 ?? ?? ?? ?? ?? ?? ?? ?? 5A 35
?? 7E ?? ?? ?? ?? 72 ?? ?? ?? ??
                28 ?? ?? ?? ?? 16 FE 01 2B ?? 17 00 13 ?? 11 ?? 2D ??
00 06 08 6F ?? ?? ?? ?? 6F ?? ??
                ?? ?? 00 00 2B ?? 08 6F ?? ?? ?? ?? 28 ?? ?? ?? ?? 2C
?? 7E ?? ?? ?? ?? 72 ?? ?? ?? ??
                28 ?? ?? ?? ?? 16 FE 01 2B ?? 17 00 13 ?? 11 ?? 2D ??
00 06 08 6F ?? ?? ?? ?? 6F ?? ??
                ?? ?? 00 00 00 DE ?? 26 00 00 DE ?? 26 00 00 DE ?? 00
00 00 11 ?? 6F ?? ?? ?? ?? 13 ??
                11 ?? 3A ?? ?? ?? ?? DE ?? 11 ?? 14 FE 01 13 ?? 11 ??
2D ?? 11 ?? 6F ?? ?? ?? ?? 00 DC
                00 00 DE ?? 26 00 00 DE ?? 26 00 00 DE ?? 00 00 11 ??
6F ?? ?? ?? ?? 13 ?? 11 ?? 3A ??
                ?? ?? ?? DE ?? 11 ?? 14 FE 01 13 ?? 11 ?? 2D ?? 11 ??
6F ?? ?? ?? ?? 00 DC 00 00 DE ??
                26 00 00 DE ?? 26 00 00 DE ?? 26 00 00 DE ?? 26 00 00
DE ?? 00 06 13 ?? 2B ?? 11 ?? 2A
        }
```

```
        $encrypt_files = {
            73 ?? ?? ?? ?? 13 ?? 11 ?? 03 7D ?? ?? ?? ?? 11 ?? 04
7D ?? ?? ?? ?? 11 ?? 05 7D ?? ??
            ?? ?? 11 ?? 0E ?? 7D ?? ?? ?? ?? 00 28 ?? ?? ?? ?? 11
?? 7B ?? ?? ?? ?? 6F ?? ?? ?? ??
            80 ?? ?? ?? ?? 02 16 9A 72 ?? ?? ?? ?? 28 ?? ?? ?? ??
16 FE 01 13 ?? 11 ?? 2D ?? 00 28
            ?? ?? ?? ?? 0A 06 8E 69 16 FE 02 16 FE 01 13 ?? 11 ??
2D ?? 00 16 0B 2B ?? 00 06 07 9A
            6F ?? ?? ?? ?? 16 FE 01 13 ?? 11 ?? 2D ?? 00 7E ?? ??
?? ?? 06 07 9A 6F ?? ?? ?? ?? 6F
            ?? ?? ?? ?? 13 ?? 11 ?? 2D ?? 00 7E ?? ?? ?? ?? 06 07
9A 6F ?? ?? ?? ?? 6F ?? ?? ?? ??
            00 00 00 00 07 17 58 0B 07 06 8E 69 FE 04 13 ?? 11 ??
2D ?? 00 00 2B ?? 00 16 0B 2B ??
            00 7E ?? ?? ?? ?? 02 07 9A 6F ?? ?? ?? ?? 13 ?? 11 ??
2D ?? 00 7E ?? ?? ?? ?? 02 07 9A
            6F ?? ?? ?? ?? 00 00 00 07 17 58 0B 07 02 8E 69 FE 04
13 ?? 11 ?? 2D ?? 00 7E ?? ?? ??
            ?? 72 ?? ?? ?? ?? 00 6F ?? ?? ?? ?? 2C ?? 7E ?? ?? ??
?? 72 ?? ?? ?? ?? 28 ?? ?? ?? ??
            16 FE 01 2B ?? 17 00 13 ?? 11 ?? 2D ?? 00 7E ?? ?? ??
?? 72 ?? ?? ?? ?? 00 6F ?? ?? ??
            ?? 26 00 00 7E ?? ?? ?? ?? 6F ?? ?? ?? ?? 13 ?? 38 ??
?? ?? ?? 14 0D 73 ?? ?? ?? ?? 13
            ?? 11 ?? 11 ?? 7D ?? ?? ?? ?? 11 ?? 12 ?? 28 ?? ?? ??
?? 7D ?? ?? ?? ?? 00 7E ?? ?? ??
            ?? 72 ?? ?? ?? ?? 28 ?? ?? ?? ?? 16 FE 01 13 ?? 11 ??
2D ?? 00 09 2D ?? 11 ?? FE 06 ??
            ?? ?? ?? 73 ?? ?? ?? ?? 0D 2B ?? 09 73 ?? ?? ?? ?? 0C
08 1A 6F ?? ?? ?? ?? 00 08 16 6F
            ?? ?? ?? ?? 00 08 6F ?? ?? ?? ?? 00 08 6F ?? ?? ?? ??
00 00 2B ?? 00 11 ?? 7B ?? ?? ??
            ?? 11 ?? 7B ?? ?? ?? ?? 11 ?? 7B ?? ?? ?? ?? 11 ?? 7B
?? ?? ?? ?? 11 ?? 7B ?? ?? ?? ??
            28 ?? ?? ?? ?? 00 00 00 12 ?? 28 ?? ?? ?? ?? 13 ?? 11
?? 3A ?? ?? ?? ?? DE ?? 12 ?? FE
            16 ?? ?? ?? ?? 6F ?? ?? ?? ?? 00 DC 00 00 2A
        }

        $remote_connection = {
            00 00 28 ?? ?? ?? ?? 72 ?? ?? ?? ?? 72 ?? ?? ?? ?? 6F
?? ?? ?? ?? 16 19 6F ?? ?? ?? ??
            72 ?? ?? ?? ?? 28 ?? ?? ?? ?? 0A 73 ?? ?? ?? ?? 0B 07
02 28 ?? ?? ?? ?? 06 28 ?? ?? ??
            ?? 6F ?? ?? ?? ?? 00 28 ?? ?? ?? ?? 06 28 ?? ?? ?? ??
0C DE ?? 26 00 00 DE ?? 00 7E ??
            ?? ?? ?? 0C 2B ?? 00 08 2A
        }
```

```
        condition:
            uint16(0) == 0x5A4D and
            (
                all of ($find_files_p*)
            ) and
            (
                $encrypt_files
            ) and
            (
                $remote_connection
            )
}
```

This rule looks for several conditions that match on specific byte code patterns, which indicate several necessary functions of the Thanos ransomware: the first looks for the methods Thanos uses to find files; the second for methods used to encrypt files; and the third looks for methods used to establish remote connectivity.

YARA's versatility allows users and security researchers alike to create custom rules for identifying various types of malware, suspicious files, or specific behavior. Yara is especially powerful because its flexibility allows it to detect malware in on the network during transfer, in a latent file at rest, or during execution—with the right rule, it can detect almost anything.

Machine learning and artificial intelligence

Machine learning and artificial intelligence (AI) techniques have grown in popularity, spurred by the increased awareness of AI due to large language model systems provided by OpenAI and others. When considering the malware detection methods discussed above, it is easy to see how machine learning and AI could be used to improve malware detection. This is especially when considering heuristic and behavioral techniques that benefit from large data sets and the identification of consistent patterns within malware families and variants, as machine learning techniques are able to automate the process of identifying and classifying malicious software based on patterns and features. These techniques enable malware detection tools to recognize more complex relationships, making them extremely effective for identifying both known and previously unseen malware.

To identify malware with machine learning, models need to be built and trained, in a process that typically consists of:[34]

- Extraction of relevant features from malware samples, such as file attributes, API calls, system behavior, code patterns, and more. These features serve as input to the machine learning model.
- Creation of labeled datasets consisting of both malicious and benign samples. Samples are associated with the features extracted (above) and a label indicating whether it is malicious or not.

- Machine learning algorithms, such as decision trees, random forests, support vector machines (SVMs), neural networks, and more, are trained using the labeled dataset. This allows the model to learn to recognize patterns that differentiate between malicious and nonmalicious samples.
- Validation and testing of the model on separate datasets allows the assessment of its performance and efficacy.

Finally, once the model is trained and validated, it can be used to detect malware. Malware evolves quickly, so machine learning models need to be updated continuously.

How much security is enough?
In an ideal world, there would be enough budget to implement dozens of network- and host-based security controls, and there would be enough resources to evaluate, test, implement, and operate those controls on an ongoing basis. In reality, budgets are shrinking, and too many security controls can actually be counter-productive and likely detrimental to the overall availability and performance of the ICS.

One of the most important factors to consider when deploying any security control is how it helps to reduce the risk of a cyber event from negatively impacting the ICS and the production assets under its control. In other words, controls should be deployed to reduce specific risk facing an individual organization. Many users are looking for a "play book" of controls that can be deployed on all ICS installations, irrespective of their impact on a particular organization's cyber risk. In these cases, it often results in not only large budgets but less than effective protection against cyber threats facing critical infrastructure and industrial facilities in general. A well thought out security program will always balance the "cost of security" versus the "cost of impact."

From theory to practice

Adding controls to production systems

With an understanding of some of the security controls that could be implemented, a challenge remains: how to implement new controls into production control system. While the specifics of implementation will depend upon the specific control you are attempting to implement (see below), there are some basic guidelines to follow to ensure maximum success with minimal interruption:[35]

- Plan ahead. Understand the specific controls being implemented, as well as the specifics of your ICS architecture, network design, and other factors. If implementing a security control directly into a control system environment, always consult your ICS equipment vendor(s) for specific guidance. If a risk assessment has been done, those findings should be carefully considered here (see Chapter 8, "Risk and Vulnerability Assessments").

- Careful progression. Implement new controls in isolated network zones, starting with lower security level zones, or zones that were identified in a risk assessment as having less potential impact in the event of a failure. Thoroughly test each new implementation before moving on to the next zone. Testing should include both testing of the new control to validate its effectiveness, and testing of the process to ensure that there has been no impact to system reliability.
- Communication. Maintain clear communication with all relevant stakeholders including operations, IT, engineering, and management. Document the implementation procedure and include any expected impact to the process control system, as well as a clear procedure for rolling back to a known good state if required. This step can make the difference between a delay in implementation and an incident.

For specific controls, Table 11.7 offers considerations and guidance.

When to implement security controls?

Cybersecurity is a journey, and all organizations will find themselves at different stages of that journey. Understanding when to implement a new cybersecurity control requires understanding where you are on that journey and what your short- and long-term goals are. This typically involves assessing the current environment: performing vulnerability assessments, risk assessments, penetration tests, red-team exercises, etc. (see Chapter 8, "Risk and Vulnerability Assessments").

The only "wrong answer" is to stop moving forward. The threat landscape continuously evolves, and a failure to continuously improving cybersecurity capabilities will degrade your ability to detect, protect, and respond to cyberattacks over time.

There are obviously some circumstances in which specific controls might be considered, however, and there are also some dependencies that make it more feasible to implement certain security controls before others (Figure 11.7).

Remember that certain cybersecurity controls and countermeasures require operationalization and may require dedicated staff and/or expertise. Security monitoring, for example, requires both up front considerations (What data needs to be collected? How is that data obtained?) and operational considerations (What needs to be done with security data once it is been aggregated? Are there analysts available to observe and react? What policies are in place if a threat is uncovered? What data need to be stored and for how long?). This is discussed in more depth in Chapter 13, "Security Monitoring of Industrial Control Systems."

Table 11.7 Implementing Various Security Controls, With Impact Considerations and Recommendations

Control	What it's for	Where is it implemented	Impact considerations	Recommendations
Perimeter enforcement • Router ACLs • Firewalls • Network IDS/IPS • Network anomaly detection	Preventing access from less-trusted networks to the industrial networks	• The demarcation point between zones. • Domain controllers/other Identity and Access management systems.	• Depending upon the infrastructure, rearchitecting the network may require physical and logical changes that will cause downtime. • Downtime could temporarily impact communications between zones.	Proper planning and testing in lab environments can minimize downtime and network disruption.
Network segmentation • L2 switches • L3 switches • Routers • Microsegmentation switches	Separation of devices into distinct networks for purpose of zone separation	On network infrastructure within and between zones.	Network downtime will take all communications offline during implementation.	Determine new network configurations, perform offline testing, and establish a clear plan before making network configuration changes. When working with redundant networks, implement changes on fully redundant systems first; once completed, fail over to the updated systems and complete implementation on remaining systems.
Network traffic inspection • Network IDS/IPS • Network anomaly detection	• Detection of unwanted, disallowed, or suspicious network activity • Active enforcement of communication policies within and between zones	On a network Span port or Tap.	• Implementation of active network controls can block unintended traffic if not implemented properly. • Port spanning or port mirroring may cause degradation of performance on older network switches and routers. • Network taps are installed in-line, which will require network downtime. Taps are largely mechanical devices, and will typically not degrade network performance, although they may introduce small amounts of network latency.	• These controls only apply to the network traffic they are able to see. Because critical communication flows should be limited to defined conduits, this means that these control will need to be implemented on each conduit that requires protection. • For greenfield deployments, building network taps into the initial design will enable future addition of network security devices with minimal impact.

Continued

Table 11.7 Implementing Various Security Controls, With Impact Considerations and Recommendations—cont'd

Control	What it's for	Where is it implemented	Impact considerations	Recommendations
OT network traffic inspection • Asset discovery • Anomaly detection (of process control traffic)	Enumeration of industrial network assets in order to detect suspicious or anomalous industrial protocol behavior.	On a network Span port or Tap, specific to each industrial network zone and/or conduit.		
Endpoint hardening • GPO policies • Disabling of unused ports & services • Removal of unused/ unneeded applications • Patching of apps & services	Minimization of attack surface	On each endpoint.	• Hardening reduces the attack surface by removing capability. Care should be taken to avoid accidently disabling or removing functionality that is actually required. • Patching assets is important, but always remember that only known vulnerabilities can be patched. Unfortunately, security research of industrial systems is still far behind that of general computing, and there are likely vulnerabilities that haven't been discovered yet. Always assume that every asset has unpatched vulnerabilities and implement compensating controls to prevent exploitation. • Always check with your vendor before changing a device configuration or applying a patch.	• Approach hardening efforts as a process, implementing and testing each change. • Entropy is the nemesis of hardening: Over time, applications and services that were disabled may be reenabled. Establish a process to periodically audit system configurations to ensure that hardened systems stay that way.
Host protection • HIDS/HIPS • Anti-malware • Application white-listing/allow-listing	Protection against malware/host intrusions	On each endpoint.	• These solutions can consumer large amounts of system resources when configured poorly, which could impact system reliability. • By design these system will prevent malicious files from being accessed or executed, and in some cases malicious files will be deleted. In the event of a false positive,	Because legitimate industrial applications can trigger a false positives from these solutions, these solutions should be implemented carefully and with the guidance and support of control system vendors or integrators.

Table 11.7 Implementing Various Security Controls, With Impact Considerations and Recommendations—cont'd

Control	What it's for	Where is it implemented	Impact considerations	Recommendations
			this could interfere with reliable operation of industrial systems.	
Removable media security • USB scanning stations/kiosks • USB device control/ authorization • Host anti-virus on-access scanning	Detection of malware and related threats from removable devices such as USB drives, phones, etc.	Scanning stations can be physically located almost anywhere. Device control and other host solutions require installation on endpoints.	If configured incorrectly, host controls could prevent or inhibit necessary access to USB devices (such as an authentication key fob) or files on a USB drive.	If scanning stations require connectivity to the Internet, ensure they do not communicate directly with protected nodes (and vice-versa).

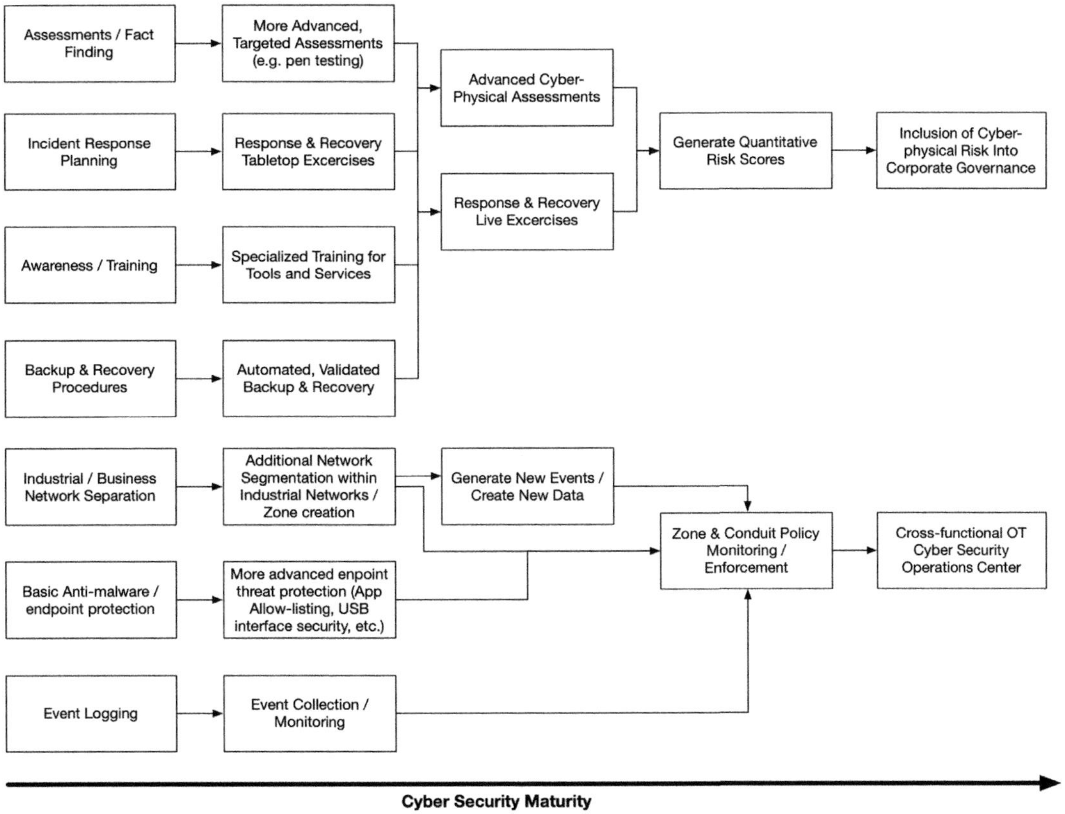

FIGURE 11.7 Cybersecurity controls and countermeasures, maturity, and dependencies.

Summary

Through the identification and isolation of functional groups, quantifiable security zones can be defined. These zones and the conduits that interconnect them can and should be secured using a variety of tools—including network- and host-based firewalls, network- and host-based intrusion detection and prevention systems (IDS/IPS), application monitoring, antivirus, and/or application allow-listing.

In addition to the direct security benefits of these various controls, each also provides useful alerting capabilities that help to improve the situational awareness within the ICS. The information collected from these and other devices can be used to identify and establish baseline behavior and thereafter to detect exceptions and anomalies (see Chapter 12, "Exception, Anomaly, and Threat Detection"). Logs and events from these zone security measures are also useful for overall activity and behavior monitoring (see Chapter 13, "Security Monitoring of Industrial Control Systems"). A solid defense-in-depth approach offers a balanced approach to not only threat prevention but also threat detection that can be used to provide early response, incident containment, and impact control.

Endnotes

1. North American Electric Reliability Corporation (NERC), Standard CIP-005-3a, "Cyber Security - Electronic Security Perimeter."
2. 2North American Electric Reliability Corporation (NERC), Standard CIP-005-4a, "Cyber Security - Electronic Security Perimeter."
3. International Society of Automation (ISA), Standard ANSI/ISA 62443-3-3-2013, "Security for industrial automation and control systems; System security requirements and security levels," SR 5.1 - Network Segmentation. Approved August 12, 2013.
4. North American Electric Reliability Corporation (NERC), Standard CIP-003-3, "Cyber Security - Electronic Security Perimeter."
5. 5International Society of Automation (ISA), Standard ANSI/ISA 62443-3-3-2013, "Security for industrial automation and control systems: System security requirements and security levels," FR 5 - Restricted Data Flow. Approved August 12, 2013.
6. Department of Homeland Security, "Risk-Based Performance Standards Guidance: Chemical Facility Anti-Terrorism Standards," May 2009.
7. U.S. Nuclear Regulatory Commission, Regulatory Guide 5.71 (New Regulatory Guide), Cyber Security Programs for Nuclear Facilities, January 2010.
8. North American Electric Reliability Corporation (NERC), Standard CIP-007-3a, "Cyber Security - Systems Security Management."
9. International Society of Automation, Standard ANSI/ISA-99.02.01-2009, "Security for Industrial Automation and Control Systems: Establishing an Industrial Automation and Control Systems Security Program," Approved January 13, 2009.
10. Department of Homeland Security, Risk-Based Performance Standards Guidance, Chemical Facility Anti-Terrorism Standards, May 2009.
11. U.S. Nuclear Regulatory Commission, Regulatory Guide 5.71 (New Regulatory Guide), Cyber Security Programs for Nuclear Facilities, January 2010.

12. National Infrastructure Security Coordination Center, NISCC Good Practice Guide on Firewall Deployment for SCADA and Process Control Networks, British Columbia Institute of Technology (BCIT), February 15, 2005.
13. Snort.org, SNORT Users Manual 2.9.0., December 2, 2010 (cited: January 19, 2011).
14. Snort. (cited: December 26, 2013).
15. Open Information Security Foundation. "Suricata" (cited: December 26, 2013).
16. Ibid.
17. NitroSecurity, Inc., Network Threat and Analysis Center, Nitrosecurity.com, January 2011.
18. K. Stouffer, J. Falco, K. Scarfone, National Institute of Standards and Technology (NIST), Special Publication 800–82 Revision 1, Guide to Industrial Control Systems (ICS) Security, Section 6.11 System and Information Integrity, May 2013.
19. Ibid.
20. Australian Dept. of Defense - Intelligence and Security, "Strategies to Mitigate Targeted Cyber Intrusion," October 2012.
21. Ibid.
22. Honeywell, Inc. Global Analysis Research and Defense Report: Industrial Cybersecurity USB Threat Report 2020. July 2020.
23. Honeywell Cybersecurity Report: USB Hardware Attack Platforms. October 2020. Honeywell, Inc.
24. Honeywell, Inc. Global Analysis Research and Defense Report: Industrial Cybersecurity USB Threat Report 2020. July 2020
25. IEC. (2021). "Industrial communication networks - Network and system security - Part 4-2: Secure Device Integration." IEC 62443-4-2.
26. "McAfee Probing Bundle That Sparked Global PC Crash," Wired, published April 22, 2010, http://www.wired.com/2010/04/mcafeebungle/, sited July 19, 2014.
27. Chirillo, J. (2003). "Hack Attacks Revealed: A Complete Reference for UNIX, Windows, and Linux with Custom Security Toolkit." Wiley.
28. Laskov, P., Rückert, C., & Kirda, E. (2011). "Detection of Malicious PDF Files and Their Analysis." In Recent Advances in Intrusion Detection (pp. 145–164). Springer.
29. Cisco. (2017). "Behavioral-Based Malware Detection." https://www.cisco.com/c/en/us/products/security/fireamp-endpoints/behavioral-based-malware-detection.html
30. Symantec. (n.d.). "Sandboxing." https://www.symantec.com/security-center/threat-report/Sandboxing
31. Mandiant. (2014). M-Trends 2014: A View from the Front Lines. https://www.fireeye.com/resources/pdfs/fireeye-mandiant-mtrends-2014.pdf
32. YARA Project. (n.d.). https://yara.readthedocs.io/en/stable/
33. Ibid.
34. olter, J. Z., & Maloof, M. A. (2006). Learning to detect and classify malicious executables in the wild. Journal of Machine Learning Research, 7(Oct), 2721-2744.
35. NIST. (2011). "Guide to Industrial Control Systems (ICS) Security." NIST Special Publication 800–82 Revision 2.
36. National Infrastructure Security Coordination Center, NISCC Good Practice Guide on Firewall Deployment for SCADA and Process Control Networks. British Columbia Institute of Technology (BCIT). February 15, 2005.

12

Exception, Anomaly, and Threat Detection

Information in this chapter

- Exception Reporting
- Behavioral Anomaly Detection
- Behavioral Whitelisting
- Advanced Threat Detection

Clear policies about what communications are allowed and what are not have already been obtained by defining zones. The operation within each zone should also be well defined and relatively predictable. This supports two important types of behavioral analysis: exception reporting and anomaly detection.

Exception reporting refers to an automated system that notifies the security administrator whenever a defined policy has been violated. In the context of zone-based security, this means a notification that the defined zone has been violated—a user, system, or service is interacting with the zone in a way that is contrary to security policies established at the perimeter and/or within the zone interior (see Chapter 9, "Establishing Zones and Conduits"). If we expect one behavior but see another, we can view this behavior as a potential threat and take action accordingly.

Anomaly detection picks up where policy-based detection ends, by providing a "ruleless" method of identifying possible threat behavior. Anomaly detection simply takes action when something out of the ordinary occurs. In an industrial system—especially if a strong defense-in-depth posture is maintained and zones are appropriately separated—the normal behavior can be determined, and variations in that behavior should be minimal. The operational behavior of an industrial network should be relatively predictable, making anomaly detection effective once all "normal" actions have been defined.

The effectiveness of anomaly detection pivots on that basic understanding of behavior. Understanding how baseline behavior can be measured is the first step to implementing a useable anomaly detection strategy.

Taken together, clearly defined policies and anomaly detection can provide an additional function called behavioral whitelisting. Behavioral whitelisting combines an understanding of what is known good/bad behavior (policies) with an understanding of expected behaviors, to define what is "known good behavior." Just as whitelists of other known good elements (IP addresses, applications, users, etc.) can be used to enforce perimeter and interior zone defenses, these higher level behavioral whitelists can help to deter broader threats, even across zones.

Industrial Network Security. https://doi.org/10.1016/B978-0-443-13737-2.00005-1

Although each method is effective on its own, attacks rarely occur in clear, direct paths (see Chapter 8 "Risk and Vulnerability Assessments"). Therefore, to detect more sophisticated threats, all anomalies and exceptions need to be assessed together, along with the specific logs and events generated by network switches, routers, security appliances, and other devices including critical industrial control system (ICS) Windows-based assets. Event correlation looks across all systems to determine larger threat patterns that can more clearly identify a security incident. Event correlation is only as good as the data that are available, requiring that all of the mentioned detection techniques be used to generate a comprehensive base of relevant security information. It also requires proper monitoring of networks and devices, as discussed in the next chapter, "Security Monitoring of Industrial Control Systems."

Caution

Automated tools for the detection of exceptions, anomalies, and advanced threats are effective measures to help notify security analysts of incidents that may need to be addressed. However, no tool should be trusted completely; the experience and insight of a human analyst is a valuable component in the security monitoring and analysis process. While tools are often sold with the promise of being "an analyst in a box," even the most well-tuned systems will still produce false positives and false negatives, therefore requiring the additional layer of human intellect to complete the assessment. At the time of publishing, several credible companies have begun offering ICS-focused managed security services that can provide the much needed 24 × 7 security coverage to industrial networks that is absent from many production environments today.

Exception reporting

In Chapter 9, "Establishing Zones and Conduits," specific policies have been developed and enforced by firewalls, intrusion prevention systems, application monitors, and other security devices. Apart from the clear examples of when a specific firewall or intrusion prevention system (IPS) rule triggers an alert, these policies can be used to assess a variety of behaviors. Exception reporting looks at all behaviors, and unlike a hard policy defined on the conduits at a zone's perimeter, which makes black-and-white decisions about what is good and bad, exception reporting can detect suspicious activities by compiling a wealth of seemingly benign security events.

This level of assessment could encompass any measurable function of a zone(s), including network traffic patterns, user access, and operational controls. At a very basic level, exception reporting might be used to inform an operator when something that should not have been allowed (based on zone perimeter policies) has occurred. The first example in Table 12.1 illustrates the concept that it should not be possible for inbound network communication to originate from an unrecognized IP address—that should have been prevented by the default Deny All firewall policy.

Table 12.1 Examples of Suspicious Exceptions

Exception	Policy being enforced	Detected by	Recommended action
Network flow originates from a different zone than the destination IP address	Network separation of functional groups/ zones	Firewall, network monitor, network IDS/IPS, etc. using $Zone_IP variables	Alert only, to create a report on all interzone communications
Network traffic originating from foreign IP addresses is seen within a secured zone	Isolation of critical zones from the internet or outside addresses	Log Manager/Analyzer, SIEM, etc. correlating !$Zone_IP variables and geolocation data	Critical alert to indicate possible penetration of a secure zone
Authorized user accessing the network from a new or different IP address	User access control policies	Log Manager/Analyzer, SIEM, etc. correlating $Zone_IP variables to user authentication activity	Alert only, to create a report on abnormal administrator activity
Unauthorized user performing administrator functions	User access control policies	Log Manager/Analyzer, SIEM, etc. correlating !$Admin_users variables to application activity	Critical alert to indicate potential unauthorized privilege escalation
Industrial protocol used in nonindustrial zones	Network separation of functional groups by protocol	Network monitor, network IDS/ IPS, Application monitor, industrial protocol monitor, etc. using !$Zone_Protocol variables	Alert only, to create a report of abnormal protocol use
Industrial protocol using WRITE function codes outside of normal business hours Identity or authentication systems indicate normal administrative shifts SIEM or other log analysis tool correlates administrative functions against expected shift hours	Administrative control policies	Application monitoring detects $Modbus_Administrator_Functions	Alert only, to create an audit trail of unexpected admin behavior
Industrial protocol using WRITE function codes is originating from a device authenticated to a nonadministrative user Authentication logs indicate a nonadministrative user SIEM or other log analysis tool correlates authentication logs with control policies and industrial protocol functions	User access control policies	Application monitoring detects $Modbus_Administrator_Functions	Critical alert to indicate possible insider threat or sabotage

Other less obvious uses for exception reporting are exemplified in the last example in Table 12.1, where two completely different detection methods (an application monitoring system and a log analysis system) indicate a policy exception that otherwise might seem benign. In this example, the function codes in question are only a concern if executed by an unauthorized user.

Exception reporting can be automated using many log analysis or security information management systems, which are designed to look at information (typically log files) from many sources, and correlate this information together (for more information on how to generate this information, see Chapter 13, "Security Monitoring of Industrial Control Systems"). Exceptions cannot be determined without an understanding of the policies that are in place. Over time, exception reporting should evolve, such that fewer exceptions occur—and therefore fewer reports—as the process matures.

Behavioral anomaly detection

Sometimes, an exception might be seen in a network's expected behavior. As with behavioral malware detection, the abnormal network behavior can indicate something malicious. In the context of network behavior, this behavior could indicate an intrusion attempt, latteral movement of an attacker once a network has been breached, communication to a command and control server, etc. These anomalies can be detected by comparing monitored behavior against known "normal" values. This can be done in a variety of ways: manually, based on real-time monitoring; manually, via log review; automatically, using a network behavior anomaly detection (NBAD) product, log analysis, or security information and event management (SIEM) tool; or automatically, by exporting data to a dedicated spreadsheet or other statistical application. Note that even with highly automated systems—such as SIEM—a degree of human analysis is still required. The value of an automation tool is in its ability to simplify the process for the human analyst, using various detection algorithms, correlation, event scoring, and other techniques to add context to the raw data. Beware of any tool that claims to eliminate the need for human cognizance, as there is no such thing as an "analyst in a box." Whether performed manually or automatically, an anomaly cannot be detected without an established baseline of activity upon which to compare. Once a baseline has been established for a given metric (such as the volume of network traffic and the number of active users), that metric must be monitored using one or more of the methods described in Chapter 13, "Security Monitoring of Industrial Control Systems."

Measuring baselines

Baselines are time-lagged calculations based on running averages. They provide a basis (base) for comparison against an expected value (line). Baselines are useful for comparing past behaviors to current behaviors but can also be used to measure network or application capacity or almost any other operational metric that can be tracked over time. A baseline should not be confused with a trend analysis—a baseline is a value; nothing more, nothing less. Using that metric in an analysis of past-observed behavior and future-predicted behavior is a trend analysis—a forward-looking application of known baselines to predict the continuation of observed trends.

A baseline can be simple or complex—anything from a gut understanding of how a system works to a sophisticated statistical calculation of hard, quantifiable data. The simplest method of establishing a baseline is to take all data collected over a period of time and use whatever metric is available to determine the average over time. This is a commonly used method that is helpful in determining whether something is occurring above or below a fixed level. In Figure 12.1, for example, it can be clearly seen that production output is either above or below the average production level for the previous 12 months. The specific peaks and valleys could represent anything from a stalled process to normal variations in process schedules. This concept is very similar to the statistical process control (SPC)/statistical quality control (SQC) $\bar{x}$ and R control chart comprising a control limit (equal to the baseline) with upper and lower control limits (UCL/LCL) that are used to signify events that are out of normal allowable tolerances.

This may or may not be useful for operations management; in a security context, this type of baseline provides little value. Knowing that 59,421,102 events over 30 days = 1,980,703 events per day average cannot tell us if the current day's event volume of 2,000,000 is meaningful or not, without some additional context. Does the yearly average include weekends and other periods of downtime? If it does, the actual per day expected values of a workday could be considerably higher. For purposes of behavioral analysis, a more applicable method would be a similar calculation that excludes known periods of downtime and creates a flat baseline that is more relevant to periods of operation. Better still are time-correlated baselines, where an observed period

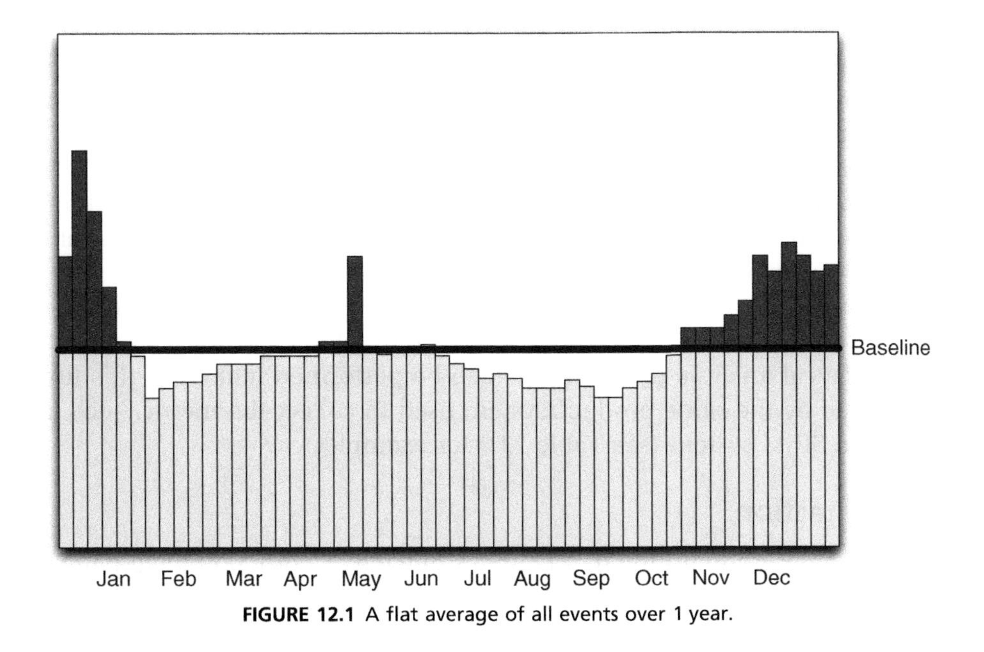

FIGURE 12.1 A flat average of all events over 1 year.

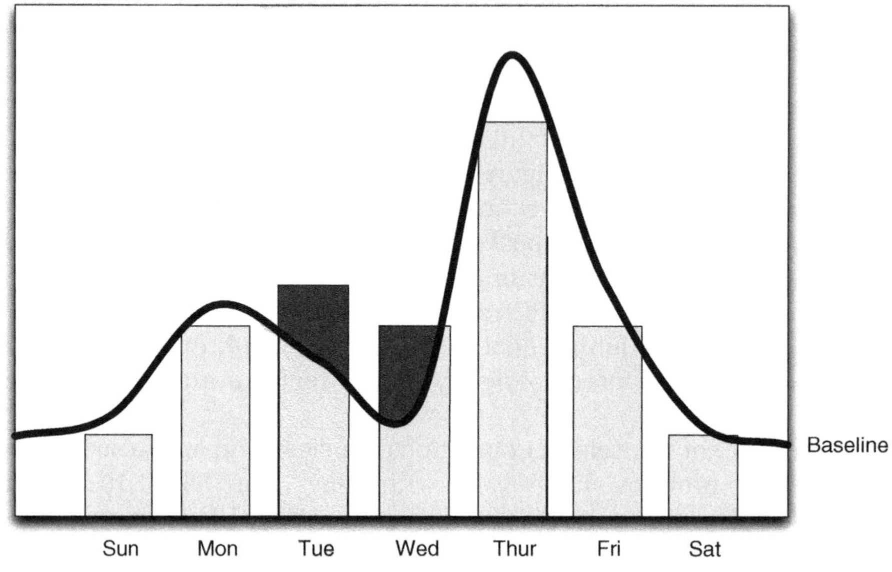

FIGURE 12.2 A time-correlated baseline shows dip on weekends, peak on Thursdays.

of activity is baselined against data samples taken over a series of similar time periods. That is, if looking at data for one (1) week, the baseline might indicate the expected patterns of behavior over a period of several weeks. Figure 12.2 illustrates how this affects the flatline average with a curved baseline that visualizes a drop in activity during weekends and shows an expected peak on Thursdays. Note that sufficient historical data are required to calculate time-correlated baselines.

Time-correlated baselines are very useful because they provide a statistical analysis of observed activity within relevant contexts of time—essentially providing historical context to baseline averages.[1] Without such a baseline, a spike in activity on Thursday might be seen as an anomaly and spur an extensive security analysis, rather than being clearly indicated as normal behavior. Consider that there may be scheduled operations at the beginning of every month, at specific times of the day, or seasonally, all causing expected changes in event volumes.

Baselines, in whatever form, can be obtained in several ways, all beginning with the collection of relevant data over time, followed by statistical analysis of that data. Although statistical analysis of any metric can be performed manually, this function is often supported by the same product/system used to collect the metric, such as a data historian or an SIEM system (see Table 12.2 for examples).

Anomaly detection

An anomaly is simply something that happens outside of normal defined parameters or boundaries of operation. Many firewalls and IDS/IPS devices may support anomaly

Table 12.2 Measurement and Analysis of Baseline Metrics

Behavior	Measured Metric(s)	Measured by	Analyzed by
Network traffic	• Total unique source IPs • Total unique destination IPs • Total unique TCP/UPD ports • Traffic volume (total flows) • Traffic volume (total bytes) • Flow duration	• Network switch/router flow logs (i.e., netFlow, jFlow, sFlow, or similar) • Network probe (i.e., IDS/IPS, network monitor, etc.)	• Network behavior anomaly detection (NBAD) system • Log management system • SIEM system
User activity	• Total unique active users • Total logons • Total logoffs • Logons by user • Logoffs by user • Activity (e.g., configuration changes) by user Note: User activity may need additional layers of correlation to consolidate multiple usernames/accounts associated with a single user	• Application logs • Database logs and/or transaction analysis • Application logs and/or session analysis • Centralized authentication (LDAP, active directory, IAM)	• Log management system • SIEM system
Process/control behavior	• Total unique function codes • Total number per individual function code • Total set point or other configuration changes	• Industrial protocol monitor • Application monitor • Data historian tags	• Data historian • SIEM system
Event/incident activity	• Total events • Total events by criticality/severity • Total events by security device	• Security device (i.e., firewall, IPS) logs	• Application monitor • Industrial protocol filter

detection directly, providing an additional detection capability at the conduits existing at a zone's perimeter. Holistically, all behaviors can be assessed for more systematic anomalies indicative of larger threats. Luckily, anomalies could be easily identified having defined expected (baseline) behaviors. In addition, many automated systems—including NBAD, log management, and SIEM systems—are available to facilitate anomaly detection across a number of different sources.

Behavioral anomaly detection is useful because there is no dependency upon a detection signature, and therefore, unknown threats or attacks that may utilize 0-day capabilities can be identified. In addition, although often thought of exclusively in terms of network anomalies, any metric that is collected over time can be statistically analyzed and used for anomaly detection.

For example, an unexpected increase in network latency—measurable by easily obtained network metrics, such as TCP errors, the size of the TCP receive window, the round-trip duration of a ping—can indicate risk to the industrial network.[2] However, as can be seen in Table 12.3, anomalies can indicate normal, benign variations in behavior as well as potential threats. In other words, the rate of false positives tends to be higher using anomaly detection techniques.

Table 12.3 Examples of Suspicious Anomalies

Normal behavior	Anomaly	Detected by	Indication
All Modbus communications to a group of PLCs originates from the same three HMI workstations	A fourth system communicates to the PLCs	• A >20% increase in the number of unique source IP addresses, from analysis of: Network flows • Security event logs from firewalls, IPS devices, etc. • Application logs • Etc.	• A new, unauthorized device has been plugged into the network (e.g., an administrator's laptop) • A rogue HMI is running using a spoofed IP address • A new system was installed and brought online
Every device has a single MAC address and a single IP address	An IP address is seen originating from two or more distinct MAC addresses	• >1 MAC addresses per IP, from analysis of: Network flows • Security event logs from firewalls, IPS devices, etc. • Application logs • Etc.	• An attacker is spoofing an IP address • A device has failed and been replaced with new hardware
Process within a control system zone is running for extended periods	Traffic increases above expected volumes	A >20% increase in the total network traffic, in bytes, from analysis of network flows	• An unauthorized service is running • A network scan or **penetration test** is being run • A shift change is underway • A new batch or process has started
Traffic decreases below expected levels	A >20% decrease in the total network traffic, in bytes, from analysis of network flows	• A service has stopped running • A networked device has failed or is offline • A batch or process has completed	
Changes to controller logic within BPCS, SIS, PLC, RTU	Industrial network monitor such as an SCADA IDS ladder logic/code review	• Any variation in the individual function codes and/or frequency of any function code, from analysis of industrial protocol monitors • Application monitors • SCADA IDS/IPS logs	• A process has been altered • A new process has been implemented • An old process has been removed • A process has been sabotaged
Control protocol traffic containing function codes that indicate changes to process operations or logic	• An operator station modifies a set point • New logic is written to a controller	Control protocol traffic containing function codes that indicate changes to process operations or logic	• Infrequent process operations have occurred (a false positive). • A planned change is being implemented by operators (a false positive). • A rogue actor has gained access to the industrial network and is making unexpected changes to the process. • An adversary is making changes to process functionality programmatically from an infected asset with the industrial network (e.g., using indestroyer or incontroller, or similar cyber-physical attack framework).
Authorized users log on to common systems at the beginning of a shift	• Unauthorized user logs on to a system normally accessed by administrators only • Authorized users log on to a system outside of normal shift hours • Authorized users log on to unknown of unexpected systems	• Any variation seen from analysis of authentication logs from active directory operating system logs • ICS application logs	• Personnel changes have been made • An administrator is on leave or absent and duties have been delegated to another user • A rogue user has authenticated to the system • An administrator account has been compromised and is in use by an attacker

Analyzing IT versus OT metrics

Up to this point, the discussion of anomaly detection has focused largely on security events derived from information technology (IT) tools. Even when looking at specialized security products for industrial network monitoring, these devices operate on the same paradigm as IT security devices to detect and block suspicious and/or "out of policy" events, subsequently generating an alert.

Anomaly detection tools

Anomaly detection can be done using anything from "gut feelings," to manual statistical analysis using a spreadsheet or mathematical application, to specialized statistics software systems, to network and security data analysis systems, such as certain log management and SIEM systems. Time-series databases, such as those used by data historians, can also be used for anomaly detection. While these systems do not typically represent anomalies within the specific context of network security, a historian configured to show comparative overlays of security events over time could easily identify dangerous anomalies that might indicate a cyberattack.

NBAD, log management, and SIEM tools are predominantly used for security-related anomaly detection. NBAD systems are focused exclusively on network activity and may or may not support the specific industrial network protocols used within an ICS environment. As such, the use of a log management or SIEM system may be better suited for anomaly detection in industrial networks. For example, Figure 12.3 shows a visual representation of anomalous authentication behavior for the administrative user (on the right) versus the same data shown without context (on the left); the security tool has done the necessary statistical analysis to show a 184% increase in administrator logins and has also brought that anomaly to the attention of the security analyst.

As shown in Table 12.3, this requires that the log management or SIEM system is used to collect relevant data over time from those systems used in perimeter and interior zone security, as well as any relevant network traffic data obtained from network switches and routers.

FIGURE 12.3 Representation of anomalous administrator logins using an SIEM system.

Tip

When selecting an analysis tool for industrial network anomaly detection, consider the greatest relevant time frame for analysis and ensure that the system is capable of automating anomaly detection over sufficient periods of time. Many systems, such as log management and SIEM systems, are not designed exclusively for anomaly detection and may have limitations as to how much information can be assessed and/or for how long.

To ensure the tool is right for the job, look at the operational lifespan of specific processes and use time-correlated baselines to determine normal activities for those processes. If a process takes 3 h, analysis of $n \times 3$ h of process data is needed for anomaly detection, where n represents the number of sampled operations. The greater the n, the more accurate the baseline and associated anomaly detection.

Tip

There are ICS network monitoring and intrusion detection systems available that automatically model normal and acceptable network behavior and generate alerts whenever some network devices perform activities that diverge from their intended operation. For adequate behavior-based detection, these systems should first analyze network communications and generate a behavioral baseline—a valuable blueprint that defines communication patterns, protocols, message types, message fields, and field values that are normal for the monitored process. A review of the "blueprint" can reveal network and system misconfigurations (e.g., rogue devices), unintended communications, and unusual field values employed in the network. Continuous monitoring is then able to detect whenever network devices perform unintended activities—or anomalies outside the normal band.

This type of continuous monitoring is also useful for reporting observed network communications—in terms of communication patterns, protocols, and protocol message types normally used by the devices in the network—to additional security analytics tools, such as SIEM or anomaly behavior analysis systems, which are then able to perform even deeper analysis over longer periods of time.

Behavioral whitelisting

Whitelisting is well understood in the context of access control and application whitelisting (AWL) for host malware prevention. However, the concept of whitelisting has many roles within control system environments, where access, communication, processes, policies, and operations are all well-defined. Using the controlled nature of these systems and the zone-based policies defined in Chapter 9, "Establishing Zones and Conduits," whitelists can be defined for a variety of network and security metrics, including users, assets, applications, and others.

Whitelists can be actively enforced via a Deny! Whitelist policy on a firewall or IPS or can be used throughout a network by combining network-wide monitoring and exception reporting with dynamic security controls. For example, if an exception is seen to a policy within a zone, a script can be run to tighten the specific perimeter defenses of that zone at all affected conduits.

User whitelists

Understanding user activity—especially of administrative users—is extremely useful for detecting cyberattacks, both by insiders (e.g., intentional actors like a disgruntled employee, or unintentional actors like the control system engineer or subcontractor/ vendor) as well as by outside attackers. Locking critical functions to administrative personnel, and then following best practices of user authentication and access control, means an attack against a critical system should have to originate from an administrative user account. In reality, enumeration is a standard process in a cyberattack because administrative accounts can be used for malicious intent (see Chapter 8, "Risk and Vulnerability Assessment"). They can be hijacked or used to escalate other rogue accounts in order to enable nonauthorized users' administrator rights.

Note

It should be pointed out that the term "administrator" does not have to mean a Windows Administrator account but could represent a special Windows Group or Organizational Unit that has been established containing users with "elevated" privileges for particular applications. Some ICS vendors have implemented this concept and facilitate the creation of separate application administrative roles from Windows administrative roles.

Note

Many ICS applications were developed and commissioned when cybersecurity was not a priority. The applications may require administrative rights to execute properly and may even require execution from an administrator interactive account. These represent a unique problem discussed not only earlier but also in Chapter 7, "Hacking Industrial Systems" due to the fact that if these applications or services can be exploited, the access level of the resulting payload is typically at the same level as the compromised component—the administrator in this case!

Tip

It is important to understand the ICS application software that is installed within a given facility, not only in terms of potential vulnerabilities within the application code base but also implementation or configuration weaknesses that can easily be exploited. It is typically not possible for a user to assess the software coding practices of their ICS vendor. The US

Department of Homeland Security (DHS) has developed the "Cyber Security Procurement Language for Industrial Control Systems"[3] guidance document that provides useful text that can be added to technical specifications and purchasing documents to expose and understand many hidden or latent potential weaknesses within the ICS components.

Fortunately, authorized users have been identified and documented (see Chapter 9, "Establishing Zones and Conduits"), and this allows us to whitelist user activities. As with any whitelist, the list of known users needs to be established and then compared to monitored activity. Authorized users can then be identified using a directory service or an Identity and Access Management (IAM) system, such as Lightweight Directory Access Protocol (LDAP) included with Microsoft Active Directory, or other commercial IAM systems from IBM, Oracle, Sun, and others.

As with exception reporting, the whitelist is first defined, and then monitored activity is compared against it. If there is an exception, it becomes a clear indicator that something outside of established policies is occurring. All known good user accounts are used as a detection filter against all login activity in the case of a user whitelist. If the user is on the list, nothing happens. If the user is not on the list, it is assumed bad and an alert is sent to security personnel. This accomplishes an immediate flag of all rogue accounts, default accounts, or other violations of the authentication policies. In early 2011, a security researcher was able to uncover hard-coded credentials within a PLC, and then used these credentials to gain shell access to the PLC.[4]

Note

In the case of hidden accounts and other hard-coded backdoor authentications, normal connections would also be flagged as an exception because those accounts would most likely not appear on the whitelist. This could generate a potential excess of false-positive alerts. However, it would also draw attention to the existence of accounts that leverage default authentication within the system so that these accounts could be more closely monitored. For example, the WinCC authentication (used as one propagation mechanism in the Stuxnet campaign) could be monitored in conjunction with baseline analysis. If the default account was then used by new malware that was developed with knowledge learned from Stuxnet, it would still be possible to detect the threat via anomaly detection.

Asset whitelists

Once an inventory of cyber assets is completed—either automatically via an appropriate soft and "friendly" network scan (see Chapter 8, "Risk and Vulnerability Assessment") or manual inventory—the resulting list of known, authorized devices can be used to whitelist known good network devices.

Unlike perimeter-based security policies that may only allow known good devices into a zone or "inter-zone," a network asset whitelist can be applied to devices within a zone or "intra-zone." If a spoofed address or rogue device appears within a zone, it can still be detected via exception reporting against the list of known good devices so that action can be taken.

A classic use case for asset whitelisting is the use of "sneaker net," which can be used to carry files (documents, databases, applications) past perimeter defenses and attached directly to a protected network, well within a secure zone. This could be benign—an employee bringing a smart phone inside a control system that has Wi-Fi enabled—or it could be a deliberate vehicle for sabotage. Either way, the IP address of the device will be detected by switches, routers, network monitors, and security devices, and will eventually be seen in logs or events that are centralized and managed, as illustrated in Figure 12.4. At this point, simple comparison against the defined whitelist will identify the presence of an unauthorized device. This example represents significant risk, as the mobile device (smart phone in this case) also connects directly to a 3G or 4G cellular network, which bypasses all defensive measures of the electronic security perimeter, and opens the zone up for attack or further exploitation.

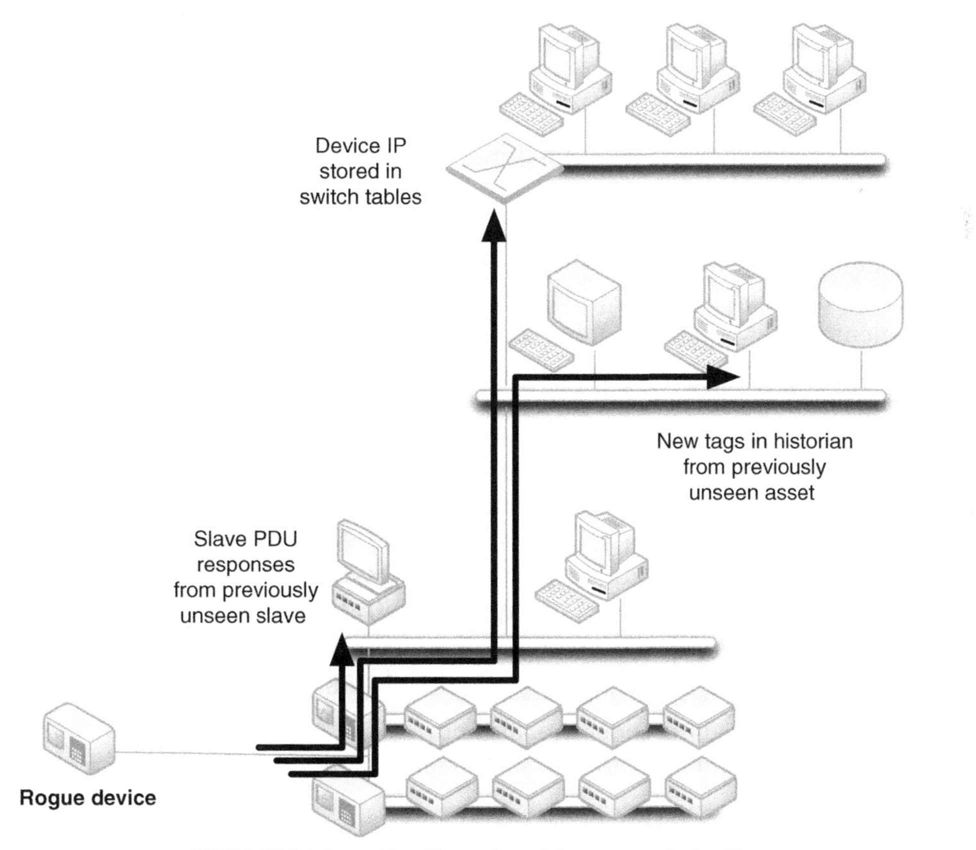

FIGURE 12.4 Information flow relevant to a rogue device IP.

Tip

One easy and effective method to prevent the introduction of unauthorized or foreign devices in a secure ICS zone is by disabling dynamic hardware addresses (e.g., media access control address) on the network switches within the zone. Default switch configurations allow dynamic creation of MAC tables within the switch effectively allowing any newly discovered device to begin forwarding and receive traffic. Disabling this feature not only secures the zone from intentional and malicious actors but also from unintentional insiders accidently connecting devices not authorized for use within the zone—as defined by the security goals of the zone (see Chapter 9, "Establishing Zones and Conduits").

The whitelists themselves would need to be generated and applied to the central management system—most likely a log management or SIEM system that is capable of looking at device metrics across the entire network. Depending upon the specific monitoring product used, the whitelist might be built through the use of a defined system variable (much like the generation of zone-specific variables in firewalls and IDS/IPS devices, as discussed in Chapter 11, "Implementing Security and Access Controls"), configurable data dictionaries, manually scripted detection signatures, and so on.

Application behavior whitelists

Applications themselves can be whitelisted per host using an AWL product. It is also possible for the application behavior to be whitelisted within the network. As with asset whitelisting, application behavior whitelists need to be defined so that good behavior can be differentiated from bad behavior. A central monitoring and management system can utilize application behavior whitelists by defining a variable of some sort within a log management or SIEM system just like asset whitelists. However, because of the nature of industrial network protocols, many application behaviors can be determined directly by monitoring those protocols and decoding them in order to determine the underlying function codes and commands being executed (see Chapter 6, "Industrial Network Protocols"). This allows for in-line whitelisting of industrial application behavior in addition to network-wide whitelisting offered by a log management or SIEM system. If in-line whitelisting is used via an industrial security appliance or application monitor, network whitelisting may still be beneficial for assessing application behavior outside of industrial control systems (i.e., for enterprise applications and ICS applications that do not utilize industrial protocols).

Some examples of application behavior whitelisting in industrial networks include

- Only read-only function codes are allowed.
- Master Protocol Data Units (PDU) or Datagrams are only allowed from predefined assets.
- Only specifically defined function codes are allowed.

- Some examples of application behavior whitelisting in enterprise networks include
- Only encrypted HTTP web traffic is allowed and only on Port 443.
- Only POST commands are allowed for web form submissions.
- Human–machine interface (HMI) applications are only allowed on predefined hosts.

  Some examples of application behavior whitelisting across both environments together include
- Write commands are only allowed in certain zones, between certain assets, or even during certain times of the day.
- HMI applications in supervisor networks are only allowed to use read functions over authorized protocols.

In other words, unlike AWL systems that only allow certain authorized applications to execute, application behavior whitelisting only allows applications authorized to execute to function in specifically defined ways on the network.

For example, an AWL system is installed on a Windows-based HMI. The AWL allows for the HMI application to execute, as well as a minimal set of necessary operating system services, and the networking services required to open Modbus/TCP network sockets so that the HMI can communicate to a series of RTUs and PLCs. However, the AWL does not control how the HMI application is used, and what commands and controls it can enforce on those RTUs and PLCs. A disgruntled employee can shut down key systems, randomly change set points, or otherwise disrupt operations using an HMI even though it is protected by AWL. Network-based application behavior whitelisting looks at how the HMI application is being used and compares that to a defined whitelist of authorized commands—in this case, a list of known good Modbus function codes. Functions that are not explicitly defined may then be actively blocked, or they may be allowed but the system may generate an alert to notify administrators of the violated policy.

Industrial protocol or application monitoring tools should possess a base understanding of industrial protocols and their functions, allowing behavioral whitelists to be generated directly within the device. For network-wide behavioral whitelisting, variables or data dictionaries need to be defined. Common variables useful in application behavioral whitelisting include these same application function codes—the specific commands used by industrial protocols, ideally organized into clear categories (read, write, system commands, synchronization, etc.).

Note

It has probably become clear that there is a great deal of similarity between application behavior whitelisting at the host-level and deep-packet inspection at the network level. Both technologies require application and/or protocol knowledge, and both provide a mechanism for an additional layer of protection beyond what or who is allowed to execute commands to what commands can be executed. These technologies should be appropriately deployed based on the target security level desired within a particular zone.

Examples of beneficial whitelists

Many whitelists can be derived using the functional groups defined in Chapter 9, "Establishing Zones and Conduits." Table 12.4 identifies some common whitelists, and how those whitelists can be implemented and enforced.

Smart-Lists

The term "Smart-Lists" was first introduced at the SANS Institute's 2010 European SCADA and Process Control Summit in London, United Kingdom. "**Smart-Listing**" combines the concept of behavioral whitelisting with a degree of deductive intelligence. Where blacklists block what is known to be bad, and whitelists only allow what is known to be good, Smart-Lists use the latter to help dynamically define the former.

For example, if a critical asset is using AWL to prevent malicious code execution, the AWL software will generate an alert when an unauthorized application attempts to execute. What can now be determined is that the application is not a known good application for that particular asset. However, it could be a valid application that is in use elsewhere and has attempted to access this asset unintentionally. A quick correlation against other whitelists can then determine if the application under scrutiny is an acceptable application on other known assets. If it is, the "Smart-Listing" process might result in an informational alert and nothing more. However, if the application under scrutiny is not defined anywhere within the system as a known good application, the smart-listing process can deduce that it is malicious in nature. It then defines it within the system as a known bad application and proactively defends against it by initiating a script or other active remediation mechanism to block that application wherever it might be detected.

Table 12.4 Examples of Behavioral Whitelists

Whitelist	Built using	Enforced using	Indications of a violation
Authorized devices by IP	• Network monitor or probe (such as a network IDS) • Network scan	• Firewall • Network monitor • Network IDS/IPS	A rogue device is in use
Authorized applications by port	• Vulnerability assessment results • Local service scan • Port scan	• Firewall • Network IDS/IPS • Application flow monitor	A rogue application is in use
Authorized applications by content	• Application monitor	An application is being used outside of policy	
Authorized function codes/ commands	• Industrial network monitor, such as an ICS IDS • Ladder logic/code review	• Application monitor • Industrial protocol monitor	A process is being manipulated outside of policy
Authorized users	• Active directory services • IAM	• Access control • Application log analysis • Application monitoring	A rogue account is in use

"Smart-Listing" therefore combines what we know from established whitelists with deductive logic in order to dynamically adapt our blacklist security mechanisms (such as firewalls and IPS devices) to proactively block newly occurring threats. This process is illustrated in Figure 12.5. First, an alert is generated that identifies a violation of an established policy. Next, the nature of that alert is checked against other system-wide behavior. Finally, a decision is made—if it is "bad" a script or other automation service may be used to dynamically update firewall, IDS/IPS, and other defenses so that they can actively block this activity. If not, the activity might generate an alert, or be ignored.

Smart-Listing is a relatively new concept that could greatly benefit zone defenses by allowing them to automatically adapt to evasive attacks as well as insider attacks. Smart-Listing is especially compelling when used with overarching security management tools (see Chapter 13, "Security Monitoring of Industrial Control Systems"), as it requires complex event association and correlation. Although it has yet to be determined how widely security analysis and information management vendors will adopt this technique and whether ICS suppliers will endorse this approach, at present, the techniques can be performed manually, using any number of log management or SIEM tools.

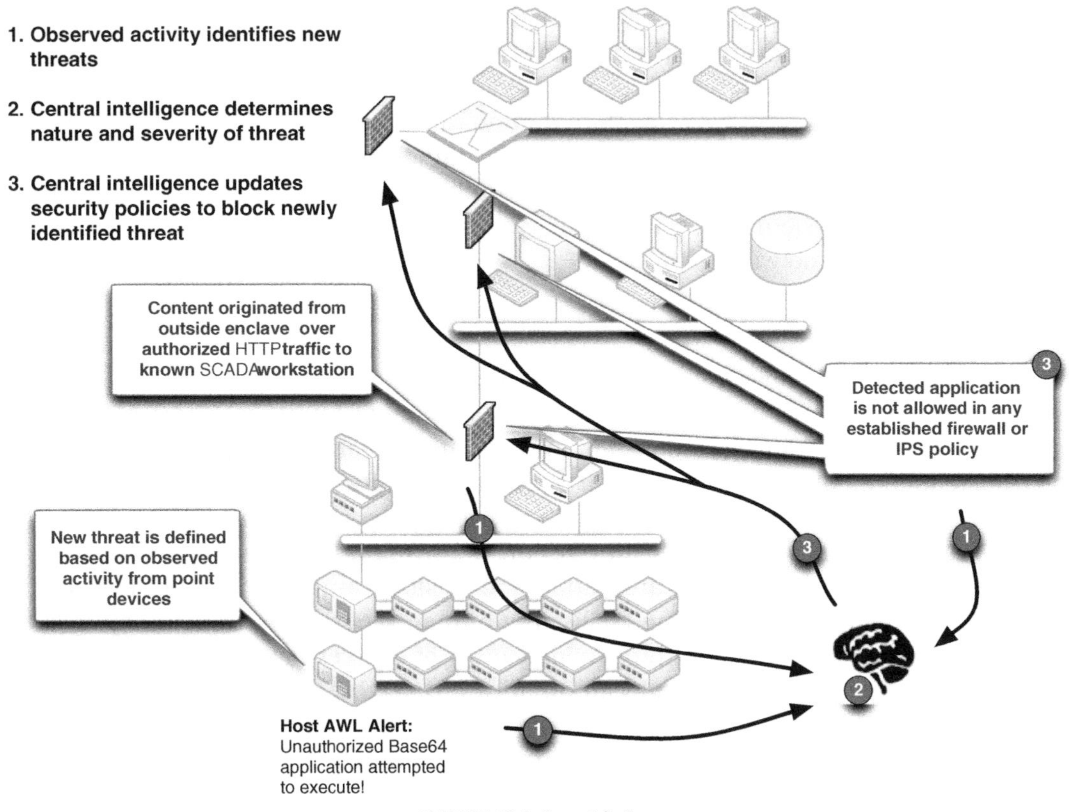

FIGURE 12.5 Smart-listing.

Advanced threat detection

Used independently, the specific detection techniques discussed up to this point—security device and application logs, network connections, specific alerts generated by exception reporting or anomaly detection, and violations of whitelists—provide valuable data points indicating events where a specific policy was violated. Even simple attacks consist of multiple steps. For the detection of an incident (vs. a discrete event), it is necessary to look at multiple events together and search for broader patterns. For example, many attacks will begin with some form of assessment of the target, followed by an enumeration technique, followed by an attempt to successfully authenticate against an enumerated account. (The remaining steps of elevating local privileges, creating persistent access, and covering tracks leave easy indicators for the numerous security controls described to this point.) This pattern might equate to firewall alerts indicating a ping sweep, followed next by access to the and files, ending with a brute force login. The detection of this larger threat pattern is known as event correlation. As cyber-attacks continue to increase in sophistication, event correlation methods have continued to expand. They consider event data from a wider network of point security devices, additional event contexts, such as user privileges or asset vulnerabilities, and search for more complex patterns.

In looking at cyber-physical threats such as Stuxnet, Industroyer, et. al., another factor is introduced that further complicates the event correlation process. Prior to Stuxnet, a threat had never before involved events from both IT and OT systems; now cyber-physical threats are increasingly common, and the correlation of events across both IT and OT systems is necessary with the evolution of threat patterns that traverse both domains. The problem is that event correlation systems were not designed to accommodate OT systems, presenting challenges in the detection of the most serious threats to industrial networks.

Event correlation

Event correlation simplifies the threat detection process by making sense of the massive amounts of discrete event data, analyzing it as a whole to find the important patterns and incidents that require immediate attention. Although early event correlation focused on the reduction of event volumes in order to simplify event management—often through filtering, compressing, or generalizing events[5]—newer techniques involve state logic to analyze event streams as they occur, performing pattern recognition to find indications of network issues, failures, attacks, intrusions, and so on.[6] Event correlation is useful in several ways, including facilitating human security assessments by making the large volumes of event data from a wide variety of sources more suitable for human consumption and comprehension, by automatically detecting clear indications of known threat patterns to easily detect incidents of cyberattack and sabotage, and by facilitating the human detection of unknown threat patterns through event normalization. The process of event correlation is depicted in Figure 12.6.

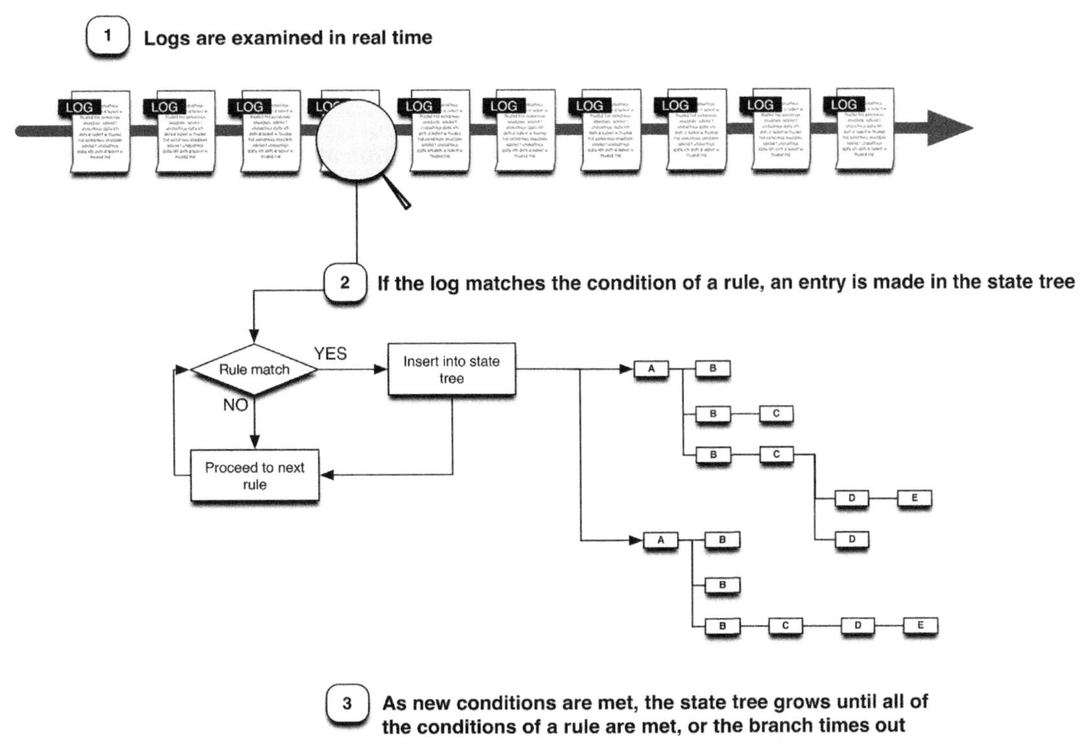
1
Logs are examined in real time
LOG
2
If the log matches the condition of a rule, an entry is made in the state tree
Rule match
YES
NO
Insert into state tree
Proceed to next rule
A
B
C
D
E
3
As new conditions are met, the state tree grows until all of the conditions of a rule are met, or the branch times out

FIGURE 12.6 The event correlation process.

Events are first compared against a defined set of known threat patterns or "correlation rules." If there is a match, an entry is made in a (typically) memory-resident state tree; if another sequence in the pattern is seen, the rule progresses until a complete match is determined. For example, if a log matches the first condition of a rule, a new entry is made in the state tree, indicating that the first condition of a rule has been met. As more logs are assessed, there may be a match for a subsequent condition of an existing branch at which point that branch is extended. A log may meet more than one condition of more than one rule, creating large and complex state trees. For example, even a simple "brute force attack" rule can create several unique branches. Consider the rule.

```
If [5 consecutive failed logins] from [the same source IP] to [the
same destination IP] within [5 minutes]
```

This example would create one branch for the first failed login event "A" from any IP address to any other IP address. The next matching login event "B" would extend that initial branch while also generating a new branch (with a new timer):

```
A+B
B
```

The third matching login event "C" would extend the first two branches while also creating a third:

```
A+B+C
B+C
C
```

This will continue ad infinitum until all of the conditions are met, or until a branch's timer expires. If a branch completes (i.e., all conditions are met), the rule triggers.

Note that events are collected from many types of information sources, such as firewalls, switches, and authentication services. They must be normalized into a common event taxonomy before they can be effectively correlated. Normalization categorizes activities into a common framework so that similar events can be correlated together even if the originating log or event formats differ.[7] Without normalization, many additional correlation rules would be required in order to check a condition (in this example a failed login) against all possible variations of that event that may be present (Windows logins, Application logins, etc.).

For purposes of threat detection, the entire event correlation process is typically performed in memory at the time the individual logs and events are collected. Correlation can also be performed manually by querying larger stores of already collected events to find similar patterns.[8]

Examples of event correlation rules are provided in Table 12.5. Event correlation may be very basic (e.g., a brute force attack) or highly complex—up to and including tiered correlation that consists of correlation rules within correlation rules (e.g., a brute force attack followed by a malware event).

Data enrichment

Data enrichment refers to the process of appending or otherwise enhancing collected data with relevant context obtained from additional sources. For example, if a username is found within an application log, that username can be referenced against a central

Table 12.5 Example Event Correlation Rules

Threat pattern	Description	Rule
Brute force attack	Passwords are guessed randomly in quick succession in order to crack the password of a known user account	A number N of failed logon events, followed by one or more successful logon events, from the same source IP
Outbound Spambot behavior	A spambot (malware designed to send spam from the infected computer) is sending bulk unsolicited e-mails to outside addresses	A large number N of outbound SMTP events, from one internal IP address, each destined to a unique email address
HTTP command and control	A hidden (covert) communication channel inside of HTTP (overt) is used as a command and control channel for malware	HTTP traffic is originating from servers that are not HTTP servers
Covert botnet, command, and control	A distributed network of malware establishing covert communications channels over applications that are otherwise allowed by firewall or IPS policy	Traffic originating from N number of $ControlSystem_Zone01_Devices to !$ControlSystem_Zone01_Devices with contents containing Base64 coding.

IAM system (or ICS application if Application Security is deployed) to obtain the user's actual name, departmental roles, privileges, and so on. This additional information "enriches" the original log with this context. Similarly, an IP address can be used to enrich a log file, referencing IP reputation servers for external addresses to see if there is known threat activity associated with that IP address, or by referencing geolocation services to determine the physical location of the IP address by country, state, or postal code (see "Additional Context" in Chapter 13, "Security Monitoring of Industrial Control Systems," for more examples of contextual information).

Caution

Many of the advanced security controls described in this chapter leverage the use of external threat intelligence data. It is always important to remember to follow strict security policies on network connectivity between trusted control zones and less-trusted enterprise and public (i.e., Internet) zones. This can be addressed by proper location of local assets requiring remote information, including the creation of dedicated "security zones" within the semi-trusted DMZ framework.

Data enrichment can occur in two primary ways. The first is by performing a lookup at the time of collection and appending the contextual information into the log. Another method is to perform a lookup at the time the event is scrutinized by the SIEM or log management system. Although both provide the relevant context, each has advantages and disadvantages. Appending the data at the time of collection provides the most accurate representation of context and prevents misrepresentations that may occur as the network environment changes. For example, if IP addresses are provided via the Dynamic Host Configuration Protocol (DHCP), the IP associated with a specific log could be different at the time of collection than at the time of analysis. Although more accurate, this type of enrichment also burdens the analysis platform by increasing the amount of stored information. It is important to ensure that the original log file is maintained for compliance purposes, requiring the system to replicate the original raw log records prior to enrichment.

The alternative, providing the context at the time of analysis, removes these additional requirements at the cost of accuracy. Although there is no hard rule indicating how a particular product enriches the data that it collects, traditional log management platforms tend toward analytical enrichment, whereas SIEM platforms tend toward enrichment at the time of collection possibly because most SIEM platforms already replicate log data for parsing and analysis, minimizing the additional burden associated with this type of enrichment.

Tip

Data enrichment is an important process when attempting to correlate cyber ("IT") events and physical ("OT") alarms. This is because the data contained within each data source is different, and there may be no single common data point to correlate on. For example, most typical events managed by cybersecurity systems will include computer names, IP address(es), MAC address(es), and similar "IT-centric" information. Most process alarms, however, will not include these data points, instead focusing on process control server IDs and domains, asset Node IDs, and other data objects specific to the industrial process. Data enrichment allows common, correlateable data points to be added to both "IT" and "OT" event data, which can then be used to correlate these events together using event management and monitoring tools.

Normalization

Event normalization is a classification system that categorizes events according to a defined taxonomy, such as the Common Event Expression Framework provided by the MITRE Corporation.[9] Normalization is a necessary step in the correlation process, due to the lack of a common log format.[10] Table 12.6 provides a comparison of authentication logs associated with logon activity from a variety of sources.

Note

In 2006, security software company ArcSight (purchased by Hewlett–Packard in 2010) saw the need to improve the interoperability of devices in terms of how event data are logged and transmitted. The problem at the time was that each vendor had their own unique format for reporting event information that was often found to lack the necessary information needed to integrate these events with other systems. This new format was called the Common Event Format (CEF) and defined a syntax for audit log records comprised of a standard header and a variable expression formatted as key-value pairs. CEF allows vendors of both security and nonsecurity devices to structure their syslog event data making it more easily parsed.[11]

Although each example in Table 12.6 is a logon, the way the message is depicted varies sufficiently such that without a compensating measure, such as event

Table 12.6 Common Logon Events Depicted by Varying log Formats[13]

Log source	Log contents	Description
Juniper firewall	Dec 17 15:45:57 10.14.93.7 ns5xp: NetScreen device_id 5 ns5xp system-warning-00515: Admin user jdoe has logged on via Telnet from 10.14.98.55:39073 (2002-12-17 15:50:53)	Successful logon
Cisco router	Dec 25 00:04:32:%SEC_LOGIN-5-LOGIN_SUCCESS:Login Success [user:jdoe] [Source:10.4.2.11] [localport:23] at 20:55:40 UTC Fri Feb 28, 2006	Successful logon
Redhat Linux	Mar 4 09:23:15 localhost sshd[27577]: Accepted password for jdoe from:ffff:192.168.138.35 port 2895 ssh2	Successful logon
Windows	Fri Mar 17 14:29:38 2006 680 security SYSTEM user failure audit ENTERPRISE account logon Logon attempt by: MICROSOFT_AUTHENTICATION_PACKAGE_V1_0 logon account: JDOE source workstation: ENTERPRISE error code: 0xC000006A 4574	Successful logon

normalization, a correlation rule looking for "logons" would need to explicitly define each known logon format. In contrast, event normalization provides the necessary categorization so that a rule can reference a "logon" and then successfully match an event against any variety of logons. Most normalization taxonomies utilize a tiered categorization structure because this level of generalization may be too broad for the detection of specific threat patterns, as illustrated in Figure 12.7.

Cross-source correlation

Cross-source correlation refers to the ability to extend correlation across multiple sources so that common events from disparate systems (such as a firewall and an IPS) may be normalized and correlated together. As correlation systems continue to mature, the availability of single-source correlation is dwindling. Cross-source correlation remains an important consideration of threat detection capability. The more types of information that can be correlated, the more effective the threat detection will be, and the fewer false positives, as shown in Table 12.7.

As more systems are monitored (see Chapter 13, "Security Monitoring of Industrial Control Systems"), the potential for expanding cross-source correlation increases accordingly—ideally with all monitored information being normalized and correlated together.

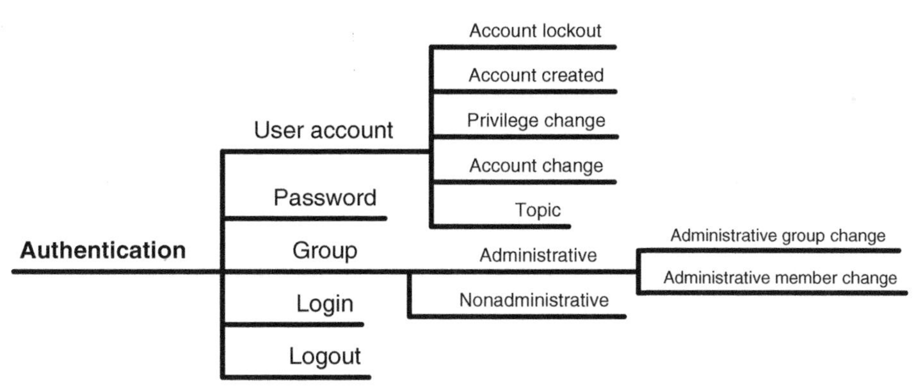

FIGURE 12.7 A partial representation of a tiered normalization taxonomy.

Table 12.7 Single-Source Versus Cross-Source Correlation

Single-source correlation example	Cross-source correlation example
Multiple failed logon followed by one or more successful logon	Multiple failed logon events by an admin user of critical assets, followed by one or more successful logon
Any successful logon to a critical asset	Any successful logon to a critical asset, by either a terminated employee or by an admin user at a time outside of normal shift hours.
HTTP traffic is originating from servers that are not HTTP servers	HTTP traffic is originating from servers that are not HTTP servers' IP addresses with a geographic location outside of the United States

Tiered correlation

Tiered correlation is simply the use of one correlation rule within another correlation rule. For example, a brute force attempt on its own may or may not be indicative of a cyberincident. If it is a cyber-attack, there is no further determination of what the attack is or its intent. By stacking correlation rules within other rules, additional rules can be enabled to target more specific attack scenarios, as shown in Table 12.8.

The third example in Table 12.8 illustrates the use of normalization within correlation by using a Malware event as a general condition of the rule. The fourth example illustrates the value of content inspection for the purposes of threat detection by exposing application authentication parameters to the correlation engine.

Correlating between IT and OT systems

Up until now, correlation has been discussed solely within the context of IT networks running standard enterprise systems and protocols. OT systems must also be analyzed, requiring that metrics within the OT network be correlated to events in the IT network. The challenge here is the disparity of the two system types and the information collection models used within each. IT systems are monitored heavily for performance and security using a wide range of available tools, whereas OT systems are monitored primarily for process efficiency and performance using a more limited range of tools consisting of data historians, process alarm management systems, spreadsheets, and statistical modeling applications (see Chapter 13, "Security Monitoring of Industrial Control Systems").

Even benign network behaviors of the IT network can impact operations, and threats do exist across both IT and OT systems. By correlating IT conditions against OT conditions, a good deal can be determined about potential cyberincidents.[12] Table 12.9 shows and example of several instances where IT systems can impact OT systems.

To fully leverage the automated correlation capability built into most IT SIEM products, OT data must first be collected into the SIEM, and then the normalization of one metric to another must be made using a common threat taxonomy.

Table 12.8 Tiered Correlation Examples

Description	Rule
Brute force attack	A number *N* of failed logon events, followed by one or more successful logon events, from the same source IP
Brute force malware injection	A number *N* of failed logon events, followed by one or more successful logon events, from the same source IP, followed by a malware event
Brute force followed by internal propagation	A number *N* of failed logon events, followed by one or more successful logon events, from the same source IP, followed by a network scan originating from the same source IP
Internal brute force enumeration using known password	A number *N* of failed logon events from the same source IP, each with a unique username but a different password

Table 12.9 Correlation of IT and OT Systems[14]

Incident	IT event	OT event	Condition
Network instability	Increased latency, measured by TCP errors, reduction of TCP receive windows, increased round-trip TTL, etc.	Reduction in efficiency, measured by historical batch comparisons	Manifestation of network condition in operational processesDeliberate cyber sabotage
Operational change	No detected event	Change to operational set points, or other process change(s)	Benign process adjustmentUndetected cyber sabotage
Network breach	Detected threat or incident using event correlation, to determine successful penetration of IT system(s)	Change to operational set points, or other process change(s)	Benign process adjustmentUndetected cyber sabotage
Targeted incident	Detected threat or incident directly targeting industrial SCADA or DCS systems connected to IT networks	Abnormal change to operational set points, unexpected PLC code writes, etc.	Potential "Stuxnet-class" cyberincident or sabotage

Caution

The ability to collect, interpret, and correlate data from disparate systems is vital to an effective security monitoring solution. The devices that comprise the network architectures must be able to communicate event data to a system that is equally capable of receiving these data. These concepts are progressive to OT networks and are a primary reason why many ICS servers, workstations, and embedded devices do not support this capability. It is not uncommon for an ICS vendor to restrict additional components that can be installed on their assets in order to maintain not only continuous performance and availability to manufacturing operations, but also the long-term support required to service these systems for years to come. At the time of publishing, there are several companies offering "SCADA SIEM" or similar packages. As SCADA and ICS systems continue to incorporate more mainstream security features, the ability of commercial monitoring and analysis tools to support industrial systems will continue to improve. Many commercial security analysis systems lack the necessary context to understand the data being collected from industrial systems, limiting the value of their analytics. This trend will change as more security solution companies partner with ICS vendors in delivering integrated OT security solutions.

Summary

A larger picture of security-related activity begins to form when zone security measures are in place. Measuring these activities and analyzing them can detect exceptions from the established security policies. In addition, anomalous activities can be identified so that they may be further investigated.

This requires well-defined policies and also that those policies be configured within an appropriate information analysis tool to ensure enforcement of those policies. Just as with perimeter defenses to a zone, carefully built variables defining allowed assets, users, applications, and behaviors can be used to aid in detection of security risks and threats. If these lists can be determined dynamically, in response to observed activity within the network, the "whitelisting" of known good policies becomes "Smart-Listing," which can help strengthen perimeter defenses through dynamic firewall configuration or IPS rule creation.

The event information can be further analyzed by event correlation systems as various threat detection techniques are used together to find larger and broader patterns that are more indicative of serious threats or incidents. Though widely used in IT network security, event correlation is now beginning to "cross the divide" into OT networks at the heels of Stuxnet and other sophisticated threats that attempt to compromise industrial network systems via attached IT networks and services.

Everything—measured metrics, baseline analysis, and whitelists—all rely on a rich base of relevant security information. Where does this security information come from? Chapter 13, "Security Monitoring of Industrial Control Systems," discusses what to monitor, and how, in order to obtain the necessary baseline of data required achieving "situational awareness" and effectively securing an industrial network.

Endnotes

1. F. Salo, Anomaly Detection Systems: Context Sensitive Analytics. NitroSecurity, Inc. Portsmouth, NH, December 2009.
2. B. Singer, Correlating Risk Events and Process Trends. Proceedings of the SCADA Security Scientific Symposium (S4). Kenexis Security Corporation and Digital Bond Press, Sunrise, FL, 2010.
3. U.S. Dept. of Homeland Security, "Cyber Security Procurement Language for Industrial Control Systems," September 2009.
4. D. Beresford, "Exploiting Siemens Simatic S7 PLCs," July 8, 2011. Prepared for Black Hat USA 2011.
5. R. Kay, QuickStudy: event correlation. Computerworld.com, July 28, 2003 (cited: February 13, 2011).
6. Softpanorama, Event correlation technologies., January 10, 2002 (cited: February 13, 2011).
7. The MITRE Corporation, About CEE (common event expression)., May 27, 2010 (cited: February 13, 2011).
8. M. Leland, Zero-day correlation: building a taxonomy. NitroSecurity, Inc., May 6, 2009 (cited: February 13, 2011).
9. The MITRE Corporation, About CEE (common event expression)., May 27, 2010 (cited: February 13, 2011).
10. A. Chuvakin, Content aware SIEM., February 2000 (cited: January 19, 2011).
11. ArcSight, "Common Event Format," Revision 16, July 22, 2010.
12. B. Singer, Correlating risk events and process trends. Proceedings of the SCADA Security Scientific Symposium (S4). Kenexis Security Corporation and Digital Bond Press, 2010, Sunrise, FL.
13. A. Chuvakin, Content aware SIEM. http://www.sans.org/security-resources/idfaq/vlan.php, February 2000 (cited: January 19, 2011).
14. B. Singer, Correlating Risk Events and Process Trends. Proceedings of the SCADA Security Scientific Symposium (S4). Kenexis Security Corporation and Digital Bond Press, 2010.

13

Security Monitoring of Industrial Control Systems

Information in this chapter

- Determining What to Monitor
- Successfully Monitoring Security Zones
- Information Management
- Log Storage and Retention

The first step of information analysis requires a certain degree of data collection so that there is a healthy body of data to assess. Collecting evidence relevant to cyber security requires knowing what to monitor and how to monitor it.

Unfortunately, there is a lot of information that could be relevant to cybersecurity, and because there are many unknown threats and exploitations, even information that may not seem relevant today may be relevant tomorrow as new threats are discovered. Even more unfortunate is that the amount of seemingly relevant data is already overwhelming—sometimes consisting of millions or even billions of events in a single day, with even higher rates of events occurring during a period of actual cyberattack.[1] It is therefore necessary to assess which events, assets, applications, users, and behaviors should be monitored—as well as any additional relevant systems that can be used to add context to the information collected, such as threat databases, user information, and vulnerability assessment results.

An additional challenge arises from the segregated nature of a properly secured industrial network. Deploying a single monitoring and information management system across multiple otherwise-separated zones violates the security goals of those zones and introduces potential risk. The methods used to monitor established zones must be considerate of the separation of those zones, and the data generated from this monitoring need to be managed accordingly as well. While there are benefits to fully centralized information management, the information being generated may be sensitive and may require "need to know" exposure to security analysts. Therefore, centralized monitoring and management needs to be overlaid with appropriate security controls and countermeasures, up to and including full separation—forgoing the efficiencies of central management so that the analysis, information management, and reporting of sensitive information remains local in order to maintain absolute separation of duties between, for example, a highly critical safety system and a less secure supervisory system.

Industrial Network Security. https://doi.org/10.1016/B978-0-443-13737-2.00010-5

In order to deal with massive volumes of log and event data that can result from monitoring established network zones, and the challenges of highly distributed and segregated zones, best practices in information management—including short- and long-term information storage—must be followed. This is necessary in order to facilitate the threat detection process, and also as a mandate for relevant compliance requirements, such as the North American Electric Reliability Corporation Critical Infrastructure Protection (NERC CIP), NRC Title 10 CFR 73.54, Chemical Facility Anti-Terrorism Standards (CFATS), and others (see Chapter 14, "Standards and Regulations").

Determining what to monitor

The trite answer to "what to monitor" is "everything and more!" Everything that we monitor, however, results in information that must be managed. Every data point results in a log record, or perhaps a security or safety alert. Assets, users, applications, and the communication channels that interconnect them all require monitoring. Because there are so many assets, users, applications, and networks that need to be monitored, the total amount of information generated every second in even a moderately sized enterprise can be staggering.[2]

While products exist to automate security event and information management, the total amount of information available can quickly overwhelm the information analysis and storage capacity of these tools. Therefore, security monitoring requires some planning and preparation in order to ensure that all necessary information is obtained, without overloading and potentially crippling the tools the information is intended to feed.

One approach is to segregate monitoring by zone. Just as the separation of functional groups into zones helps minimize risk, it also helps to minimize the total information load that is generated by that zone. In other words, there are limited assets and activities within a zone, and therefore there are less total logs and events.

To further complicate matters, operational technology (OT) activities and metrics must also be considered when securing industrial networks—representing new data types from yet another potentially overwhelming source of new assets such as remote terminal units (RTUs), programmable logic controllers (PLCs), intelligent electronic devices (IEDs), and other industrial assets; applications such as human–machine interfaces (HMIs), and Historians; and networks such as fieldbus and smart grid networks.

Tip

When considering network monitoring and information management, it is helpful to benchmark the information load currently being produced in both IT and OT networks. IT networks require identifying which devices need to be monitored. This means understanding what servers, workstations, firewalls, routers, proxies, and so on (almost every IT device is capable of producing logs of some sort) are important—the process of determining critical assets described in Chapter 2, "About Industrial Networks," and Chapter 9, "Establishing Zones and Conduits," is helpful here. Once it has been determined which devices need to be monitored, the event load generated by these devices needs to be calculated. One method is to measure the event load of a period of time that contains both normal and peak activity and divide the total number of events by the time period (in seconds) to determine the average event per second (EPS) load of the network. Alternately, a worst-case calculation can be based entirely on peak event rates, which will result in a higher EPS target.[3]

Most assets in OT networks, mainly the embedded device types, like PLCs, RTUs, and IEDs, which make up the majority of network-attacked assets, do not produce events or logs at all, and therefore they cannot be measured. However, they do produce information. This can be easily derived by looking at historized data from the control plants and/or through the use of specialized industrial protocol monitors or alarm managers. Determine which assets you wish to monitor and use the Data Historian system to determine the amount of information collected from these assets over time. This information will need to be normalized and centralized—either automatically via an SIEM or similar product, or manually via human time and effort—so it may be prudent to limit the amount of historized data that need to be exposed for security assessment. Some Historian tags—especially system tags concerning authentication, critical alarm tags concerning point or operational changes, stopped or failed processes, and so on—are obvious choices, while others may have little relevance to security. This step is effectively a form of security event "rationalization," similar to the process performed on the process event systems of ICS to improve operational effectiveness.

Once the initial benchmark is obtained, add room for growth, and room for headroom—perhaps 10% (this will vary by situation). When sizing the IT network, it is also prudent to plan for "peak averages" where peak traffic rates occur for extended periods of time (i.e., the peak becomes the average), as this condition can occur during an extended attack or as a result of a successful breach and subsequent infection with malware.[4] Unusual peak averages may also occur on OT systems during abnormal events, such as plant startups and shutdowns, or during system patching or on-process migrations and upgrades. OT systems may report different conditions but are less likely to report higher numbers of conditions unless the control process being historized has been significantly altered.

So what really needs to be monitored? The following guidelines help to identify what systems should be monitored.

Security events

Security events are those events generated by security and infrastructure products: network- or host-based firewalls, network routers and switches, malware prevention systems, intrusion detection and prevention systems, application monitors, and so on. Ideally, any event generated by a security device should be relevant, and therefore, these devices should be used for promiscuous monitoring. Realistically, false positives can dilute the relevance of valid security events.

Note

The term "false positive" is often misused. False positives are often associated with what are seemingly irrelevant security data because security logs and events originate from many sources and are often generated quickly and in large quantities. When an alert is generated because a benign activity matches a detection signature of an intrusion detection system (IDS), the result is a false positive. Similarly, if an antivirus system falsely indicates that a file is infected, the result is a false positive. False positives make security analysis more difficult by generating extra data points that need to be assessed, potentially clouding real incidents from detection.

False positives can be minimized through tuning of the faulty detection signatures—a process that should be performed regularly to ensure that detection devices are operating as efficiently as possible. While false positives often result in large amounts of unnecessary or irrelevant data, not all irrelevant data are false positives. Many security analysts and even security vendors are tempted to overly tune devices to eliminate any alert that occurs in large numbers because of this common misconception. The issue with overly aggressive tuning is that while it will make incidents easier to manage in day-to-day operations, it can introduce *false negatives*—that is, when a real threat fails to create an alert or when a correlation rule fails to trigger because a necessary condition was suppressed by over-tuning (see Chapter 12, "Exception, Anomaly, and Threat Detection"). Remembering that event correlation signatures are signature-matching rules that detect known threat patterns, the elimination of smaller seemingly irrelevant events can prevent detection of the larger pattern. Similarly, as security researchers discover new patterns, event data that seem irrelevant today may become relevant in the future (see Figure 13.1).

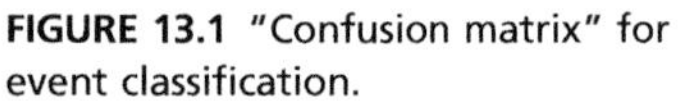

FIGURE 13.1 "Confusion matrix" for event classification.

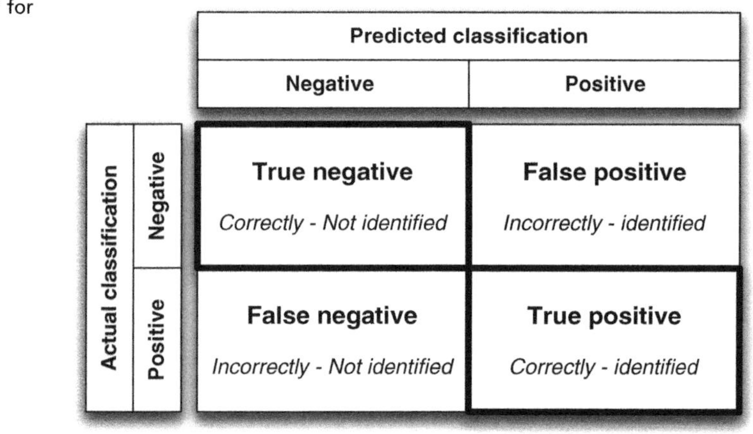

To ensure accurate threat detection and correlation, all legitimately produced events should be retained short-term for live analysis (i.e., kept on-line) and long-term for forensic and compliance purposes (i.e., kept off-line) regardless of how irrelevant they may seem at the time of collection. Only true false positives—the events generated due to a false signature match—should be eliminated via tuning or filtering.

When considering the relevance of security events in industrial networks, consider the source of the event and its relevance to the specific zone being monitored. For example, all zones should have at least one perimeter security device, such as a firewall or IPS, but there may also be multiple host-based security devices capable of generating events, such as antivirus, application whitelisting, intrusion detection and prevention systems (HIDS/HIPS), firewalls, or other security devices (see Chapter 9, "Establishing Zones and Conduits"). One example is industrial security appliances that use industrial protocol and application monitoring to enforce how industrial protocols are used.

These logs might provide much more specific data to a zone than do general security events, as seen in the example below from a Tofino industrial security appliance that provides detailed information pertaining to the unauthorized use of an industrial protocol (Modbus/TCP) function code (6 = "write single register"):

```
May 20 09:25:50 169.254.2.2 Apr 14 19:47:32 00:50:C2:B3:23:56
CEF:1|Tofino Security Inc|Tofino SA|02.0.00|300008|Tofino Modbus/
TCP Enforcer: Function Code List Check|6.0|msg = Function code 6
is not in permitted function code list TofinoMode = OPERATIONAL
smac = 9c:eb:02:a6:22 src = 192.168.1.126 spt = 32500
dmac = 00:00:bc:cf:6b:08 dst = 192.168.1.17 dpt = 502 proto = TCP
TofinoEthType = 800 TofinoTTL = 64 TofinoPhysIn = eth0
```

In contrast, a generic Snort IDS might produce a syslog event string identifying a perimeter policy violation, such as the attempted Windows update shown below, but cannot provide the context of application function codes within the industrial network (see Chapter 6, "Industrial Network Protocols").

```
Jan 01 00:00:00 [69.20.59.59] snort: [1:2002948:6] ET POLICY
External Windows Update in Progress [**] [Classification: Potential
Corporate Privacy Violation] [Priority: 1] {TCP} 10.1.10.33:1665
-> 192.168.25.35:80
```

An often-overlooked step prior to commissioning any device that will generate security events is to "tune" or validate that normal traffic does not trigger events. Figure 13.2 illustrates how a complete rule set for a Tofino Security Appliance might look once commissioned. Note that only the last rule (as indicated by the arrow) is actually enforcing segregation on the conduit by performing deep-packet inspection on Modbus/TCP (502/tcp) traffic originating in the ICS Host zone and destined for the ICS Controllers zone. There are many other types of valid traffic that is generated to support functionality like the Network Neighborhood used in Windows operating systems and Neighboring Switches/Routers typical in both IT and OT network devices that is commonly sent to broadcast and multicast addresses. This valid traffic, if not properly handled with "drop-no log" entries in the rule set would generate "false positives" in terms of the security events within an industrial network. Some of the traffic that must be considered include

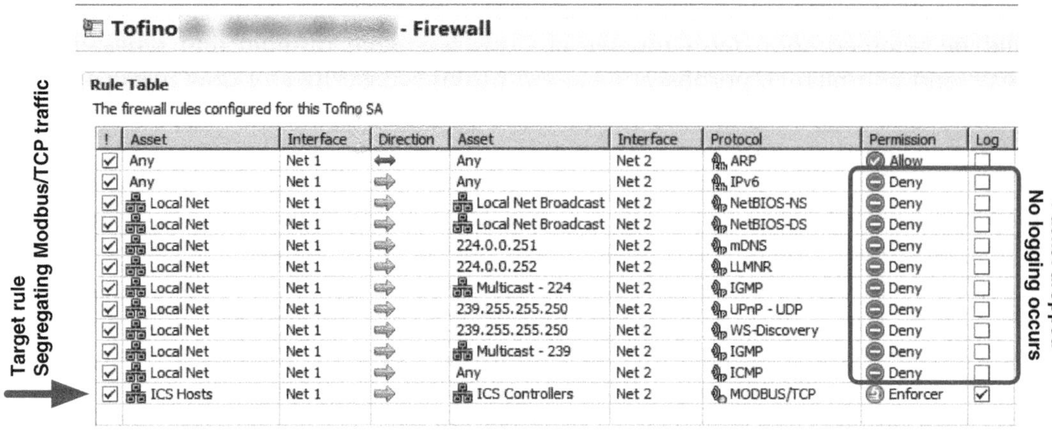

Tofino - Firewall

Rule Table

The firewall rules configured for this Tofino SA

!	Asset	Interface	Direction	Asset	Interface	Protocol	Permission	Log
☑	Any	Net 1	⟷	Any	Net 2	ARP	Allow	☐
☑	Any	Net 1	→	Any	Net 2	IPv6	Deny	☐
☑	Local Net	Net 1	→	Local Net Broadcast	Net 2	NetBIOS-NS	Deny	☐
☑	Local Net	Net 1	→	Local Net Broadcast	Net 2	NetBIOS-DS	Deny	☐
☑	Local Net	Net 1	→	224.0.0.251	Net 2	mDNS	Deny	☐
☑	Local Net	Net 1	→	224.0.0.252	Net 2	LLMNR	Deny	☐
☑	Local Net	Net 1	→	Multicast - 224	Net 2	IGMP	Deny	☐
☑	Local Net	Net 1	→	239.255.255.250	Net 2	UPnP - UDP	Deny	☐
☑	Local Net	Net 1	→	239.255.255.250	Net 2	WS-Discovery	Deny	☐
☑	Local Net	Net 1	→	Multicast - 239	Net 2	IGMP	Deny	☐
☑	Local Net	Net 1	→	Any	Net 2	ICMP	Deny	☐
☑	ICS Hosts	Net 1	→	ICS Controllers	Net 2	MODBUS/TCP	Enforcer	☑

Target rule
Segregating Modbus/TCP traffic

Packet dropped
No logging occurs

FIGURE 13.2 Tuning an industrial network security appliance.

- Multicast DNS (5353/udp)
- Link-Layer Multicast Name Resolution (5355/udp)
- Universal Plug 'n Play (1900/udp and 2869/tcp)
- Web Services Discovery Protocol (3702/udp)
- Cisco Discovery Protocol
- Link Layer Discovery Protocol
- Internet Control Message Protocol (IP Protocol 1)
- Internet Group Management Protocol (IP Protocol 2)
- Internet Protocol Version 6 (IPv6).

Assets

Assets—the physical devices connected to the network—can also provide security data, typically in the form of logs. Assets can produce logs that track activity on a variety of levels. The operating system itself produces many logs, including system logs, application logs, and file system logs.

System logs are useful for tracking the status of devices and the services that are (or are not) running, as well as when patches are (or are not) applied. Logs are useful for determining the general health of an asset, as well as validating that approved ports and services are running. These logs are valuable in tracking which users (or applications) have authenticated to the asset, satisfying several compliance requirements. The following represents individual records from a Redhat Linux system log showing a successful user login and a Windows failed authentication:

```
<345> Mar 17 11:23:15 localhost sshd[27577]: Accepted password
for knapp from ::ffff:10.1.1.1 port 2895 ssh2
<345> Fri Mar 17 11:23:15 2011 680 Security SYSTEM User Failure
Audit ENTERPRISE Account Logon attempt by:
MICROSOFT_AUTHENTICATION_PACKAGE_V1_0 Logon account: KNAPP Source
Workstation: ENTERPRISE Error Code: 0xC000006A 4574
```

Although syslog is ubiquitously used across a variety of systems, other event logging systems are used as well—the most notable of which is the Windows Management Instrumentation (WMI) framework. WMI produces auditable events in a structured data format that can be used against scripts (for automation) as well as by other Windows operating system functions.[5] Because syslog is so widely supported, WMI events are often logged using a Windows syslog agent, such as Snare for Windows to stream WMI events over syslog. It is also possible to configure log forwarding between Windows hosts when restrictions prohibit the installation of agents on critical assets using the Windows Event Collector functionality.

The following WMI event example indicates the creation of a new process on a Windows server:

```
Computer Name: WIN-0Z6H21NLQ05
Event Code: 4688
Type: Audit Success (4)
User Name:
Category: Process Creation
Log File Name: Security
String[%1]: S-1-5-19
String[%2]: LOCAL SERVICE
String[%3]: NT AUTHORITY
String[%4]: 0x3e5
String[%5]: 0xc008
String[%6]: C:\Windows\System32\RacAgent.exe
String[%7]: %%1936
String[%8]: 0xc5e4
Message: A new process has been created. Subject: Security ID:
S-1-5-19 Account Name: LOCAL SERVICE Account Domain: NT AUTHORITY
Logon ID: 0x3e5 Process Information: New Process ID: 0xc008 New
Process Name: C:\Windows\System32\RacAgent.exe Token Elevation
Type: TokenElevationTypeDefault (1) Creator Process ID: 0xc5e4
Token Elevation Type indicates the type of token that was assigned
to the new process in accordance with User Account Control policy.
Type 1 is a full token with no privileges removed or groups
disabled. Afull token is only used if User Account Control is
disabled or if the user is the built-in Administrator account or
a service account. Type 2 is an elevated token with no privileges
removed or groups disabled. An elevated token is used when User
Account Control is enabled and the user chooses to start the
program using Run as administrator. An elevated token is also used
when an application is configured to always require administrative
privilege or to always require maximum privilege, and the user is
a member of the Administrators group. Type 3 is a limited token
with administrative privileges removed and administrative groups
disabled. The limited token is used when User Account Control is
enabled, the application does not require administrative privilege,
and the user does not choose to start the program using Run as
administrator.
```

The same event, when collected via syslog using a WMI agent, such as Snare, might look like this:

```
<12345> Fri Mar 17 11:23:15 2011||WIN-0Z6H21NLQ05||4688||Audit
Success (4)||||Process Creation||Security||S-1-5-19||LOCAL
SERVICE||NT AUTHORITY||0x3e5||0xc008||C:\Windows\System32\RacAgent.
exe||%%1936||0xc5e4
```

Application logs (covered in more detail under the section "Applications") provide a record of application-specific details, such as logon activities to an HMI, configuration changes, and other details that indicate how an application is being used. These Application Logs are an important component in the security associated with many ICS applications since these applications commonly utilize a single Windows logon authentication account and manage individual user actions via local application accounts and security settings.

File system logs typically track when files are created, changed, or deleted, when access privileges or group ownerships are changed, and similar details. File system logging is included in Windows using the Windows File Protection (WFP) within WMI, which is an "infrastructure for management data and operations on Windows-based operating systems."[6] File monitoring in Unix and Linux systems is performed using **auditd**. For Windows systems, commercial file integrity monitoring (FIM) products are available, such as Tripwire (www.tripwire.com). In addition, many commercial and open source security monitoring tools include file integrity monitoring features. This is extremely valuable for assuring the integrity of important files stored on an asset—such as configuration files (ensuring that the asset's configurations remain within policy), and the asset's log files themselves (ensuring that logged activities are valid and have not been tampered with to cover up indications of illicit behavior).

Note

Not all assets can be monitored. Unfortunately, sometimes assets do not produce logs with relevant security context, while some assets don't produce logs at all. For many industrial assets—PLCS, RTUs, IEDs, etc.—the operating systems are highly optimized in order to support real-time operation and maximum reliability. For this reason, vendors may minimize or eliminate logging. This obviously makes monitoring these assets extremely difficult. To compensate, consider alternative monitoring methods. For example, many of these types of assets will be fully visible to DCS alarm management systems. Exposing process alarms and/or historian data to cybersecurity analysts can be an extremely effective way to detect potential cyber-physical threats, although without proper training these alarms can be easily misinterpreted and the process can become counterproductive. Alternatively, monitoring industrial protocol traffic on network segments reachable by unmonitored assets can help determine—and log—what these assets are doing and how they are behaving.

Process data and alarms

Cyber-physical attacks intend to manipulate the physical outcome of a process. While this can be extremely difficult to monitor using traditional cybersecurity monitoring tools, an industrial control system is typically engineered to provide precise and real-time visibility into all aspects of process control. Two primary sources of information that are relevant to cybersecurity are data historians and alarm management systems.

Historians maintain historical records of all data associated with a process. This information was traditionally used to analyze how a process behaved over time, in order to identify ways to improve performance and efficiency. Historians have grown in capability over the years and now include many features to extend their analytical capability to support many additional use cases, including alarm management, safety monitoring, asset management, and more.[7]

Alarm management functionality can be provided by historian systems and are also available by some vendors as standalone software packages that focus exclusively on process alarms. By nature, a process alarm focuses on condition(s) that are of interest and so have a higher value to cybersecurity monitoring efforts. See "Data Historians and Process Alarms" below for more information on how to use historian data with cybersecurity monitoring tools.

Configurations

Configuration monitoring refers to the process of monitoring baseline configurations for any indications of change,[8] Recommended Security Controls for Federal Information Systems and Organizations, August, 2009, and is only a small part of configuration management (CM). Basic configuration monitoring can be done at a rudimentary level through a combination of host configuration file monitoring (to establish the baseline), system and application log monitoring (to look for change actions), and FIM (to ensure that configurations are not altered). While this does not provide true CM, it does provide an indication as to when established configurations are altered, providing a valuable security resource.

Full CM systems provide additional key functions, typically mapping at least partially to the security controls outlined in NIST SP 800-53 under the section "Configuration Management," which provides a total of nine CM controls:[9]

- CM policy and procedures—establishes a formal, documented CM policy.
- Baseline configurations—identifying and documenting all aspects of an asset's configurations to create a secure template against which all subsequent configurations are measured.
- Change control—monitoring for changes and comparing changes against the established baseline.
- Security impact analysis—the assessment of changes to determine and test how they might impact the security of the asset.

- Access restrictions for change—limiting configuration changes to a strict subset of administrative users.
- Configuration settings—identification, monitoring, and control of security configuration settings and changes thereto.
- Least functionality—the limitation of any baseline configuration to provide the least possible functionality to eliminate unnecessary ports and services.
- Information service (IS) component (asset) inventory—establishing an asset inventory to identify all assets that are subject to CM controls, as well as to detect rogue or unknown devices that may not meet baseline configuration guidelines.
- Establishment of a CM plan—assigning roles and responsibilities around an established CM policy to ensure that CM requirements are upheld.

Configuration management tools may also offer automated controls to allow batch configurations of assets across large networks, which is useful for ensuring that proper baselines are used in addition to improving desktop management efficiencies. For the purposes of security monitoring, it is the monitoring and assessment of the configuration files themselves that is a concern. This is because an attacker will often attempt to either escalate user privileges in order to obtain higher levels of access or alter the configurations of security devices in order to penetrate deeper into secured zones—both of which are detectable with appropriate CM controls in place.

The logs produced by the CM are therefore a useful component of overall threat detection by using change events in combination with other activities, such as an event correlation system. For example, a port scan, followed by an injection attempt on a database, followed by a configuration change on the database server is indicative of a directed penetration attempt. Change logs are also highly beneficial (and in some cases mandatory) for compliance and regulatory purposes, with configuration and change management being a common requirement of most industrial security regulations (see Chapter 14, "Standards and Regulations").

Tip

The problem with CM within ICS is that a large portion of the critical configuration information is retained in embedded devices often running proprietary or closed operating systems using nonstandard communication protocols. These devices (PLCs, RTUs, IEDs, SIS, etc.) represent the true endpoint with a connection to the physical process under control, making their configuration details (control logic, hardware configuration, firmware, etc.) one of the most critical components pertaining to the operational integrity of the ICS. While several available IT products, such as Tripwire, Solarwinds, and What'sUpGold, can provide configuration and change management for servers, workstations, and network devices, specialized products, such as Cyber Integrity by PAS and the Industrial Defender Automation Systems Manager from Lockheed Martin, provide not only the necessary database components to identify and track configuration changes, but an extensive library of system and device connectors necessary to extract configuration data from ICS components.

Applications

Applications run on top of the operating system and perform specific functions. While monitoring application logs can provide a record of the activities relevant to those functions, direct monitoring of applications using a dedicated application monitoring product or application content firewall will likely provide a greater granularity of all application activities. Application logs can indicate when an application is executed or terminated, who logs into the application (when application-level security is implemented), and specific actions performed by users once logged in. The information contained in application logs is a summary, as it is in all log records. A sample application log record generated by an Apache web server is provided here:

```
Jan 01 00:00:00 [69.20.32.12] 93.80.237.221 - - [24/
Feb/2011:01:56:33 -0000] "GET/spambot/spambotmostseendownload.
php HTTP/1.0" 500 71224 "http://yandex.ru/yandsearch?text=video.
krymtel.net" "Mozilla/4.0 (compatible; MSIE 6.0; Windows NT 5.1;
MRA 4.6 (build 01425))"
```

A corresponding application log entry from an ICS illustrating a local access level change is shown here:

```
Jan 01 00:00:00 ICSSERVER1 HMI1 LEVEL Security Level Admin
Jan 01 00:00:00 ICSSERVER1 HMI1 LEVEL Security Level Oper
```

For a more detailed accounting of application activity, an application monitoring system can be used. For example, while it is possible that malware might be downloaded over HTTP and be indicated in a log file, such as the first example shown earlier, monitoring an application's contents across a session could indicate malware that is embedded in a file being downloaded from an otherwise normal-seeming website, as shown in Figure 13.3.

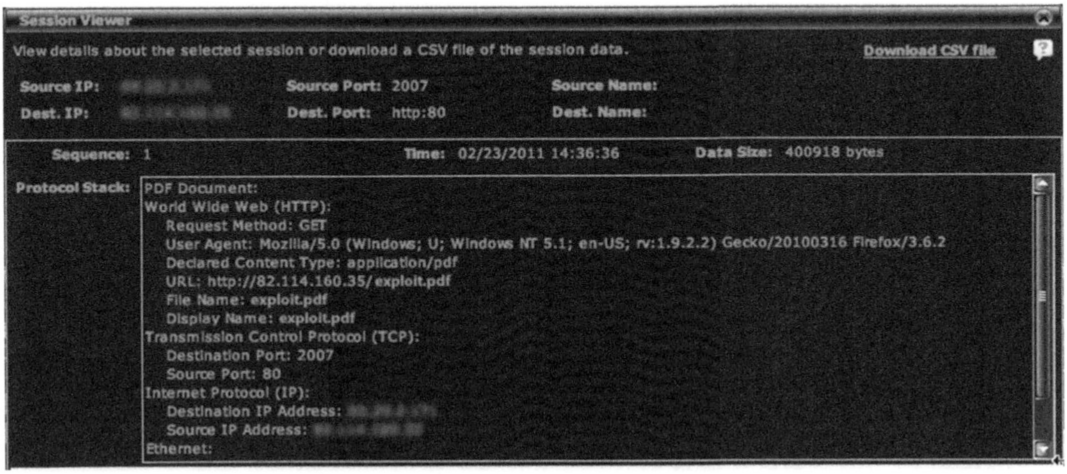

FIGURE 13.3 Application session details from an application monitor.

Networks

Network flows are records of network communications, from a source to one or more destinations. Network infrastructure devices, such as switches and routers, usually track flows. Flow collection is typically proprietary to the network device manufacturer (e.g., Cisco supports NetFlow and Juniper supports J-Flow), although many vendors also support the sFlow standard (see Table 13.1).

Monitoring flows provides an overview of network usage over time (for trending analysis, capacity planning, etc.) as well as at any given time (for impact analysis, security assessment, etc.) and can be useful for a variety of functions, including:[10]

- Network diagnosis and fault management.
- Network traffic management or congestion management.
- Application management, including performance management, and application usage assessments.
- Application and/or network usage accounting for billing purposes.
- Network security management, including the detection of unauthorized devices, traffic, and so on.

Table 13.1 Network Flow Details

Flow detail	What it indicates	Security ramifications
SNMP interface indices (ifIndex in IF-MIB)	The size of the flow in terms of traffic volume (bytes, packets, etc.), as well as errors, latency, discards, physical addresses (MAC addresses), etc.	SNMP details can provide indications of abnormal protocol operation that might indicate a threatMore germane to industrial networks, the presence of interface errors, latency, etc. can be directly harmful to the correct operation of many industrial protocols (see Chapter 6, "Industrial Network Protocols")
Flow start time	When a network communication was initiated and when it ended	Essential for the correlation of communications against security events
Flow end time	Collectively, the start and stop timestamps also indicate the duration of a network communications	
Number of bytes/ packets	Indicates the "size" of the network flow, indicative of how much data is being transmitted	Useful for the detection of abnormal network access, large file transfers, as might occur during information theft (e.g., retrieving a large database query result, downloading sensitive files, etc.)
Source and destination IP addresses	Indicates where a network communication began and where it was terminated	Essential for the correlation of related logs and security events (which often track IP address details)
Source and destination port	Note that in non-IP industrial networks, the flow may terminate at the IP address of an MI or PLC even though communications may continue over specialized industrial network protocols	IP addresses may also be used to determine the physical switch or router interface of the asset, or even the geographic location of the asset (through the use of a geo-location service)

Network flow analysis is extremely useful for security analysis because it provides the information needed to trace the communications surrounding a security incident back to its source. For example, if an application whitelisting agent detects malware on an asset, it is extremely important to know where that malware came from, as it has already breached the perimeter defenses of the network and is now attempting to move laterally and infect adjacent machines. By correlating the malware attempt to network flows, it may be possible to trace the source of the malware and may also provide a path of propagation (i.e., where else did the virus propagate).

Network flow analysis also provides an indication of network performance for industrial network security. This is important because of the negative impact that network performance can have on process quality and efficiency, as shown in Table 13.1. An increase in latency can cause certain industrial protocols to fail, halting industrial processes.[11]

Caution

It is important to verify with the ICS supplier that network flow functionality can be enabled on the industrial network without negatively impacting the performance and integrity of the network and its connected devices. Many industrial protocols include real-time extensions (see Chapter 6, "Industrial Network Protocols") that see switch performance issues when available forwarding capacity has been altered. Network vendors like Cisco have addressed this with special "lite" capabilities for netflow reporting. Always consult the ICS supplier before making modifications to recommended or qualified network topologies and operating parameters.

User identities and authentication

Monitoring users and their activities is an ideal method for obtaining a clear picture of what is happening on the network and who is responsible. User monitoring is also an important component of compliance management, as most compliance regulations require specific controls around user privileges, access credentials, roles, and behaviors. This requirement is enforced more so on systems that must comply with requirements, such as 21 CFR Part 11 and similar standards common in "FDA-regulated industries," such as pharmaceutical, food, and beverage.

Unfortunately, the term "user" is vague—there are user account names, computer account names, domain names, host names, and of course the human user's identity. While the latter is what is most often required for compliance management (see Chapter 14, "Standards and Regulations"), the former are what are typically provided within digital systems. Authentication to a system typically requires credentials in the form of a username and password, from a machine that has a host name, which might be one of several hosts in a named domain. The application itself might then authenticate to another backend system (such as a database), which has its own name and to which the application authenticates using yet another set of credentials. To further complicate things, the same human operator might need to authenticate to several systems, from several different machines, and may use a unique username on each. As mentioned earlier, ICS users may utilize a "common" Windows account shared by many, while each possesses a unique "application" account used for authentication and authorization within the ICS applications.

It is therefore necessary to normalize users to a common identity, just as it is necessary to normalize events to a common taxonomy. This can be done by monitoring activities from a variety of sources (network, host, and application logs), extracting whatever user identities might be present, and correlating them against whatever clues might be preset within those logs. For example, if a user authenticates to a Windows machine, launches an application and authenticates to it, and then the application authenticates to a backend system, it is possible to track that activity back to the original username by looking at the source of the authentications and the time at which they occurred. It can be assumed that all three authentications were by the same user because they occurred from the same physical console in clear succession.

As the systems become more complex and distributed, and as the number of users increases, each with specific roles and privileges, this can become cumbersome, and an automated identity management mechanism may be required.

This process is made simpler through the use of common directories, such as Microsoft Active Directory and/or the **Lightweight Directory Access Protocol** (LDAP), which act as identity directories and repositories. However, there may still be several unique sets of credentials per human operator that are managed locally within the applications versus centrally via a directory service. The difficulty lies in the lack of common log formats, and the corresponding lack of universal identities between diverse systems. User monitoring therefore requires the extraction of user information from a variety of network and application logs, followed by the normalization of that identity information. John Doe might log into a Windows domain using the username j.doe, have an e-mail address of jdoe@company.com, and log into a corporate intranet or Content Management System (CMS) as johnnyd, and so on. To truly monitor user behavior, it is necessary to recognize j.doe, jdoe, and johnnyd as a single identity.

Several commercial identity and access management (IAM) systems (also sometimes referred to as identity and authentication management systems) are available to facilitate this process. Some commercially available IAM systems include: NetIQ (formerly Novell and spun off as part of the merger with Attachmate), Oracle Identity Management (also encompassing legacy Sun Identity Management prior to Oracle's acquisition of Sun Microsystems), and IBM's Tivoli Identity. Other third-party identity solutions, such as Securonix Identity Matcher, offer features of both a centralized directory and IAM by mining identity information from other IAMs and normalizing everything back to a common identity.[12] More sophisticated SIEM and log management systems might also incorporate identity correlation features to provide user normalization. An authoritative source of identity is provided by managing and controlling authentications to multiple systems via a centralized IAM irrespective of the method used, as shown in Figure 13.4.

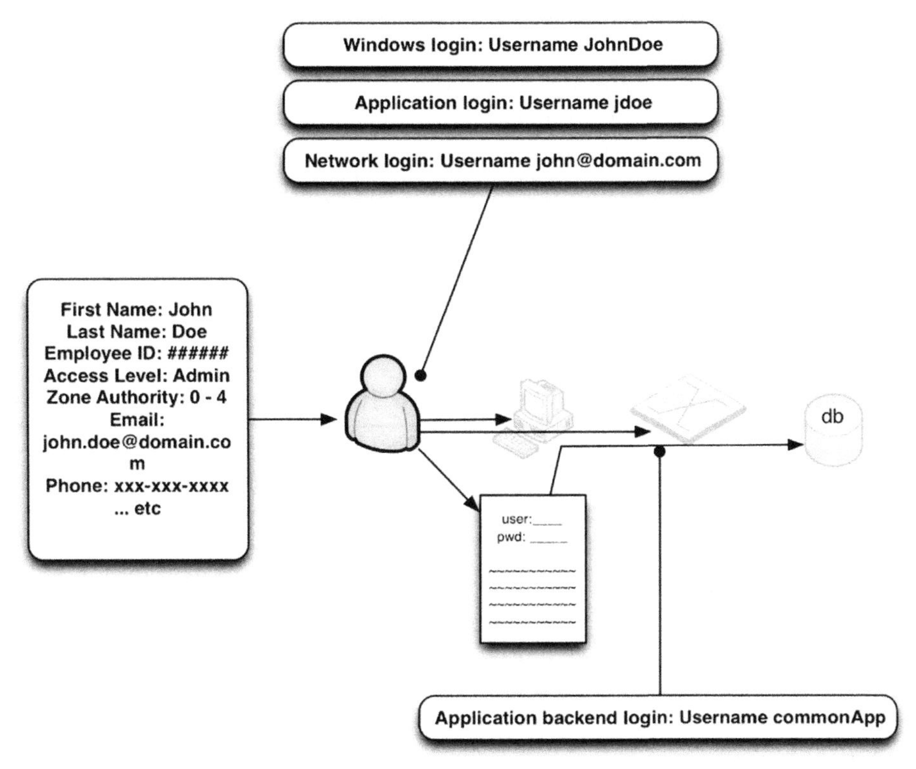

FIGURE 13.4 Normalization of user identity.

Once the necessary identity context has been obtained, it can be utilized in the information and event management process to cross-reference logs and events back to users. A SIEM dashboard shows both network and event details associated with their source users in Figure 13.5.

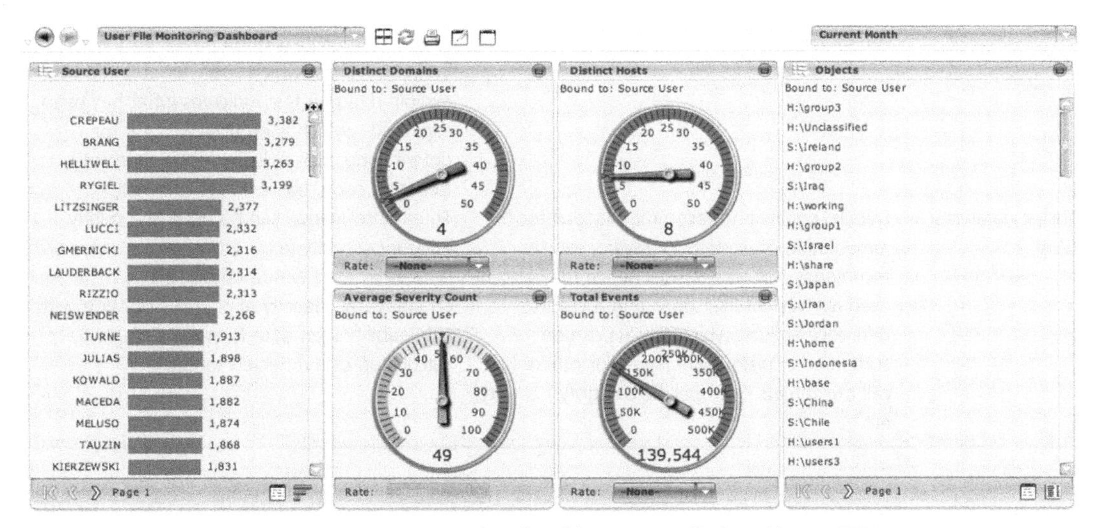

FIGURE 13.5 User activity related to file access as displayed by an SIEM.

Additional context

While user identity is one example of contextual information, there is a wealth of additional information available that can provide context. This information—such as vulnerability references, IP reputation lists, and threat directories—supplements the monitored logs and events with additional valuable context. Examples of contextual information are provided in Table 13.2.

Contextual information is always beneficial, as the more context is available for any specific event or group of events, the easier it will be to assess relevance to specific

Table 13.2 Contextual Information Sources and Their Relevance

Information source	Provided context	Security implications
Directory services (e.g., active directory)	User identity information, asset identity information, and access privileges	Provides a repository of known users, assets, and roles that can be leveraged for security threat analysis and detection, as well as for compliance
Identity and authentication management systems	Detailed user identity information, usernames and account aliases, access privileges, and an audit trail of authentication activity	Enables the correlation of users to access and activities based upon privilege and policy. When used to enrich security events, provides a clear audit trail of activity versus authority that is necessary for compliance auditing
Vulnerability scanner	Asset details including the operating system, applications in use (ports and services), patch levels, identified vulnerabilities, and related known exploits	Enables security events to be weighted based upon the vulnerability of their target (i.e., a Windows virus is less concerning if it is targeting a Linux workstation) Also provides valuable asset details for use in exception reporting, event correlation, and other functions
Penetration tester	Exploitation success/failure, method of exploitation, evasion techniques, etc.	Like with a vulnerability scanner, pen test tools provide the context of an attack vector. Unlike VA scan results, which show what could be exploited, a pen test indicates what has been exploited—which is especially useful for determining evasion techniques, detecting mutating code, etc.
Threat database/ CERT	Details, origins and recommendations for the remediation of exploits, malware, evasion techniques, etc.Threat intelligence may also be used as "watchlists," providing a cross-reference against which threats can be compared in order to highlight or otherwise call out threats of a specific category, severity, etc.	Threat intelligence can be used in a purely advisory capacity (e.g., providing educational data associated with a detected threat), or in an analytical capacity (e.g., in association with vulnerability scan data to weight the severity calculation of a detected threat)

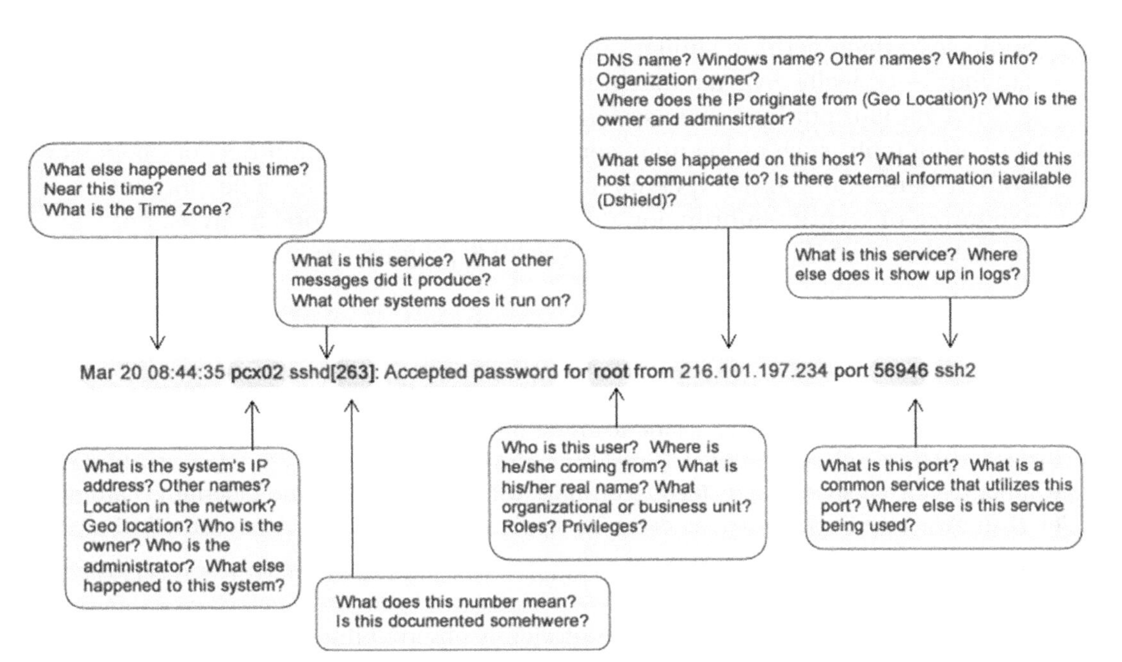

FIGURE 13.6 A log file, illustrating the lack of context image.

security and business policies. This is especially true because the logs and events being monitored often lack the details that are most relevant, such as usernames (see Figure 13.6).[13]

It is important to know that contextual information adds to the total volume of information already being assessed. It is therefore most beneficial when used to enrich other security information in an automated manner (see section "Information Management").

Behavior

Behavior can be directly monitored or derived. Direct behavior monitoring includes looking for specific behavior patterns of an application, such as the loading of a specific dynamic linked library, or communication to a specific external IP address or domain. Derived behavior refers to a behavior that seems suspicious based on previously observed behavior or by the analysis of a specific monitored metric over time. The result is an indication of expected versus unexpected activity, which is extremely useful for a wide range of security functions, including anomaly-based threat detection, as well as capacity or threshold-based alarming. Behavior is also a useful condition in security event correlation (see Chapter 12, "Exception, Anomaly, and Threat Detection").

- Behavioral analysis is provided by a variety of tools, including:
- Many antimalware tools that analyze application behavior as an application is executing

- Sandboxes, which perform similar detection of malicious behavior, but in a safe "sandbox"—typically a virtual machine or container—so that the behavior can be detected without putting a host system at risk.
- Monitoring tools, such as log management systems, SIEMs, and network behavior anomaly detection (NBAD) systems (if the system used for the collection and monitoring of security information does not provide behavioral analysis, an external tool, such as a spreadsheet or statistics program, may be required).

Note

As mentioned in Chapter 11, machine learning (ML) is commonly used to facilitate threat detection. ML, a subset of artificial intelligence (AI), is able to analyze large sets of data without relying on specific, provided algorithms. ML systems are able to determine patterns on their own based upon observed data. The applications for ML in cybersecurity include:

- Helping to detect threats based on code similarity or behavioral similarity ("this looks suspicious").
- Help to detect attack patterns based on previously observed threat lifecycles /kill chains (see Chapter 10, "OT Attack and Defense Lifecycles").
- Helping to automate tasks associated with threats once they are detected, e.g., helping to prioritize events, detect threat patterns, eliminate false positives, etc. ("This looks important").

Machine learning and AI are evolving rapidly at the time of publication, identified applications for ML in cybersecurity are being identified seemingly every day. For the purposes of industrial cybersecurity, it is important to recognize that ML and AI are a means to an end: they can automate many cybersecurity tasks, make those tasks more efficient, and perhaps even perform some tasks better; however, the fundamentals of cybersecurity have not changed, and the use of ML and AI does not negate them.

Successfully monitoring security zones

Understanding what to monitor is only the first step—actually monitoring all of the users, networks, applications, assets, and other activities still needs to happen. The discussion of what to monitor focused heavily on logs because log files are designed to describe activities that have occurred, are fairly ubiquitous, and are well understood. Log files are not always available however and may not provide sufficient detail in some instances. Therefore, monitoring is typically performed using a combination of methods, including the following:

- Log collection and analysis
- Direct monitoring or network inspection
- Inferred monitoring via tangential systems.

Except in pure log-collection environments, where logs are produced by the assets and network devices that are already in place, specialized tools are required to monitor the various network systems. The results of monitoring (by whatever means) needs to be dealt with because while manual logs and event reviews are possible (and allowed by most compliance regulations), automated tools are available and are recommended.

The central analysis of monitored systems is contrary to a security model built upon functional isolation. This is true because industrial networks should be separated into functional security zones, and centralized monitoring requires that log and event data either remain within a functional group (limiting the value for overall situation awareness of the complete system) or be shared between zones (potentially putting the security of the zone at risk). In the first scenario, logs and events are not allowed across the zone perimeter where they may be collected, retained, and analyzed only by local systems within that zone. In the second scenario, special considerations must be made for the transportation of log and event data across zone perimeters to prevent the introduction of a new inbound attack vector. A common method is to implement special security controls (such as a data diode, unidirectional gateway, or firewall configured to explicitly deny all inbound communications) to ensure that the security data are only allowed to flow toward the centralized management system. A hybrid approach may be used in industrial networks where critical systems in remote areas need to operate reliably. This provides local security event and log collection and management so that the zone can operate in total isolation while also pushing security data to a central location to allow for more complete situational awareness across multiple zones.

Log collection

Log collection is simply the collection of logs from whatever sources produce them. This is often a matter of directing the log output to a log aggregation point, such as a network storage facility and/or a dedicated log management system. Directing a log is often as simple as directing the syslog event data service to the IP address of the aggregator. In some cases, such as WMI, events are stored locally within a database rather than as log files. These events must be retrieved, either directly (by authenticating to Windows and querying the event database via the Windows Event Collector functionality) or indirectly (via a software agent, such as Snare, which retrieves the events locally and then transmits them via standard syslog transports).

Direct monitoring

Direct monitoring refers to the use of a "probe" or other device to passively examine network traffic or hosts by placing the device in-line with the network. Direct monitoring is especially useful when the system being monitored does not produce logs natively (as is the case with many industrial network assets, such as RTUs, PLCs, and IEDs). It is also useful as a verification of activity reported by logs, as log files can be altered deliberately in order to hide evidence of malicious activities. Common monitoring devices include firewalls, IDSs, **database activity monitors** (**DAMs**), application monitors, and network probes. These are often available commercially as software or appliances, or via open-source distributions, such as Snort (IDS/IPS), Wireshark (network sniffer and traffic analyzer), and Kismet (wireless sniffer).

Often, network monitoring devices produce logs of their own, which are then collected for analysis with other logs. Network monitoring devices are sometimes referred to as "passive logging" devices because the logs are produced without any direct interaction with the system being monitored. DAMs, for example, monitor database activity on the network—often on a span port or network tap. The DAM decodes network packets and then extracts relevant SQL transactions in order to produce logs. There is no need to enable logging on the database itself resulting in no performance impact to the database servers.

In industrial networks, it is similarly possible to monitor industrial protocol use on the network by providing "passive logging" to those industrial control assets that do not support logging. Passive monitoring is especially important in these networks, as many industrial protocols operate in real time and are highly susceptible to network latency and jitter. This is one reason why it is difficult to deploy logging agents on the devices themselves (which would also complicate asset testing policies), making passive network logging an ideal solution in these cases. Special consideration to any industrial network redundancy should also be considered when deploying network-based monitoring solutions.

In some instances, the device may use a proprietary log format or event streaming protocol that must be handled specially. Cisco's Security Device Event Exchange protocol (SDEE) (used by most Cisco IPS products) requires a username and password in order to authenticate with the security device so that events can be retrieved on demand and/or "pushed" via a subscription model. While the end result is the same, it is important to understand that syslog is not absolutely ubiquitous.

Inferred monitoring

Inferred monitoring refers to situations where one system is monitored in order to infer information about another system. Many applications connect to a database. So as an example, monitoring the database in lieu of the application itself will provide valuable information about how the application is being used, even if the application itself is not producing logs or being directly monitored by an application monitor.

Note

Network-based monitoring inevitably leads to the question, "Is it possible to monitor encrypted network traffic?" Many industrial network regulations and guidelines recommend the encryption of control data when these data are transferred between trusted security zones via untrusted conduits ... so how can these data be monitored via a network probe? There are a few options, each with benefits and weaknesses. The first is to monitor the sensitive network connection between the traffic source and the point of encryption. That is, encrypt network traffic externally using a network-based encryption appliance, such as the Certes Networks Enforcement Point (CEP) variable speed encryption appliances, and place the network probe immediately between the asset and the encryption. The second option is to utilize a dedicated network-based decryption device, such as the Netronome SSL Inspector. These devices perform deliberate, hardware-based man-in-the-middle attacks in order to break encryption and analyze the network contents for security purposes. A third option is not to monitor the encrypted traffic at all, but rather to monitor for instances of data that should be encrypted (such as industrial protocol function codes) but are not producing exception alerts indicating that sensitive traffic is not being encrypted.

To determine which tools are needed, start with your zone's perimeter and interior security controls (see Chapter 9, "Establishing Zones and Conduits") and determine which controls can produce adequate monitoring and which cannot. If they can, start by aggregating logs from the absolute perimeter (the demarcation between the least critical zone and any untrusted networks—typically the business enterprise LAN) to a central log aggregation tool (see the section "Information Collection and Management Tools"). Begin aggregating logs from those devices protecting the most critical zones and work outward until all available monitoring has been enabled, or until the capacity of your log aggregation has become saturated. At this point, if there are remaining critical assets that are not being effectively monitored, it may be necessary to increase the capacity of the log aggregation system.

Tip

Adding capacity does not always mean buying larger, more expensive aggregation devices. Distribution is also an option—keep all log aggregation local within each zone (or within groups of similar zones) and then aggregate subsets of each zone to a central aggregation facility for centralized log analysis and reporting. While this type of event reduction will reduce the effectiveness of threat detection and will produce less comprehensive reports from the centralized system, all the necessary monitoring and log collection will remain intact within the zones themselves, where they can be accessed as needed.

This concept is particularly well-suited for industrial networks in that it allows the creation of a local "dashboard" where relevant events for nearby assets can be displayed and responded to quickly by a "first responder" that may reside in the operational or plant environment while offering the ability to export these events to upper-level aggregators that have a much broader view of more assets and can focus more on event correlation and threat analysis typically performed in a security operations center.

If all logs are being collected and there are still critical assets that are not adequately monitored, it may be necessary to add additional network monitoring tools to compensate for these deficiencies. This process is illustrated in Figure 13.7.

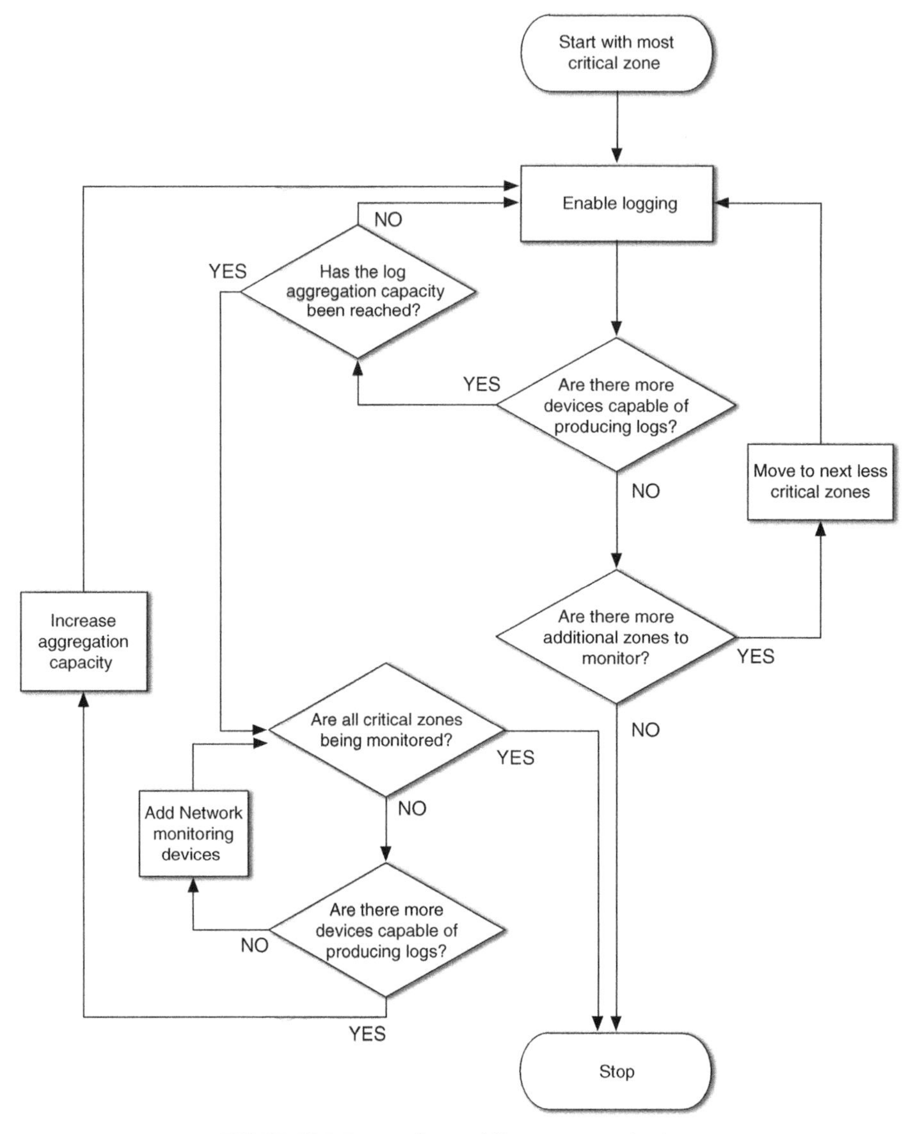

FIGURE 13.7 Process for enabling zone monitoring.

> **Caution**
>
> Remember that when aggregating logs, it is still necessary to respect the boundaries of all established security zones. If logs need to be aggregated across zones (which is helpful for the detection of threats as they move between zones), make sure that the zone perimeter is configured to only allow the movement of logs in one direction; otherwise, the perimeter could potentially be compromised. In most instances, simply creating a policy that explicitly

Caution—cont'd

states the source (the device producing logs) and the destination (the log aggregation facility) for the specified service (e.g., syslog, port 514) is sufficient in order to enforce a restricted one-way transmission of the log files. For critical zones, physical separation using a data diode or unidirectional gateway may be required to assure that all log transmissions occur in one direction, and that there is no ability for malicious traffic to enter the secure zone from the logging facility.

Additional monitoring tools might include any asset or network monitoring device, including host-based security agents, or external systems, such as an IDS, an application monitor, or an industrial protocol filter. Network-based monitoring tools are often easier to deploy, because they are by nature nonobtrusive and, if configured to monitor a spanned or mirrored interface, typically do not introduce latency.

Information collection and management tools

The "log collection facility" is typically a log management system or a security information and event management (SIEM) system. These tools range from very simple to very complex and include free, open-source, and commercial options. Some options include syslog aggregation and log search, commercial log management systems, the open source security information management (**OSSIM**) system, and commercial SIEM systems.

Syslog aggregation and log search

Syslog allows log files to be communicated over a network. By directing all syslog outputs from supported assets to a common network file system, a very simple and free log aggregation system can be established. While inexpensive (essentially free), this option provides little added value in terms of utilizing the collected logs for analysis, requiring the use of additional tools, such as open source log search or IT search tools, or through the use of a commercial log management system or SIEM. If logs are being collected for compliance purposes as well as for security monitoring, additional measures will need to be taken to comply with log retention requirements. These requirements include nonrepudiation and chain of custody, as well as ensuring that files have not been altered, or accessed by unauthorized users. This can be obtained without the help of commercial systems, although it does require additional effort by IT managers.

Log management systems

Log management systems provide a commercial solution for log collection, analysis, and reporting. Log management systems provide a configuration interface to manage log collection, as well as options for the storage of logs—often allowing the administrator to configure log retention parameters by individual log source. At the time of collection, log management systems also provide the necessary nonrepudiation features to ensure the integrity of the log files, such as "signing" logs with a calculated hash that can be later compared to the files as a checksum. Once collected, the logs can then also be analyzed and searched, with the ability to produce prefiltered reports in order to present log data relevant to a specific purpose or function, such as compliance reports, which produce log details specific to one or more regulatory compliance controls, as shown in Figure 13.8.

Security information and event management systems

Security information and event management systems, or SIEMs, extend the capabilities of log management systems with the addition of specific analytical and contextual functions. According to security analysts from Gartner, the differentiating quality of an SIEM is that it combines the log management and compliance reporting qualities of a log management or legacy security information management (SIM) system with the real-time monitoring and incident management capabilities of a security event manager (SEM).[14] A SIEM must also support "data capture from heterogeneous data sources, including network devices, security devices, security programs, and servers,"[15] making the qualifying SIEM an ideal platform for providing situational awareness across security zone perimeters and interiors.

Many SIEM products are available, including several open-source variants in addition to the many commercial SIEMs. Because an SIEM is designed to support real-time

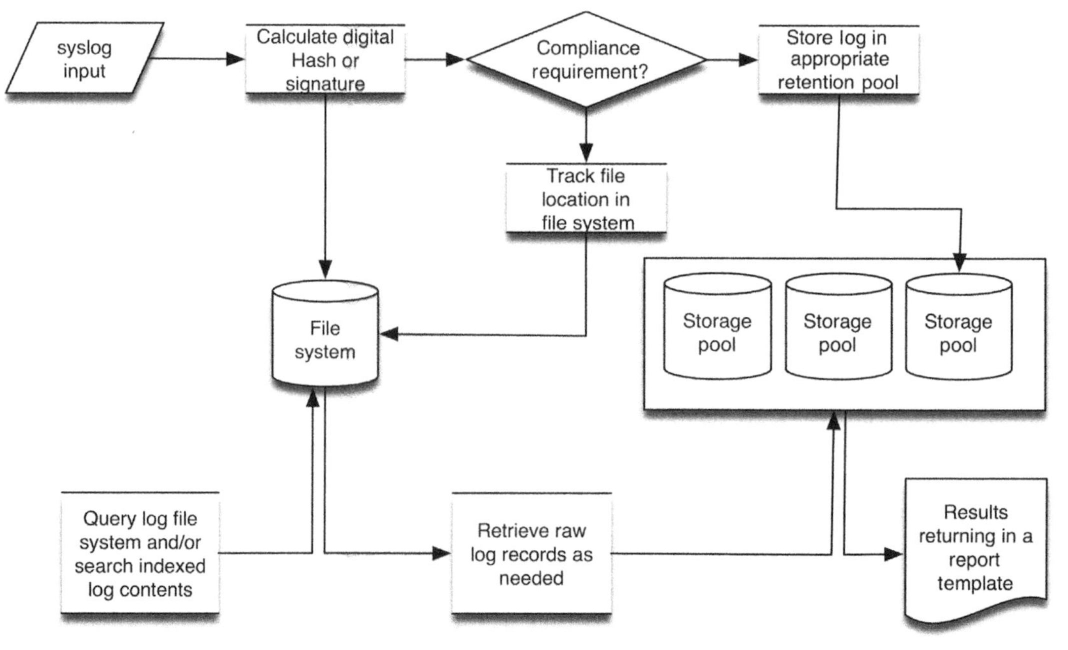

FIGURE 13.8 Typical log management operations.

monitoring and analytical functions, it will parse the contents of a log file at the time of collection, storing the parsed information in some sort of structured data store, typically a database or a specialized flat-file storage system. By parsing out common values, they are more readily available for analytics, helping to support the real-time goals of the SIEM, as shown in Figure 13.9. The parsed data are used for analytics, while a more traditional log management framework that will hash the logs and retain them for compliance. Because the raw log file may be needed for forensic analysis, a logical connection between the log file and the parsed event data is typically maintained within the data store.

SIEM platforms are often used in security operations centers (SOCs), providing intelligence to security operators that can be used to detect and respond to security concerns. Typically, the SIEM will provide visual dashboards to simplify the large amounts of disparate data into a more human-readable form. Figure 13.10 illustrates how a custom dashboard is created within Splunk to visual ICS-related security events. Figure 13.11 shows how this dashboard can be expanded to provide more application-layer event information pertaining to industrial protocol security events (e.g., use of invalid function codes).

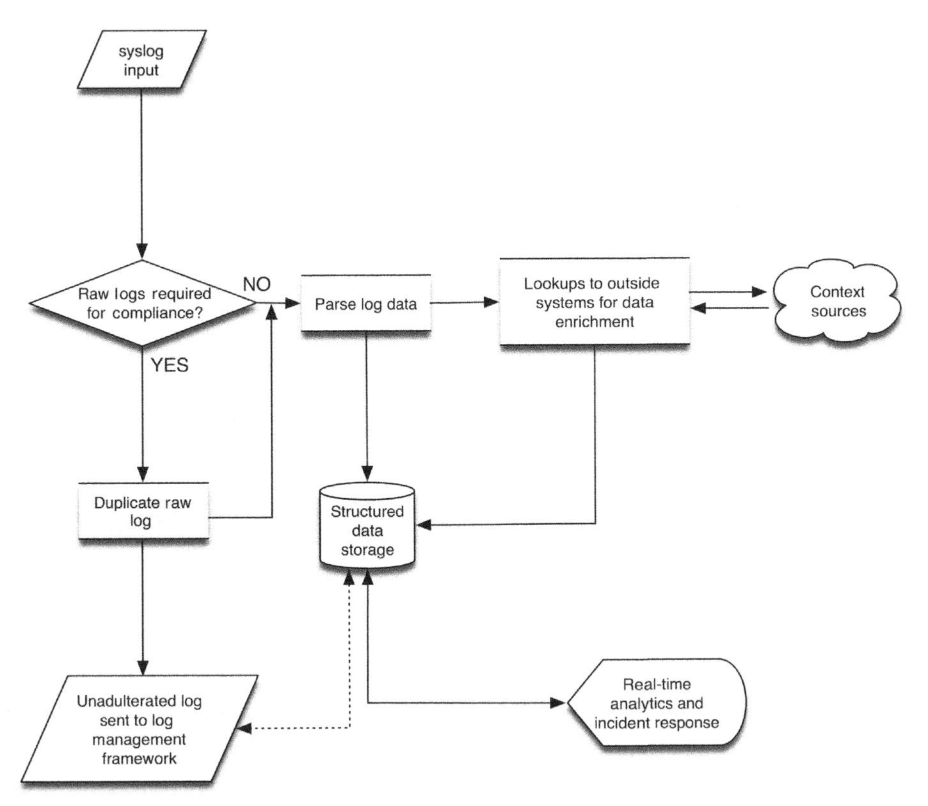

FIGURE 13.9 Typical SIEM operations.

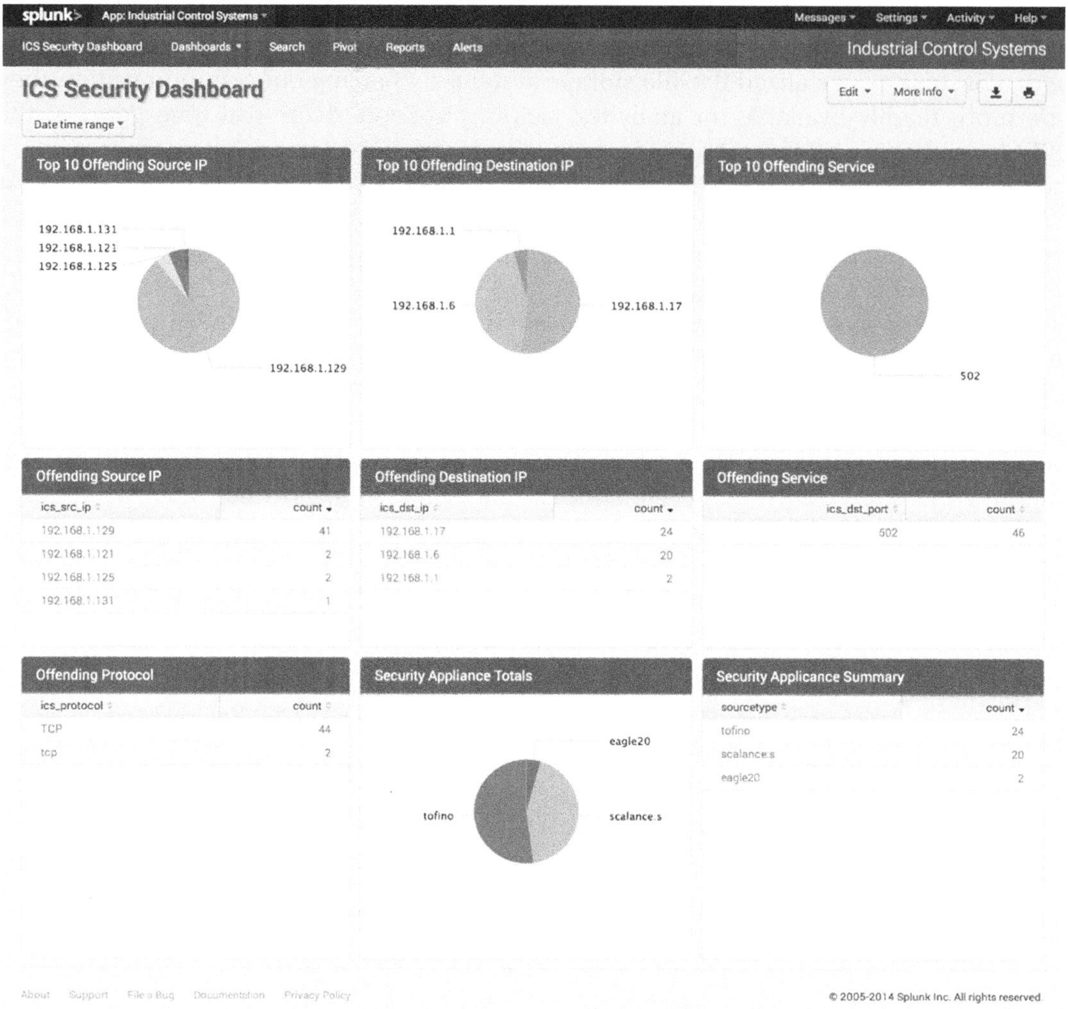

FIGURE 13.10 ICS security dashboard for Splunk.

> **Note**
>
> Log management and SIEM platforms are converging as information security needs become more closely tied to regulatory compliance mandates. Many traditional log management vendors now offer SIEM features, while traditional SIEM vendors are offering log management features.

Data historians and process alarms

Data historians are not security monitoring products, but they do monitor activity (see Chapter 4, "Introduction to Industrial Control Systems and Operations") and can be a useful supplement to security monitoring solutions in several ways, including

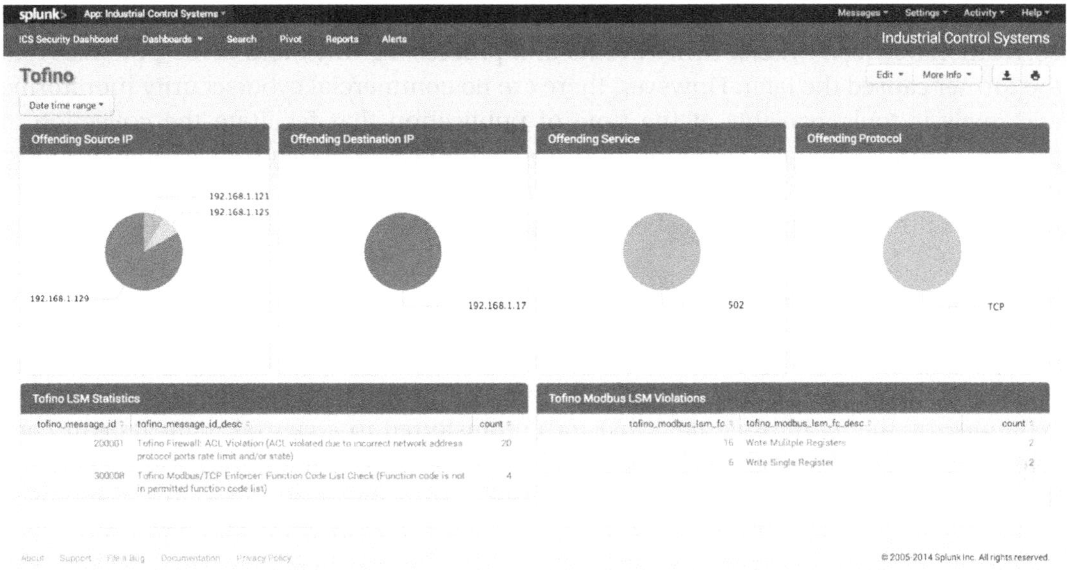

FIGURE 13.11 ICS security dashboard—application layer event analysis.

- Providing visibility into control system assets that may not be visible to typical network monitoring tools.
- Providing process efficiency and reliability data that can be useful for security analysis.

Because most security monitoring tools are designed for enterprise network use, they are typically restricted to TCP- and UDP-based IP networks and therefore have no visibility into large portions of most industrial plants that may utilize serial connectivity or other nonroutable protocols. Many industrial protocols are evolving to operate over Ethernet using TCP and UDP transports over IP, meaning these processes can be impacted by enterprise network activities. The security analysis capabilities of SIEM are made available to operational data by using the operational data provided by the ICS, allowing threats that originate in IT environments but target OT systems (i.e., Stuxnet, Industroyer, Incontroller, and other cyber-physical attacks) to be more easily detected and tracked by security analysts. Those activities that could impact the performance and reliability of industrial automations systems can be detected as well by exposing IT network metrics to operational processes, including network flow activity, heightened latency, or other metrics that could impact the proper operation of industrial network protocols (see Chapter 6, "Industrial Network Protocols").

Process alarms are by nature more focused and typically represent a condition or abnormality within the process that an operator should be concerned about (hence the generation of an alarm). This makes process alarms a valuable data point when mapping cybersecurity events to any potential impact on the physical process: with a successful correlation between cybersecurity events and process alarms, there is the potential that the former caused the later. However, there are no commercial cybersecurity monitoring and analysis tools available at the time of publication that facilitate the collection of process alarms. Manual integration of historized process data and/or alarms into most cybersecurity tools is possible but will require input from subject matter experts in both process control and industrial cybersecurity.

Monitoring across secure boundaries

As mentioned in the section "Successfully monitoring security zones," it is sometimes necessary to monitor systems across secure zone boundaries via defined conduits. This requires zone perimeter security policies that will allow the security logs and events generated by the monitoring device(s) to be transferred to a central management console. Data diodes are ideal for this application as they force the information flow in one direction—away from the zones possessing higher security levels and toward the central management system. If a firewall is used, any "hole" provided for logs and events represents a potential attack vector. The configuration must therefore explicitly limit the communication from the originating source(s) to the destination management system, by IP (Layer 3), Port (Layer 4), and preferably application content (Layer 7), with no allowed return communication path. Ideally, this communication would be encrypted as well, as the information transmitted could potentially be sensitive in nature.

Information management

The next step in security monitoring is to utilize the relevant security information that has been collected. Proper analysis of this information can provide the situational awareness necessary to detect incidents that could impact the safety and reliability of the industrial network.

Ideally, the SIEM or Log Manager will perform many underlying detection functions automatically—including normalization, data enrichment, and correlation (see Chapter 12, "Exception, Anomaly, and Threat Detection")—providing the security analyst with the following types of information at their disposal:

- The raw log and event details obtained by monitoring relevant systems and services, normalized to a common taxonomy.
- The larger "incidents" or more sophisticated threats derived from those raw events that may include correlation with external global threat intelligence sources.
- The associated necessary context to what has been observed (raw events) and derived (**correlated events**).

Typically, an SIEM will represent a high-level view of the available information on a dashboard or console, as illustrated in Figure 13.12, which shows the dashboard of the

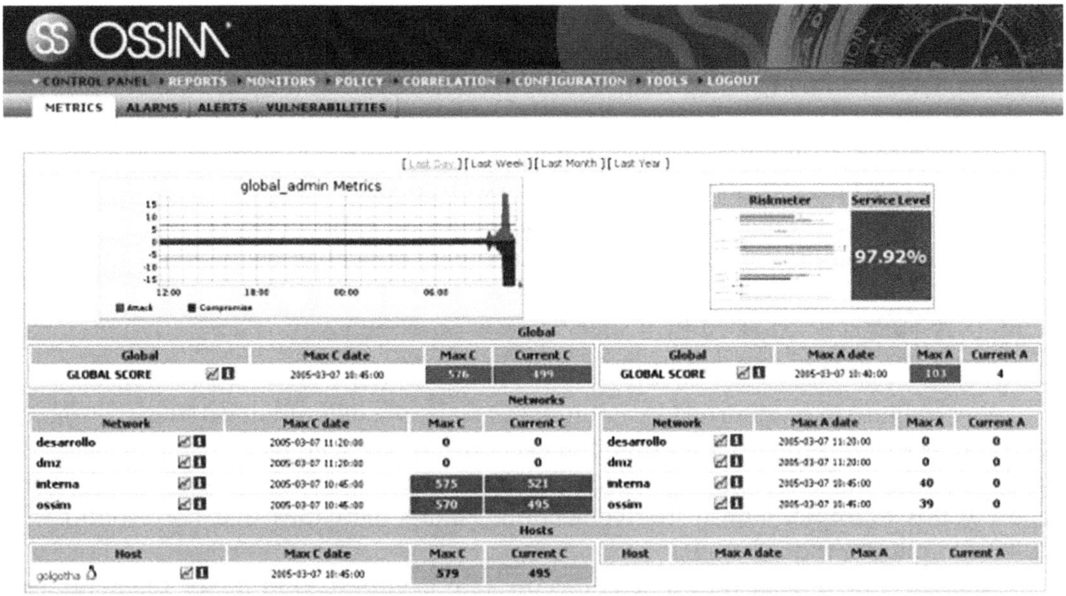

FIGURE 13.12 The open source security information management project.

Open Source Security Information Management (OSSIM) platform. With this information in hand, automated and manual interaction with the information can occur. This information can be queried directly to achieve direct answers to explicit questions. It can also be formulated into a report to satisfy specific business, policy, or compliance goals, or it can be used to proactively or reactively notify a security or operations officer of an incident. The information is available to further investigate incidents that have already occurred.

Queries

The term "query" refers to a request for information from the centralized data store. This can sometimes be an actual database query, using structured query language (SQL), or it may be a plain-text request to make the information more accessible by users without database administration skills (although these requests may use SQL queries internally, hidden from the user). Common examples of initial queries include the following:

- Top 10 talkers (by total network bandwidth used)
- Top talkers (by unique connections or flows)
- Top events (by frequency)
- Top events (by severity)
- Top events over time
- Top applications in use
- Open ports.

These requests can be made against any or all data that are available in the data store (see the section "data availability"). By providing additional conditions or filters, queries can be focused yielding results more relevant to a specific situation. For example

- Top 10 talkers during nonbusiness hours
- Top talkers using specific industrial network protocols
- All events of a common type (e.g., user account changes)
- All events targeting a specific asset or assets (e.g., critical assets within a specific zone)
- All ports and services used by a specific asset or assets
- Top applications in use within more than one zone.

Query results can be returned in a number of ways: via delimited text files, a graphical user interface or dashboard, preformatted executive reports, an alert that is delivered by SMS or e-mail, and so on. Figure 13.13 shows user activity filtered by a specific event type—in this example, administrative account change activities that correspond with NERC compliance requirements.

A defining function of an SIEM is to correlate events to find larger incidents (see Chapter 12, "Exception, Anomaly, and Threat Detection"). This includes the ability to define correlation rules, as well as present the results via a dashboard. Figure 13.14 shows a graphical event correlation editor that allows the logical conditions (such as "if A and B then C"), while Figure 13.15 shows the result of an incident query—in this case the selected incident (an HTTP Command and Control Spambot) being derived from four discrete events.

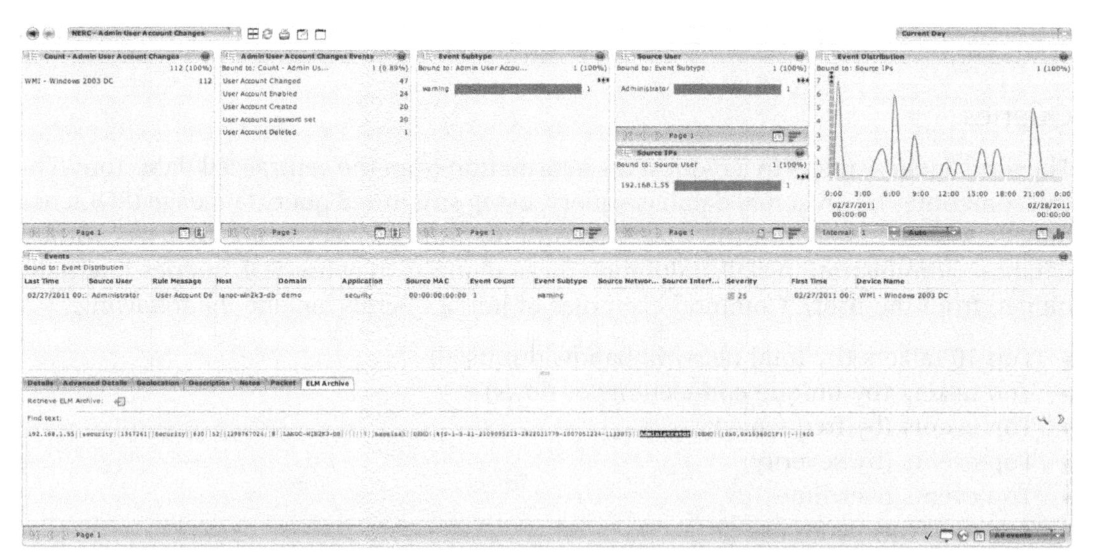

FIGURE 13.13 An SIEM dashboard showing administrative account changes.

FIGURE 13.14 An example of a graphical interface for creating event correlation rules.

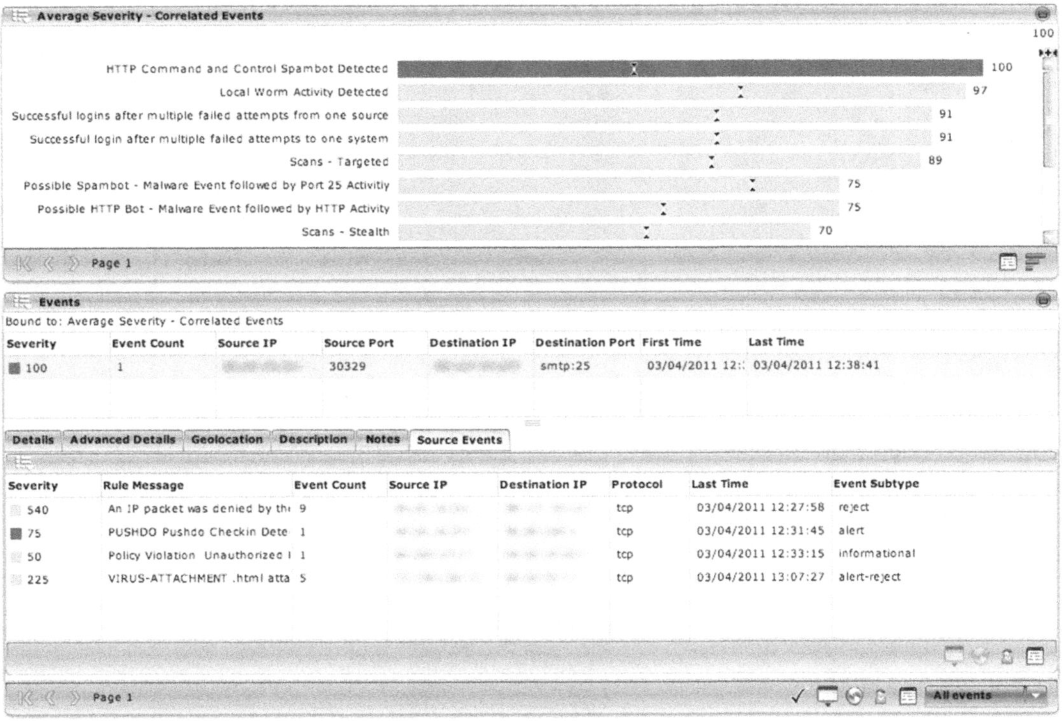

FIGURE 13.15 An SIEM dashboard showing a correlated event and its source events.

Reports

Reports select, organize, and format all relevant data from the enriched logs and events into a single document. Reports provide a useful means to present almost any data set. Reports can summarize high-level incidents for executives or include precise and comprehensive documentation that provides minute details for internal auditing or for compliance. An example of a report generated by an SIEM is shown in Figure 13.16 showing a quick summary of the OSIsoft PI Historian authentication failures and point change activity.

Industrial Incidents
Report Generated: Mar 4, 2011 1:58 PM
Time Zone: Greenwich Mean Time : Dublin, Edinburgh, Lisbon, London GMT+00:00
Report Period: 2011/01/01 00:00:00 to 2011/04/01 00:00:00
Device Count: 49

Incident overview

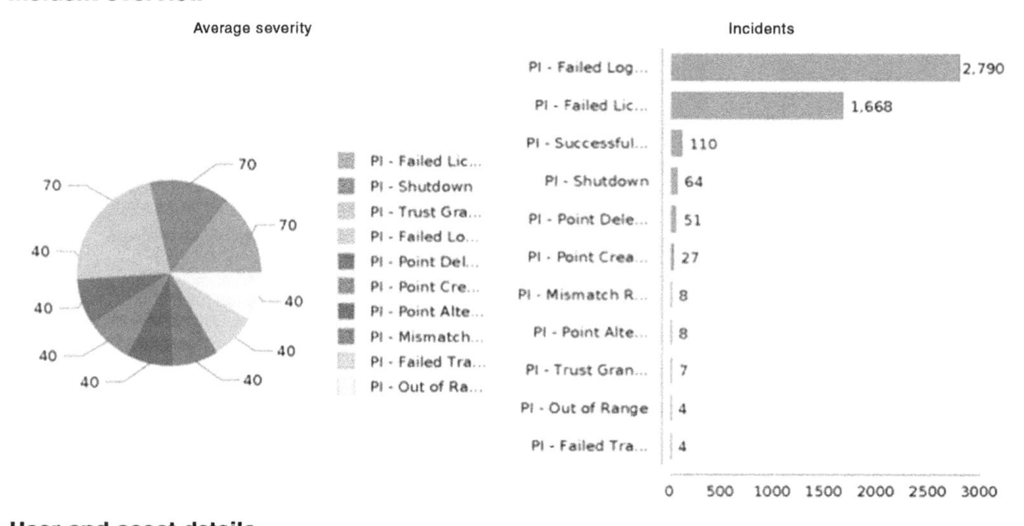

User and asset details

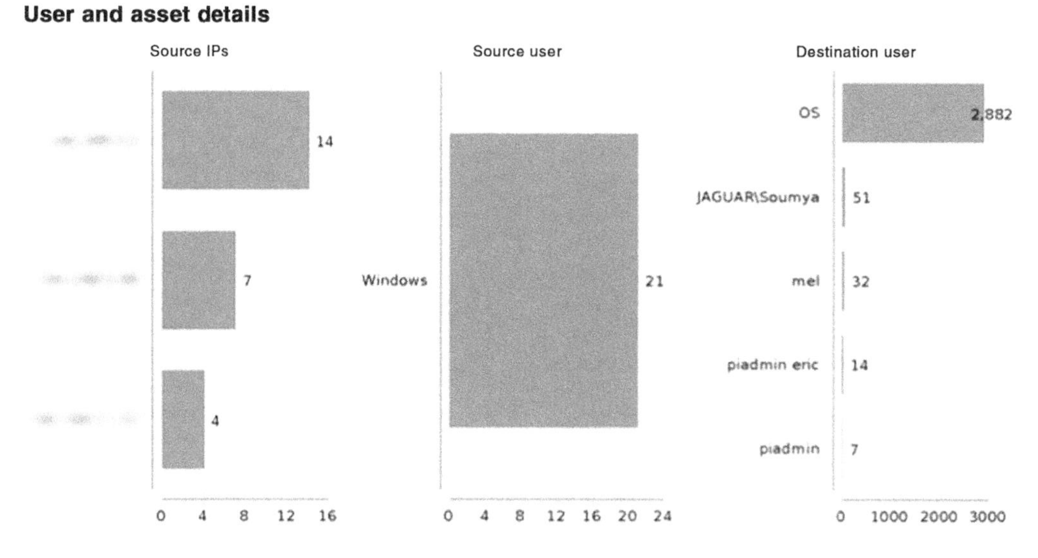

Page 1 of 3

FIGURE 13.16 An SIEM report showing industrial activities.

Alerts

Alerts are active responses to observed conditions within the SIEM. An alert can be a visual notification in a console or dashboard, a direct communications (e-mail, page, SMS, etc.) to a security administrator, or even the execution of a custom script. Common alert mechanisms used by commercial SIEMs include the following:

- Visual indicators (e.g., red, orange, yellow, green)
- Direct notification to a user or group of users
- Generation and delivery of a specific report(s) to a user or group of users
- Internal logging of alert activity for audit control
- Execution of a custom script or other external control
- Generation of a ticket in a compatible help desk or incident management system.

Several compliance regulations, including NERC CIP, CFATS, and NRC RG 5.71, require that incidents be appropriately communicated to proper authorities inside and/or outside of the organization. The alerting mechanism of an SIEM can facilitate this process by creating a useable variable or data dictionary with appropriate contacts within the SIEM and automatically generating appropriate reports and delivering them to key personnel.

Incident investigation and response

SIEM and log management systems are useful for incident response because the structure and normalization of the data allow an incident response team to drill into a specific event to find additional details (often down to the source log file contents and/or captured network packets) and to pivot on specific data fields to find other related activities. For example, if there is an incident that requires investigation and response, it can be examined quickly providing relevant details, such as the username and IP address. The SIEM can then be queried to determine what other events are associated with the user, IP, and so on.

In some cases the SIEM may support active response capabilities, including

- Allowing direct control over switch or router interfaces via SNMP, to disable network interfaces.
- Executing scripts to interact with devices within the network infrastructure, to reroute traffic, isolate users, and so on.
- Execute scripts to interact with perimeter security devices (e.g., firewalls) to block subsequent traffic that has been discovered to be malicious.
- Execute scripts to interact with directory or IAM systems to alter or disable a user account in response to observed malicious behavior.

These responses may be supported manually or automatically, or both.

Caution

While automated response capabilities can improve efficiencies, they should be limited to noncritical security zones and/or to zone perimeters. As with any control deployed within industrial networks, all automated responses should be carefully considered and tested prior to implementation. A false positive could trigger such a response and cause the failure of an industrial operation, with potentially serious consequences.

Log storage and retention

The end result of security monitoring, log collection, and enrichment is a large quantity of data in the form of log files, which must be stored for audit and compliance purposes (in the cases where direct monitoring is used in lieu of log collection, the monitoring device will still produce logs, which must also be retained). This represents a few challenges, including how to ensure the integrity of the stored files (a common requirement for compliance), how and where to store these files, and how they can be kept readily available for analysis.

Nonrepudiation

Nonrepudiation refers to the process of ensuring that a log file has not been tampered with, so that the original raw log file can be presented as evidence, without question of authenticity, within a court of law. This can be achieved in several ways, including digitally signing log files upon collection as a checksum, utilizing protected storage media, or the use of third-party FIM systems.

A digital signature is typically provided in the form of a hash algorithm that is calculated against the log file at the time of collection. The result of this calculation provides a checksum against which the files can be verified to ensure they have not been tampered with. If the file is altered in any way, the hash will calculate a different value and the log file will fail the integrity check. If the checksum matches, the log is known to be in its original form.

The use of appropriate storage facilities can ensure nonrepudiation as well. For example, by using write once read many (WORM) drives, raw log records can be accessed but not altered, as the write capability of the drive prevents additional saves. Many managed storage area network (SAN) systems also provide varying levels of authentication, encryption, and other safeguards.

An FIM may already be in use as part of the overall security monitoring infrastructure, as described in the section "Assets." The FIM observes the log storage facility for any sign of changes or alterations, providing an added level of integrity validation.

Data retention/storage

The security monitoring tools just mentioned all require the collection and storage of security-related information. The amount of information that is typically required could easily surpass 170 GB over an 8-h period for a medium-sized enterprise collecting information at approximately 20,000 events per second.[16] It is worth mentioning that event generation within an industrial network is typically a small fraction of this number, and when properly tuned, presents a manageable amount of information storage.

Data retention refers to the amount of information that is stored long-term and can be measured in volume (the size of the total collected logs in bytes) and time (the number of months or years that logs are stored for). The length of time a log is retained is important, as this metric is often defined by compliance regulations—NERC CIP requires that logs are retained for anywhere from 90 days to up to 3 years, depending upon the nature of the log.[17] The amount of physical storage space that is required can be calculated by determining which logs are needed for compliance and for how long they must be kept. Some of the factors that should be considered include the following:

- Identifying the quantity of inbound logs
- Determining the average log file size
- Determining the period of retention required for logs
- Determining the supported file compression ratios of the log management or SIEM platform being used.

Table 13.3 illustrates how sustained log collection rates map to total log storage requirements over a retention period of 7 years, resulting in a few terabytes (10^{12}) of storage up to hundreds of terabytes or even petabytes (10^{15}) of storage.

There may be a requirement to retain an audit trail for more than one standard or regulation depending upon the nature of the organization, often with each regulation mandating different retention requirements. As with NERC CIP, there may also be a change in the retention requirements depending upon the nature of the log, and whether an incident has occurred. All of this adds up to even greater, long-term storage requirements.

Table 13.3 Log Storage Requirements Over Time

Logs per Second	Logs per Day (in Billions)	Logs per Year (in Billions)	Average Bytes per Event	Retention Period in Years	Raw log Size (TB)	Compressed Bytes (TB) 5:1	Compressed Bytes (TB) 10:1
100,000	8.64	3154	508	7	10,199	2040	1020
50,000	4.32	1577	508	7	5100	1020	510
25,000	2.16	788	508	7	2550	510	255
10,000	0.86	315	508	7	1020	204	102
5000	0.43	158	508	7	510	102	51
1000	0.09	32	508	7	102	21	11
500	0.04	16	508	7	51	11	6

Tip

Make sure that the amount of available storage has sufficient headroom to accommodate spikes in event activity because event rates can vary (especially during a security incident).

Data availability

Data availability differs from retention, referring to the amount of data that is accessible for analysis. Also called "live" or "online" data, the total data availability determines how much information can be analyzed concurrently—again, in either volume (bytes and/or total number of events) or time. Data retention affects the ability of an SIEM to detect "low and slow" attacks (attacks that purposefully occur over a long period of time in order to evade detection), as well as to perform trend analysis and anomaly detection (which by definition requires a series of data over time—see Chapter 12, "Exception, Anomaly, and Threat Detection").

Tip

In order to meet compliance standards, it may be necessary to produce a list of all network flows within a particular security zone that originated from outside of that zone, for the past 3 years. For this query to be successful, 3 years of network flow data need to be available to the SIEM at once. There is a work-around if the SIEM's data availability is insufficient (for example, it can only keep 1 year of data active). The information can be stored in volumes consistent with the SIEM's data availability by archiving older data sets. A partial result is obtained by querying the active data set. Two additional queries can be run by then restoring the next-previous backup or archive, producing multiple partial result sets of 1 year each. These results can then be combined to obtain the required 3-year report. Note that this requires extra effort on the part of the analyst. The archive/retrieval process on some legacy SIEMs may interfere with or interrupt the collection of new logs until the process is complete.

Unlike data retention, which is bound by the available volume of data storage (disk drive space), data availability is dependent upon the structured data that are used by the SIEM for analysis. Depending upon the nature of the data store, the total data availability of the system may be limited to a number of days, months, or years. Typically, one or more of the following limits databases:

- The total number of columns (indices or fields)
- The total number of rows (discreet records or events)
- The rate at which new information is inserted (i.e., collection rate)
- The rate at which query results are required (i.e., retrieval rates).

Depending upon the business and security drivers behind information security monitoring, it may be necessary to segment or distribute monitoring and analysis into zones to meet performance requirements. Some factors to consider when calculating the necessary data availability include:

- The total length of time over which data analysis may be required by compliance standards.
- The estimated quantity of logs that may be collected in that time based on event estimates.
- The incident response requirements of the organization—certain governmental or other critical installations may require rapid-response initiatives that necessitate fast data retrieval.
- The desired granularity of the information that is kept available for analysis (i.e., are there many vs. few indices).

Summary

A larger picture of security-related activity begins to form once zone security measures are in place. Exceptions from the established security policies can then be detected by measuring these activities and further analyzing them. Anomalous activities can also be identified so that they may be further investigated.

This requires well-defined policies with those policies configured within an appropriate information analysis tool. Just as with perimeter defenses to the security zone, carefully built variables defining allowed assets, users, applications, and behaviors can be used to aid in detection of security risks and threats. If these lists can be determined dynamically, in response to observed activity within the network, the "whitelisting" of known-good policies, becomes "smart-listing." This helps further strengthen perimeter defenses through dynamic firewall configuration or IPS rule creation.

The event information can be further analyzed as various threat detection techniques are used together by event correlation systems that find larger patterns more indicative of serious threats or incidents. Widely used in IT network security, event correlation is beginning to "cross the divide" into OT networks, at the heels of Stuxnet and other sophisticated threats that attempt to compromise industrial network systems via attached IT networks and services.

Everything (measured metrics, baseline analysis, and whitelists) rely on a rich base of relevant security information. Where does this security information come from? The networks, assets, hosts, applications, protocols, users, and everything else that is logged or monitored contributes to the necessary base of data required to achieve "situational awareness" and effectively secure an industrial network.

Endnotes

1. J.M. Butler. Benchmarking Security Information Event Management (SIEM). The SANS Institute Analytics Program, February 2009.
2. Ibid.
3. Ibid.
4. Ibid.
5. Microsoft. Windows Management Instrumentation., January 6, 2011 (cited: March 3, 2011).
6. Ibid.
7. Wayne Matthews. New roles for process historians. International Society of Automation. Intech Magazine. November/December 2017.
8. National Institute of Standards and Technology, Special Publication 800-53 Revision 3.
9. Ibid.
10. Flow.org. Traffic Monitoring using sFlow. http://www.sflow.org/sFlowOverview.pdf. 2003 (cited: March 3, 2011).
11. B. Singer, Kenexis Security Corporation; D. Peterson (Ed.), Proceedings of the SCADA Security Scientific Symposium, 2: Correlating Risk Events and Process Trends to Improve Reliability, Digital Bond Press, 2010.
12. Securonix, Inc., Securonix Indentity Matcher: Overview. http://www.securonix.com/identity.htm, 2003 (cited: March 3, 2011).
13. A. Chuvakin, Content Aware SIEM. http://www.sans.org/security-resources/idfaq/vlan.php. February 2000 (cited: January 19, 2011).
14. K.M. Kavanagh, M. Nicolett, O. Rochford, "Magic quadrant for security information and event management," Gartner Document ID Number: G00261641, June 25, 2014.
15. Ibid.
16. J.M. Butler, Benchmarking Security Information Event Management (SIEM). The SANS Institute Analytics Program, February 2009.
17. North American Electric Reliability Corporation. NERC CIP Reliability Standards, version 4. http://www.nerc.com/page.php?cid=2/20. February 3, 2011 (cited: March 3, 2011).

14 Standards and Regulations

Information in this chapter

- Common Cybersecurity Standards and Regulations
- ISA/IEC-62443
- Mapping Industrial Network Security to Compliance
- Mapping Compliance Controls to Network Security Functions
- Industry Best Practices for Conducting ICS Assessments
- Common Criteria and FIPS Standards

There are many cybersecurity standards, guidelines, and regulations imposed by governments and industry, which provide everything from "best practices" to hard requirements that are enforced through penalties and fines. Many of these standards are general information security documents; however, the number of industry-related documents focused on industrial control systems (ICSs) is growing. In the United States, common standards include the North American Electric Reliability Corporation's (NERC's) Critical Infrastructure Protection (CIP) Reliability Standards, the US Department of Homeland Security's (DHS) Chemical Facility Anti-Terrorism Standards (CFATS), the Regulated Security of Nuclear Facilities by the US Nuclear Regulatory Commission (NRC), and general ICS security recommendations published by the National Institute of Standards and Technology (NIST) in Special Publication 800-82. In Europe, standards and guidelines include the EU M/490 and the SGCG, which provide guidance for modern power, and the many publications of the European Union Agency for Network and Information Security (ENISA). Global standards include the International Organization for Standardization (ISO)/ International Electrotechnical Commission (IEC) 27000 series of standards, of which ISO-27002:2013 "Code of practice for information security controls" is widely adopted.

Arguably, the standard most relevant to industrial security is International Society of Automation (ISA) 62443 (formerly ISA 99), which is the product of the International Society of Automation. ISA 62443 is concerned with the security of industrial automation and control systems and is applicable to any organization or industry that uses these systems. ISA 62443 also aligns with international standard IEC 62443 and is under revision and restructuring for acceptance by the ISO as ISO 62443.

Regardless of which standard you are working with, it is important to remember that standards are designed for a large and sometimes diverse audience, and so caution should be taken when applying them to an industrial architecture. These guidelines will make recommendations or requirements for specific cybersecurity controls, which have been vetted for general use by the target audience of the standard. However, even when the target audience is suppliers, integrators, and end-users of ICS—as is the case with

Industrial Network Security. https://doi.org/10.1016/B978-0-443-13737-2.00015-4

ISA 62443—there is no way for a standard to address the intricacies and nuances of an individual company or facility. No two networks are identical—even the same process within the same company will have subtle differences from site-to-site due to commissioning dates, system updates/migrations, and general lifecycle support. Therefore, each recommendation should be given careful consideration taking into account the specifics of your own unique industrial network environment.

This chapter attempts to map specific controls referenced in common standards to the relevant topics and discussions that are covered in this book (see Table 14.1). Please note that in many instances, policies and procedures may be the right answer; however, these are not covered in any detail in this book. You may realize, having made it to Chapter 14 that this book focuses largely on technology. This is not to suggest that people and process are less important to technology; only to explain that there are many additional security controls to consider beyond what is covered here. On a similar note, we will not attempt to focus on any one standard in detail within this book because efforts to maintain compliance with just one of these regulations can be challenging and complex enough to fill entire books. Because of slight variations in terminology and methodology, complying with multiple standards can be a nightmare. However, it can often be valuable for someone who is attempting to follow a particular standard to utilize both the normative and informative text of other standards to gain additional insight and understanding that may be absent from the original document. "Crosswalks" between standards can be a valuable asset in mapping between the various standards and their particular requirements.

There are also standards and regulations that do not apply to industrial networks at all, but rather to the products that might be utilized by an industrial network operator to help secure (see Chapter 9, "Establishing Zones and Conduits") and monitor (see Chapter 13, "Security Monitoring of Industrial Control Systems") the network. Among these are the international Common Criteria standards and various Federal Information Processing Standards (FIPS) including the FIPS 140-2 Security Requirements for Cryptographic Modules.

Common standards and regulations

As mentioned in Chapter 2, "About Industrial Networks," industrial networks are of interest to several national and international regulatory and standards organizations. In the United

Table 14.1 International Society of Automation (ISA) 62443 Security Levels[35]

Security level	Description
1	Prevent the unauthorized disclosure of information via eavesdropping or casual exposure
2	Prevent the unauthorized disclosure of information to an entity actively searching for it using simple means with low resources, generic skills, and low motivation
3	Prevent the unauthorized disclosure of information to an entity actively searching for it using sophisticated means with moderate resources, IACS-specific skills, and moderate motivation
4	Prevent the unauthorized disclosure of information to an entity actively searching for it using sophisticated means with extended resources, IACS-specific skills, and high motivation

States and Canada, NERC is well known because of the NERC CIP reliability standards, which heavily regulate security within the North American bulk electric system. NERC operates independently under the umbrella of the Federal Energy Regulatory Commission (FERC), which regulates interstate transmission of natural gas, oil, and electricity. FERC also reviews proposals to build liquefied natural gas (LNG) terminals, interstate natural gas pipelines, and licensing for hydropower projects. The Department of Energy (DoE) and DHS also produce several security recommendations and requirements, including the CFATS, the Federal Information Security Management Act (FISMA), and Homeland Security Presidential Directive Seven, which all refer back to several special publications of the NIST, particularly SP 800-53 "Recommended Security Controls for Federal Information Systems and Organizations" and SP 800-82 "Guide to Industrial Control Systems (ICS) Security." The ISA's standard for the Security for Industrial Automation and Control Systems (ISA 62443) provides security recommendations that are applicable to industrial control networks. ISO also has published the ISO-27033 standard for network security and is considering the release of industry-specific standard ISO-27013 for manufacturing systems.

NERC CIP

It is hard to discuss critical infrastructure security without referring to the NERC CIP reliability standards, which has gained wide notoriety due to its heavy penalties for noncompliance. Although NERC CIP standards are only enforceable within North American bulk electric systems, the standards represented are technically sound and in alignment with other standards and are presented in the spirit of improving the security and reliability of the electric industry.[1] Furthermore, the critical infrastructures of the electric utilities—specifically the distributed control systems responsible for the generation of electricity and the stations, substations, and control facilities used for transmission of electricity—utilize common industrial network assets and protocols, making the standards relevant to a wider base of industrial network operators.

CFATS

The Risk-Based Performance Standards (RBPS) for the CFATS outline various controls for securing the cyber systems of chemical facilities. Specifically, RBPS Metric 8 ("Cyber") outlines controls for (1) security policy, (2) access control, (3) personnel security, (4) awareness and training, (5) monitoring and incident response, (6) disaster recovery and business continuity, (7) system development and acquisition, (8) configuration management, and (9) audits.

Controls of particular interest are Cyber Metric 8.2.1, which requires that system boundaries are identified and secured using perimeter controls, which supports the zone-based security model. Metric 8.2 includes perimeter defense, access control (including password management), the limiting of external connections, and "least-privilege" access rules.[2]

Metric 8.3 (Personnel Security) also requires that specific user access controls be established, primarily around the separation of duties, and the enforcement thereof by using unique user accounts, access control lists, and other measures.[3]

Metric 8.5 covers the specific security measures for the monitoring of asset security (primarily patch management and antimalware), network activity, log collection and alerts, and incident response, whereas Metric 8.8 covers the ongoing assessment of the architecture, assets, and configurations to ensure that security controls remain effective and in compliance.[4]

Of particular note are RBPS 6.10 (cyber security for potentially dangerous chemicals), RBPS 7 (sabotage), RBPS 14 (specific threats, vulnerabilities, and risks), and RBPS 15 (reporting)—all of which include cybersecurity controls outside of the RBPS 8 recommendations for cybersecurity. RBPS 6.10 implicates ordering and shipping systems as specific targets for attack that should be protected according to RBPS 8.[5] RBPS 7 indicates that cyber systems are targets for sabotage and that the controls implemented "deter, detect, delay, and respond" to sabotage.[6] RBPS 14 requires that measures be in place to address specific threats, vulnerabilities, and risks, inferring a strong security and vulnerability assessment (SVA) plan,[7] whereas RBPS 15 defines the requirements for the proper notification of incidents when they do occur.[8]

Update

CFATS expired in July 2023, during the writing of this text. If reauthorized, CISA will follow up with facilities. Per CISA, "CISA cannot enforce compliance with the CFATS regulations at this time. This means that CISA will not require facilities to report their chemicals of interest or submit any information in CSAT, perform inspections, or provide CFATS compliance assistance, among other activities. CISA can no longer require facilities to implement their CFATS Site Security Plan or CFATS Alternative Security Program.

CISA encourages facilities to maintain security measures. CISA's voluntary ChemLock resources are available on the ChemLock webpages."[9]

ISO/IEC 27002

The ISO/IEC 27002:2022 Standard is part of the ISO/IEC 27000 series of international standards published by the ISO, the IEC, and the American National Standards Institute (ANSI). Figure 14.1 illustrates the organization of the ISO 27000 series. ISO 27002 was previously published as ISO 17799 and later renamed, outlines hundreds of potential security controls that may be implemented according to the guidance outlined in ISO 27001. Although ISO/IEC 27002 provides less guidance for the specific protection of industrial automation and control, it is useful in that it maps directly to additional national security standards in Australia and New Zealand, Brazil, Chile, Czech Republic, Denmark, Estonia, Japan, Lithuania, the Netherlands, Poland, Peru, South Africa, Spain, Sweden, Turkey, the United Kingdom, Uruguay, Russia, and China.[10]

As with NERC CIP and CFATS, ISO/IEC 27002 focuses on risk assessment and security policies in addition to purely technical security controls.[11] The 2013 revision includes 114 security controls that are discussed including asset management and configuration

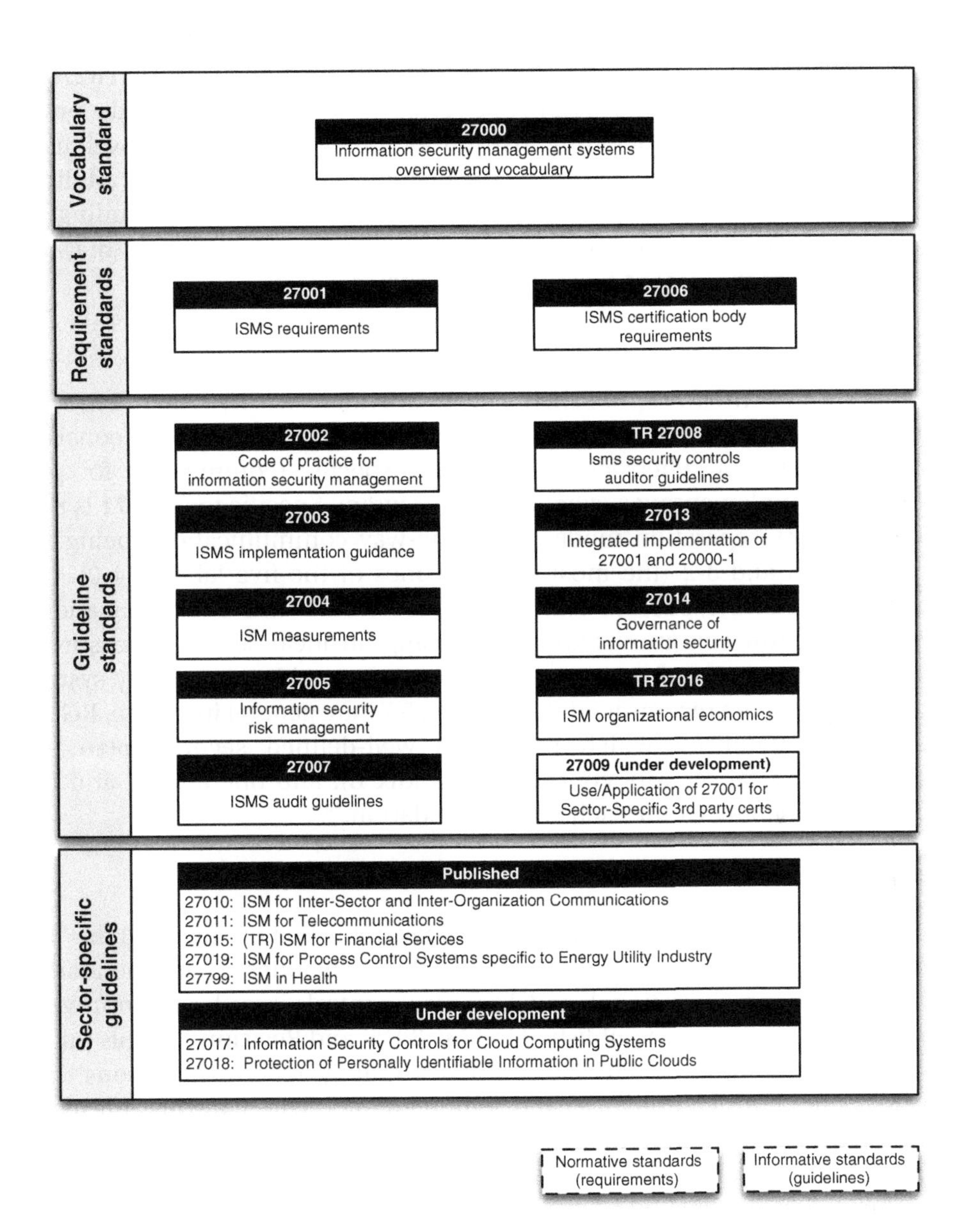

FIGURE 14.1 ISO 27000 organizational structure.[33]

management controls, separation and security controls for network communications, specific host security controls regarding access control, and antimalware protection. Of particular interest is a group of controls around security incident management—the first of the standards discussed in this book to specifically mention the anticipation of a security breach using anomaly detection. Specifically, ISO/IEC mentions "malfunctions or other anomalous system behavior may be an indicator of a security attack or actual security breach."[12]

In 2013, ISO/IEC released the energy-sector specific technical report TR27019:2013, which was updated in 2017 to TR27019:2017. This document expands on the requirements of NERC CIP by including distribution of electric power, as well as storage and distribution of gas and heat. The report includes 42 sector-specific additions and recommendations outside the current content of ISO/IEC 27002, including security controls for (potentially insecure) legacy systems, data communications, malware protection, and patch management for industrial systems.

NRC Regulation 5.71

NRC Regulation 5.71 (RG 5.71) published in 2010 provides security recommendations for complying with Title 10 of the Code of Federal Regulations (CFR) 73.54. It consists of the general requirements of cybersecurity, including specific requirements for planning, establishing, and implementing a cybersecurity program. Specific to RG 5.71 is the use of a five-zone network separation model, with one-way communications being required between levels 4-3 and 3-2 (the most critical zones of the five labeled 4-0). One-way communication gateways, such as data diodes, allow outbound communications while preventing any return communications, promising an ideal security measure for the transmission of information from a secure zone to an outside supervisory system.

Although many of the recommendations in RG 5.71 are general in nature, RG 5.71 also includes three appendices, which provide a well-defined security plan template (Appendix A), technical security controls (Appendix B), and operational and management controls (Appendix C) for each recommendation.[13]

NIST SP 800-82

The National Institute of Standards and Technology published in May 2013 the "Guide to Industrial Control Systems (ICS) Security," which includes recommendations for Security, Management, Operational, and Technical controls in order to improve control system security. Revision 2 was published in May 2015, and Revision 3 of this publication is currently in draft form. SP 800-82 comprises mainly recommendations, not hard regulations subject to compliance and enforcement. The controls presented are comprehensive and map well to additional NIST recommendations, such as those provided in Special Publication (SP) 800-53 ("Recommended Security Controls for Federal Information Systems and Organizations") and SP 800-92 ("Guide to Computer Security Log Management").[14]

ISA/IEC-62443

ISA 62443 is actually a series of standards, organized into four groups that address a broad range of topics necessary for the implementation of a secure Industrial Automation and Control System (IACS). The standard, which originated as ISA 99 when developed by the

Standards and Practices Committee 99 (SP99), is now being aligned with IEC 62443. At the time of this writing, several of the documents produced under ISA 62443 have been published and adopted by IEC, while others remain in various stages of genesis. Due to timing, there is no guarantee that what is referenced here within this book will fully align with what is eventually published, so as always it is a good idea to reference the documents directly via ISA.org. The document number for each identifies the standard (62443), the Group Number, and the Document Number (e.g., ISA 62443-1-1 is document number "1," belonging to group "1" of the ISA 62443 standard). Figure 14.2 illustrates the organizational structure of the ISA 62443 series.

ISA 62443 Part 1: "General"

ISA 62443 Part 1 (ISA 62443-1-x) focuses on the standardization of terminology and consistency of references, metrics, and models, with the goal of establishing a baseline of the fundamentals that are then referenced within the other groups. At this time, there are four documents actively being developed, including a master glossary (62443-1-2) and definitions of an IACS security lifecycle (62443-1-4). Of particular interest is 62443-1-3, which defines conformance metrics that are extremely useful in quantifying compliance to IACS security practices. These metrics are also extremely valuable to cybersecurity information analytics platforms, exception reporting, and other useful security monitoring tools (see Chapter 13, "Security Monitoring of Industrial Control Systems").

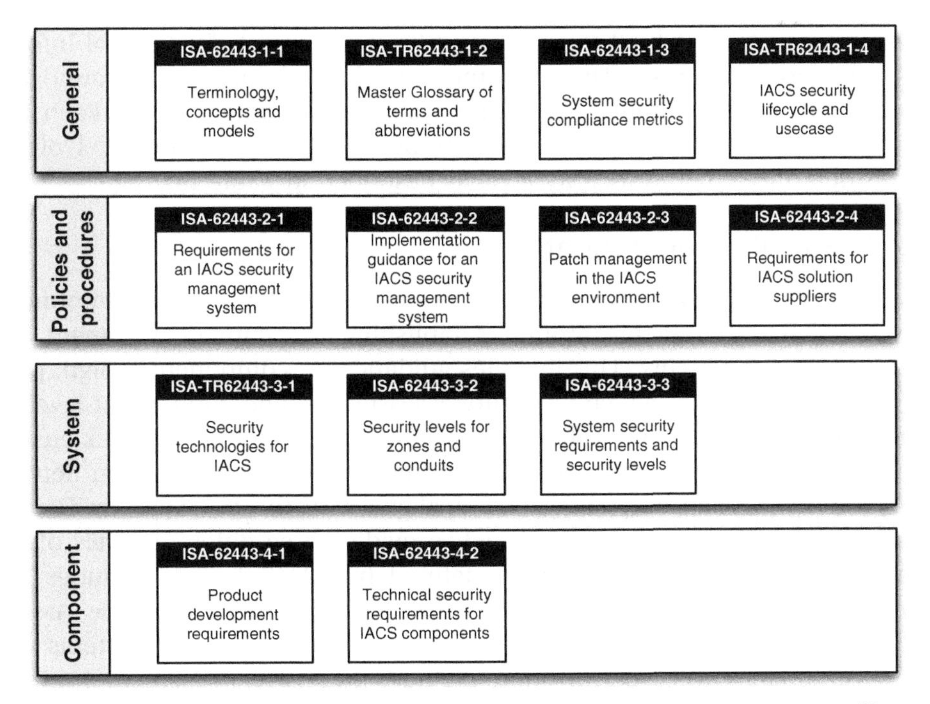

FIGURE 14.2 International Society of Automation (ISA) 62443 organizational structure.[34]

ISA 62443 Part 2: "Policies and procedures"

ISA 612443 Part 2 (ISA 62443-2-x) focuses on the necessary policies and procedures for the creation of an effective IACS security program. Group 2 includes 62443-2-1, which was one of the first standards published in the series, and details the requirements necessary for an IACS security management system. 62443-2-3 addresses patch management within industrial architectures (see Chapter 8, "Risk and Vulnerability Assessments"). 62443-2-4 has been adapted from guideline document "Process Control Domain Security Requirements for Vendors" originally developed by the Process Automation Users' Association (WIB) in Europe, and provides requirements for the certification of IACS suppliers.

ISA 62443 Part 3: "System"

ISA 62443 Part 3 (ISA 62443-3-x) focuses on cybersecurity technologies and includes documents covering available technologies, assessment and design methodologies, and security requirements and assurance levels. 62443-3 is where information and guidance on network zones and conduits will be found (along with reference models defined in 62443-1-1), as well as ISA's methodologies for risk assessments (these topics are also covered in Chapter 8, "Risk and Vulnerability Assessments," Chapter 9, "Establishing Zones and Conduits," and Chapter 11, "Implementing Security and Access Controls"). 62443-3-3 represents the security controls catalog applicable to IACS, in much the same manner as ISO 27002 "Security Techniques - Code of Practice for Information Security Management" and NIST 800-53 "Security and Privacy Controls for Federal Information Systems and Organizations." This document is divided into seven foundation requirements (FR), each containing multiple system requirements (SR). Each SR then contains zero or more requirement enhancements (RE) where the level of security required is determine by the security level as described in Table 14.1.

ISA 62443 Part 4: "Component"

ISA 62443 Part 4 (ISA 62443-4-x) focuses on the secure development of components and includes detailed requirements around establishing a secure development lifecycle (SDLC) for IACS components. This includes guidance for component design, planning, code development and review, vulnerability assessments, and component level testing. 62443-4 supports the test and validation of component "robustness" to ensure that components used within an IACS are not unduly vulnerable to common network aberrations, anomalies, and excesses. 62443-4 aligns with the ISA Security Compliance Institute's (ISCI) ISASecure program, which provides three different levels of security certification aligned with the standards defined by 62443-4. This includes supplier product development for ICS systems (Security Development Lifecycle Assurance), embedded devices (Embedded Device Security Assurance), and systems (System Security Assurance). Device certification includes extensive robustness testing using

ISCI-validated test tools including the Wurldtech (a GE company) Achilles Test Platform, Codenomicon's Defensics X test platform, and FFRI's Raven for ICS test platform. The result from the testing and certifications defined by 62443-4 is the establishment of a particular "capability" Security Level as described in Chapter 9, "establishing zones and conduits" necessary to align the capabilities of ICS components with the design "target" established earlier in the automation project lifecycle.

Mapping industrial network security to compliance

Again, there are many security regulations, guidelines, and recommendations that are published globally. Many are applicable to industrial networks; some are enforced, some not; some are regional; some are applicable to all industrial networks, while some (such as NERC CIP) apply to specific industries. Although most standards and regulations focus on a variety of general security measures (including physical security, security policy development and planning, training, and awareness), each has specific controls and measures for cybersecurity.

Tip

Many enforced compliance regulations (e.g., NERC CIP) require that "**compensating controls**" be used where a requirement cannot be feasibly met. Using additional compliance standards as a guide, alternate "compensating controls" may be identified. Therefore, even if the compliance standard is not applicable to a particular organization, the recommendations made within may prove useful.

These cybersecurity measures often overlap, although there are differences (both subtle and strong) among them. Efforts to normalize all the available controls to a common "compliance taxonomy" are being led by organizations, such as the Unified Compliance Framework (UCF), which has currently mapped close to 500 authority documents to a common framework consisting of thousands of individual controls.[15] The advantages of a common mapping are significant and include the following:

- Facilitating compliance efforts for organizations that are responsible for multiple sets of compliance controls. For example, a nuclear energy facility that must track industrial regulations, such as NRC Title 10 CFR 73.54, NRC RG 5.71, and **NEI** 08/09 requirements, as well as business regulations, such as Sarbanes-Oxley (SOX). Understanding which specific controls are common among all regulations prevents the duplication of efforts and can significantly reduce the costs of collecting, maintaining, storing, and documenting the information necessary for compliance.
- Facilitating the implementation of specific security controls by providing a comprehensive list of controls that must be implemented across all relevant standards and regulations.

This chapter begins to map the security and compliance requirements for this purpose; however, owing to the extensive nature of most regulations, as well as the changing nature of specific compliance control documents, only a select sample of common controls has been included in this text.

Industry best practices for conducting ICS assessments

There are several documents published that discuss various methodologies for testing and assessing IT architectures. This number is greatly reduced when an attempt is made to identify documents that understand the unique nature of industrial networks and offer any guidance in safely, accurately, and reliably performing these assessments. Table 14.2 provides a listing of most of the documents published on industrial security assessments. Table 14.3 provides a mapping of regulations to cybersecurity controls.

Department of Homeland Security (USA)/Center for Protection of National Infrastructure (UK)

The US Department of Homeland Security co-authored a guidance document in November 2010,[16] which the UK Centre for the Protection of National Infrastructure (CPNI) also published in April 201[17] as a "Good Practice Guide." This guideline is comprehensive in content and provides a well-documented assessment methodology or process flow chart for the testing process. The coverage of the testing process is extensive and can form the foundation for any organization's internal methodology.

The guide discusses the uniqueness associated with industrial networks and addresses the differences between assessing industrial environments and traditional IT architectures. In particular, it describes the differences between an "assessment" and a "penetration test" and how the goals desired from a particular exercise should be used to drive the overall process. The guide also provides a list of alternate methodologies that can be used to address specific requirements or constraints that may exist, including

- Lab assessments
- Component testing

Table 14.2 Industry Best Practices for Conducting ICS Assessments

Publishing organization	Description
American Petroleum Institute/National Petrochemicals and Refiners Association (USA)	Security vulnerability assessment methodology for the petroleum and petrochemicals industries
Centre for the Protection of National Infrastructure (UK)	Cyber security assessments of industrial control systems—A good practice guide
Department of Homeland Security (USA)	Can be used to test ability to exploit vulnerabilities (ethical hacking)
Institute for Security and Open Methodologies	Open-source security testing methodology manual
National Security Agency (NSA)	A framework for assessing and improving the security posture of industrial control systems (ICS)

Table 14.3 Sample Mappings of Regulations and Guidelines to Cybersecurity Controls

Example requirements	Recommendations	Chapter to reference

• Establish electronic security perimeter (NERC CIP)
• Establish system boundaries (CFATS)
• Establish secure conduit (ISA-62443)
• Segregation of networks (ISO/IEC 27002:2005)
• Sensitive system Isolation (ISO/IEC 27002:2005)
• Cybersecurity controls (CFATS)
• Access control lists (CFATS)
• Network connection control (ISO/IEC 27002:2005)
• Network routing control (ISO/IEC 27002:2005)
• Information flow enforcement (NRC)
• Network architecture control/Firewall between corporate network and control network (NIST 800-82)
• Security control, intrusion detection and prevention (NIST 800-82)
• Network access control (NRC)
• Information flow enforcement (NRC)
	• Implement network segmentation at Layer 2 (VLANs), or Layer 3 (subnets). If segmentation is not supported due to ICS requirements (e.g., multicast messaging), filter traffic at the switch to control traffic
	• Add network security to control traffic between segments. This can include:
	  • NAC
	  • ACLs
	  • firewalls
	  • NGFW
	  • IPS
	  • Application filters
	  • UTM
		Chapter 5, "Industrial Network Design and Architecture"
		Chapter 9, "Establishing Zones and Conduits"
		Chapter 11, "Implementing Security Controls"

• Electronic access control (NERC CIP) • User authentication for external connections (ISO/IEC 27002:2005) • Password requirements (NRC) • Password management (CFATS) • Unique accounts (CFATS) • User registrations (ISO/IEC 27002:2005) • Access enforcement (NRC) • User identification and authentication (NRC)	• Require authentication to access all privileged network zones and all data contained therein • Maintain least-privilege and separation of duties on all user accounts • Maintain strong password management on all user accounts • Monitor all user activity for indicators of inappropriate data access • Implement identity access management (IAM) tools to manage user accounts and ensure strong authentication and authorization practices	Chapter 11, "Implementing Security Controls" Chapter 13, "Security Monitoring of Industrial Control Systems"

• Monitoring electronic access (NERC CIP)
• Network monitoring (CFATS)
	• Monitor network flows to validate network segmentation and ensure that network configurations and implemented security controls are functioning as intended. This can include the use of:
	  • Network management (NMS)
	  • Network Behavior Anomaly Detection (NBAD)
	  • Log Management system (LMS)
	• Security Information and Event Management system (SIEM)
		Chapter 12, "Exception, Anomaly and Threat Detection"
		Chapter 13, "Security Monitoring of Industrial Control Systems"

• Denial of service protection (NRC)	• Ensure that proper zoning is in place and that industrial systems are not exposed to the Internet • Implement anti-DoS technology in outer perimeters (e.g., between business networks and the Internet) • Validate critical network, security and ICS components are robust (i.e., test for resiliency during traffic anomalies and floods)	Chapter 11, "Implementing Security Controls" Chapter 8, "Risk and Vulnerability Assessments"

Continued

Table 14.3 Sample Mappings of Regulations and Guidelines to Cybersecurity Controls—cont'd

Example requirements	Recommendations	Chapter to reference
• Remote diagnostic and configuration port protection (ISO/IEC 27002:2005)	• Maintain a protected network zone for all external connectivity and remote communication, and control access into and out of this zone	Chapter 5, "Industrial Network Design and Architecture" Chapter 9, "Establishing Zones and Conduits" Chapter 11, "Implementing Security Controls"

• Change control and configuration management (NERC CIP. NRC)
• Change management (ISO/IEC 27002:2005)
• Changes to file system and operating system permissions (NRC)
	• Host configuration monitoring using built-in Windows security audit tools and/or linux *auditd* tool
	• Additional host cyber security controls for file integrity monitoring (FIM) and configuration management
	• Host cybersecurity controls to prevent file tampering or changes, including Host Intrusion Detection Systems (HIDS) and application whitelisting (AWL)
	• Monitor hosts for indications of file tampering or unauthorized changes. This can include the use of:
	  • Log Management system (LMS)
	  • Security Information and Event Management system (SIEM)
		Chapter 11, "Implementing Security Controls"
		Chapter 13, "Security Monitoring of Industrial Control Systems"

• Ports and services (NERC CIP) • Removal of unnecessary services and programs (NRC) • Open and insecure protocol restrictions (NRC)	• Monitor hosts for open ports and services using asset management or configuration management tools • Monitor network and log behavior for indicators of unauthorized ports and services that may be in use, using SIEM and similar tools	Chapter 13, "Security Monitoring of Industrial Control Systems"
• Patch management (NERC CIP) • Control of technical vulnerabilities (ISO/IEC 27002:2005) • Cyber vulnerability assessment (NERC CIP) • Vulnerability scans and assessments (NRC)	• Perhaps the most difficult challenge in industrial cyber security, patching is fundamental to maintaining a strong security posture • The most important ingredient to good patch management is knowledge: keep informed of the latest vulnerabilities and threats, and keep your patch management procedure fluid enough to accommodate urgent patching requirements • Automated solutions can ease this burden (e.g., using WSUS for Windows system and security patches).	Chapter 8, "Risk and Vulnerability Assessments"
• Cyberasset identification (CFATS)	• Implement access management either procedurally or through the use of asset management tools • Implement security monitoring tools such as SIEM, preferably with integrated asset management capabilities	Chapter 8, "Risk and Vulnerability Assessments" Chapter 12, "Exception, Anomaly and Threat Detection" Chapter 13, "Security Monitoring of Industrial Control Systems"

• Malicious software prevention (NERC CIP) • Cybersecurity controls (CFATS) • Controls against malicious code (ISO/IEC 27002:2005) • Host intrusion detection system (NRC) • Malicious code detection (NIST 800-82) • Antivirus • Malware protection • Endpoint hardening to minimize the vulnerability of devices to malware • Antivirus, application whitelisting and/or HIDS to prevent the effectiveness of malware • Network • Network cyber security controls including: • Segment the network to minimize the propagation or spread of malware if/when it occurs. • Implement network traffic inspection (DPI) using IPS to prevent known exploits and malware from traversing the network.	• To protect against malware, both host-based and network-based security controls should be used. Because malware changes often, multiple layers of defense are recommended, and all antimalware efforts should be well-managed, and kept current with any necessary patches or updates • Host cybersecurity controls including:	Chapter 5, "Industrial Network Design and Architecture" Chapter 9, "Establishing Zones and Conduits" Chapter 11, "Implementing Security Controls"
• Incident reporting (CFATS, NERC CIP) • Audit logging (ISO/IEC 27002:2005) • Reporting information security events (ISO/IEC 27002:2005) • Collection of evidence (ISO/IEC 27002:2005) • Records retention and handling (NRC)	• While incident reporting can be largely procedural, a good log management or SIEM solution can assist with the auditing of evidence and activities surrounding an incident, produce supporting documentation, and store the records (in this case, the event logs) in a secure, nonrepudiated manner	Chapter 13, "Security Monitoring of Industrial Control Systems"
• Monitoring electronic access (NERC CIP) • Security status monitoring (NERC CIP) • Network monitoring (CFATS) • Monitoring system Use (ISO/IEC 27002:2005) • Security alerts and advisories (NRC) • Continuous monitoring and assessment (NRC)	• Again, a good log management or SIEM solution will collect data from the network in addition to security events, providing a continuous monitoring solution needed to support a variety of standards. Most solutions will include standard-specific report templates as well, further easing compliance efforts	Chapter 13, "Security Monitoring of Industrial Control Systems"

- Functionality review
- Configuration review
- Risk assessments.

National Security Agency (USA)

The National Security Agency (NSA) published their framework in August 2010.[18] As the case with many of the documents, this framework is broad in nature and provides a high-level approach to conducting security assessments specifically for industrial systems. This document provides guidance that can be very helpful in assisting with risk assessments for ICS by helping assess the threats and understanding the resulting impacts or consequences.

The framework provides valuable information on the system characterization activity defined in the text as a "network connectivity assessment." This is an important first step in understanding the complete system under consideration (SuC) and can be applied to any methodology as an early activity. The document also provides information on loss assessments, and how to calculate metrics that help to identify important services within the architecture and consequences to the overall system operation should these services fail to perform as designed.

This framework provides guidance of assessment of threats by first identifying the roles and responsibilities of authorized users. The potential attack vectors that target these users is introduced along with the concept of "attack difficulty," which provides a more qualitative means of measuring the "likelihood" of a cyber-event occurring. This framework also stands out from others reviewed in that it provides steps on prioritization of the defense efforts in order to address weaknesses discovered during the assessment process.

American Petroleum Institute (USA)/National Petrochemical and Refiners Association (USA)

The American Petroleum Institute (API) and the National Petrochemical and Refiners Association (NPRA), both from the USA, were among the earliest publishers of security guidance material releasing their document in May 2003. The second edition of this document was released in October 2004.[19] This document does not contain any specific reference to industrial systems but rather provides the most comprehensive approach in terms of a complete security analysis called a security vulnerability assessment (SVA). This document is industry-specific, but the examples provided and the associated process applies to a broad range of process and industrial sectors. It discusses the concepts of an SVA in terms of risk including the concept of "asset attractiveness" that offers a different approach to the underlying motivation that a potential attacker may have for a given target. This factor is then combined with the other common risk components (threat, vulnerability, consequences) to provide a form of risk screening that can be used to understand how risk differs from industry to industry.

Sample forms and checklists are part of the methodology, which have not been included in any of the other documents reviewed. Several real-world assessments are provided, covering petroleum refining, petroleum pipeline, and transportation and distribution systems for truck and rail.

Institute for Security and Open Methodologies (Spain)

The Institute for Security and Open Methodologies is an open community and nonprofit organization that first published version 1.0 of the Open-Source Security Testing Methodology Manual in January 2001. The current version 3.0 was released in 2010.[20] The OSSTMM is generic in nature and does not include any specific reference to industrial networks. The terminology used in the methodology is inconsistent with other ICS-related documents. So why is this methodology included?

This document provides valuable reference information that may be useful as a methodology is customized to a particular organization's unique needs. The document provides assistance in utilizing "quantitative" methods and metrics of assessing security over the more traditional "qualitative" approach. One area that is addressed within the methodology that is not covered in the other documents focuses on "human security testing," and the processes that can be used to assess the involvement of operational personnel within the overall assessment framework extending beyond simple social engineering measures. The methodology provides a valuable discussion on analyzing trust and using this to identify and correct security weaknesses.

The OSSTMM provides an extensive section on compliance, including not only standards-based requirements, but also a list of countries and legislative requirements within these countries.

Common Criteria and FIPS standards

Unlike other standards, Common Criteria and Federal Information Processing Standards (FIPS) aim to certify security *products*, rather than security *policies* and *processes*. The Common Criteria for Information Technology Security Evaluation ("Common Criteria" or "CC") is an international framework that is currently recognized by Australia/New Zealand, Canada, France, Germany, Japan, the Netherlands, Spain, the United Kingdom, and the United States.[21] FIPS is defined by NIST in FIPS PUBs. Although there are several standards in FIPS, it is the FIPS 140-2 Standard that validates information encryption that is most relevant to information security products.

Common Criteria

Common Criteria's framework defines both functional and assurance requirements that security vendors can test against in order to validate the security of the product in question.[22] Certification by an authorized Common Criteria testing facility provides a high level of assurance that specific security controls have been appropriately specified and implemented into the product.

The evaluations required prior to certification are extensive and include

- Protection profiles (PP)
- Security target (ST)
- Security functional requirements (SFRs)
- Security assurance requirements (SARs)
- Evaluation assurance level (EAL).

The Security Target defines what is evaluated during the certification process, providing both the necessary guidance during evaluation as well as high-level indication of what has been evaluated after an evaluation is complete.[23]

The Security Targets are translated to the more specific Security Functional Requirements, which provide the detailed requirements against which the various STs are evaluated. The SFRs provide a normalized set of terms and requirements designed so that different STs for different products can be evaluated using common tests and controls, to provide an accurate comparison.

When common requirements are established for a particular product type or category, typically by a standards organization, they can be used to develop a common Protection Profile that is similar to an ST in that it provides a high-level indication of the assessment, but different in that the specific targets are predefined within the PP.[24] For example, there is a Common Criteria Protection Profile for Intrusion Detection and Prevention Systems that defines the specific STs that an intrusion detection system (IDS) or intrusion prevention system (IPS) must meet to earn certification.

Perhaps, the most commonly identified CC metric is the Evaluation Assurance Level (EAL). EALs measure Development (ADV), Guidance Documents (AGD), Lifecycle Support (ALC), Security Target Evaluation (ASE), Tests (ATE), and Vulnerability Assessment (AVA).[25] There are seven total assurance levels, EAL 1 through EAL 7, each of which indicates a more extensive degree of evaluation against a more exhaustive set of requirements for each of these components. For example, to compare just one of the evaluation requirements (AVA-Vulnerability Assessment), CC EAL 1 provides a basic level of assurance using a limited security target, and a vulnerability assessment consisting only of a search for potential vulnerabilities in the public domain.[26] In contrast, EAL 3 requires a "vulnerability analysis ... demonstrating resistance to penetration attackers with a basic attack potential,"[27] and EAL 4 requires a "vulnerability analysis ... demonstrating resistance to penetration attackers with an Enhanced-Basic attack potential" (i.e., more sophisticated attack profiles for a more thorough vulnerability assurance level).[28] At the most extensive end of the certification assurance spectrum is EAL 7, which requires "complete independent confirmation of the developer test results, and an independent vulnerability analysis demonstrating resistance to penetration attackers with a high attack potential."[29]

It is important to understand that the EAL level does not measure the level of security of the product that is under evaluation, but rather measures the degree to which the product's security is tested. Therefore, a higher EAL does not necessarily indicate a more secure system. It is the specific STs being evaluated that indicate the functional requirements of the system. When comparing like systems that are tested against identical targets, the higher EAL indicates that those targets were more thoroughly tested and evaluated, and therefore, the higher EAL provides additional confidence or assurance in the proper and secure function of the system.

FIPS 140-2

The Federal Information Processing Standards Publication (FIPS PUB) 140-2 establishes the requirements for the "cryptographic modules" that are used within a cyber asset or system. There are four qualitative levels of FIPS validation, Levels 1 through 4, which like Common Criteria's EALs intend to validate increasingly thorough assurance. With FIPS 140-2, this assurance is in the form of cryptographic integrity; basically, how resistant encrypted boundaries are to penetration.[30] FIPS 140-2 covers the implementation and use of Symmetric and Asymmetric Keys, the Secure Hash Standard, Random Number Generators, and Message Authentication.[31] The specific validation levels represent increasingly more stringent controls to prevent physical access to information with the encrypted boundary. For example, FIPS 140-2 Level 2 requires that data cannot be accessed physically, even through the removal of disk drives or direct access to system memory. Level 3 provides stronger physical controls to prevent access to and tampering, even through ventilation holes, whereas Level 4 even accommodates environmental failures to protect the encrypted data against recovery during or following a failure.[32]

Caution

FIPS 140-2 defines what are called security assurance "levels," numbered 1–4 with 1 represented the lowest level of security requirements and 4 the highest allowing appropriate solutions be deployed based on unique local requirements. These security levels are not the same as those defined by ISA 62443 and cannot be used interchangeably when working with the various standards.

Summary

Understanding how regulatory standards and regulations can impact the security of a network or system will help at all stages of industrial network security planning and implementation. Specific compliance controls might dictate the use of certain products or services to improve security and/or how to configure those security products.

The security products themselves are subject to regulation as well, of course. The Common Criteria standards provide a means for evaluating the function and assurance of a product in a manner designed to facilitate the comparison of similar products, whereas standards in FIPS, such as FIPS 140-2, can provide further validation of specific security functions (in this case, encryption) used by a products.

Endnotes

1. M. Asante, NERC, Harder questions on CIP compliance update: ask the expert, 2010 SCADA and Process Control Summit, The SANS Institute, March 29, 2010.
2. Department of Homeland Security, Risk-Based Performance Standards Guidance; Chemical Facility Anti-Terrorism Standards, May 2009.
3. Ibid.
4. Ibid.
5. Ibid.
6. Ibid.
7. Ibid.
8. Ibid.
9. CISA. CFATS Announcement. July 28, 2023. https://www.cisa.gov/resources-tools/programs/chemical-facility-anti-terrorism-standards-cfats/cfats-risk-based-performance-standards
10. International Standards Organization/International Electrotechnical Commission (ISO/IEC), About ISO. http://www.iso.org/iso/about.htm (cited: March 21, 2011).
11. "Information technology – Security techniques – Information security management systems – Overview and vocabulary," ISO/IEC 27000:2014, third Edition, January 15, 2014.
12. International Standards Organization/International Electrotechnical Commission (ISO/IEC), International ISO/IEC Standard 27002:2005 (E), Information Technology—Security Techniques—Code of Practice for Information Security Management, first edition 2005-06-15.
13. U.S. Nuclear Regulatory Commission, Regulatory Guide 5.71 (New Regulatory Guide) Cyber Security Programs for Nuclear Facilities, January 2010.
14. K. Stouffer, J. Falco, K. Scarfone, National Institute of Standards and Technology, Special Publication 800-82 (Final Public Draft), Guide to Industrial Control Systems (ICS) Security, September 2008.
15. The Unified Compliance Framework, What is the UCF? (cited: March 21, 2011).
16. "Cyber Security Assessments of Industrial Control Systems," U.S. Dept. of Homeland Security, November 2010.
17. "Cyber Security Assessments of Industrial Control Systems – A Good Practice Guide," Center for the Protection of National Infrastructure, April 2011.
18. "A Framework for Assessing and Improving the Security Posture of Industrial Control Systems (ICS)," National Security Agency, August 2010.
19. "Security Vulnerability Assessment Methodology for the Petroleum and Petrochemical Industries," API SVA-2004, American Petroleum Institute/National Petroleum Refiners Association, second Edition, October 2004.
20. "Open-Source Security Testing Methodology Manual," Version 3.0, Institute for Security and Open Methodologies, 2010.
21. The Common Criteria Working Group, Common Criteria for Information Technology Security Evaluation, Part 1: Introduction and General Model, Version 3.1, Revision 3 Final, July 2009.
22. Ibid.
23. Ibid.
24. Ibid.
25. Ibid.
26. Ibid.
27. Ibid.
28. Ibid.
29. Ibid.
30. National Institute of Standards and Technology, Information Technology Laboratory, Federal Information Processing Standards Publication 140-2, Security Requirements for Cryptographic Modules, May 25, 2001.

31. Ibid.
32. Ibid.
33. "Information technology – Security techniques – Information security management systems – Overview and vocabulary," ISO/IEC 27000:2014, third Edition, January 15, 2014.
34. ISA99 Committee on Industrial Automation and Control Systems Security, sited July 21, 2014.
35. "Security for industrial automation and control systems: System security requirements and security levels," ISA 62443-3-3:2013.

15

Common Pitfalls and Mistakes

Information in this chapter

- The Basics
- Lack of Operationalization
- Lack of Awareness
- Misunderstanding Vulnerability
- Worlds are Colliding!
- The Mistake that You are Making Right Now

Even with best of intentions, a qualified staff, a strong budget, and time it can be difficult to implement strong security measures into any network, and even more so into an industrial network … but who realistically has all of these things? In reality, most industrial cybersecurity experts are trying to do their best with insufficient resources. Therefore, it should be clear that the intention of sharing these common pitfalls and mistakes is to learn, and maybe laugh a little. The intent is not to shame anyone, even though these issues are all derived from actual conversations with me, the author. However, I have heard these often enough that I felt it important to discuss them here, to help others avoid making the same mistakes … and perhaps to end this book on a lighter note.

The basics

The first edition of this book was written over a decade ago, and yet some industrial control operators are still making basic mistakes. There's simply no excuse for this anymore: if this is you it's time to get motivated and improve!

The KISS of death

"Our network is pretty basic …"

The acronym "Keep is Simple, Stupid" is often used to extol the virtues of avoiding complexity. However, this is a bad idea when it means your industrial network consists of a single subnet that is connected directly to the internet. Yes, network segmentation can be complicated. Yes, firewalls can be difficult to properly configure and manage. But if you choose simplicity in these circumstances … consider it a Fail.

Industrial Network Security. https://doi.org/10.1016/B978-0-443-13737-2.00018-X

Password123

"I don't think we ever changed it ..."

Everything is a target. If something comes with a default password, you can bet that the attackers know what it is, which makes it an *easy* target. If the industrial network is also connected directly to the internet, you're a few Internet searches away from being part of a botnet.

People are people

"We don't need that particular security control, because we've told our people they're not allowed to do that."

Not to sound like a pessimist, but people don't always do what they're told. An employee might understand that they aren't allowed to use USB devices, but if they *really* need to charge their phone, or if they *really* need to move that one file ... you never know what someone might do. Not all misuse is intentional or malicious, but it is inevitable.

The Air Gap myth

"We're safe — we're fully air-gapped."

Open networking protocols and wireless networks are ubiquitous, yet many still believe that a true Air Gap exists, protecting critical industrial systems because they somehow can't be reached.

In reality, even a real Air Gap (if one truly does exist) is of little use in defending against cyber attacks, because cyber attacks have evolved past physical wires. Many assets that were not designed or intended to support wireless network communications include embedded Wi-Fi capabilities at the microprocessor level,[1] which can be exploited by attackers ranging from the skilled cyber terrorist, to a disgruntled worker with an understanding of wireless technologies.[2]

In addition, there is the high possibility that a threat could be walked into a critical network, stepping across the Air Gap with the aid of a human carrier. Only strong security awareness and strong technical security controls can truly "gap" a networked system.

The future is now

"We aren't allowed to send any data outside of our plant."

You can replace 'plant' with 'region', 'country', 'organization', or whatever other arbitrary identifier you want. This complaint, ironically, often comes via email: which in most cases is a clear example of data being sent outside of the supposed boundary.

In reality, there are established and legitimate data flows into and out of almost any environment. The restriction might be real, but take the time to determine what data is being transmitted and what the specific risks and regulations are in a given circumstance. It might not be an issue at all, or there might be specific controls or countermeasures that are required.

IIoT is not spelled with a "d" in it

"IT and IoT are the same thing!"

This is, of course, simply not true. However it highlights a growing problem in an industry with a real labor shortage (at least at the time of publication): people don't know what they don't know. If based on cursory Internet searches, without prior experience to guide you, it might be easy to believe that Operational Technology (OT) and the Internet of Things (IoT) are the same. While this example is fairly harmless, misinformation can be counterproductive or even dangerous. Luckily you've read Chapter 3 already, so this isn't you.

Lack of proper operationalization

Cybersecurity requires people, process, and technology. However, it remains common for some companies to invest in one without the others. Unless all three are considered, it is impossible to fully operationalize your cybersecurity efforts, and you'll be left with highly trained people with no tools, or (more commonly) an abundance of tools without trained personnel to utilize them, nor the processes in place to make the most of what these tools provide.

Schrödingers event logs

"We collect tons of security logs. When I log into one of the firewalls, I can see if there are any alerts there."

Security controls create event logs, and some of those event logs could be extremely important. If you have tools creating events, make sure there are people looking at them, who have the right tools at their disposal to manage them effectively. It doesn't always require a huge investment (there are many open source options available), but event logs that aren't being managed are useless. Maybe useful in certain cirmstances? Both.

Planning versus practice

"We have an extremely thorough cybersecurity response plan. It fills three binders."

This is great except the person explaining this also admitted to never once having practiced that plan. If the first step required in the event of incident is to find and read three binders full of documentation, don't expect the rest to go well.

Inadequate staffing

> *"I just made a huge investment in cybersecurity controls. It's on that shelf over there because there's no one to use it!"*

I have witnessed "Security operations centers" with fancy dashboards on screens, but without anyone looking at them. In one case, there wasn't even a desk or chair in the room, indicating that no one *ever* looked at them. In another case, I saw firewalls still their boxes, covered in dust. They weren't spares; they had never been deployed. Most cybersecurity tools have an operational requirement. Those that are more straightforward (such as security events from basic controls like firewalls and anti-malware software) can be automated to a degree, but someone still has to make sense of them all. Those controls that are more prone to false positives may require several dedicated team members. Those that are never unboxed? They dont require a lot of effort to use, at least.

Lack of awareness

Cybersecurity is a journey, and like all journeys it is important to know where you are and where you are trying to get to. Without some degree of self-awareness, it's easy to get lost along the way. And without some awareness of the world around you, you're likely to crash into something unexpected.

Driving without a map

> *"We have a flat network that's connected directly to the Internet, so we need the most advanced monitoring tools that money can buy!"*

Not everyone's cybersecurity journey is the same, but as a rule it's best to implement the basics first. If you're connecting your network directly to the Internet, that should probably be solved *before* implementing something more advanced like an industrial network anomaly detection product or an advanced SIEM or XDR solution. Otherwise, there's going to be a lot more to analyze and a lot more event data to sift through as your network gets continuously pummeled by outside traffic.

One-and-done

> *"We were told we had to put a firewall in. We did that years ago, so what's the problem?"*

Every step on the cybersecurity journey is important, but no step is the last one. If the next step that you need to take isn't obvious, consider bringing in an outside agency to perform an assessment of where you and where you need to be.

We're not at risk

> *"This is all 'nation-state' level stuff ... we don't have to worry about any of this, we're not that important."*

Unfortunately, the threat has evolved in many ways. First, the threat of ransomware has put *all* industrial operators at risk, because any company represents a potential pay-out for the bad guys. Second, the techniques available by adversaries — even those that aren't nation-state actors — are the same techniques that were considered "nation state" level threats just a few years ago. Third, if you defend against a larger threat than you expect you need, you'll be that much more likely to succeed.

Driving too slow in the fast lane

> *"We finished a project this spring to map out security levels and create zones and conduits. In next year's budget cycle we'll put in the request for the segmentation project, and if we get the funding, we'll start procurement in the following year ..."*

The threat landscape is always evolving. To use the "cybersecurity journey" analogy once again, it's important to try and at least keep pace with adversaries — because they're on a journey too and if you can't keep up you will fall behind.

Passing on the right

> *"We need to implement this security control ASAP!"*

Usually in response to an incident, a very specific security control will become suddenly urgent. While urgency isn't a bad thing — as just stated, there is a need to keep up — I've seen at least half of the pitfalls mentioned here made while rushing into a response without proper consideration.

Misunderstanding vulnerability

Vulnerabilities are often a top consideration of cybersecurity professionals. As we learned in Chapter 8, however, vulnerabilities and risk in industrial systems (like most things in industrial systems) require special consideration.

Paralysis by vulnerability analysis

> *"We haven't been able to start the segmentation project, because we're tasked with implementing a new patching process and we're being held up patching our OT assets by uncooperative vendors."*

If there's something in your plan that is difficult, move ahead on other projects while sorting it out. Don't let one stubborn process (yes, patching OT assets can be difficult) get in the way of other important work.

We've patched all the vulnerabilities

> *"We invested a lot into a patching process that lets us keep everything up to date, within 24 hours of an available patch. We're safe now!"*

In reality, there are unknown threats that cannot be accounted for. Therefore, no security plan is fully complete without some method of accounting for unknown attacks. Assume all critical assets are vulnerable, and plan accordingly. Many industrial control assets are still unavailable to the broader cybersecurity research community; it is reasonable to assume that there are vulnerabilities that have not yet been identified and disclosed. If a vulnerability is unknown, a vulnerability scan is not going to identify it ... yet the vulnerability still exists.

Software versus systems

> *"Patching our industrial assets is the absolutely most important thing to do."*

Similar to the belief that being fully patched equals being fully secure, there are many who feel that patching industrial assets such as PLCs is absolutely the most important thing to keep an ICS safe from attack. In reality, an ICS is by design a command and control infrastructure; if you get access to the ICS, the whole system is vulnerable to misuse. Patched or not, an adversary that has breached the ICS will be able to manipulate that PLC. Yes, it's important to patch systems to make direct manipulation of assets more difficult, but industrial cybersecurity requires a holistic approach.

Worlds are colliding!

> *"IT" and "OT" have been converging, and will likely continue to do so. Along the way, there's been a lot mistakes made — from both sides.*

Cybersecurity for OT is just like IT

"OT uses Windows servers and they're no different than the ones used in finance or other business functions"

The hardware is often the same, and the operating system is often the same. Although both might be older (even eold enough to be unsupported). Some aspects of the opporating system — such as network drivers — may be proprietary in order to support the low-latency, real-time communication, and redundancy requirements of an industrial control system. And of course the control system software itself is highly specialized, to the degree that vendors will typically only provide support for tested and certified configurations. In short, even though two computers might look the same on the surface, the operational and cybersecurity needs are very different.

All processes are fragile

"If you breath on an industrial network wrong, you could tip the whole thing over!"

This is true in some very specific scenarios (e.g., unexpected network traffic that introduces latency or consumes unnecessary capacity in real-time networks), but it is untrue in others (e.g., control systems that are designed for robustness and resiliency). If a cybersecurity control is needed (e.g., as determined by a risk assessment), test it. If it impacts reliability, instead of abandoning the control, consider how to improve reliability in order to accommodate it.

My process is resilient: It's unhackable!

"A properly designed ICS will keep running smoothly no matter what you throw at it."

The inverse use case is to have too much confidence in the design of the system. Even extremely resilient environments can be manipulated by a determined attacker. It might be more difficult, but an adversary only needs to find one weakness to potentially disrupt an otherwise robust system.

The mistake that you are making right now

Too much reading, not enough practice

"I can suggest a good book on the subject ..."

I'm fond of recommending this book to those looking to learn industrial cybersecurity. While this book is huge, and you are at the end of it, it unfortunately isn't enough. You will never be able to know all there is to know. Industrial network security is a broad and complex subject, and you've done great to get this far, but cybersecurity is not a purely academic discipline: everything you just learned now needs to be put into practice. If you're reading this as an industrial operator who needs to understand cybersecurity better: get some hands on training to get a feel for what "hacking" really means, and to get comfortable with the tools of the trade. If you're reading this as a cybersecurity expert who is looking to understand what "OT" is all about: complete your safety training, put on your PPE, and get your boots dirty. Good luck!

Summary

With the proper intentions, a well informed network security administrator can plan, implement and execute best-in class security measures for any industrial network .... But mistakes can happen, and they often do. Laugh, and learn!

Endnotes

1. Jason Larson, Idaho National Laboratories. Control Systems at Risk: Sophisticated Penetration Testers Show How to Get Through the Defenses. In: Proc. 2009 SANS European SCADA and Process Control Security Summit; October 2009.
2. Jacob Brodsky, Anthony McConnell, Marco Cajina, and Dale Peterson. Security and reliability of wireless LAN Protocol Stacks Used in Control Systems. Proceedings of the SCADA Security Scientific Symposium (S4). Kenexis Security Corporation, 2010. Digital Bond Press.

Glossary

Active Directory Microsoft's Active Directory (AD) is a centralized directory framework for the administration of network devices and users, including user identity management and authentication services. AD utilizes the Lightweight Directory Access Protocol (LDAP) along with domain and authentication services.

Advanced Persistent Threat The Advanced Persistent Threat (APT) refers to a class of cyber threat designed to infiltrate a network and remain persistent through evasion and propagation techniques. APTs are typically used to establish and maintain an external command and control channel through which the attacker can continuously exfiltrate data.

Allowlist, Whitelist Whitelists or allowlists refer to defined lists of "known good" items: users, network addresses, applications, and so on, typically for the purpose of exception-based security where any item not explicitly defined as "known good" results in a remediation action (e.g. alert and block). Allowlists or whitelists contrast blocklists or blacklists, which define "known bad" items.

Allowlisting, Whitelisting Allowlisting, also referred to as whitelisting, refers to the act of comparing an item against a list of approved items for the purpose of assessing whether it is allowed or should be blocked. Typically referred to in the context of Application Allowlisting, which prevents unauthorized applications from executing on a host by comparing all applications against a list of authorized applications.

Antivirus Antivirus (AV) systems inspect network and/or file content for indications of infection by malware. Signature-based AV works by comparing file contents against a library of defined code signatures; if there is a match, the file is typically quarantined to prevent infection, at which point the option to clean the file may be available. Most modern AV solutions also offer other detection techniques including file reputation matching, behavioral analysis, sandboxing, and other techniques.

Application Control, Application Allowlisting, Application Whitelisting A method of controlling which executable files (applications) are allowed to operate. These systems typically work by first establishing a list of allowed applications, after which point any attempt to execute code will be compared against that list. If the application is not allowed, it will be prevented from executing. These controls often operate at low levels within the kernel of the host operating system.

Application Monitor/Application Data Monitor An application content monitoring system that functions much like an intrusion detection system, only performing deep inspection of a session rather than of a packet, so that application contents can be examined at all layers of the OSI model, from low level protocols through application documents, attachments, and so on. Application Monitoring is useful for examining industrial network protocols for malicious content (malware).

APT See Advanced Persistent Threat.

Assessment (see also "Risk Assessment", "Vulnerability Assessment") A cybersecurity assessment refers to any evaluation of an organization's digital security profile. Common assessments include vulnerability assessments and risk assessments, but any aspect of cybersecurity could be evaluated and measured. Assessments are a fundamental tool for obtaining the awareness needed to identify areas of improvement in an organization's overall cybersecurity posture.

Asset An asset is any device used within an industrial network.

ATT&CK The ATT&CK framework is a publicly available knowledge base of observed cyber attack tactics and techniques, maintained by MITRE.

Attack Surface The attack surface of a system or asset refers to the collectively exposed portions of that system or asset. A large attack surface means that there are many exposed areas that an attack could target, while a small attack surface means that the target is relatively unexposed.

Attack Vector An attack vector is the direction(s) through which an attack occurs, often referring to specific vulnerabilities that are used by an attacker at any given stage of an attack.

auditd Auditd is the auditing component of the Linux Auditing System, responsible for writing audit events to disk. The Linux Auditing System is a useful tool for monitoring file access and file integrity in Linux systems.

AV See Antivirus.

AWL See Application Whitelisting.

Backchannel A backchannel typically refers to a communications channel that is hidden or operates "in the background" to avoid detection, but is also used in reference to hidden or covert communications occurring back toward the originating sender, that is, malware hidden in the return traffic of a bidirectional communication.

BCS See Building Control System.

Behavioral Analysis Behavioral malware analysis evaluates the actions and interactions of software in an attempt to identify malicious functions, evasion techniques, and other suspicious behaviors that are indicative of malware.

Blacklisting, Blocklisting (Also see "Allowlisting, Whitelisting") Blacklisting or Blocklisting refers to the technique of defining known malicious behavior, content, code, and so on. Blocklists are typically used for threat detection, comparing network traffic, files, users, or some other quantifiable metric against a relevant blacklist. For example, an intrusion prevention system (IPS) will compare the contents of network packets against blocklists of known malware, indicators of exploits, and other threats so that offending traffic (i.e. packets that match a signature within the blocklist) can be blocked.

Building Control System A Building Control System (BCS) is a centralized management system used to monitor, regulate, and control various aspects of a building's infrastructure, such as heating, ventilation, air conditioning, lighting, access control, etc. BCS systems much like ICS systems, using a combination of sensor readings, physical actuations, and programmable logic.

CDA See Critical Digital Asset.

CFATS The Chemical Facility Anti-Terrorism Standard is established by the US Department of Homeland Security to protect the manufacture, storage, and distribution of potentially hazardous chemicals.

ChemITC The Chemical Information Technology Center (ChemITC) is an industry group within the American Chemistry Council (ACC) that focuses on the use of information technology for the advancement of chemical products and services.

Cloud, Cloud Computing Cloud computing is a technology that allows users to access and utilize computing resources (servers, storage, databases, networking, software, etc) over the internet, rather than leveraging traditional data centers. Cloud computing provides benefits such as scalability, flexibility, and cost-efficiency. Cloud computing also makes data and applications available on the Internet, and therefore special considerations and controls are required when using cloud computing in industrial environments.

Compensating Controls The term "compensating controls" is typically used within regulatory standards or guidelines to indicate when an alternative method than those specifically addressed by the standard or guideline is used.

Control Center A control center typically refers to an operations center where a control system is managed. Control centers typically consist of SCADA and HMI systems that provide interaction with industrial/automated processes.

Correlated Event A correlated event is a larger pattern match consisting of two or more regular logs or events, as detected by an event correlation system. For example, a combination of a network scan event (as reported by a firewall) followed by an injection attempt against an open port (as reported by an IPS) can be correlated together into a larger incident; in this example, an attempted reconnaissance and exploit. Correlated events may be very simple or very complex, and can be used to detect a wide variety of more sophisticated attack indicators.

Critical Cyber Asset A critical cyber asset is a cyber asset that is itself responsible for performing a critical function, or directly impacts an asset that performs a critical function. The term "critical cyber asset" is used heavily within NERC reliability standards for Critical Infrastructure Protection.

Critical Digital Asset A "critical digital asset" is a digitally connected asset that is itself responsible for performing a critical function, or directly impacts an asset that performs a critical function. The term "critical digital asset" is used heavily within NRC regulations and guidance documents. Also See: Critical Cyber Asset.

Critical Infrastructure Any infrastructure whose disruption could have severe impact on a nation or society. In the United States, Critical Infrastructures are defined by the Homeland Security Presidential Directive Seven as: Agriculture and Food; Banking and Finance; Chemical; Commercial Facilities; Critical Manufacturing; Dams; Defense Industrial Base; Drinking Water and Water Treatment Systems; Emergency

Services; Energy; Government Facilities; Information Technology; National Monuments and Icons; Nuclear Reactors, Materials, and Waste; Postal and Shipping; Public Health and Healthcare; Telecommunications; and Transportation Systems.

CsHAZOP Cybersecurity HAZOP studies extend the methods of a traditional HAZOP to include risk in the overall assessment of reliability. csHAZOP is an extensive assessment of cyber-physical risk.

Cyber Asset A digitally connected asset; that is, an asset that is connected to a routable network or a Host. The term Cyber Asset is used within the NERC reliability standards, which defines a Cyber Asset as any Asset connected to a routable network within a control system; any Asset connected to a routable network outside of the control system; and/or any Asset reachable via dial-up.

Cyber-Physical Attack, Cyber-Physical Threat An attack that utilizes digital manipulation (i.e., "cyber") to create an impact to the physical world (i.e., "physical"), usually through the manipulation of a process control system.

Cyber-Physical Threat Model Cyber-physical threat modeling refers to the practice of identifying threats to a process that could potentially be initiated by cyber activity, and that might not be considered within a normal HAZOP. Identifying cyber threats that could impact physical systems is also a component in understanding the overall cyber-physical risk (ie., the risks identified by csHAZOP).

DAM See Database Activity Monitor.

Data Diode A data diode is a "one-way" data communication device, often consisting of a physical-layer unidirectional limitation. Using only 1/2 of a fiber optic "transmit/receive" pair would enforce unidirectional communication at the physical layer, while proper configuration of a network firewall could logically enforce unidirectional communication at the network layer.

Database Activity Monitor A Database Activity Monitor (DAM) monitors database transactions, including SQL, DML, and other database commands and queries. A DAM may be network- or host based. Network-based DAMs monitor database transactions by decoding and interpreting network traffic, while host-based DAMs provide system-level auditing directly from the database server. DAMs can be used for indications of malicious intent (e.g. SQL injection attacks), fraud (e.g. the manipulation of stored data), and/or as a means of logging data access for systems that do not or cannot produce auditable logs.

Database Monitor See Database Activity Monitor

DCS See Distributed Control System.

Deep-Packet Inspection The process of inspecting a network packet all the way to the application layer (Layer 7) of the OSI model. That is, past datalink, network, or session headers to inspect all the way into the payload of the packet. Deep-packet inspection is used by most intrusion detection and prevention systems (IDS/IPS), newer firewalls, and other security devices.

Distributed Control System An industrial control system deployed and controlled in a distributed manner, such that various distributed control systems or processes are controlled individually. See also Industrial Control System.

DPI See Deep Packet Inspection.

Electronic Security Perimeter An Electronic Security Perimeter (ESP) refers to the demarcation point between a secured enclave, such as a control system, and a less trusted network, such as a business network. The ESP typically includes those devices that secure that demarcation point, including firewalls, IDS, IPS, industrial protocol filters, application monitors, and similar devices.

Enclave A logical grouping of assets, systems, and/or services that defines and contains one (or more) functional groups. Enclaves represent network "zones" that can be used to isolate certain functions in order to more effectively secure them.

Enumeration Enumeration is the process of identifying valid identities of devices and users in a network, typically as an initial step in a network attack process. Enumeration allows an attacker to identify valid systems and/or accounts that can then be targeted for exploitation or compromise.

ESP See Electronic Security Perimeter.

EtherNet/IP EtherNet/IP is a real-time Ethernet protocol supporting the Common Industrial Protocol (CIP), for use in industrial control systems.

Event An event is a generic term referring to any datapoint of interest, typically alerts that are generated by security devices, logs produced by systems and applications, alerts produced by network monitors, and so on.

Finger The finger command is a network tool that provides detailed information about a user.

Function Code Function Codes refer to various numeric identifiers used within industrial network protocols for command and control purposes. For example, a function code may represent a request from a Master device to a Slave device(s), such as a request to read a register value, to write a register value, or to restart the device.

HAZOP HAZOP (Hazard and Operability) refers to a systematic approach to identifying potential hazards, deviations from normal operations, and operational vulnerabilities in a system or process, with the goal of improving safety and operability.

HIDS Host IDS. A Host Intrusion Detection System, which detects intrusion attempts via a software agent running on a specific host. A HIDS detects intrusions by inspecting packets and matching the contents against defined patterns or "signatures" that indicate malicious content, and produce an alert.

HIPS Host IPS. A Host Intrusion Prevention System, which detects and prevents intrusion attempts via a software agent running on a specific host. Like a HIDS, a HIPS detects intrusions by inspecting packets and matching the contents against defined patterns or "signatures" that indicate malicious content. Unlike a HIDS, a HIPS is able to perform active prevention by dropping the offending packet(s), resetting TCP/IP connections, or other actions in addition to passive alerting and logging actions.

HMI A human–machine interface (HMI) is the user interface to the processes of an industrial control system. An HMI effectively translates the communications to and from PLCs, RTUs, and other industrial assets to a human-readable interface, which is used by control systems operators to manage and monitor processes.

Homeland Security Presidential Directive Seven The United States Homeland Security Presidential Directive Seven (HSPD-7) defines the 18 critical infrastructures within the United States, as well as the governing authorities responsible for their security.

Host A host is a computer connected to a network, that is, a Cyber Asset. The term differs from an Asset in that hosts typically refer to computers connected to a routable network using the TCP/IP stack—that is, most computers running a modern operating system and/or specialized network servers and equipment—while an Asset refers to a broader range of digitally connected devices, and a Cyber Asset refers to any Asset that is connected to a routable network.

HSPD-7 See Homeland Security Presidential Directive Seven.

IACS Industrial Automation Control System. See Industrial Control System.

IAM See Identity Access Management.

ICCP See Inter Control Center Protocol.

ICS See Industrial Control System.

Identity Access Management Identity access management refers to the process of managing user identities and user accounts, as well as related user access and authentication activities within a network, and a category of products designed to centralize and automate those functions.

IDS Intrusion Detection System. Intrusion detection systems perform deep-packet inspection and pattern matching to compare network packets against known "signatures" of malware or other malicious activity in order to detect a possible network intrusion. IDS operates passively by monitoring networks either in-line or on a tap or span port, and providing security alerts or events to a network operator.

IEC See International Electrotechnical Commission.

IED See Intelligent Electronic Device.

IIoT See "Industrial Internet of Things."

Industrial Control System An industrial control system (ICS) refers to the systems, devices, networks, and controls used to operate and/or automate an industrial process. See also Distributed Control System.

Industrial Internet of Things Industrial Internet of Things, or IIoT, refers to the interconnection of industrial devices, sensors, machines, and systems that collect, exchange, and analyze data in manufacturing and other industrial settings.

Intelligent Electronic Device An intelligent electronic device (IED) is an electronic component (such as a regulator and circuit control) that has a microprocessor and is able to communicate, typically digitally using fieldbus, real-time Ethernet, or other industrial protocols.

Inter-Control Center Protocol The Inter-Control Center Protocol (ICCP) is a real-time industrial network protocol designed for wide-area intercommunication between two or more control centers. ICCP is an internationally recognized standard published by the International Electrotechnical Commission (IEC) as IEC 60870-6. ICCP is also referred to as the Telecontrol Application Service Element-2 or TASE.2.

International Electrotechnical Commission The International Electrotechnical Commission (IEC) is an international standards organization that develops standards for the purposes of consensus and conformity among international technology developers, vendors, and users. The IEC 62443 standard aligns with the ISA99/62443 standard for cybersecurity of industrial automation and control systems.

International Standards Organization The International Standards Organization (ISO) is a network of standards organizations from over 160 countries, which develops and publishes standards covering a wide range of topics.

Internet of Things Internet of Things, or IoT, refers to the interconnection of "things"—any device with internet connectivity, that are typically able to collect and exchange data with each other and central systems.

IoT See "Internet of Things."

IPS Intrusion Prevention System. Intrusion protection systems perform the same detection functions of an IDS, with the added capability to block traffic. Traffic can typically be blocked by dropping the offending packet(s), or by forcing a reset of the offending TCP/IP session. IPS works in-line, and therefore may introduce latency.

ISA The International Society of Automation (ISA) is a nonprofit organization dedicated to industrial automation, and the development of standards and certifications associated with industrial automation. The ISA99 standard aligns with the IEC 62243 defines standards for cybersecurity of industrial automation and control systems.

ISO See International Standards Organization.

Kill Chain A cybersecurity concept that describes the stages of a cyberattack, from initial reconnaissance to the final objective, which is often data theft or disruption. These stages typically include reconnaissance, weaponization, delivery, exploitation, installation, command and control, and actions on objectives.

Kinetic Attack See Cyber-Physical Attack.

LDAP See Lightweight Directory Access Protocol.

Lightweight Directory Access Protocol The Lightweight Directory Access Protocol (LDAP) is a standard published under IETF RFC 4510, which defines a standard process for accessing and utilizing network-based directories. LDAP is used by a variety of directories and IAM systems.

Log A log is a file used to record activities or events, generated by a variety of devices, including computer operating systems, applications, network switches and routers, and virtually any computing device. There is no standard for the common format or structure of a log.

Log Management Log management is the process of collecting and storing logs for purposes of log analysis and data forensics, and/or for purposes of regulatory compliance and accountability. Log management typically involves collection of logs, some degree of normalization or categorization, and short-term (for analysis) and long-term storage (for compliance).

Log Management System A system or appliance designed to simplify and/or automate the process of log management. See also Log Management.

Master Station A master station is the controlling asset or host involved in an industrial protocol communication session. The master station is typically responsible for timing, synchronization, and command and control aspects of an industrial network protocol.

Metasploit Metasploit is a commercial exploit package, used for penetration testing.

Modbus Modbus is the Modicon Bus protocol, used for intercommunication between industrial control assets. Modbus is a flexible master/slave command and control protocol available in several variants including Modbus ASCII, Modbus RTU, Modbus TCP/IP, and Modbus Plus.

Modbus ASCII A Modbus variant that uses ASCII characters rather than binary data representation.

Modbus Plus A Modbus extension that operates at higher speeds, which remains proprietary to Schneider Electric.

Modbus RTU A Modbus variant that uses binary data representation.

Modbus TCP A Modbus variant that operates over TCP/IP.

NAC See Network Access Control.

NEI The Nuclear Energy Institute is an organization dedicated to and governed by the United States nuclear utility companies.

NERC See North American Electric Reliability Corporation.

NERC CIP The North American Electric Reliability Corporation reliability standard for Critical Infrastructure Protection.

Network Access Control Network Access Control (NAC) provides measures of controlling access to the network, using technologies, such as 802.1X (port network access control), to require authentication for a network port to be enabled, or other access control methods.

Network Whitelisting See Allowlisting, Whitelisting.

NIDS Network IDS. A network intrusion detection system detects intrusion attempts via a network interface card, which connects to the network either in-line or via a span or tap port.

NIPS Network IPS. A network intrusion prevention detection system detects and prevents intrusion attempts via a network-attached device using two or more network interface cards to support inbound and outbound network traffic, with optional bypass interfaces to preserve network reliability in the event of a NIPS failure.

NIST The National Institute of Standards and Technology. NIST is a nonregulatory federal agency within the United States Department of Commerce, whose mission is to promote innovation through the advancement of science, technology, and standards. NIST provides numerous research documents and recommendations (the "Special Publication 800 series") around information technology security.

NIST Cybersecurity Framework A set of guidelines and best practices developed by the National Institute of Standards and Technology (NIST). The NIST Cybersecurity framework is designed to help organizations continuously improve their cybersecurity posture. It consists of five core functions: Identify, Protect, Detect, Respond, and Recover.

nmap Nmap or "Network Mapper" is a popular network scanner distributed under GNU General Public License GPL-2 by nmap.org.

North American Electric Reliability Corporation The North American Electric Reliability Corporation is an organization that develops and enforces reliability standards for and monitors the activities of the bulk electric power grid in North America.

NRC See Nuclear Regulatory Commission.

Nuclear Regulatory Commission The United States Nuclear Regulatory Commission (NRC) is a five-member presidentially appointed commission responsible for the safe use of radioactive materials including but not limited to nuclear energy, nuclear fuels, radioactive waste management, and the medical use of radioactive materials.

Operational Technology Operational Technology refers generally to the assets and systems of industrial automation and control systems, and is analogous to "Information Technology" or "IT." The term is often misused and is discussed in detail in Chapter 2.

OSSIM OSSIM is an Open Source Security Information Management project, whose source code is distributed under GNU General Public License GPL-2 by AlienVault.

OT See Operational Technology.

Outstation An outstation is the DNP3 slave or remote device. The term outstation is also used more generically as a remote SCADA system, typically interconnected with central SCADA systems by a Wide Area Network.

PCS Process Control System. See Industrial Control System.

Pen test A Penetration Test. A method for determining the risk to a network by attempting to penetrate its defenses. Pen-testing combines vulnerability assessment techniques with evasion techniques and other attack methods to simulate a "real attack."

PLC See Programmable Logic Controller.

Profibus Profibus is an industrial fieldbus protocol defined by IEC standard 61158/IEC 61784-1.

Profinet Profinet is an implementation of Profibus designed to operate in real time over Ethernet.

Programmable Logic Controller A programmable logic controller (PLC) is an industrial device that uses input and output relays in combination with programmable logic in order to build an automated control loop. PLCs commonly use Ladder Logic to read inputs, compare values against defined set points, and (potentially) write to outputs.

Project Aurora A research project that demonstrated how a cyber-attack could result in the explosion of a generator.

RBPS Risk-Based Performance Standards are recommendations for meeting the security controls required by the Chemical Facility Anti-Terrorism Standard (CFATS), written by DHS.

Red Network A "red network" typically refers to a trusted network, in contrast to a "black network," which is less secured. When discussing unidirectional communications in critical networks, traffic is typically only allowed outward from the red network to the black network, to allow supervisory data originating from critical assets to be collected and utilized by less secure SCADA systems. In other use cases, such as data integrity and fraud prevention, traffic may only be allowed from the black network into the red network, to prevent access to classified data once they have been stored.

Remote Terminal Unit A remote terminal unit (RTU) is a device combining remote communication capabilities with programmable logic for the control of processes in remote locations.

Risk Assessment Risk assessments evaluate the likelihood of a cybersecurity incident against the impact of such an incident. Risk assessments leverage threat and vulnerability awareness to quantify risk in order to make informed decisions.

RTU See Remote Terminal Unit.

SCADA See Supervisory Control and Data Acquisition.

SCADA-IDS SCADA aware Intrusion Detection System. An IDS designed for use in SCADA and ICS networks. SCADA-IDS devices support pattern matching against the specific protocols and services used in control systems, such as Modbus, ICCP, DNP3, and others. SCADS-IDS is passive, and is therefore suitable for deployment within a control system, as it does not introduce any risk to control system reliability.

SCADA-IPS SCADA aware Intrusion Prevention System. An IPS system designed for use in SCADA and ICS networks. SCADA-IPS devices support pattern matching against the specific protocols and services used in control systems, such as Modbus, ICCP, DNP3, and others. SCADA-IPS is active and can block or blacklist traffic, making it most suitable for use at control system perimeters. SCADA-IPS is not typically deployed within a control system for fear of a false positive disrupting normal control system operations.

Security Information and Event Management Security information and event management (SIEM) combines security information management (SIM or log management) with security event management (SEM) to provide a common centralized system for managing network threats and all associated information and context.

SERCOS III SERCOS III is the latest version of the Serial Realtime Communications System, a real-time Ethernet implementation of the popular SERCOS fieldbus protocols.

Set Points Set points are defined values signifying a target metric against which programmable logic can operate. For example, a set point may define a high temperature range, or the optimum pressure of a container, and so on. By comparing set points against sensory input, automated controls can be established. For example, if the temperate in a furnace reaches the set point for the maximum temperature ceiling, reduce the flow of fuel to the burner.

SIEM See Security Information and Event Management.

Situational Awareness Situational Awareness is a term used by the National Institute of Standards and Technology (NIST) and others to indicate a desired state of awareness within a network in order to identify and respond to network-based attacks. The term is a derivative of the military command and control process of perceiving a threat, comprehending it, making a decision and taking an action in order to maintain the security of the environment. Situational Awareness in network security can be obtained through network and security monitoring (perception), alert notifications (comprehension), security threat analysis (decision-making), and remediation (taking action).

Smart-listing A term referring to the use of blacklisting and whitelisting technologies in conjunction with a centralized intelligence system, such as a SIEM in order to dynamically adapt common blacklists in response to observed security event activities. See also: Allowlisting, Whitelisting, Blacklisting, Blocklisting.

Stuxnet An advanced cyber-attack against an industrial control system, consisting of multiple zero-day exploits used for the delivery of malware that then targeted and infected specific industrial controls for the purposes of sabotaging an automated process. Stuxnet is widely regarded as the first cyber-attack to specifically target an industrial control system.

Supervisory Control And Data Acquisition Supervisory Control and Data Acquisition (SCADA) refers to the systems and networks that communicate with industrial control systems to provide data to operators for supervisory purposes, as well as control capabilities for process management.

TASE.1 See Telecontrol Application Service Element-1.

TASE.2 See Telecontrol Application Service Element-2.

Technical Feasibility/Technical Feasibility Exception (TFE) The term "Technical Feasibility" is used in the NERC CIP reliability standard and other compliance controls to indicate where a required control can be reasonably implemented. Where the implementation of a required control is not technically feasible, a Technical Feasibility Exception can be documented. In most cases, a TFE must detail how a compensating control is used in place of the control deemed to not be feasible.

Telecontrol Application Service Element-1 The initial communication standard used by the ICCP protocol. Superseded by Telecontrol Application Service Element-2

Telecontrol Application Service Element-2 The Telecontrol Application Service Element-2 standard or TASE.2 refers to the ICCP protocol. See also Inter Control Center Protocol.

Unidirectional Gateway A network gateway device that only allows communication in one direction, such as a Data Diode. See also Data Diode.

User Whitelisting, User Allowlisting The process of establishing a list of known valid user identities and/or accounts, for the purpose of detecting and/or preventing rogue user activities. See also Application Whitelisting.

VA See Vulnerability Assessment.

Vulnerability A vulnerability refers to a weakness in a system that can be utilized by an attacker to damage the system, obtain unauthorized access, execute arbitrary code, or otherwise exploit the system.

Vulnerability Assessment The process of scanning networks to find hosts or assets, and probing those hosts to determine vulnerabilities. Vulnerability assessment can be automated using a vulnerability assessment scanner, which will typically examine a host to determine the version of the operating system and all running applications, which can then be compared against a repository of known software vulnerabilities to determine where patches should be applied.

Whitelist, Allowlist Whitelists or allowlists refer to defined lists of "known good" items: users, network addresses, applications, and so on, typically for the purpose of exception-based security where any item not explicitly defined as "known good" results in a remediation action (e.g. alert and block). Whitelists contrast blacklists, which define "known bad" items.

xIoT A term that most commonly eludes to a broad spectrum of IoT devices used across multiple industries and applications, where the 'x' acts as a variable when defining IoT. See also "Internet of Things."

Zone A zone refers to a logical boundary or enclave containing assets of like function and/or criticality, for the purposes of facilitating the security of common systems and services. See also Enclave.

Index

'*Note:* Page numbers followed by "f" indicate figures, "t" indicate tables and "b" indicate boxes.'

D

J

K

9780443137372